Coding for Erasure Channels

Understand how to make wireless communication networks, digital storage systems, and computer networks robust and reliable in the first unified, comprehensive treatment of erasure correcting codes. Data loss is unavoidable in modern computer networks; as such, data recovery can be crucial, and these codes can play a central role. Through a focused and detailed approach, you will gain a solid understanding of the theory and the practical knowledge to analyze, design, and implement erasure codes for future computer networks and digital storage systems. Starting with essential concepts from algebra and classical coding theory, this book provides specific code descriptions and efficient design methods, with practical applications and advanced techniques stemming from cutting-edge research. This is an accessible and self-contained reference, invaluable to both theorists and practitioners in electrical engineering, computer science, and mathematics.

Enrico Paolini is Associate Professor in the Department of Electrical, Electronic and Information Engineering at the University of Bologna, Italy. He is also IEEE Senior Member and served as an associate editor for the *IEEE Transaction on Communications* and the *IEEE Communications Letters*.

Gianluigi Liva is Leader of the Information Transmission Group at the German Aerospace Center. He lectures on graph-based coding theory at the Technical University of Munich, Germany. He is also IEEE Senior Member and serves as an associated editor for the *IEEE Transactions on Communications*.

Balázs Matuz is currently an Optical Research Engineer at Huawei's Munich Research Center. From 2007 to 2023, he was a member of the technical staff at the German Aerospace Center, where he conducted research on forward error correction FEC for satellite links, among other topics. Since 2014, he also serves as a lecturer at the Technical University of Munich.

Marco Chiani is Professor of Information Theory at the University of Bologna, Italy. He is also IEEE Fellow and Research Affiliate at the Massachusetts Institute of Technology. He received the 2011 IEEE Abraham Prize, the 2012 IEEE Ellersick Prize, and the 2012 IEEE Rice Prize in the fields of communications systems and theory.

Coding for Erasure Channels

ENRICO PAOLINI
University of Bologna

GIANLUIGI LIVA
German Aerospace Center, Wessling

BALÁZS MATUZ
Huawei Munich Research Center, Munich

MARCO CHIANI
University of Bologna

CAMBRIDGE
UNIVERSITY PRESS

CAMBRIDGE
UNIVERSITY PRESS

Shaftesbury Road, Cambridge CB2 8EA, United Kingdom

One Liberty Plaza, 20th Floor, New York, NY 10006, USA

477 Williamstown Road, Port Melbourne, VIC 3207, Australia

314–321, 3rd Floor, Plot 3, Splendor Forum, Jasola District Centre,
New Delhi – 110025, India

103 Penang Road, #05–06/07, Visioncrest Commercial, Singapore 238467

Cambridge University Press is part of Cambridge University Press & Assessment,
a department of the University of Cambridge.

We share the University's mission to contribute to society through the pursuit of
education, learning and research at the highest international levels of excellence.

www.cambridge.org
Information on this title: www.cambridge.org/9781107131552

DOI: 10.1017/9781316443354

When citing this work, please include a reference to the DOI 10.1017/9781316443354

First published 2026

Cover image: Roman characters. ©THEPALMER / E+ / Getty Images

A catalogue record for this publication is available from the British Library

A Cataloging-in-Publication data record for this book is available from the Library of Congress

ISBN 978-1-107-13155-2 Hardback

Contents

Preface

Th? e?asur? ch?nne?, wh?re som? parts of mess?ges can be l?st during transm?ssion, is dafferent form the arror channel, where chatacters are misinterpreled rether than simpty lost.

This book provides a self-contained introduction to both the theoretical foundations and practical applications of coding for erasure channels. While many excellent texts cover coding theory broadly, none are devoted specifically to this fundamental and practically significant topic. Erasure channels arise naturally in the real-world systems – such as wireless links, data storage devices, and distributed networks – where information is not corrupted but lost. This book aims to fill that gap by offering a focused yet comprehensive treatment.

We begin in Chapter 2 with the foundations: basic tools for error correction, operations in finite fields, decision rules, classical block codes, and Reed–Solomon codes – one of the cornerstones of erasure correction.

Chapter 3 introduces the memoryless erasure channel and derives its capacity and finite-length performance bounds. This chapter also provides an overview of erasure channels with memory, offering a more general modeling framework for real communication scenarios.

Chapters 4–6 are devoted to low-density parity-check (LDPC) codes – one of the most effective families of codes for the erasure channel. We introduce their graphical structure, discuss both unstructured and structured ensembles (such as protograph-based codes), analyze their weight distributions, and detail decoding algorithms. Special emphasis is placed on iterative decoding and maximum-likelihood decoding.

The performance analysis of these decoding algorithms is the focus of Chapters 7 and 8. We explore threshold phenomena in the infinite-length regime, finite-length effects such as stopping sets and error floors, and analyze both iterative and maximum-likelihood decoding, using both asymptotic and finite-blocklength tools.

Chapter 9 generalizes the LDPC framework, introducing spatially coupled, generalized LDPC, and low-density generator matrix codes – each bringing design advantages and decoding insights. We also describe erasure decoding methods for these code ensembles.

We then turn to other code families in Chapters 10 and 11. Chapter 10 covers polar codes, which leverage channel polarization to enable efficient erasure decoding. Chapter 11 explores fountain codes. We describe LT and Raptor codes and highlight their performance versus complexity trade-offs.

This book is intended for graduate students, researchers, and professionals in electrical engineering, computer science, and related fields. A basic understanding of probability theory, linear algebra, and information theory is assumed. We have aimed to provide both rigorous foundations and practical insights, guiding the reader through both classical concepts and recent advances.

On a personal note, this book reflects years of research, collaboration, and intellectual discovery. We are grateful to our colleagues – especially Mark Flanagan, Francisco Lázaro Blasco, and Giuliano Garrammone – for their valuable contributions. Our thanks also go to the editorial staff at Cambridge University Press and to our families for their constant support.

We hope this book will serve not only as a reference but also as an entry point to the rich and growing field of coding for erasure channels.

Abbreviations

3GPP	3rd Generation Partnership Project
ACE	approximate cycle extrinsic message degree
APP	a posteriori probability
ARJA	accumulate-repeat jagged-accumulate
ARQ	automatic retransmission query
ATSC	Advanced Television Systems Committee
AWGN	additive white Gaussian noise
BEC	binary erasure channel
BLEC	block erasure channel
BP	belief propagation
BSC	binary symmetric channel
c.d.f.	cumulative distribution function
CCSDS	Consultative Committee for Space Data Systems
CN	check node
CRC	cyclic redundancy check
DFT	discrete Fourier transform
DMC	discrete memoryless channel
DVB	Digital Video Broadcasting
DVB-H	Digital Video Broadcasting – Handheld
EXIT	extrinsic information transfer
FEC	forward error correction
GE	Gaussian elimination
GEC	Gilbert erasure channel
GLDPC	generalized LDPC
HDFS	Hadoop file system
IDFT	inverse discrete Fourier transform
IETF	Internet Engineering Task Force
i.i.d.	independent and identically distributed
IP	internet protocol
IRA	irregular repeat-accumulate
IT	iterative
LDGM	low-density generator matrix
LDPC	low-density parity-check
LEO	low Earth orbit

LLR	log-likelihood ratio
LR	likelihood ratio
LRF	linear random fountain
LTE	long-term evolution
MAP	maximum a posteriori
MBMS	multimedia broadcast multicast services
MDS	maximum distance separable
MI	mutual information
MIMO	multi-input multi-output
ML	maximum likelihood
MPE	multi-protocol encapsulation
MSE	mean-square error
MTBL	maximum tolerable burst length
p.m.f.	probability mass function
PEG	progressive edge growth
PIR	private information retrieval
PSS	pivot search and swap
QEC	q-ary erasure channel
r.v.	random variable
RAID	redundant array of independent disks
RJA	repeat jagged-accumulate
RM	Reed–Muller
RS	Reed–Solomon
SBUEC	single burst erasure channel
SC	successive cancellation
SC-LDPC	spatially coupled low-density parity-check
SCL	successive cancellation list
SPA	sum–product algorithm
SPC	single parity-check
SVN	separated variable nodes
VN	variable node

1 Introduction

Communication theory, as founded by Claude Shannon in 1948, is the branch of science which studies the fundamental limits and the methods of transmitting information between a source and a destination, regardless of their spatial or temporal separation. For example, data storage can be seen as a way to send information in the future. Transmission of information is accomplished by changing, over time and/or space, the state of some physical entity, such as the electromagnetic field, the air pressure, the current in a circuit, or the spin of a particle, thus obtaining a communication channel. The state changes need to be properly interpreted at the destination by a set of procedures put in place by a "receiver."

The channel is described by a statistical model derived from the laws of physics: It has a set of possible inputs, a set of possible outputs, and a probabilistic relation between input and output. Channel models should not only be sufficiently accurate to capture the behavior of the real communication channels, but also as simple as possible, in order to be useful for system design and performance evaluation.

In this book, we will deal with discrete-time channels, where the input and output are time series, indicated with $x_1, x_2, \ldots$, and $y_1, y_2, \ldots$, respectively. The channel input and output alphabets are $\mathcal{X}$ and $\mathcal{Y}$, so that $x_i \in \mathcal{X}$ and $y_i \in \mathcal{Y}$. In many cases, the channel can be considered as memoryless, so that at any time i, the response y_i given the input x_i is conditionally independent on all other inputs and outputs. If the input and output sets $\mathcal{X}$ and $\mathcal{Y}$ are also discrete (usually with finite cardinality), we get a discrete memoryless channel (DMC).

One of the simplest and most widely employed DMC models is the so-called binary symmetric channel (BSC), which is depicted in Figure 1.1. At any given instant, the BSC has a binary input and a binary output, so one binary digit is sent per unit of time. The channel is not reliable since there is a fixed probability, p, that the output bit will differ from the input. The equal probability of error for both input bits 0 and 1 justifies the term "symmetric."

This model is simple and also very useful, as in many cases we have communication systems where, from an end-to-end perspective, bits are processed by the channel in a memoryless way, with some fixed probability of error. Consequently, there is a vast literature on coding techniques and performance analysis for the BSC.

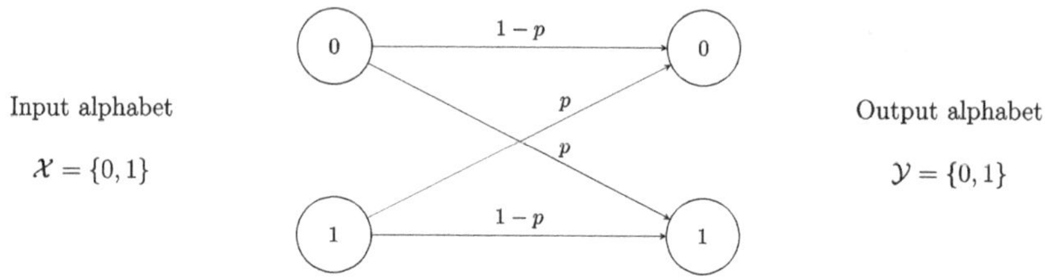

Figure 1.1 The binary symmetric channel.

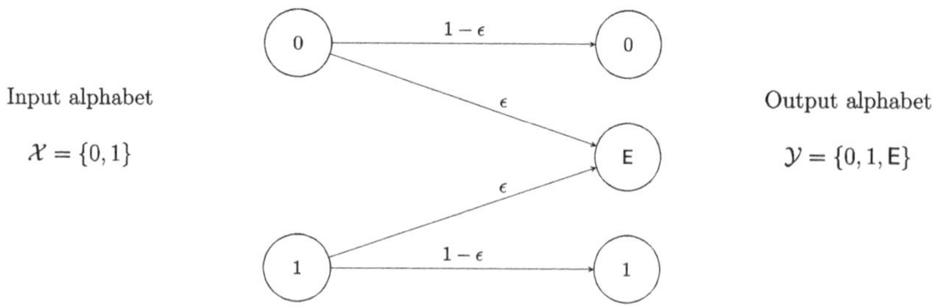

Figure 1.2 The binary erasure channel with erasure probability ϵ.

Another simple yet important model is the *erasure channel*, where the information is either received correctly or marked as "garbage." More precisely, the binary erasure channel (BEC) accepts binary symbols as input and produces ternary symbols as output (Figure 1.2). Each input symbol has probability $\epsilon < 1$ of being erased, and a probability $1 - \epsilon$ of being received correctly. An erasure is indicated to the receiver with E. If the receiver receives a 0 or a 1, it confirms the correct transmission of that symbol, but the reception of an erasure gives the receiver no information about the transmitted symbol.

Generalizing, the q-ary erasure channel (QEC) accepts q-ary symbols: The output is coincident with the input with probability $1 - \epsilon$ and is an erasure E with probability ϵ. Therefore, the output alphabet comprises $q + 1$ symbols. The binary erasure channel (BEC) was introduced by Elias in 1955 and was subsequently used as a simple model to derive information theoretic results. Since the 1990s, it has gained large popularity as a simple model to obtain fundamental results for modern coding schemes (following the spirit of Elias' approach) and as a realistic model for packet-based communications.

In fact, with the advent of data networks, the erasure channel has become the fundamental model for the design and performance analysis of packet error-correcting control strategies. Indeed, in network architecture, where the objective is to reliably exchange data packets among nodes, protocols are organized in layers, where each layer offers a service to the layer above. The internet protocol stack is reported in Figure 1.3: The physical layer is in charge of moving bits from one node to another; the role of the link layer is to move packets of bits (link-layer packets or frames) from one node to the next node along a route; the network layer is responsible for moving network-layer packets (datagrams) from one host to another; the transport layer moves transport-layer segments, which are used by the application

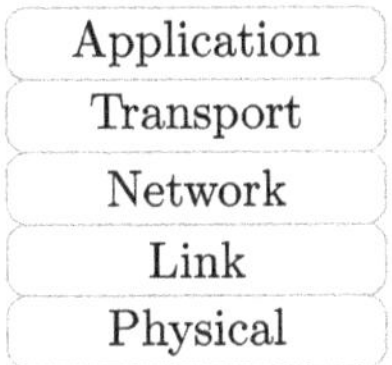

Figure 1.3 Five-layer internet protocol stack.

layer to transport application-layer messages between application endpoints. For any layer above the physical one, packets from the lower layer are usually either correctly received or lost. A packet can be marked as lost if there are errors in it as detected by an error-detecting code, or if it never arrived due to network congestion or unavailability.

Error control can in general be pursued by forward error correction (FEC), that is, by employing error-correcting codes to correct errors at the receiver as described later, or using error detection and retransmission, called automatic retransmission query (ARQ), or with a combination in the so-called hybrid ARQ. In ARQ, when errors are detected, the receiver sends a request to the transmitter to resend the message, and this continues until the message is received correctly. Hence, ARQ requires a feedback link from the receiver to the transmitter. On the other hand, forward error correction (FEC) can be used also for one-way systems, where the transmission is unidirectional. One advantage of ARQ over FEC is that error detection is much simpler than error correction. Also, ARQ is inherently adaptive, as retransmissions occur only when errors are detected. On the other hand, FEC allows for immediate error correction at the receiver, so avoiding the round-trip propagation delay needed by ARQ for the retransmissions. This is particularly important for real-time network applications, or in scenarios, such as deep-space satellite communications, where the propagation delays are long. Moreover, in one-to-many communication scenarios, such as distributing live video over the internet, implementing the ARQ strategy, with a single transmitter managing the individual retransmission requests from many users, can be difficult. The FEC strategy, where correction is accomplished directly at each receiver, can be preferred in these situations.

In this book, we mainly refer to FEC for transmission systems where individual bits, or entire packets, are occasionally erased by the communication channel.

A simple example presents the parallelism between the case where a bit is erased and the case where an entire packet is erased. Assume we have a way to send blocks of four bits through a channel, which however sometimes erases one bit out of the four. To protect our information from these erasures, we can simply send through the channel three information bits c_1, c_2, and c_3 plus an additional "parity bit" c_4, chosen so that the total number of bits of value "1" is even. Mathematically, it must be $c_1 + c_2 + c_3 + c_4 = 0$, where + stands for binary sum (i.e., the sum modulo 2, see Chapter 2). Since the binary difference is coincident with the binary sum, we also have $c_4 = c_1 + c_2 + c_3$. So, if any of the four transmitted bits is not received (erased), its value can be calculated from the other three. For example, if c_1 is erased, we can

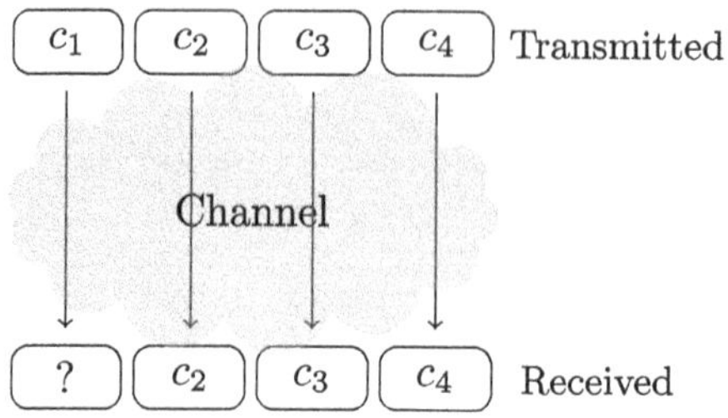

Figure 1.4 Three data packets (c_1, c_2, c_3) with one additional parity packet.

calculate it as $c_1 = c_2 + c_3 + c_4$. More in general, it is clear that by adding a parity bit to a block of bits, we can recover one erasure occurring on any of the bits.

Next, consider one layer of the protocol stack which is exchanging packets $c_1, c_2, \ldots$, each of 1 000 bits, with the lower layer (note that here each c_i represents one entire packet). Due to network issues, sometimes a packet might get lost. We can adopt a simple strategy to recover a lost packet as follows: With every three information packets c_1, c_2, and c_3, we send an additional "parity packet" c_4 (still of 1 000 bits), with $c_4 = c_1 + c_2 + c_3$ where the sum is the element-wise binary sum (i.e., the sum modulo 2 bit by bit). So, if any of the four transmitted packets is not received, its value can be calculated from the other three. For example, if the packet c_1 is lost, we can reconstruct it (all 1 000 bits) as $c_1 = c_2 + c_3 + c_4$ where again the sum is the element-wise binary sum (see Figure 1.4). Thus, we can design erasure correcting codes both at the single bit level, where one symbol in the channel is one bit, or for whole packets, where one symbol is an entire packet. While in the former case, the symbol operations are the usual binary sum and difference, in the latter, we need to properly define an algebraic structure of group to operate with packets. When packets have the same number of bits, we can use simply the element-wise binary sum.

In general, we can interpret the erasures as *unknowns*: The constraints imposed on the transmitted symbols should then be suitably designed to allow solving for the unknowns in the largest possible number of situations. In the abovementioned example, the constraint is described by the linear equation $c_1 + c_2 + c_3 + c_4 = 0$, which expresses a "parity" constraint among the symbols c_i. Since we have one equation, we can solve for at most one unknown. The same constraint can be represented in matrix form as

$$\begin{bmatrix} 1 & 1 & 1 & 1 \end{bmatrix} \begin{bmatrix} c_1 \\ c_2 \\ c_3 \\ c_4 \end{bmatrix} = 0. \tag{1.1}$$

Another way to represent this constraint is by means of a bipartite (Tanner) graph, indicating the symbols c_i as circles and the equation as a square (Figure 1.5).

If we want to be able to recover more erasures occurring over n symbols, we need more equations. With $c_i \in \{0, 1\}$ and m linear equations, the constraints would be in the form:

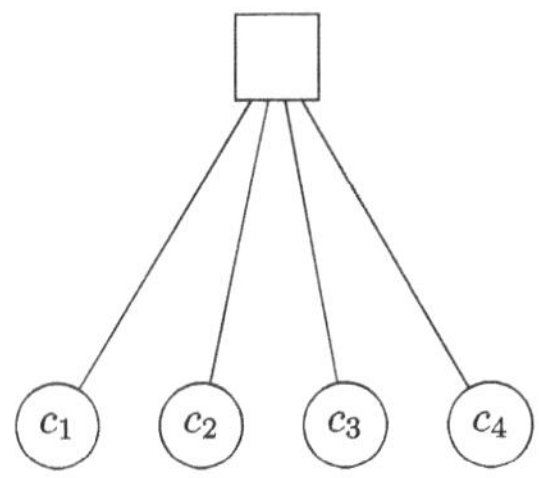

Figure 1.5 Bipartite graph for the equation $c_1 + c_2 + c_3 + c_4 = 0$.

$$
\begin{aligned}
h_{1,1}c_1 + h_{1,2}c_2 + \cdots + h_{1,n}c_n &= 0 \\
h_{2,1}c_1 + h_{2,2}c_2 + \cdots + h_{2,n}c_n &= 0 \\
&\vdots \\
h_{m,1}c_1 + h_{m,2}c_2 + \cdots + h_{m,n}c_n &= 0,
\end{aligned}
\tag{1.2}
$$

where $h_{i,j} \in \{0,1\}$ and the operations are the binary sum and multiplication. Here, $k = n - m$ symbols will be data, and the other m symbols, determined uniquely by the set of m linearly independent equations, can be seen as parity symbols. An n-tuple of symbols $c_1, c_2, \ldots, c_n$ which satisfies all constraints is a *codeword*. The set of all codewords is a *code*, which in our case is used to *correct erasures*. Assuming some symbols have been erased and the others are known, recovering (decoding) consists of solving, if possible, the system of linear equations for the unknowns; with the Gaussian elimination method, this would take a time proportional to the cube of the number of unknowns.

In this binary case, a weight $h_{i,j} = 0$ implies that the symbol c_j is not present in the ith equation, while $h_{i,j} = 1$ means that it participates in the equation. Thus, the same set of constraints can also be represented with a bipartite graph, with m equations (so m squares, also called *check nodes*), where the symbol c_j is connected through an *edge* to the check node i if $h_{i,j} = 1$. For example, the set of equations

$$
\begin{aligned}
c_1 + c_2 + c_3 + c_5 &= 0 \\
c_2 + c_3 + c_4 + c_6 &= 0 \\
c_1 + c_2 + c_4 + c_7 &= 0
\end{aligned}
\tag{1.3}
$$

for the seven variables $c_1, c_2, \ldots, c_7$, can be expressed in matrix form as

$$
\begin{bmatrix} 1 & 1 & 1 & 0 & 1 & 0 & 0 \\ 0 & 1 & 1 & 1 & 0 & 1 & 0 \\ 1 & 1 & 0 & 1 & 0 & 0 & 1 \end{bmatrix}
\begin{bmatrix} c_1 \\ c_2 \\ c_3 \\ c_4 \\ c_5 \\ c_6 \\ c_7 \end{bmatrix}
=
\begin{bmatrix} 0 \\ 0 \\ 0 \end{bmatrix}
\tag{1.4}
$$

or represented by the bipartite graph in Figure 1.6. It can be verified that the code described by the relation (1.4), or, equivalently, by the graph in Figure 1.6, can be

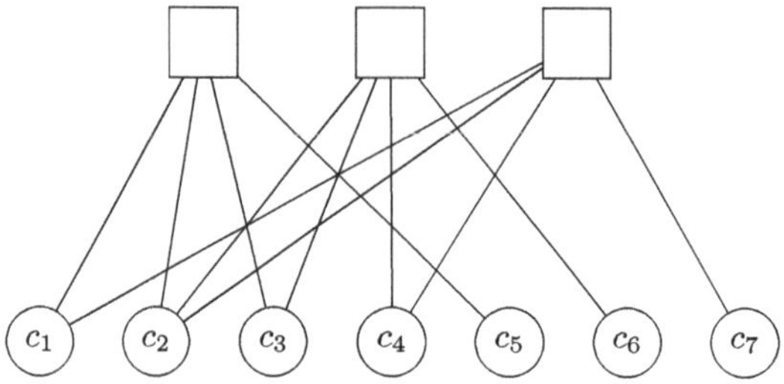

Figure 1.6 Bipartite graph for the code defined by equation (1.4).

interpreted as four data symbols with additional three parity symbols. This is known as the Hamming code (see Chapter 2). A close look to the structure of the matrix in (1.4) reveals that the code allows recovering the seven symbols of a codeword (so, in particular the four data symbols) if any one or any two of them are erased (see Chapter 2). So, besides the case of no erasures, the code corrects the seven patterns consisting of one erasure, and the $\binom{7}{2} = 21$ patterns of two erasures. There are also 28 configurations of 3 erasures which are correctable, all other situations with 3 or more erasures being not correctable. The erasure correction capability is achieved at the cost of a redundancy of three additional parity symbols for each group of four data symbols. We say that the *code rate* is $R = 4/7 \approx 57\%$: This is the fraction of data symbol per codeword symbol, that is, the fraction of data we send per "use of the channel." Clearly, with a higher code rate we are sending more data, but on the other side, we are in general more prone to errors.

With reference to the abovementioned example, when a codeword is transmitted through a BEC with erasure probability ϵ, the probability that the erasures cannot be resolved (block error probability) is then

$$P_B = 1 - \left[(1-\epsilon)^7 + 7\epsilon(1-\epsilon)^6 + 21\epsilon^2(1-\epsilon)^5 + 28\epsilon^3(1-\epsilon)^4\right]. \tag{1.5}$$

The curve reporting the block error probability versus the channel erasure probability is shown in Figure 1.7.

We could wonder if, with the same code rate, it is possible to do better, specifically, to correct a larger fraction of erasures. To answer this question, we can recall the fundamental channel coding theorem due to Shannon: He proved that there are codes able to correct almost all errors, provided that the code rate is below the *channel capacity*, C. This general theorem applies also to the BEC, for which the capacity is simply $\mathsf{C} = 1 - \epsilon$ (see Chapter 3). Thus, the theorem assures that there are, at least for blocklengths n sufficiently large, codes with code rate R correcting almost all erasures introduced by a BEC, provided that

$$R < 1 - \epsilon\,. \tag{1.6}$$

So, a capacity-limit approaching code with rate R will give a negligible block error probability for any $\epsilon < 1 - R$. In the previous example, for $R = 4/7$ the theorem says that we can work with BEC having erasure probability up to $\epsilon_{\max} = 1 - R = 3/7 \approx 0.43$, which is also reported in Figure 1.7. However, although the Shannon's theorem

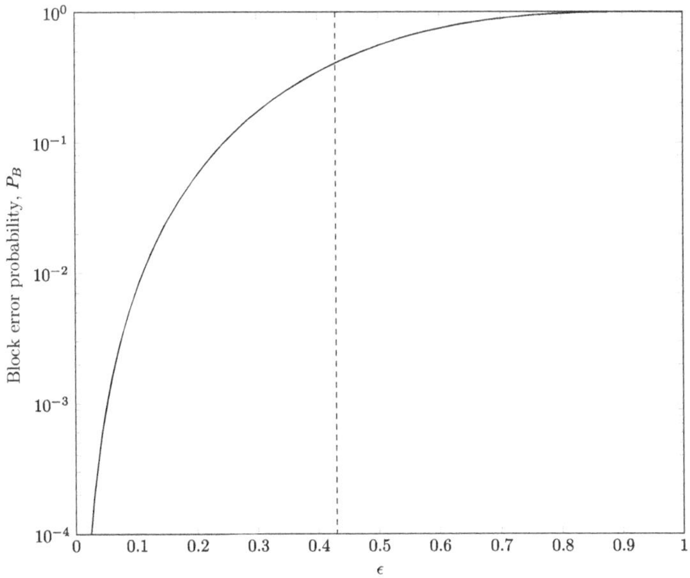

Figure 1.7 Block error probability for the $n = 7$ symbols, $R = 4/7$ code described by the graph in Figure 1.6 over the BEC. The dashed line is at $\epsilon_{\text{max}} = 0.43$ giving a BEC capacity $\mathsf{C} = 4/7$: It represents the performance of a code with rate $R = 4/7$ approaching the Shannon limit.

assures that there exist codes with much better performance than the one in Figure 1.6, it does not tell us how to find practical codes to get close to the capacity.

Here there are two problems: (a) We should find codes (bipartite graphs) with good properties and (b) practical methods for recovering the unknowns which have been erased by the channel. There are many classes of codes in the literature; they have all advantages and disadvantages. In this book, we will mention the most important ones and focus on codes with sparse graph representation. In general, longer codes perform better, so we will have to design large *graph codes*, and therefore we will need low-complexity methods to correct erasures.

There is therefore a trade-off between complexity and performance, so that in some applications, short codes are to be preferred to better-performing longer codes. For example, some of the current applications of erasure correcting codes include packet loss recovery in computer networks, distributed storage systems, distributed computing, network coding, and privacy, where the requirements in terms of latency, reliability, redundancy, and complexity are quite different. The design and analysis of erasure correcting codes described by graphs and of methods for low-complexity decoding are the subjects of this book.

Notation

Throughout the book, we will use lowercase bold to indicate vectors and uppercase bold for matrices. The general block diagram of a communication link is reported in Figure 1.8: From the source data $\boldsymbol{u}$ (also referred to as information message), the channel *encoder* produces a codeword $\boldsymbol{c}$. The *mapper* maps the vector $\boldsymbol{c}$ into a vector $\boldsymbol{x}$ suitable for the transmission over the communication *channel*. The output of the

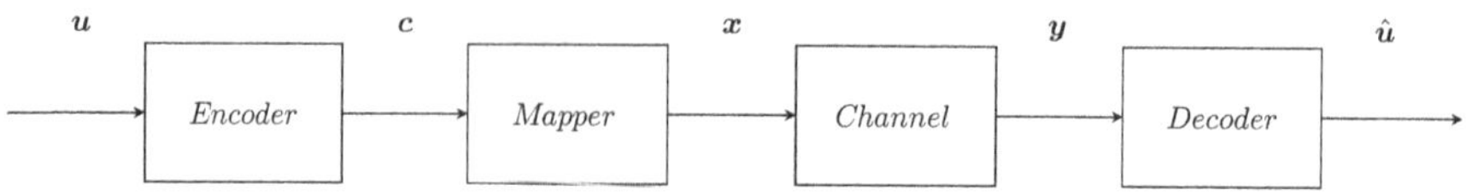

Figure 1.8 Block diagram of the reference communication system.

channel, $\boldsymbol{y}$, is used by the *decoder* to produce the estimated data vector $\hat{\boldsymbol{u}}$. In most of the cases analyzed in the book, the mapper is included in the channel model, so that, if not explicitly indicated, the notations $\boldsymbol{c}$ and $\boldsymbol{x}$ will be used interchangeably to denote codewords.

The book is organized as follows.

- Basic definitions and tools for error correction. In Chapter 2, we provide the basic elements of classical error-correcting codes, how to perform operations in finite fields, the decision rules, the structure and properties of classical block codes, and finally a description of the Reed–Solomon codes that are particularly important for the erasure channel.
- Erasure channel models and bounds. In Chapter 3, we introduce the memoryless erasure channel. The performance limits of the memoryless erasure channel are derived in terms of capacity, as well as in the form of finite-length upper and lower bounds on the block error probability. The chapter continues with a survey of models for erasure channels with memory.
- Low-density parity-check (LDPC) codes. In Chapter 4, we introduce low-density parity-check (LDPC) codes, a class of powerful and efficient channel codes that admit a graphical representation based on sparse bipartite graphs. The chapter introduces fundamental concepts in LDPC coding theory and guides the reader through unstructured and structured LDPC code ensembles, including the important case of protograph-based LDPC code ensembles.
- Weight distribution of LDPC codes. More advanced topics on LDPC codes are covered by Chapter 5 that focuses on structural properties of these codes related to their weight distributions. The chapter addresses weight distributions of unstructured and partially structured LDPC code ensembles as well as weight distribution exponents. It also addresses ensemble expurgation and minimum distance analysis for LDPC codes.
- Decoding of LDPC codes on the erasure channel. In Chapter 6, we illustrate different decoding algorithms for LDPC codes over erasure channels, namely, iterative (IT) and maximum likelihood (ML) decoding. Decoding on an erasure channel can be strongly simplified with respect to decoding on other channels, since whenever a symbol is not erased, its value is known with full certainty. We illustrate that iterative decoding can be done by a peeling process that resolves one unknown per iteration. ML decoding can be performed efficiently by solving a sparse system of equations by variants of the Gaussian elimination algorithm.
- Performance analysis for iterative decoders. In Chapter 7, we discuss the behavior of LDPC codes under iterative erasure decoding. When the blocklength goes to

infinity, for many LDPC codes the symbol error probability exhibits a so-called threshold phenomenon, that is, there exists a certain channel erasure probability below which error-free communication is possible while this is not guaranteed above it. We discuss how to compute this threshold for ensembles of LDPC codes on memoryless erasure channels. In the finite-length setting, one may observe a flattening of the symbol error rate curve owing to stopping sets – specific structures in the code's bipartite graph. Knowing their number and size allows predicting this so-called error floor. Based on our findings, we discuss how to design a good LDPC for memoryless erasure channels with extension to channels with memory.

- ML LDPC decoder analysis. In Chapter 8, the performance of LDPC codes under ML decoding is analyzed. ML decoding is intended here either as the block-wise or the symbol-wise decoding criterion (see Section 2.2). More specifically, the asymptotic analysis on the ML decoding threshold addresses the performance in terms of symbol-wise ML decoding, whereas finite-length bounds are provided for the block error probability under block-wise ML decoding. While the focus is on unstructured LDPC code ensembles, the results in this chapter can be considered to a large extent valid for other LDPC code ensembles.
- Generalizations of LDPC codes. In Chapter 9, we present code ensembles that may be regarded as special instances, generalizations, or modifications of LDPC code ensembles introduced in Chapter 4. The chapter starts with spatially coupled LDPC codes, introduced within a protograph-based framework, then addresses generalized LDPC codes, where some of the check nodes impose multiple linear constraints, and finally describes low-density generator matrix codes. Erasure decoding algorithms are described for all code classes.
- Polar codes. In Chapter 10, one of the most recent proposals for error-correcting codes is described, which are called polar codes. They are block codes designed for simplifying the implementation of the decoder. Specifically, polar codes are designed assuming a successive cancellation (SC) decoder. Channel polarization and subchannel ranking are also discussed in this chapter.
- Fountain codes. In Chapter 11, we address an alternative paradigm for erasure coding that exploits feedback from (multiple) receivers to the transmitter. Here, the receivers are assumed to be interested in the same content, and the transmission takes place over a broadcast channel. The feedback is employed to adapt "on the fly" the coding rate to the channel conditions. This approach, which shares several similarities with the framework of rate compatible codes with hybrid ARQ, relies on the so-called fountain codes. Two well-established classes of fountain codes are discussed, namely, the class of LT codes (strongly related to LDPC codes) and the class of Raptor codes.

Finally, some of the current applications of erasure correcting codes, such as reliable data transmission in computer networks, distributed storage systems, and distributed computing, are described in Appendix A.

1.1 Historical Notes

The fundamental limits for communication systems were established in Shanon's landmark paper in 1948 [1], where he introduced the concept of channel capacity and proved the existence of error-correcting codes able to cope with noisy channels. The BSC channel model and its capacity were also analyzed by Shannon in [1]. The BEC was introduced by Elias in 1955 [2]. For this channel, a simple decoding algorithm based on solving simultaneous linear binary equations was investigated by Elias and Epstein in the 1950s [2, 3], where it was observed that the computational effort needed to decode a random linear code grows at most as the cube of the number of erasures that need to be corrected. The erasure channel was subsequently used as a simple model to derive information theoretic results [4–6]. More recently, it has gained large popularity to obtain fundamental results and design guidelines for modern coding schemes following the spirit of Elias' approach [7–11] and as a realistic model for packet-based communications, multiple access, and distributed storage [12–18].

References

[1] C. Shannon, "A mathematical theory of communication," *Bell Syst. Tech. J.*, vol. 27, no. 3, pp. 379–423, 623–656, July 1948.
[2] P. Elias, "Coding for two noisy channels," in *Third London Symposium on Information Theory*, London, UK, Sep. 1955, pp. 61–74.
[3] M. A. Epstein, "Algebraic decoding for a binary erasure channel," Research Laboratory of Electronics, Massachusetts Institute of Technology, RLE Technical Report 340, 1958.
[4] R. Gallager, "Low-density parity-check codes," *IRE Trans. Inf. Theory*, vol. 8, no. 1, pp. 21–28, Jan. 1962.
[5] R. Gallager, *Information theory and reliable communication.* New York, NY: Wiley, 1968.
[6] E. Berlekamp, "The technology of error-correcting codes," *Proc. IEEE*, vol. 68, no. 5, pp. 564–593, May 1980.
[7] T. Richardson, A. Shokrollahi, and R. Urbanke, "Design of capacity-approaching irregular low-density parity-check codes," *IEEE Trans. Inf. Theory*, vol. 47, no. 2, pp. 619–637, Feb. 2001.
[8] C. Di, D. Proietti, I. Telatar, T. Richardson, and R. Urbanke, "Finite-length analysis of low-density parity-check codes on the binary erasure channel," *IEEE Trans. Inf. Theory*, vol. 48, no. 6, pp. 1570–1579, June 2002.
[9] A. Ashikhmin, G. Kramer, and S. ten Brink, "Extrinsic information transfer functions: Model and erasure channel properties," *IEEE Trans. Inf. Theory*, vol. 50, no. 11, pp. 2657–2673, Nov. 2004.
[10] T. Richardson and R. Urbanke, *Modern Coding Theory.* New York, NY, USA: Cambridge University Press, 2008.
[11] E. Arıkan, "Channel polarization: A method for constructing capacity-achieving codes for symmetric binary-input memoryless channels," *IEEE Trans. Inf. Theory*, vol. 55, no. 7, pp. 3051–3073, July 2009.

[12] L. Rizzo, "Effective erasure codes for reliable computer communication protocols," *ACM Comput. Commun. Rev.*, vol. 27, no. 2, pp. 24–36, Apr. 1997.

[13] M. Luby, M. Mitzenmacher, A. Shokrollahi, D. Spielman, and V. Stemann, "Practical loss-resilient codes," in *Proc. 29th Symp. Theory Computing*, El Paso, TX, USA, May 1997, pp. 150–159.

[14] J. Byers, M. Luby, M. Mitzenmacher, and A. Rege, "A digital fountain approach to reliable distribution of bulk data," in *Proc. ACM SIGCOMM*, vol. 28, no. 4, Vancouver, Canada, Oct. 1998, pp. 56–67.

[15] J. Byers, M. Luby, and M. Mitzenmacher, "A digital fountain approach to asynchronous reliable multicast," *IEEE J. Sel. Areas Commun.*, vol. 20, no. 8, pp. 1528–1540, Oct. 2002.

[16] G. Liva, "Graph-based analysis and optimization of contention resolution diversity slotted ALOHA," *IEEE Trans. Commun.*, vol. 59, no. 2, pp. 477–487, Feb. 2011.

[17] E. Paolini, G. Liva, and M. Chiani, "Coded slotted ALOHA: A graph-based method for uncoordinated multiple access," *IEEE Trans. Inf. Theory*, vol. 61, no. 12, pp. 6815–6832, Dec 2015.

[18] A. Dimakis, V. Prabhakaran, and K. Ramchandran, "Decentralized erasure codes for distributed networked storage," *IEEE Trans. Inf. Theory*, vol. 52, no. 6, pp. 2809–2816, June 2006.

2 Basics

This chapter provides essential concepts from algebra and classical coding theory, useful to understand the subsequent chapters. A basic knowledge of linear algebra is assumed. The treatment is rigorous but limited: More details can be found in the recommended textbooks given at the end of the chapter.

2.1 Finite Fields and Vector Spaces

We assume that digital data are organized in n-tuples (sequences of length n), whose elements can take value within a finite set, $F = \{\alpha_0, \alpha_1, \ldots, \alpha_{q-1}\}$, composed of q elements. One n-tuple is represented in bold as follows:

$$\boldsymbol{c} = \begin{bmatrix} c_0 & c_1 & \ldots & c_{n-1} \end{bmatrix} \qquad c_i \in F. \tag{2.1}$$

To efficiently manipulate n-tuples, we define an algebraic structure, where elements of F are called scalars, and n-tuples represent vectors. We have to first properly define addition and multiplication of two elements in the set F, such that the results of these operations are still in F (in other words, F must be closed both under addition and under multiplication), as well as the inverse operations of difference and division. Then, we will define the addition of two n-tuples and the multiplication of an element of F by an n-tuple, to interpret n-tuples as vectors in a vector space.

Definition 2.1 (Groups) A group is a set together with an operation which associates an element to each ordered pair of elements in the set. The group is commutative (or Abelian) if the order of the pair is irrelevant. The operation must be associative, there must be an identity element, and for each element, there must exist an inverse.

Definition 2.2 (Fields) A field is a set F together with a pair of operations called addition "+" and multiplication "·," satisfying the following conditions:

i. The addition forms an additive commutative group, where the identity element is indicated as $\alpha_0 = 0$. The additive inverse of an element a is denoted by $-a$.

ii. The elements, exclusive of 0, form a multiplicative commutative group, where the identity element is $\alpha_1 = 1$. The multiplicative inverse of an element $a \neq 0$ is denoted by a^{-1}.

iii. The two operations are connected by the distributive law.

For $a, b \in F$, the difference is then naturally defined as $a - b = a + (-b)$, and the division as $a/b = a \cdot b^{-1}$, $b \neq 0$.

There are a number of properties which can be derived from the above definition, making it possible to add, subtract, multiply, and divide (except that division by 0 is not defined) elements in a generic field similar to the usual operations in the real number system (which is itself an example of a field).

Definition 2.3 (Finite Fields) A field $(F, +, \cdot)$ with a finite number of elements $q = |F|$ is called a *finite field*. The integer q is called the *order* of the field. The set of elements is often indicated as F_q. Finite fields are also called Galois fields after their discoverer, and indicated as $\mathrm{GF}(q)$.

It can be proved that all finite fields of the same order are isomorphic (i.e., they only differ in the way the elements are named), and that finite fields exist if and only if q is a prime (prime fields) or an integer power of a prime number, that is, $q = p^m$, with p a prime number and m a positive integer (extension fields). The finite field with $q = 2^m$ elements, $\mathrm{GF}(2^m)$, is of particular interest, as in this case each field element can be interpreted as an m-tuple of bits. For example, using $\mathrm{GF}\left(2^8\right)$, we can add, subtract, multiply, and divide two bytes, producing one byte as output.

Before explaining how to perform operations on finite fields, it is useful to remind some facts from elementary arithmetic.

Theorem 2.1 (Euclidean division) *Given any two integers a and $b \neq 0$, there exist two unique integers q,r with $0 \leq r < |b|$ such that*

$$a = b\,q + r\,. \tag{2.2}$$

The integers a,b,q, and r are called dividend, divisor, quotient, and remainder, respectively. We also say that r is a modulo b and write

$$r = a \bmod b\,. \tag{2.3}$$

Starting from the Euclidean division, it is possible to prove that, if $\gcd(a,b)$ is the greatest common divisor of two integers a and b, then there exist integers p,q such that (Bézout's identity)

$$\gcd(a,b) = a\,p + b\,q. \tag{2.4}$$

Note that the couple p,q is not unique. A couple can be obtained by the so-called extended Euclidean algorithm.

In particular, if two integers a,b are relatively prime (no common prime factors), then there exist two integers p,q such that $a\,p + b\,q = 1$.

2.1.1 Implementation of Operations on Finite Fields

2.1.1.1 Prime Fields: GF(p), p Prime

For each prime number p, the field set can simply be taken as $F_p = \{0, 1, \ldots, p-1\}$, and the operations in the field $\mathrm{GF}(p)$ are just the usual addition and multiplication

Table 2.1 GF(3) operations

+	0	1	2
0	0	1	2
1	1	2	0
2	2	0	1

·	0	1	2
0	0	0	0
1	0	1	2
2	0	2	1

between integers, computed modulo p. Thus, the sum and product of $a, b \in F_p$ give the elements $c, d \in F_p$ according to

$$c = (a + b) \bmod p \tag{2.5}$$

$$d = a\,b \bmod p, \tag{2.6}$$

where $a+b$ and $a\,b$ are the usual sum and multiplication of integers. The multiplicative inverse a^{-1} of the element $a \neq 0$ must give $a\,a^{-1} = 1 \bmod p$, and may be computed by using the extended Euclidean algorithm. From (2.6), we can prove that p must be a prime. In fact, when p is a prime, for $a < p$, there is always the multiplicative inverse of a since, from (2.4), there always exist two integers b, q such that $a\,b + p\,q = \gcd(p, a) = 1$, that is, $a\,b = 1 \bmod p$, and $b = a^{-1}$ is therefore the reciprocal of a. On the contrary, if p is not a prime, then there is no multiplicative inverse under modulo p multiplication for the integers $a > 1$ dividing p. In fact, if $p = ak$ for some integers a, k, then the inverse element b of a should be such that $ab = pq + 1$ or $a(b - kq) = 1$, which has no solution for integer numbers.

It is possible to show that there is always at least one element $\alpha \in \{1, \ldots, p-1\}$ whose successive $p - 1$ powers $\alpha^0, \alpha^1, \ldots, \alpha^{p-1}$, reduced modulo p, generate all nonzero field elements. The multiplicative group above is thus a cyclic group, and an element α with this property is called a primitive element of the field.

Example 2.1 The smallest prime field is the well-known binary field GF(2), with $0 + 0 = 0, 0 + 1 = 1, 1 + 1 = 0, 0 \cdot 0 = 0, 0 \cdot 1 = 0, 1 \cdot 1 = 1$. It follows also that in GF(2), the difference is the same as the sum. As a less trivial example, we describe in Table 2.1 the operations for GF(3), for which 2 is the primitive element (in this case, there is only one).

2.1.1.2 Extension Fields: GF(p^m), p Prime, $m \geq 2$ Integer

When the field order $q = p^m$ is nonprime, the set and operations are defined by using polynomials, whose coefficients are from the prime field GF(p). We here refer specifically to the important practical case $q = 2^m$, but the presented results given in the following may be extended to the general case GF(p^m) with p being prime, $p > 2$.

Definition 2.4 A *polynomial over GF*(2) is a polynomial with coefficients from GF(2):

$$f(X) = f_0 + f_1 X + f_2 X^2 + \cdots + f_n X^n, \tag{2.7}$$

where $f_i \in \{0,1\}$ for $i = 0,\ldots,n$. For nonzero polynomials, the largest i for which $f_i = 1$ is called the degree of the polynomial. The polynomial may also be represented as the binary tuple $(f_0\ f_1\ \ldots\ f_n)$.

Polynomials over GF(2) can be multiplied, added, subtracted, and divided in the usual way, where the sum and product of coefficients are performed modulo 2. For example, given two polynomials $f(X)$ and $g(X)$ over GF(2), their sum (coincident with their difference) is as follows:

$$f(X) + g(X) = (f_0 + g_0) + (f_1 + g_1)X + (f_2 + g_2)X^2 + \cdots + (f_n + g_n)X^n, \quad (2.8)$$

where $f_i + g_i$ is the addition performed modulo 2.

The field GF(2^m) can be implemented by considering as set elements the 2^m distinct polynomials of degree less than m, whose coefficients are from the prime field GF(2):

$$F_{2^m} = \left\{0, 1, X, 1 + X, X^2, \ldots, 1 + X + X^2 + \cdots X^{m-1}\right\}. \quad (2.9)$$

With this definition of field set, we now need to define the operations between two elements. The addition and subtraction between elements is simply that between polynomials as in (2.8). For multiplication and division, there are essentially three equivalent approaches, presented below.

2.1.1.2.1 *First Method to Implement Multiplications/Divisions in GF(2^m)*

To define multiplication between elements in GF(2^m), we first choose a polynomial $p(X)$ over GF(2) of degree m, which is irreducible over GF(2).

Definition 2.5 A polynomial is *irreducible over a field* if it cannot be expressed as the product of two polynomials of lower degree in the field.

Loosely speaking, an irreducible polynomial plays the role of a prime number, in that it has no nontrivial factors. Such an irreducible polynomial always exists for any positive degree. Given an irreducible polynomial over GF(2) of degree m, similarly to the case of prime fields, the product of two elements $a(X), b(X) \in F_{2^m}$ is the polynomial $c(X) \in F_{2^m}$ obtained by the usual multiplication between polynomials, computed modulo $p(X)$:

$$c(X) = a(X)b(X) \bmod p(X). \quad (2.10)$$

The multiplicative inverse of a nonzero element $a(X) \in F_{2^m}$ is the unique polynomial $a^{-1}(X) \in F_{2^m}$ giving $a(X)a^{-1}(X) \bmod p(X) = 1$. As for integers, it may be computed with the extended Euclidean algorithm.

For example, the polynomial $p(X) = 1 + X^2 + X^3$ is irreducible over GF(2) and can be used for calculating multiplications and divisions in GF$\left(2^3\right)$ (see Example 2.2).

2.1.1.2.2 *Second Method to Implement Multiplications/Divisions in GF(2^m)*

Since in general for a given m there are multiple irreducible polynomials over GF(2), we may ask if there could be some advantage in choosing one or another. Actually, all irreducible polynomials yield the same field up to an isomorphism. However, some could be more suitable for implementing operations. Indeed, it is possible to show that there exists a special subset of the irreducible polynomials, called *primitive polynomials*, for which:

1. The remainders

$$X^\ell \bmod p(X) \qquad \ell = 0, 1, \ldots, q-2 \tag{2.11}$$

   are all the nonzero $q - 1 = 2^m - 1$ distinct polynomials over GF(2) of degree less than m;
2. We also have[1]

$$X^{q-1} \bmod p(X) = 1\,. \tag{2.12}$$

It follows cyclically $X^q \bmod p(X) = X$, $X^{q+1} \bmod p(X) = X^2$, and so on. Requiring that a polynomial satisfies the two conditions above is equivalent to the following definition.

Lemma 2.2 *An irreducible polynomial of degree m over GF(2) is primitive if the smallest positive integer n for which it divides* $X^n - 1$ *is* $n = 2^m - 1$.

Proof To have $X^\ell \bmod p(X) \neq X^k \bmod p(X)$ for $0 \leq k < \ell \leq q-2$ it must be $(X^\ell - X^k) \bmod p(X) \neq 0$, that is, $X^k(X^{\ell-k} - 1) \bmod p(X) \neq 0$. Since X^k is not a factor of $p(X)$, we have $(X^{\ell-k} - 1) \bmod p(X) \neq 0$, which is like to say that $p(X)$ does not divide $X^n - 1$ for $n \leq q-2 = 2^m - 2$. Moreover, requiring that $X^{q-1} \bmod p(X) = 1$ is equivalent to $(X^{q-1} - 1) \bmod p(X) = 0$, that is, $p(X)$ must divide $X^{q-1} - 1$. □

The advantage in using a primitive polynomial in (2.10) is that the nonzero elements of GF(q) with $q = 2^m$ can also be expressed, keeping in mind the equivalence under reduction modulo $p(X)$, as the *integer powers* of the element X, called the primitive element of the field, that is, $F_{2^m} = \left\{0, 1, X, X^2, X^3, \ldots, X^{q-2}\right\}$. It is common to denote the *primitive element* by α, and thus

$$F_{2^m} = \left\{0, 1, \alpha, \alpha^2, \alpha^3, \ldots, \alpha^{q-2}\right\}, \tag{2.13}$$

where the elements are all distinct, and $\alpha^{q-1} = 1$ (in algebra language, the element α has order $q-1$). With this notation, the multiplication between two elements α^ℓ and α^k can be calculated either by (2.10), that is, $\left(\alpha^\ell \alpha^k\right) \bmod p(\alpha)$, or simply as $\alpha^{(\ell+k) \bmod (q-1)}$, and the multiplicative inverse of α^ℓ as $\alpha^{q-1-\ell}$. For the addition we have $\left(\alpha^\ell + \alpha^k\right) \bmod p(\alpha) = \alpha^\ell \bmod p(\alpha) + \alpha^k \bmod p(\alpha)$.

In summary, operations in GF(2^m) algebra are the usual polynomial operations, computed modulo a primitive polynomial $p(\alpha)$. Note that, since $p(\alpha) \bmod p(\alpha) = 0$, we can say that, in GF(q), the primitive element α is a root of $p(\alpha)$.

2.1.1.2.3 *Third Method to Implement Multiplications/Divisions in GF(2^m)*

A third possibility is to associate suitable matrices with the field elements, so that addition and multiplication in the field are the usual operations on the corresponding matrices. This can be obtained by introducing the $m \times m$ *companion matrix* of a primitive monic polynomial $p(X) = p_0 + p_1 X + \cdots + p_{m-1}X^{m-1} + X^m$:

[1] It is possible to show that every irreducible polynomial satisfies condition 2.

$$A = \begin{bmatrix} 0 & 1 & 0 & \cdots & 0 \\ 0 & 0 & 1 & \cdots & 0 \\ \vdots & \vdots & \vdots & \vdots & \vdots \\ 0 & 0 & 0 & 0 & 1 \\ -p_0 & -p_1 & -p_2 & \cdots & -p_{m-1} \end{bmatrix} \tag{2.14}$$

with elements in GF(2). Now, it is easy to verify that $\det(X\boldsymbol{I} - \boldsymbol{A}) = p(X)$, that is, $p(X)$ is the characteristic polynomial of the companion matrix $\boldsymbol{A}$, and hence $p(\boldsymbol{A}) = \det(\boldsymbol{A} - \boldsymbol{A}) = \boldsymbol{0}$. Thus, similarly to the previous case, the matrix polynomials $\boldsymbol{A}^\ell \bmod p(\boldsymbol{A})$, $\ell \in \{0, 1, \ldots, q-2\}$, are all distinct, and $\boldsymbol{A}^{q-1} \bmod p(\boldsymbol{A}) = \boldsymbol{I}$. Therefore, the field set can be taken as $F_{2^m} = \{\boldsymbol{0}, \boldsymbol{I}, \boldsymbol{A}, \boldsymbol{A}^2, \boldsymbol{A}^3, \ldots, \boldsymbol{A}^{q-2}\}$, multiplication and division follow the exponential rules, and the sum of elements is just the element-wise sum in GF(2) of the corresponding matrices. Moreover, it can be verified that the product between the binary m-tuple corresponding to α^ℓ with the companion matrix $\boldsymbol{A}^k$ provides directly the m-tuple corresponding to $\alpha^\ell \alpha^k$.

Example 2.2 The different representations of the elements for $\mathrm{GF}(2^3)$ are reported in Table 2.2 for the primitive polynomial $p(X) = 1 + X^2 + X^3$. Note that, for example, α^5 and $1 + \alpha$ represent the same element under reduction modulo $p(\alpha) = 1 + \alpha^2 + \alpha^3$. Consider the elements $\alpha^3 = \begin{bmatrix} 1 & 0 & 1 \end{bmatrix}$ and $\alpha^5 = \begin{bmatrix} 1 & 1 & 0 \end{bmatrix}$. Their product can be computed as

$$\alpha^3 \alpha^5 = \alpha^8 = \alpha = \begin{bmatrix} 0 & 1 & 0 \end{bmatrix} \tag{2.15}$$

or

$$((1 + \alpha^2)(1 + \alpha)) \bmod (1 + \alpha^2 + \alpha^3) = \alpha. \tag{2.16}$$

or

$$\begin{bmatrix} 1 & 0 & 1 \\ 1 & 1 & 1 \\ 1 & 1 & 0 \end{bmatrix} \begin{bmatrix} 1 & 1 & 0 \\ 0 & 1 & 1 \\ 1 & 0 & 0 \end{bmatrix} = \begin{bmatrix} 0 & 1 & 0 \\ 0 & 0 & 1 \\ 1 & 0 & 1 \end{bmatrix} = \boldsymbol{A} \tag{2.17}$$

or

$$\begin{bmatrix} 1 & 0 & 1 \end{bmatrix} \begin{bmatrix} 1 & 1 & 0 \\ 0 & 1 & 1 \\ 1 & 0 & 0 \end{bmatrix} = \begin{bmatrix} 0 & 1 & 0 \end{bmatrix} = \alpha\,. \tag{2.18}$$

The addition $\alpha^3 + \alpha^5$ is computed as $[1\ 0\ 1] + [1\ 1\ 0] = [0\ 1\ 1] = \alpha^6$, or

$$(\alpha^3 + \alpha^5) \bmod (1 + \alpha^2 + \alpha^3) = (1 + \alpha^2) + (1 + \alpha) = \alpha + \alpha^2 = \alpha^6 \tag{2.19}$$

or

$$\begin{bmatrix} 1 & 0 & 1 \\ 1 & 1 & 1 \\ 1 & 1 & 0 \end{bmatrix} + \begin{bmatrix} 1 & 1 & 0 \\ 0 & 1 & 1 \\ 1 & 0 & 0 \end{bmatrix} = \begin{bmatrix} 0 & 1 & 1 \\ 1 & 0 & 0 \\ 0 & 1 & 0 \end{bmatrix} = \boldsymbol{A}^6\,. \tag{2.20}$$

Table 2.2 GF(2^3) elements representations, $p(X) = 1 + X^2 + X^3$

Power α^ℓ	**Polynomial** $\alpha^\ell \bmod p(\alpha)$	**3-tuple**	**Matrix** A^ℓ
0	0	$\begin{bmatrix}0 & 0 & 0\end{bmatrix}$	$\begin{bmatrix}0 & 0 & 0\\0 & 0 & 0\\0 & 0 & 0\end{bmatrix}$
1	1	$\begin{bmatrix}1 & 0 & 0\end{bmatrix}$	$\begin{bmatrix}1 & 0 & 0\\0 & 1 & 0\\0 & 0 & 1\end{bmatrix}$
α	α	$\begin{bmatrix}0 & 1 & 0\end{bmatrix}$	$\begin{bmatrix}0 & 1 & 0\\0 & 0 & 1\\1 & 0 & 1\end{bmatrix}$
α^2	α^2	$\begin{bmatrix}0 & 0 & 1\end{bmatrix}$	$\begin{bmatrix}0 & 0 & 1\\1 & 0 & 1\\1 & 1 & 1\end{bmatrix}$
α^3	$1 + \alpha^2$	$\begin{bmatrix}1 & 0 & 1\end{bmatrix}$	$\begin{bmatrix}1 & 0 & 1\\1 & 1 & 1\\1 & 1 & 0\end{bmatrix}$
α^4	$1 + \alpha + \alpha^2$	$\begin{bmatrix}1 & 1 & 1\end{bmatrix}$	$\begin{bmatrix}1 & 1 & 1\\1 & 1 & 0\\0 & 1 & 1\end{bmatrix}$
α^5	$1 + \alpha$	$\begin{bmatrix}1 & 1 & 0\end{bmatrix}$	$\begin{bmatrix}1 & 1 & 0\\0 & 1 & 1\\1 & 0 & 0\end{bmatrix}$
α^6	$\alpha + \alpha^2$	$\begin{bmatrix}0 & 1 & 1\end{bmatrix}$	$\begin{bmatrix}0 & 1 & 1\\1 & 0 & 0\\0 & 1 & 0\end{bmatrix}$

2.1.2 Vectors in GF(q)^n

Having defined the operations between scalars (elements of a finite field), we can now define the vector space of n-tuples.

Definition 2.6 (The vector space GF$(q)^n$) The set of all n-tuples (vectors) of elements from F_q

$$\boldsymbol{c} = \begin{bmatrix}c_1 & c_2 & \dots & c_n\end{bmatrix} \qquad c_i \in F_q \tag{2.21}$$

under component-wise vector addition

$$\boldsymbol{c} + \boldsymbol{d} = \begin{bmatrix}c_1 + d_1 & c_2 + d_2 & \dots & c_n + d_n\end{bmatrix} \quad c_i, d_i \in F_q, \tag{2.22}$$

where the sum $c_i + d_i$ is the sum in GF(q), and under component-wise scalar multiplication

$$a\,\boldsymbol{c} = a\begin{bmatrix} c_1 & c_2 & \dots & c_n \end{bmatrix} = \begin{bmatrix} a \cdot c_1 & a \cdot c_2 & \dots & a \cdot c_n \end{bmatrix} \qquad a, c_i \in F_q, \quad (2.23)$$

where the product $a \cdot c_i$ is the product in GF(q), form the *vector space* GF$(q)^n$, also denoted with $\mathcal{S}$. The total number of vectors in $\mathcal{S}$ is $|\mathcal{S}| = q^n$, and the dimension of the vector space (maximum number of linearly independent vectors) is

$$\dim(\mathcal{S}) = n\,. \qquad (2.24)$$

We also define the *inner product* between two vectors as the scalar

$$\langle \boldsymbol{c}, \boldsymbol{d} \rangle = c_1 \cdot d_1 + c_2 \cdot d_2 + \cdots + c_n \cdot d_n, \qquad (2.25)$$

where all operations are in GF(q). If $\langle \boldsymbol{c}, \boldsymbol{d} \rangle = 0$, we say that the vectors are orthogonal. Note that in finite fields a nonzero vector can be orthogonal to itself.[2] For example, in GF$(2)^3$, we have $\langle \begin{bmatrix} 1 & 0 & 1 \end{bmatrix}, \begin{bmatrix} 1 & 0 & 1 \end{bmatrix} \rangle = 1 + 1 = 0$.

If we select $k \le n$ linearly independent vectors $\boldsymbol{c}_1, \boldsymbol{c}_2, \dots, \boldsymbol{c}_k$, each of length n, and form all possible linear combinations of them, we obtain a *subspace* $\mathcal{C}$, which is itself a vector space. The number of vectors in $\mathcal{C}$ is $|\mathcal{C}| = q^k$, and its dimension is

$$\dim(\mathcal{C}) = k\,. \qquad (2.26)$$

The set of vectors in $\mathcal{S}$ which are orthogonal to all vectors in $\mathcal{C}$ (it suffices that they are orthogonal to a base of $\mathcal{C}$) is also a subspace of $\mathcal{S}$, called *null space* or *orthogonal space* of $\mathcal{C}$, indicated as $\mathcal{C}^\perp$. We have

$$\langle \boldsymbol{c}, \boldsymbol{d} \rangle = 0 \qquad \forall \boldsymbol{c} \in \mathcal{C}, \boldsymbol{d} \in \mathcal{C}^\perp\,. \qquad (2.27)$$

It can also be proved that $\dim(\mathcal{C}) + \dim(\mathcal{C}^\perp) = n$, and therefore

$$\dim\left(\mathcal{C}^\perp\right) = n - k\,. \qquad (2.28)$$

The number of vectors in the null space is $|\mathcal{C}^\perp| = q^{n-k}$.

2.1.2.1 Vectors as Polynomials

Vectors in GF$(q)^n$ can also be represented by polynomials over GF(q). By indicating with X the indeterminate, the vector $\boldsymbol{c} = \begin{bmatrix} c_0 & c_1 & \cdots & c_{n-1} \end{bmatrix}$ is associated with the polynomial

$$c(X) = c_0 + c_1 X + c_2 X^2 + \cdots + c_{n-1} X^{n-1} = \begin{bmatrix} c_0 & c_1 & \cdots & c_{n-1} \end{bmatrix} \begin{bmatrix} 1 \\ X \\ X^2 \\ \vdots \\ X^{n-1} \end{bmatrix}, \qquad (2.29)$$

where the c_i and all operations are in GF(q). In the polynomial representation, the sum of two vectors $\boldsymbol{c} + \boldsymbol{d}$, and the scalar multiplication $a\,\boldsymbol{c}$, become $c(X) + d(X)$ and $a\,c(X)$, respectively.

[2] Note that, on the contrary, for $\mathbb{R}^n$ the product $\langle \boldsymbol{c}, \boldsymbol{c} \rangle$ is positive definite, that is, $\langle \boldsymbol{c}, \boldsymbol{c} \rangle = 0 \Leftrightarrow \boldsymbol{c} = \boldsymbol{0}$.

Denoting again by α a primitive element of GF(q), the polynomial (2.29) evaluated at the nonzero field elements is

$$c(\alpha^\ell) = \sum_{i=0}^{n-1} c_i \alpha^{\ell i}, \quad \ell = 0, \dots, q-2. \tag{2.30}$$

2.1.2.2 Discrete Fourier Transform

In the previous equation, assume the vector $\boldsymbol{c}$ has length $n = q - 1$. In this case, we have $\alpha^n = \alpha^{q-1} = 1$, and thus the vector $[c(1)\ c(\alpha)\ \cdots\ c(\alpha^{n-1})]$ can be interpreted as the discrete Fourier transform (DFT) of the vector $\boldsymbol{c} = [c_0\ c_1\ \cdots\ c_{n-1}]$, where α plays the role that $e^{-j2\pi/n}$ has for the DFT in the complex field.[3]

We can hence define the DFT of the polynomial $c(X) = c_0 + c_1 X + c_2 X^2 + \cdots + c_{n-1}X^{n-1}$ as the polynomial $C(X)$

$$C(X) = C_0 + C_1 X + \cdots + C_{n-1}X^{n-1} \tag{2.31}$$

with coefficients $C_\ell = c(\alpha^\ell)$.

The inverse discrete Fourier transform (IDFT) is in this case

$$c_i = \frac{1}{n}\sum_{\ell=0}^{n-1} C_\ell\, \alpha^{-\ell i} = \frac{1}{n} C(\alpha^{-i}), \quad i = 0, \dots, n-1, \tag{2.32}$$

where $1/n = n^{-1}$ and $n = 1 + 1 + \cdots + 1$ in GF(q). Note that for $q = 2^m$, we have $n = 1 + 1 + \cdots + 1 = 1$ and therefore $n^{-1} = 1$. So, similarly to the complex field case, it is possible to use fast algorithms to compute the DFT and IDFT defined above.

Remark 2.1 (Fields and field sets). To facilitate readability, the distinction between the notion of a field and a field set will be omitted in the remainder of the book.

2.2 Communication Channel and Decision Rules

When data is transmitted through a noisy channel, the output of the channel is related to the transmitted data only in a probabilistic manner. Assume for example the scheme in Figure 1.8, where an n-tuple of symbols $\boldsymbol{c}$ from a finite field is transmitted through a noisy channel. The output of the channel is indicated as $\boldsymbol{y}$, and the relation between input and output is described by a statistical model, for example, by the conditional probability $P(\boldsymbol{y}|\boldsymbol{c})$. To allow error control, among the q^n possible n-tuples, only a subset of them is used, so as to make as more reliable as possible recognition of the transmitted n-tuple. The subset is called *code* and its elements are called *codewords*.

Definition 2.7 (Communication channel) Assume that, with a probability $P(\boldsymbol{c})$, the codeword $\boldsymbol{c}$ is selected from the set $\mathcal{C} = \{\boldsymbol{c}_1, \boldsymbol{c}_2, \dots, \boldsymbol{c}_M\}$, and transmitted at a given

[3] In the complex field, an nth root of unity is $e^{-j2\pi/n}$, where $j = \sqrt{-1}$. In GF(q), from (2.12), it is $\alpha^n = 1$ (modulo $p(\alpha)$) for any n multiple of $q - 1$, and in particular for $n = q - 1$.

time through a *communication channel*, whose output is indicated as $\boldsymbol{y}$. We assume the channel output $\boldsymbol{y}$ does not depend on previous or future inputs. Also, due to noise and interference, the channel output is a nondeterministic function of the input. This statistical behavior is completely characterized by the *transition probability* $P(\boldsymbol{y}|\boldsymbol{c})$. When seen as a function of $\boldsymbol{c}$ with $\boldsymbol{y}$ fixed, the function $P(\boldsymbol{y}|\boldsymbol{c})$ is called a *likelihood function*. The channel output $\boldsymbol{y}$ is referred to as the *observable*.

Definition 2.8 (Decoder) Denote by $\mathcal{Y}$ the set of all possible channel outputs, and by $\mathcal{Y}^n$ the set of n-tuples with elements in $\mathcal{Y}$. A *decoder* is a mapping $D\colon \mathcal{Y}^n \to \mathcal{C}$, associating to the channel output $\boldsymbol{y}$ the hypothesized codeword $\hat{\boldsymbol{c}} \in \mathcal{C}$ (decision). We have a *decoding error* if $\hat{\boldsymbol{c}} \neq \boldsymbol{c}$.

Definition 2.9 (Block error probability) The error probability for the specific codeword $\boldsymbol{c}_\ell$ is $P(\hat{\boldsymbol{c}} \neq \boldsymbol{c}|\boldsymbol{c} = \boldsymbol{c}_\ell)$. The *word error probability* or *block error probability* is $P_B = P(\hat{\boldsymbol{c}} \neq \boldsymbol{c}) = \sum_\ell P(\hat{\boldsymbol{c}} \neq \boldsymbol{c}|\boldsymbol{c} = \boldsymbol{c}_\ell)P(\boldsymbol{c}_\ell)$, where $P(\boldsymbol{c}_\ell)$ is the a priori probability of the ℓth codeword.

Definition 2.10 (Pairwise error probability) Assume the vector $\boldsymbol{c}_\ell$ has been transmitted, and that the decoder has just two possible choices, the correct one $\boldsymbol{c}_\ell$, and an alternative codeword $\boldsymbol{c}_m$. The *pairwise error probability* $P(\boldsymbol{c}_\ell \to \boldsymbol{c}_m)$ is the probability that, having only these two choices, the decoder makes the wrong decision, choosing $\boldsymbol{c}_m$ instead of $\boldsymbol{c}_\ell$.

Definition 2.11 (MAP and ML decoding criteria) If the decoder knows only the a priori probabilities $P(\boldsymbol{c}_\ell)$, $\ell = 1,\dots,M$, without knowing $\boldsymbol{y}$, then the block error probability is minimized by choosing the most probable codeword, that is, $\hat{\boldsymbol{c}} = \arg\max_{\boldsymbol{c}\in\mathcal{C}} P(\boldsymbol{c})$. If the decoder has also access to the observable $\boldsymbol{y}$ and knows the channel model $P(\boldsymbol{y}|\boldsymbol{c})$, then it can update its knowledge to the a posteriori probabilities

$$P(\boldsymbol{c}_\ell|\boldsymbol{y}) = \frac{P(\boldsymbol{y}|\boldsymbol{c}_\ell)P(\boldsymbol{c}_\ell)}{P(\boldsymbol{y})}, \quad \ell = 1,2,\dots,M\,. \tag{2.33}$$

Thus, for a given $\boldsymbol{y}$, the block error probability is minimized by choosing the codeword with the largest a posteriori probability. The maximum a posteriori (MAP) decoder decides for the codeword maximizing the a posteriori probability $P(\boldsymbol{c}_\ell|\boldsymbol{y})$:

$$\hat{\boldsymbol{c}} = \arg\max_{\boldsymbol{c}\in\mathcal{C}} P(\boldsymbol{c}|\boldsymbol{y}) = \arg\max_{\boldsymbol{c}\in\mathcal{C}} P(\boldsymbol{y}|\boldsymbol{c})P(\boldsymbol{c})\,. \tag{2.34}$$

The ML decoder decides for the codeword maximizing the likelihood $P(\boldsymbol{y}|\boldsymbol{c})$:

$$\hat{\boldsymbol{c}} = \arg\max_{\boldsymbol{c}\in\mathcal{C}} P(\boldsymbol{y}|\boldsymbol{c})\,. \tag{2.35}$$

If the codewords have the same a priori probabilities, that is, if $P(\boldsymbol{c}_\ell) = 1/M$, $\ell = 1,2,\dots,M$, then the MAP and the ML decoders are equivalent.

In the previous definitions we implicitly assumed that there is exactly one codeword maximizing the a posteriori probability or the likelihood. While for continuous output channels, like for instance the unquantized Gaussian additive channel, this is true with probability one, for discrete output channels, such as the BEC, this may not be the case as there could be multiple codewords maximizing the a posteriori probability or the likelihood (tie).

Remark 2.2 (Incomplete decoding). In case of tie we can either: (1) assume that the MAP or ML decoder picks randomly one of the tied codewords; or that (2) the decoder raises a red flag and declares a decoding failure. We will indicate the decoder (2) as an *incomplete* decoder. In the following, incomplete decoding will be always assumed when analyzing the code performance over erasure channels.

The previous decoder strategies are designed to minimize the block error probability and are sometimes called *block-wise* decoders. If the aim is to minimize the symbol error probability, we obtain the following:

Definition 2.12 (Symbol-wise MAP decoder) The symbol-wise MAP decoder decides for the symbol c_i maximizing the a posteriori probability $P(c_i|\boldsymbol{y})$:

$$\hat{c}_i = \arg\max_{c_i \in \mathrm{GF}(q)} P(c_i|\boldsymbol{y}) = \arg\max_{c_i \in \mathrm{GF}(q)} P(\boldsymbol{y}|c_i)P(c_i)\,. \tag{2.36}$$

Definition 2.13 (Symbol-wise ML decoder) The symbol-wise ML decoder decides for the symbol c_i maximizing the likelihood $P(\boldsymbol{y}|c_i)$:

$$\hat{c}_i = \arg\max_{c_i \in \mathrm{GF}(q)} P(\boldsymbol{y}|c_i)\,. \tag{2.37}$$

If the symbols have the same a priori probabilities, the symbol-wise MAP and ML decoders are equivalent.

2.3 Linear Block Codes

The mathematical structure of vector space allows to simplify the analysis, design, and implementation of codes.

Definition 2.14 (Linear block code) A linear block code $\mathcal{C}(n,k)$ over $\mathrm{GF}(q)^n$ is a subspace of dimension k of the vector space $\mathrm{GF}(q)^n$. The q^k vectors in $\mathcal{C}(n,k)$ are the *codewords*, the *blocklength* is n, and k is called the *dimension* of the code. The ratio $R = k/n$ is called the *code rate*.

A linear block code is therefore a set of q^k n-tuples over $\mathrm{GF}(q)$ (codewords), such that the sum of two codewords is a codeword, and the product of any codeword with a field element is a codeword. Since $\mathcal{C}(n,k)$ is a vector space, the all-zero n-tuple is always a codeword.

Definition 2.15 (Hamming weight, Hamming distance) The *Hamming weight* (or just *weight*) $w_{\mathrm{H}}(\boldsymbol{c})$ of the n-tuple $\boldsymbol{c}$ is the number of nonzero components in $\boldsymbol{c}$. The *Hamming distance* $d_{\mathrm{H}}(\boldsymbol{c},\boldsymbol{d})$ between the two n-tuples $\boldsymbol{c},\boldsymbol{d}$ is the number of positions in which they differ. It follows that $d_{\mathrm{H}}(\boldsymbol{c},\boldsymbol{d}) = w_{\mathrm{H}}(\boldsymbol{c}-\boldsymbol{d})$.

Example 2.3 For vectors with integer elements, the weight of $\begin{bmatrix}0 & 1 & 0 & 0 & 2\end{bmatrix}$ is 2, the weight of $\begin{bmatrix}2 & 1 & 0 & 1 & 1\end{bmatrix}$ is 4, and their Hamming distance is

$$d_{\mathrm{H}}\left(\begin{bmatrix}0 & 1 & 0 & 0 & 2\end{bmatrix}, \begin{bmatrix}2 & 1 & 0 & 1 & 1\end{bmatrix}\right) = 3 \tag{2.38}$$

which is also the weight of the difference

$$\begin{bmatrix}0 & 1 & 0 & 0 & 2\end{bmatrix} - \begin{bmatrix}2 & 1 & 0 & 1 & 1\end{bmatrix} = \begin{bmatrix}-2 & 0 & 0 & -1 & 1\end{bmatrix}. \tag{2.39}$$

Definition 2.16 (Minimum Hamming distance of a code) The *minimum Hamming distance* $d_{\min}$ of a code $\mathcal{C} = \left\{\boldsymbol{c}_1, \boldsymbol{c}_2, \ldots, \boldsymbol{c}_{q^k}\right\}$ is the minimum Hamming distance between any two different codewords

$$d_{\min} = \min_{i \neq j} d_{\mathrm{H}}(\boldsymbol{c}_i, \boldsymbol{c}_j). \qquad \boldsymbol{c}_i, \boldsymbol{c}_j \in \mathcal{C} \tag{2.40}$$

Definition 2.17 (Weight distribution of a code) Let A_w denote the number of codewords of weight w of a code $\mathcal{C} = \left\{\boldsymbol{c}_1, \boldsymbol{c}_2, \ldots, \boldsymbol{c}_{q^k}\right\}$. The numbers $A_0, A_1, A_2, \ldots, A_n$ form the *weight distribution* or *weight (distance) spectrum* of the code, with associated *weight enumerating function* $A(W) = A_0 + A_1 W + \cdots + A_n W^n$. For linear codes we always have $A(0) = A_0 = 1$ and $A(1) = A_0 + A_1 + \cdots + A_n = q^k$.

For a linear block code, any difference $\boldsymbol{c}_i - \boldsymbol{c}_j$ is also a codeword. It follows that, for any linear block code $\mathcal{C} = \left\{\boldsymbol{c}_1, \boldsymbol{c}_2, \ldots, \boldsymbol{c}_{q^k}\right\}$, the minimum distance is coincident with the smallest weight among all nonzero codewords. By indicating the all-zero codeword as $\boldsymbol{c}_1 = \begin{bmatrix}0 & 0 & \cdots & 0\end{bmatrix}$, we have for a linear code

$$d_{\min} = \min_{i \neq j} d_{\mathrm{H}}(\boldsymbol{c}_i, \boldsymbol{c}_j) = \min_{i > 1} w_{\mathrm{H}}(\boldsymbol{c}_i) \tag{2.41}$$

which is also the minimum $w > 0$ such that $A_w > 0$.

Example 2.4 Consider the subspace of $\mathcal{S} = \mathrm{GF}(2)^5$ generated by the vectors $[10001], [01111]$. Taking all linear combinations of these two vectors gives the linear block code

$$\begin{aligned}\mathcal{C}(5,2) &= \left\{\begin{bmatrix}0 & 0 & 0 & 0 & 0\end{bmatrix}, \begin{bmatrix}1 & 1 & 1 & 1 & 0\end{bmatrix},\right.\\ &\qquad \left.\begin{bmatrix}1 & 0 & 0 & 0 & 1\end{bmatrix}, \begin{bmatrix}0 & 1 & 1 & 1 & 1\end{bmatrix}\right\}\\ &= \{\boldsymbol{c}_1, \boldsymbol{c}_2, \boldsymbol{c}_3, \boldsymbol{c}_4\}\end{aligned} \tag{2.42}$$

with $|\mathcal{C}(5,2)| = 2^2 = 4$, $w_{\mathrm{H}}(\boldsymbol{c}_1) = 0, w_{\mathrm{H}}(\boldsymbol{c}_2) = 4, w_{\mathrm{H}}(\boldsymbol{c}_3) = 2, w_{\mathrm{H}}(\boldsymbol{c}_4) = 4$. The weight distribution is $A_0 = 1, A_2 = 1, A_4 = 2$, and the minimum distance of the code is $d_{\min} = 2$.

In the scheme of Figure 1.8, the source is assumed to produce symbols in $\mathrm{GF}(q)$. The *encoder* has the aim to associate, with a one-to-one mapping, all possible source

k-tuples in GF(q) to the q^k distinct codewords in $\mathcal{C}(n,k)$. The simplest mapping is linear, which results in a linear encoder, specified by its generator matrix.

Definition 2.18 (Generator matrix) Consider k linearly independent vectors $\mathbf{g}_1, \mathbf{g}_2, \ldots, \mathbf{g}_k$ in GF$(q)^n$, where $\mathbf{g}_\ell = \begin{bmatrix} g_{\ell,1} & g_{\ell,2} & \cdots & g_{\ell,n} \end{bmatrix}$ and construct the $k \times n$ matrix

$$\boldsymbol{G} = \begin{bmatrix} g_{11} & g_{12} & \cdots & g_{1n} \\ g_{21} & g_{22} & \cdots & g_{2n} \\ \vdots & \vdots & & \vdots \\ g_{k1} & g_{k2} & \cdots & g_{kn} \end{bmatrix} = \begin{bmatrix} \mathbf{g}_1 \\ \mathbf{g}_2 \\ \vdots \\ \mathbf{g}_k \end{bmatrix}. \tag{2.43}$$

Then, the rowspace (all possible linear combinations of the rows) of $\boldsymbol{G}$ is a linear code $\mathcal{C}(n,k)$, and $\boldsymbol{G}$ is called a *generator matrix* of the code.

A linear encoder associates the source k-tuple $\boldsymbol{u}_m = \begin{bmatrix} u_{m1} & u_{m2} & \cdots & u_{mk} \end{bmatrix}$ (*information symbols* or *message*) to the n-tuple $\boldsymbol{c}_m = \begin{bmatrix} c_{m1} & c_{m2} & \cdots & c_{mn} \end{bmatrix}$ (*codeword*), via the linear transformation

$$\boldsymbol{c}_m = \boldsymbol{u}_m \boldsymbol{G} = u_{m1}\mathbf{g}_1 + u_{m2}\mathbf{g}_2 + \cdots + u_{mk}\mathbf{g}_k \,. \tag{2.44}$$

Definition 2.19 (Parity-check matrix) Given the subspace $\mathcal{C}(n,k)$ with generator matrix $\boldsymbol{G}$, its null space $\mathcal{C}^\perp$ is also a subspace and therefore can be interpreted as a linear block code. The code $\mathcal{C}^\perp(n, n-k)$ has dimension $n-k$, and is called *dual code* of $\mathcal{C}$. Let's call $\boldsymbol{H}$ an $(n-k)\times n$ generator matrix of $\mathcal{C}^\perp$. Then, a vector $\boldsymbol{c}$ is a codeword in $\mathcal{C}(n,k)$ if and only if

$$\boldsymbol{H}\,\boldsymbol{c}^{\mathsf{T}} = \mathbf{0} \tag{2.45}$$

or equivalently, indicating with h_{ij} the elements of $\boldsymbol{H}$,

$$\begin{aligned} h_{11}c_1 + h_{12}c_2 + \cdots + h_{1n}c_n &= 0 \\ h_{21}c_1 + h_{22}c_2 + \cdots + h_{2n}c_n &= 0 \\ &\cdots \\ h_{(n-k)1}c_1 + h_{(n-k)2}c_2 + \cdots + h_{(n-k)n}c_n &= 0\,. \end{aligned} \tag{2.46}$$

These are called *parity-check equations*, since in the binary case they are simply checks for even parity on certain sets of symbols in the codeword.[4] The matrix $\boldsymbol{H}$ is called the *parity-check matrix* of the code $\mathcal{C}(n,k)$.

Equation (2.45) holds for every vector in $\mathcal{C}$, and, in particular, it holds for the k basis vectors of the matrix $\boldsymbol{G}$. It follows that

$$\boldsymbol{H}\boldsymbol{G}^{\mathsf{T}} = \mathbf{0}\,. \tag{2.47}$$

Definition 2.19 considers a full-rank matrix, that is rank$(\boldsymbol{H}) = n-k$. Sometimes, a parity-check matrix (in the sense of a matrix satisfying (2.45), (2.47)) is used with

[4] For the binary field, an equation like $a+b+c=0$ is satisfied by any (a,b,c) with an even number of 1. From here the word "parity," which, with a little abuse, is extended also to nonbinary codes.

rank($\boldsymbol{H}$) $= n - k$, but more than $n - k$ rows (which are necessarily linearly dependent). Unless otherwise stated, we will consider full-rank parity-check matrices in the following, so with rank($\boldsymbol{H}$) $= n - k$ rows.

Definition 2.20 (Error syndrome) Assume a codeword $\boldsymbol{c} \in \mathcal{C}(n,k)$ is transmitted over a channel which, due to noise and disturbances, may change some elements with other elements in $\mathrm{GF}(q)$. Thus, the received vector $\boldsymbol{y} \in \mathrm{GF}(q)^n$ can be written as $\boldsymbol{y} = \boldsymbol{c} + \boldsymbol{e}$, where $\boldsymbol{e} \in \mathrm{GF}(q)^n$ is the error vector. Then, the vector

$$\boldsymbol{s} = \boldsymbol{y}\,\boldsymbol{H}^{\mathsf{T}} \in \mathrm{GF}(q)^{n-k} \tag{2.48}$$

which can be computed at the receiver, depends only on $\boldsymbol{e}$ and not on the particular codeword transmitted, since $\boldsymbol{s}^{\mathsf{T}} = \boldsymbol{H}\,(\boldsymbol{c}^{\mathsf{T}} + \boldsymbol{e}^{\mathsf{T}}) = \boldsymbol{H}\,\boldsymbol{c}^{\mathsf{T}} + \boldsymbol{H}\,\boldsymbol{e}^{\mathsf{T}} = \boldsymbol{H}\,\boldsymbol{e}^{\mathsf{T}}$. The vector $\boldsymbol{s}$ is the error pattern *syndrome* of the vector $\boldsymbol{y}$. Clearly, the error syndrome is $\boldsymbol{0}$ if and only if $\boldsymbol{y}$ is a codeword, that is, if and only if $\boldsymbol{e}$ is a codeword (including the all-zero vector).

When an error correcting code is used over a channel with independent errors, its error correction capability depends on the distances among codewords, or equivalently, for linear codes, on its weight distribution. This distribution does not change if we apply row operations on $\boldsymbol{G}$, since these change only the base of the subspace, but do not change the code itself. The code weight distribution is also preserved, leaving unaltered the error correction, if we permute all symbols for all codewords with any given permutation.

Definition 2.21 (Equivalent codes) Two codes that differ only for symbols permutation (same permutation for all codewords), are called *equivalent codes*. Then, if $\boldsymbol{G}'$ is obtained by row operations and/or column permutation of a generator matrix $\boldsymbol{G}$, the two generated codes are *equivalent codes*.

Definition 2.22 (Systematic encoder) Suppose $\mathcal{C}$ is a code, with an encoder which maps the message $\begin{bmatrix} u_1 & u_2 & \cdots & u_k \end{bmatrix}$ into the codeword $\begin{bmatrix} c_1 & c_2 & \cdots & c_n \end{bmatrix}$. The encoder is said to be systematic if each codeword is composed by appending $n - k$ symbols to the k message symbols.

Thus, for linear codes, a *systematic encoder* is obtained with the *generator matrix in systematic form*

$$\boldsymbol{G} = \begin{bmatrix} 1 & 0 & \cdots & 0 & p_{11} & \cdots & p_{1,n-k} \\ 0 & 1 & \cdots & 0 & p_{21} & \cdots & p_{2,n-k} \\ \vdots & \vdots & & \vdots & & & \\ 0 & 0 & \cdots & 1 & p_{k1} & \cdots & p_{k,n-k} \end{bmatrix} = \begin{bmatrix} \boldsymbol{I}_k | \boldsymbol{P} \end{bmatrix}. \tag{2.49}$$

With this generator matrix, the input k-tuple $\boldsymbol{u} = [u_1\, u_2 \cdots u_k]$ is mapped to

$$\begin{bmatrix} c_1 & c_2 & \cdots & c_n \end{bmatrix} = \boldsymbol{u}\,\boldsymbol{G} = \begin{bmatrix} u_1 & u_2 & \cdots & u_k & c_{k+1} & c_{k+2} & \cdots & c_n \end{bmatrix} \tag{2.50}$$

that is, the codeword is composed by the k information symbols followed by $n-k$ additional symbols, called *redundant* or *check symbols*, obtained by linear combinations of the input symbols (through the $k \times (n - k)$ submatrix $\boldsymbol{P}$).

It is possible then to verify that, by row operations and column permutations of the generator matrix, every encoder is equivalent to an encoder in systematic form. We also observe that, if $\boldsymbol{G} = \left[\boldsymbol{I}_k|\boldsymbol{P}\right]$ is in systematic form, then a possible parity-check matrix is simply $\boldsymbol{H} = \left[-\boldsymbol{P}^{\mathsf{T}}|\boldsymbol{I}_{n-k}\right]$. In fact, we have

$$\boldsymbol{G}\boldsymbol{H}^{\mathsf{T}} = \left[\boldsymbol{I}_k|\boldsymbol{P}\right]\left[\frac{-\boldsymbol{P}}{\boldsymbol{I}_{n-k}}\right] = -\boldsymbol{P} + \boldsymbol{P} = \boldsymbol{0}. \tag{2.51}$$

Example 2.5 Consider the code of Example 2.4. A generator matrix is

$$\boldsymbol{G} = \begin{bmatrix} 1 & 0 & 0 & 0 & 1 \\ 0 & 1 & 1 & 1 & 1 \end{bmatrix} \tag{2.52}$$

which is already in systematic form. A parity-check matrix is then

$$\boldsymbol{H} = \begin{bmatrix} 0 & 1 & 1 & 0 & 0 \\ 0 & 1 & 0 & 1 & 0 \\ 1 & 1 & 0 & 0 & 1 \end{bmatrix}. \tag{2.53}$$

Besides through row operations and column permutations of $\boldsymbol{G}$, another approach to build a systematic encoder given a nonsystematic generator matrix is to first partition it as $\boldsymbol{G} = [\boldsymbol{G}_1|\boldsymbol{G}_2]$, with $\boldsymbol{G}_1$ the $k \times k$ submatrix formed by the first k columns. Then, if $\boldsymbol{G}_1$ is nonsingular,[5] we build $\boldsymbol{G}_{\text{syst}} = \boldsymbol{G}_1^{-1}\boldsymbol{G} = [\boldsymbol{I}_k|\boldsymbol{P}]$, where $\boldsymbol{P} = \boldsymbol{G}_1^{-1}\boldsymbol{G}_2$. For example, if a nonsystematic encoder $\boldsymbol{G}$ is available, we can precode the information vector $\boldsymbol{u}$ to obtain $\boldsymbol{u}' = \boldsymbol{u}\,\boldsymbol{G}_1^{-1}$, which is then encoded as $\boldsymbol{c} = \boldsymbol{u}'\boldsymbol{G}$. In other words, the matrix $\boldsymbol{G}_1^{-1}$ is used as a precoding matrix to obtain a systematic encoder from a nonsystematic one.

Definition 2.23 (Information set) For any set of k independent columns of $\boldsymbol{G}$ with indexes $j_1 < j_2 < \ldots < j_k$, the codeword elements $c_{j_1}, c_{j_2}, \ldots, c_{j_k}$ are sufficient to recover the information vector $\boldsymbol{u}$ (and therefore the whole codeword $\boldsymbol{c}$). Since there exists a one-to-one correspondence between $c_{j_1}, c_{j_2}, \ldots, c_{j_k}$ and $\boldsymbol{u}$ (hence $\boldsymbol{c}$), the set $\{j_1, j_2, \ldots, j_k\}$ is called an *information set.*

Definition 2.24 ($(n, 1)$ repetition code) Consider an encoder which takes one input symbol ($k = 1$) in GF(q) and repeats it n times. The code rate is $R = 1/n$. The $1 \times n$ generator matrix is

$$\boldsymbol{G} = \begin{bmatrix} 1 & 1 & 1 & \cdots & 1 \end{bmatrix} \tag{2.54}$$

[5] Otherwise, some columns of $\boldsymbol{G}$ should be permuted so that $\boldsymbol{G}_1$ is full rank.

which is in systematic form. An $(n-1) \times n$ parity-check matrix is

$$\boldsymbol{H} = \begin{bmatrix} 1 & 0 & \cdots & 0 & -1 \\ 0 & 1 & \cdots & 0 & -1 \\ \vdots & \vdots & & \vdots & \vdots \\ 0 & 0 & \cdots & 1 & -1 \end{bmatrix} \tag{2.55}$$

and the minimum Hamming distance is $d_{\min} = n$. When used in GF(2), this code has only two codewords, $\mathcal{C} = \left\{\begin{bmatrix} 0 & 0 & \cdots & 0 \end{bmatrix}, \begin{bmatrix} 1 & 1 & \cdots & 1 \end{bmatrix}\right\}$, and its weight distribution is $A_0 = 1, A_n = 1$.

Definition 2.25 ($(k+1,k)$ single parity-check (SPC) code) Consider a systematic encoder which takes k information symbols $u_1, \ldots, u_k$ in GF(q) and adds one check symbol $c_n = -u_1 - u_2 - \ldots - u_k$. Therefore $n = k+1$, and we have $c_1 = u_1, c_2 = u_2, \ldots, c_k = u_k, c_{k+1} = -\sum_{\ell=1}^{k} u_\ell$. Then, for all codewords

$$c_1 + c_2 + \cdots + c_n = 0\,. \tag{2.56}$$

The code rate is $R = k/(k+1)$. The generator matrix is

$$\boldsymbol{G} = \begin{bmatrix} 1 & 0 & \cdots & 0 & -1 \\ 0 & 1 & \cdots & 0 & -1 \\ \vdots & \vdots & & \vdots & \vdots \\ 0 & 0 & \cdots & 1 & -1 \end{bmatrix} \tag{2.57}$$

which is in systematic form. A parity-check matrix is thus

$$\boldsymbol{H} = \begin{bmatrix} 1 & 1 & 1 & \cdots & 1 \end{bmatrix} \tag{2.58}$$

and the minimum Hamming distance is $d_{\min} = 2$. From $\boldsymbol{H}$ we see that the SPC code is the dual of the repetition code.

In particular, in GF(2) the SPC code takes k input bits and adds one check bit, so that the total number of 1 in a codeword is even. In other words, the check bit is the modulo 2 sum of the information bits. The codewords are in this case all binary n-tuples with an even number of 1. The weight distribution is

$$A_0 = 1,\ A_1 = 0,\ A_2 = \binom{n}{2},\ A_3 = 0,\ A_4 = \binom{n}{4}, \ldots\,. \tag{2.59}$$

The weight enumerating function of a binary $(n, n-1)$ SPC code is therefore

$$P_{2,n}(W) = \sum_{\substack{t=0 \\ t \bmod 2=0}}^{n} \binom{n}{t} W^t = \frac{1}{2}\left[(1+W)^n + (1-W)^n\right]. \tag{2.60}$$

In the general case $q \geq 2$, the weight enumerating function of an $(n, n-1)$ SPC code is

$$P_{q,n}(W) = \sum_{i=0}^{n} \binom{n}{i} p_{q,i} W^i, \tag{2.61}$$

where

$$p_{q,i} = \frac{1}{q}\left[(q-1)^i + (-1)^i(q-1)\right]. \tag{2.62}$$

Definition 2.26 (MacWilliams identity, weight enumerator of the dual code) Call $A(W)$ the weight enumerating function for an (n,k) linear code $\mathcal{C}$ over GF(q), and $B(W)$ that of its dual $\mathcal{C}^{\perp}$. The "MacWilliams identity" states that

$$A(W) = \frac{1}{q^{n-k}}(1+(q-1)W)^n B\left(\frac{1-W}{1+(q-1)W}\right). \tag{2.63}$$

Definition 2.27 ((7,4) Hamming code) Consider in GF$(2)^7$, the parity-check matrix whose columns are all possible nonzero 3-tuples

$$\boldsymbol{H} = \begin{bmatrix} 1 & 1 & 1 & 0 & 1 & 0 & 0 \\ 0 & 1 & 1 & 1 & 0 & 1 & 0 \\ 1 & 1 & 0 & 1 & 0 & 0 & 1 \end{bmatrix} \tag{2.64}$$

which can be interpreted as $\boldsymbol{H} = \left[-\boldsymbol{P}^{\mathsf{T}}|\boldsymbol{I}_3\right]$. A generator matrix is then

$$\boldsymbol{G} = \begin{bmatrix} 1 & 0 & 0 & 0 & 1 & 0 & 1 \\ 0 & 1 & 0 & 0 & 1 & 1 & 1 \\ 0 & 0 & 1 & 0 & 1 & 1 & 0 \\ 0 & 0 & 0 & 1 & 0 & 1 & 1 \end{bmatrix}. \tag{2.65}$$

This code has $n = 7$ and $k = 4$, and is known as the binary (7,4) *Hamming code*. The rowspace of $\boldsymbol{G}$ gives $2^4 = 16$ codewords

$$\mathcal{C} = \left\{\begin{bmatrix}0 & 0 & 0 & 0 & 0 & 0 & 0\end{bmatrix}, \begin{bmatrix}1 & 0 & 0 & 0 & 1 & 0 & 1\end{bmatrix}, \cdots\right\}. \tag{2.66}$$

The weight distribution is $A_0 = 1, A_3 = 7, A_4 = 7, A_7 = 1$, the weight enumerating function is thus $A(W) = 1 + 7W^3 + 7W^4 + W^7$, and the minimum Hamming distance is $d_{\min} = 3$. It is possible to generalize the construction to parity-check matrices of dimension $m \times n$ for an arbitrary $m > 1$, where the $n = 2^m - 1$ columns are all possible nonzero m-tuples. The resulting $(n, n-m)$ Hamming code has always $d_{\min} = 3$.

The minimum Hamming distance of a code can also be derived from the following theorem.

Theorem 2.3 *The minimum Hamming distance of a code with parity-check matrix* $\boldsymbol{H}$ *is the minimum number of linearly dependent columns of* $\boldsymbol{H}$.

Proof Let us indicate as $\boldsymbol{h}_j^{\mathsf{T}}$, $j = 1, \cdots, n$, the columns of $\boldsymbol{H} = \left[\boldsymbol{h}_1^{\mathsf{T}}\boldsymbol{h}_2^{\mathsf{T}}\cdots\boldsymbol{h}_n^{\mathsf{T}}\right]$. Then, for each codeword it must be

$$\boldsymbol{H}\boldsymbol{c}^{\mathsf{T}} = c_1\boldsymbol{h}_1^{\mathsf{T}} + c_2\boldsymbol{h}_2^{\mathsf{T}} + \ldots c_n\boldsymbol{h}_n^{\mathsf{T}} = \boldsymbol{0}^{\mathsf{T}} \tag{2.67}$$

which can be read as a linear combination of the columns of $\boldsymbol{H}$ producing a zero vector. A minimum weight codeword picks the minimum number of columns to give a zero vector. □

For the Hamming code, since all columns of $\boldsymbol{H}$ are distinct, and are all the $2^m - 1$ nonzero binary m-tuples, it is possible to find three linearly dependent columns of $\boldsymbol{H}$, while there are not two linearly dependent columns. Thus, the minimum Hamming distance is $d_{\min} = 3$. Theorem 2.3 has the following corollary.

Corollary 2.1 *A code with generator matrix and parity-check matrix in systematic form* $\boldsymbol{G} = \left[\boldsymbol{I}_k|\boldsymbol{P}\right]$, $\boldsymbol{H} = \left[-\boldsymbol{P}^{\mathsf{T}}|\boldsymbol{I}_{n-k}\right]$, *where* $\boldsymbol{P}$ *is a* $k \times (n-k)$ *matrix, has minimum distance* $d_{min} = n - k + 1$ *if and only if every square submatrix of* $\boldsymbol{P}$ *is nonsingular.*

Linear codes are thus defined by specifying n, k, q, and a generator matrix or a parity-check matrix. A given code can be modified in several ways, as described below.

Definition 2.28 (Modifications of linear codes) Given an (n, k) linear code, there are several possible modifications that allow obtaining new codes with different parameters. The most important ones are the following:

- *Extended codes.* These are codes obtained by adjoining to all codewords one or more parity-check symbols. The code rate decreases. For example, in the binary case, appending to all codewords one overall parity-check bit (such that the overall number of 1s is even) gives an $(n + 1, k)$ binary code with only even weight codewords. If $\boldsymbol{H}$ is the original parity-check matrix, that of the extended is obtained by adding one all-zero column, and then an all-one row. If $d_{\min}$ of the original binary code is odd, appending one overall parity-check bit increases the minimum distance to $d_{\min} + 1$. Otherwise, $d_{\min}$ remains unchanged.
- *Punctured codes.* These are codes obtained by dropping symbols in some fixed coordinates from all codewords. The code rate increases, while $d_{\min}$ may be reduced.
- *Shortened codes.* A code with a systematic encoder can be shortened by forcing ℓ information symbols to zero. The resulting codewords have then the leading ℓ symbols equal to zero; these symbols are not transmitted (since the receiver knows they are zero). The resulting is an $(n - \ell, k - \ell)$ code. The code rate decreases, and $d_{\min}$ may increase.
- *Expurgated codes.* These are codes obtained by decreasing the number of information symbols per codeword, without changing n. This can be obtained by eliminating rows of the generator matrix $\boldsymbol{G}$, which corresponds to adding rows to the parity-check matrix $\boldsymbol{H}$. For example, appending one row of all 1's to $\boldsymbol{H}$, and assuming that the resulting matrix is still full rank, we retain only the codewords with $c_1 + \cdots + c_n = 0$, and we obtain an $(n, k - 1)$ code. The code rate decreases, and $d_{\min}$ may increase.
- *Lengthened codes.* These are codes obtained by increasing the number of information symbols per codeword, with corresponding increase in the length n. This can be obtained by increasing of the same amount the number of rows and columns of the generator matrix $\boldsymbol{G}$. The code rate increases.
- *Augmented codes.* These are codes obtained by increasing the number of information symbols per codeword, without changing the length n. This

corresponds to increasing the number of rows of the generator matrix $\boldsymbol{G}$. The code rate increases.

Theorem 2.4 (Singleton bound) *For each code (linear or nonlinear)*

$$d_{min} \leq n - k + 1 . \quad (2.68)$$

Proof Assume an (n, k) code on GF(q) has a given minimum distance $d_{\min}$. This implies that if we delete up to a number $e = d_{\min} - 1$ of coordinates in all q^k codewords, the resulting $(n - d_{\min} + 1)$-tuples are still all different. All possible $(n - d_{\min} + 1)$-tuples are $q^{n-d_{\min}+1}$, so it must be $q^k \leq q^{n-d_{\min}+1}$, which gives $k \leq n - d_{\min} + 1$.

For the special case of linear codes, an alternative proof is the following: Since the parity-check matrix $\boldsymbol{H}$ is $(n - k) \times n$, its maximum rank is $n - k$. Thus, from Theorem 2.3, the minimum distance cannot exceed $n - k + 1$. Moreover, in order to have $d_{\min} = n - k + 1$, any $n - k$ columns of $\boldsymbol{H}$ must be linearly independent. □

Definition 2.29 (Maximum Distance Separable codes) If a code achieves the Singleton bound, that is, $d_{\min} = n - k + 1$, it is called a maximum distance separable (MDS) code.

It follows that a linear code is MDS if and only if any $n - k$ columns of its parity-check matrix are linearly independent. The $(n, 1)$ repetition codes ($d_{\min} = n$), and the $(n, n - 1)$ SPC codes ($d_{\min} = 2$) are MDS in every field. In the binary field, no other MDS codes exist, and all other binary codes have $d_{\min} < n - k + 1$ (typically $d_{\min} \ll n - k + 1$). Nontrivial MDS codes exist in nonbinary fields, namely, the Reed-Solomon codes. Their shortened, extended by 1 or 2 symbols, or punctured versions are also MDS (see Section 2.4.1).

Definition 2.30 (Cyclic codes) A code is cyclic if the cyclic shift of any codeword is also a codeword.

With cyclic codes it is customary to work with polynomials. Also, instead of indexing the elements of vectors or polynomial coefficients from 1 to n, when dealing with cyclic codes it is convenient to use indexes from 0 to $n - 1$. Then, using a polynomial representation for the codewords $c(X) = c_0 + c_1 X + c_2 X^2 + \cdots + c_{n-1} X^{n-1}$, where the c_i are in GF(q), it is possible to show that a code is cyclic if and only if all codewords can be written as

$$c(X) = q(X) g(X) . \quad (2.69)$$

In this equation $g(X)$, which characterizes the code and is called the *generator polynomial* of the cyclic code, has degree $n - k$ and is a factor of $X^n - 1$, while the $q(X)$ are all polynomials with maximum degree $k - 1$. All possible q^k polynomials $q(X)$, multiplied by $g(X)$, generate all possible codewords through (2.69).

A simple encoder for cyclic codes is obtained by using (2.69) with $q(X) = u(X)$, where $u(X) = u_0 + u_1 X + \cdots + u_{k-1} X^{k-1}$ is the information polynomial. However, this encoder is nonsystematic. A systematic encoder is obtained by calculating the $n - k$ parity-check symbols as the remainder $r(X) = X^{n-k} u(X) \bmod g(X)$, then constructing the codeword as $c(X) = X^{n-k}u(X) - r(X)$. Since the minimum degree of $X^{n-k}u(X)$ is $n-k$ and the maximum degree of $r(X)$ is $n-k-1$, it follows that $c(X)$ is in systematic form. Moreover $X^{n-k}u(X) = q(X)g(X) + r(X)$ and therefore $c(X)$ is in the form (2.69).

Definition 2.31 (Quasi-Cyclic codes) A $(Q\,n_0, Q\,k_0)$ code is quasi-cyclic if the cyclic shift of n_0 symbols of any codeword is also a codeword.

To simplify the manipulation of the generator and parity-check matrices of a quasi-cyclic (Qn_0, Qk_0) code, it is convenient to rearrange the code symbols by grouping together every other n_0th column of the generator matrix. This produces a total of n_0 groups. Each group forms circulant matrices of order Q because of the quasi-cyclic property of the code. This equivalent generator matrix has the form

$$\boldsymbol{G} = \begin{bmatrix} \mathbf{C}_{1,1} & \mathbf{C}_{1,2} & \cdots & \mathbf{C}_{1,n_0} \\ \vdots & \vdots & \cdots & \vdots \\ \mathbf{C}_{k_0,1} & \mathbf{C}_{k_0,2} & \cdots & \mathbf{C}_{k_0,n_0} \end{bmatrix}, \tag{2.70}$$

where $\mathbf{C}_{i,j}$ is a $Q \times Q$ circulant matrix of the form

$$\begin{bmatrix} c_0 & c_1 & \cdots & c_{Q-1} \\ c_{Q-1} & c_0 & \cdots & c_{Q-2} \\ \vdots & \vdots & \cdots & \vdots \\ c_1 & c_2 & \cdots & c_0 \end{bmatrix} \tag{2.71}$$

with $c_i \in \mathrm{GF}(q)$.

As for cyclic codes, quasi-cyclic codes appear attractive from an implementation standpoint. In particular, quasi-cyclic codes can be encoded in exactly the same manner as cyclic codes using feedback shift registers.

2.4 Classical Codes for Erasure Channels

Assume we use a linear (n, k) code over an erasure channel. Denote by $\boldsymbol{c}$ the transmitted codeword, and by $\boldsymbol{y}$ the received vector, with elements $y_i \in \{c_i, \mathsf{E}\}$, where $y_i = \mathsf{E}$ in case of a symbol erasure, else $y_i = c_i$. Denote by $i_1, i_2, \ldots, i_e$ the indexes of the e erased symbols, and $j_1, j_2, \ldots, j_{n-e}$ the indexes of the correctly received symbols. To reconstruct the erased symbols $\mathbf{e} = \begin{bmatrix} e_1 & e_2 & \ldots & e_e \end{bmatrix}$, where $e_\ell = c_{i_\ell}$, we recall that it must be

$$\boldsymbol{H}\boldsymbol{c}^{\mathsf{T}} = c_1\boldsymbol{h}_1^{\mathsf{T}} + \cdots + c_n\boldsymbol{h}_n^{\mathsf{T}} = \mathbf{0} \tag{2.72}$$

which can be reformulated as $e_1\boldsymbol{h}_{i_1}^{\mathsf{T}} + e_2\boldsymbol{h}_{i_2}^{\mathsf{T}} + \cdots + e_e\boldsymbol{h}_{i_e}^{\mathsf{T}} = -\sum_k y_{j_k}\boldsymbol{h}_{j_k}^{\mathsf{T}}$. By indicating with $\boldsymbol{H}_{\mathcal{E}}$ the $(n-k) \times e$ submatrix of $\boldsymbol{H}$ consisting of the columns with indexes $i_1, i_2, \ldots, i_e$, and with $\boldsymbol{s}^{\mathsf{T}} = -\sum_k y_{j_k}\boldsymbol{h}_{j_k}^{\mathsf{T}}$ the known contribution of the correctly received symbols, we have the linear system

$$\boldsymbol{H}_{\mathcal{E}}\boldsymbol{e}^{\mathsf{T}} = \boldsymbol{s}^{\mathsf{T}}. \tag{2.73}$$

This linear system has a unique solution if the rank of $\boldsymbol{H}_{\mathcal{E}}$ is e, that is, if the e columns of $\boldsymbol{H}_{\mathcal{E}}$ are linearly independent. It follows that:

- an (n,k) code with minimum distance $d_{\min}$ can correct all patterns of $e \le d_{\min} - 1$ erasures;
- no patterns with $e > n - k$ erasures can be corrected;
- depending on the erasure positions, some patterns with $d_{\min} \le e \le n - k$ can be corrected.

An equivalent approach is based on the code generator matrix. Recall that the transmitted codeword is $\boldsymbol{c} = \boldsymbol{u}\boldsymbol{G}$. Keeping only the correctly received symbols $\tilde{\boldsymbol{y}} = \begin{bmatrix} y_{j_1} & y_{j_2} & \cdots & y_{j_{n-e}} \end{bmatrix}$ and introducing the corresponding $k \times (n-e)$ submatrix $\boldsymbol{G}_{\mathcal{E}} = [\boldsymbol{g}_{j_1}^{\mathsf{T}} \quad \boldsymbol{g}_{j_2}^{\mathsf{T}} \quad \cdots \quad \boldsymbol{g}_{j_{n-e}}^{\mathsf{T}}]$, we can write[6]

$$\tilde{\boldsymbol{y}} = \boldsymbol{u}\boldsymbol{G}_{\mathcal{E}} \,. \tag{2.74}$$

The linear system (2.74) has $n-e$ equations and k unknowns. For $e \le n-k$, the system can thus be solved for $\boldsymbol{u}$ if $\boldsymbol{G}_{\mathcal{E}}$ is full rank. If all possible $k \times (n-e)$ submatrices of $\boldsymbol{G}$ are full rank (which is the case, for example, for Reed–Solomon codes), then it follows that all patterns of no more than e erasures can be solved.

The dimension of the system to solve is $e \le n - k$ for the approach based on the parity-check matrix (2.73), and k for that based on the generator matrix (2.74). Note that if in (2.74) there is a column with just one nonzero element, the corresponding input symbol is directly determined. If the encoder is systematic, there are k columns of $\boldsymbol{G}$ which have weight one, and at most $n-k$ with greater weight, so the two approaches are equivalent in terms of complexity.

2.4.1 Reed–Solomon Codes

Definition 2.32 (Vandermonde matrix) Given $x_1, x_2, \ldots, x_n$, the $n \times n$ matrix

$$\mathbf{V}(x_1, x_2, \ldots, x_n) = \left\{x_j^{i-1}\right\}_{i=1:n}^{j=1:n} = \begin{bmatrix} 1 & 1 & \cdots & 1 \\ x_1 & x_2 & \cdots & x_n \\ x_1^2 & x_2^2 & \cdots & x_n^2 \\ \vdots & \vdots & & \vdots \\ x_1^{n-1} & x_2^{n-1} & \cdots & x_n^{n-1} \end{bmatrix} \tag{2.75}$$

is called a Vandermonde matrix. Its determinant is

$$\det \mathbf{V}(x_1, x_2, \ldots, x_n) = \prod_{1 \le i < j \le n} \left(x_j - x_i\right). \tag{2.76}$$

It follows that, in any field, given n arbitrary distinct elements $x_1, x_2, \ldots, x_n$, the Vandermonde matrix $\mathbf{V}(x_1, x_2, \ldots, x_n)$ is full rank.

[6] Here $\boldsymbol{g}_j^{\mathsf{T}}$ is the jth column of $\boldsymbol{G}$.

Similarly, for *distinct nonzero* $x_1, \ldots, x_n$, the square matrix

$$A = \begin{bmatrix} x_1 & x_2 & \cdots & x_n \\ x_1^2 & x_2^2 & \cdots & x_n^2 \\ \vdots & \vdots & & \vdots \\ x_1^n & x_2^n & \cdots & x_n^n \end{bmatrix} \tag{2.77}$$

is full rank, since

$$\det A = x_1\, x_2 \cdots x_n \;\det \mathbf{V}(x_1, x_2, \ldots, x_n) = x_1\, x_2 \cdots x_n \prod_{1 \le i < j \le n} \left(x_j - x_i\right). \tag{2.78}$$

Definition 2.33 (Cyclic Reed-Solomon codes) Let α be a primitive element of the finite field $\mathrm{GF}(q) = \{0, 1, \alpha, \alpha^2, \ldots, \alpha^{q-2}\}$. For a given integer $1 \le r < q$, construct the $r \times (q-1)$ parity-check matrix

$$H = \left\{\alpha^{i(j-1)}\right\}_{i=1:r}^{j=1:q-1} = \begin{bmatrix} 1 & \alpha & \alpha^2 & \cdots & \alpha^{q-2} \\ 1 & \alpha^2 & \alpha^4 & \cdots & \alpha^{2(q-2)} \\ \vdots & \vdots & \vdots & & \vdots \\ 1 & \alpha^r & \alpha^{2r} & \cdots & \alpha^{r(q-2)} \end{bmatrix}, \tag{2.79}$$

so that the codewords are all vectors $c = [c_0 \;\ldots\; c_{n-1}]$ with $c_i \in \mathrm{GF}(q)$ satisfying

$$H\, c^{\top} = \mathbf{0}, \tag{2.80}$$

where all operations are in $\mathrm{GF}(q)$. The parity-check matrix (2.79) specifies a cyclic (n, k) code in $\mathrm{GF}(q)$ with

$$n = q - 1, \quad k = n - r \tag{2.81}$$

called Reed-Solomon (RS) code and indicated as $RS(n, k)$. A generator matrix is obtained easily in the similar form

$$G = \left\{\alpha^{(i-1)(j-1)}\right\}_{i=1:k}^{j=1:n} = \begin{bmatrix} 1 & 1 & 1 & \cdots & 1 \\ 1 & \alpha & \alpha^2 & \cdots & \alpha^{q-2} \\ 1 & \alpha^2 & \alpha^4 & \cdots & \alpha^{2(q-2)} \\ \vdots & \vdots & \vdots & & \vdots \\ 1 & \alpha^{k-1} & \alpha^{2(k-1)} & \cdots & \alpha^{(q-2)(k-1)} \end{bmatrix}. \tag{2.82}$$

More in general, an $RS(n, k)$ cyclic code is a code in $\mathrm{GF}(q)$ with $n = q - 1$ and $k < n$, having parity-check and generator matrices which can be put into the form[7]

$$H = \left\{\alpha^{(i+h)(j-1)}\right\}_{i=1:n-k}^{j=1:n} \qquad G = \left\{\alpha^{(i-h-1)(j-1)}\right\}_{i=1:k}^{j=1:n}, \tag{2.83}$$

where h is an arbitrary integer.

[7] In fact, for any integers v, u such that $\alpha^{v+u} \neq 1$, we have

$$\begin{aligned} \langle [1\; \alpha^v\; \alpha^{2v}\; \cdots\; \alpha^{v(q-2)}], [1\; \alpha^u\; \alpha^{2u}\; \cdots\; \alpha^{u(q-2)}] \rangle &= \\ &= 1 + \alpha^{v+u} + \alpha^{2(v+u)} + \cdots + \alpha^{(q-2)(v+u)} \\ &= \left(1 - \alpha^{(q-1)(v+u)}\right) / \left(1 - \alpha^{v+u}\right) = 0, \end{aligned}$$

where we used the equality $\alpha^{q-1} = 1$. It follows that $HG^{\top} = \mathbf{0}$.

The two more used versions for RS codes are $h = 0$, yielding (2.79) and (2.82), and $h = -1$, for which the parity-check matrix is

$$\boldsymbol{H} = \begin{bmatrix} 1 & 1 & 1 & \cdots & 1 \\ 1 & \alpha & \alpha^2 & \cdots & \alpha^{q-2} \\ 1 & \alpha^2 & \alpha^4 & \cdots & \alpha^{2(q-2)} \\ \vdots & \vdots & \vdots & & \vdots \\ 1 & \alpha^{r-1} & \alpha^{2(r-1)} & \cdots & \alpha^{(q-2)(r-1)} \end{bmatrix}. \tag{2.84}$$

Comparing (2.82) with (2.83), it results that the dual of an $RS(n,k)$ code is also an $RS(n,n-k)$ code.

The previous matrix formulation has a simple polynomial interpretation: Recalling (2.29), the parity-check equations (2.80) with (2.79) are equivalent to say that, for all codewords, the code polynomial $c(X) = c_0 + c_1X + \cdots + c_{n-1}X^{n-1}$ must satisfy

$$c(\alpha) = c(\alpha^2) = \cdots = c(\alpha^r) = 0\,. \tag{2.85}$$

Moreover, denoting the information polynomial by $u(X) = u_0 + u_1X + \cdots + u_{k-1}X^{k-1}$, the relation $\boldsymbol{c} = \boldsymbol{u}\,\boldsymbol{G}$ with (2.82) gives

$$\begin{bmatrix} c_0 & c_1 & \cdots & c_{n-1} \end{bmatrix} = \begin{bmatrix} u(1) & u(\alpha) & \cdots & u(\alpha^{q-2}) \end{bmatrix}. \tag{2.86}$$

More in general, for (2.83), the parity-check equations in polynomial form are

$$c(\alpha^{1+h}) = c(\alpha^{2+h}) = \cdots = c(\alpha^{r+h}) = 0 \tag{2.87}$$

and the codeword symbols are calculated from the input polynomial as

$$\begin{bmatrix} c_0 & c_1 & \cdots & c_{n-1} \end{bmatrix} = \alpha^{-h}\begin{bmatrix} u(\alpha^{-h}) & u(\alpha^{1-h}) & \cdots & u(\alpha^{q-2-h}) \end{bmatrix}. \tag{2.88}$$

Theorem 2.5 *The $RS(n,k)$ codes defined by* (2.83) *have minimum Hamming distance $d_{min} = n - k + 1$, and are therefore MDS.*

Proof As the elements $1, \alpha, \alpha^2, \ldots, \alpha^{q-2}$ are nonzero and all distinct, if we pick r arbitrary columns of $\boldsymbol{H}$, we have a square matrix like (2.75) or (2.77) (depending on the chosen h), with nonzero determinant. It follows that any arbitrary $e \le r$ columns of $\boldsymbol{H}$ are linearly independent. Due to Theorem 2.3, this code has minimum distance $d_{\min} = r + 1 = n - k + 1$, and is therefore MDS. □

Another simple proof is based on (2.86). In fact, since the polynomial $u(X)$ has degree $k-1$, it has at most $k-1$ zeros.[8] The information polynomials giving exactly $k-1$ zeros when evaluated in $\mathrm{GF}(q) \setminus \{0\} = \{1, \alpha, \cdots, \alpha^{n-1}\}$ are of the form $u(X) = \beta(X - a_1)\cdots(X - a_{k-1})$, where $\beta \in \mathrm{GF}(q) \setminus \{0\}$ and $a_1, a_2, \ldots, a_{k-1}$ are distinct elements in $\mathrm{GF}(q) \setminus \{0\}$. There are exactly

$$A_{d_{\min}} = (q-1)\binom{n}{k-1} \tag{2.89}$$

[8] This is related to the fact that every polynomial can be factored into irreducible polynomials in a unique way.

such polynomials. Therefore the minimum codeword weight is $d_{\text{min}} = n - (k - 1)$, and the number of codewords with weight d_{min} is given by (2.89). Actually it can be proved that (2.89) generalizes to all MDS codes.

Theorem 2.6 *The $RS(n, k)$ codes defined by* (2.83) *are cyclic.*

Proof From $\boldsymbol{H}\,\boldsymbol{c}^{\top} = \boldsymbol{0}$ and (2.83), we already observed that each codeword polynomial must satisfy $c(\alpha^{1+h}) = c(\alpha^{2+h}) = \cdots = c(\alpha^{r+h}) = 0$ where h is arbitrary. This means that the codewords are all proportional to the polynomial of degree r

$$g(X) = \left(X - \alpha^{1+h}\right)\left(X - \alpha^{2+h}\right)\cdots\left(X - \alpha^{r+h}\right) \tag{2.90}$$

which is called *generator polynomial* of the RS code. The codewords are therefore all polynomials $c(X) = q(X)g(X)$ where $q(X)$ are all q^k polynomials of maximum degree $n - r = k$. Since $g(X)$ divides $X^n - 1$, the code is also cyclic.[9] □

Example 2.6 For $m = 3$, $\mathrm{GF}(2^m) = \mathrm{GF}(8)$ we can use Table 2.2 for the parity-check matrix. With $q = 8$, setting for example $r = 2$ in (2.79) we have an $RS(7,5)$ code with parity-check matrix

$$\boldsymbol{H} = \begin{bmatrix} 1 & \alpha & \alpha^2 & \alpha^3 & \alpha^4 & \alpha^5 & \alpha^6 \\ 1 & \alpha^2 & \alpha^4 & \alpha^6 & \alpha & \alpha^3 & \alpha^5 \end{bmatrix}. \tag{2.91}$$

The codewords have $c(\alpha) = c(\alpha^2) = 0$, and the generator polynomial is $g(X) = (X - \alpha)(X - \alpha^2)$. The encoder takes $3 \cdot 5 = 15$ bits and produces a codeword of $3 \cdot 7 = 21$ bits.

Definition 2.34 (Extended RS codes) The $RS(n, k)$ codes above are cyclic and have $n = 2^m - 1$, arbitrary $k < n$, and $d_{\text{min}} = n - k + 1$. They can be lengthened by adding one or two symbols, without changing the minimum distance, to obtain $RS(n+1, k+1)$ and $RS(n+2, k+2)$, with the same minimum distance, with the parity-check matrices

$$\boldsymbol{H}_1 = \left[\ \boldsymbol{H} \ \begin{vmatrix} 1 \\ 0 \\ \vdots \\ 0 \end{vmatrix}\ \right] \qquad \boldsymbol{H}_2 = \left[\ \boldsymbol{H} \ \begin{vmatrix} 1 & 0 \\ 0 & 0 \\ \vdots & \vdots \\ 0 & 0 \\ 0 & 1 \end{vmatrix}\ \right], \tag{2.92}$$

respectively, where $\boldsymbol{H}$ is the one in (2.84). In fact, it can be verified that any $n - k$ columns of $\boldsymbol{H}_1$ or $\boldsymbol{H}_2$ are linearly independent. The resulting codes are still MDS, but no longer cyclic. However, cyclic MDS codes with the same lengths have been proved to exist.

Definition 2.35 (Shortened RS codes) The $RS(n, k)$ codes with $n = 2^m - 1$ are sometimes called primitive RS codes. They can be shortened, obtaining $RS(n - \ell, k - \ell)$,

[9] We recall that $\alpha^{q-1} - 1$, thus for $n = q - 1$ the field element α is a root of $X^n - 1$. Also, $(\alpha^j)^{q-1} = 1^j = 1$ for all j, so any power of α is also a root of $X^n - 1$.

where $0 \leq \ell < k$ is arbitrary. A possible parity-check matrix is composed in this case by the first $n - \ell$ columns and $n - k - \ell$ rows of (2.79). The resulting code has the same $d_{\min} = n - k + 1$, and is therefore still MDS. Indeed, if we use just $n - \ell$ columns in (2.79), any $r < n - \ell$ columns remain linearly independent. Clearly, the shortened codes are no longer cyclic.

2.4.2 Other Forms of MDS Codes

Theorem 2.7 *A code with generator matrix in systematic form* $\mathbf{G} = \left[\boldsymbol{I}_k | \boldsymbol{P}\right]$, *where* $\boldsymbol{P}$ *is a* $k \times (n-k)$ *matrix, is MDS if and only if every square submatrix of* $\boldsymbol{P}$ *is nonsingular.*

Proof The theorem is a direct consequence of the definition of MDS codes and Corollary 2.1. □

Definition 2.36 (Cauchy matrix) Given $x_1, x_2, \ldots, x_n$ and $y_1, y_2, \ldots, y_n$, the $n \times n$ matrix

$$\mathbf{C} = \left\{\frac{1}{x_i + y_j}\right\}_{1 \leq i,j \leq n} = \begin{bmatrix} \frac{1}{x_1+y_1} & \frac{1}{x_1+y_2} & \cdots & \frac{1}{x_1+y_n} \\ \vdots & \vdots & & \vdots \\ \frac{1}{x_n+y_1} & \frac{1}{x_n+y_2} & \cdots & \frac{1}{x_n+y_n} \end{bmatrix} \tag{2.93}$$

is called a Cauchy matrix. Its determinant is

$$\det \mathbf{C} = \frac{\prod_{1 \leq i < j \leq n} \left(x_j - x_i\right)\left(y_j - y_i\right)}{\prod_{1 \leq i,j \leq n} \left(x_i + y_j\right)}. \tag{2.94}$$

It follows that, in any field, if $x_i \neq x_j$ and $y_i \neq y_j$ $\forall i \neq j$, and also $x_i + y_j \neq 0$ $\forall i, j$, the Cauchy matrix $\mathbf{C}$ is full rank. Moreover, any square submatrix of a Cauchy matrix is nonsingular.

As a consequence, systematic encoders with $\boldsymbol{G} = \left[\boldsymbol{I}_k | \boldsymbol{P}\right]$, $\boldsymbol{H} = \left[-\boldsymbol{P}^{\mathsf{T}} | \boldsymbol{I}_{n-k}\right]$, can be constructed by using for $\boldsymbol{P}$ a $k \times (n - k)$ submatrix of a Cauchy matrix. Actually, it has been shown that systematic encoders for RS codes can be built by a generalized version of the Cauchy matrices given above. Other types of MDS codes (array codes) are discussed in Appendix A.

2.4.3 Erasure Correction with RS Codes

When an RS code is used, the matrices appearing in (2.73) or (2.74) are of Vandermonde type. For example, (2.73) becomes

$$\begin{bmatrix} \alpha^{i_1} & \alpha^{i_2} & \cdots & \alpha^{i_e} \\ \alpha^{2i_1} & \alpha^{2i_2} & \cdots & \alpha^{2i_e} \\ \vdots & \vdots & & \vdots \\ \alpha^{(n-k)i_1} & \alpha^{(n-k)i_2} & \cdots & \alpha^{(n-k)i_e} \end{bmatrix} \begin{bmatrix} e_1 \\ e_2 \\ \vdots \\ e_e \end{bmatrix} = \begin{bmatrix} y_1 \\ y_2 \\ \vdots \\ y_{n-k} \end{bmatrix}. \tag{2.95}$$

An alternative approach to solve efficiently the linear system (2.95) is provided by the following theorem.

Theorem 2.8 (RS decoding when at least k codeword symbols are known) *Consider an $RS(n,k)$ code with codewords $[c_0\ c_1\ \cdots\ c_{n-1}] = [u(1)\ u(\alpha)\cdots u(\alpha^{q-2})]$, where $u(X)$ is the message polynomial. Given any k symbols $c_{J_0}, c_{J_1}, \ldots, c_{J_{k-1}}$ the complete message polynomial $u(X)$ can be reconstructed as*

$$u(X) = \sum_{i=0}^{k-1} c_{J_i} \prod_{\substack{0\le \ell \le k-1\\ \ell\ne i}} \frac{X - \alpha^{J_\ell}}{\alpha^{J_i} - \alpha^{J_\ell}} \tag{2.96}$$

and therefore the missing codeword symbols as

$$c_j = \sum_{i=0}^{k-1} c_{J_i} \prod_{\substack{0\le \ell \le k-1\\ \ell\ne i}} \frac{\alpha^{j} - \alpha^{J_\ell}}{\alpha^{J_i} - \alpha^{J_\ell}}, \quad j \notin \{J_0, J_1, \ldots, J_{k-1}\}. \tag{2.97}$$

Proof By the Lagrange interpolator polynomial for $u(X)$. □

The erasure decoding formula (2.97) can be implemented by following the steps:

- Compute and store the integers

$$\mu_{i,j} = \log_\alpha\left(\alpha^i - \alpha^j\right), \qquad i \ne j \tag{2.98}$$

  such that $\alpha^i - \alpha^j = \alpha^{\mu_{i,j}}$, for $i, j = 0, \ldots, n-1$, $i \ne j$. Note that if $q = 2^m$ we have $\mu_{i,j} = \mu_{j,i}$;
- In the positions of the missing symbols $j \notin \{J_0, J_1, \ldots, J_{k-1}\}$, evaluate the integers

$$g_{i,j} = \left(\sum_{\ell\ne i} \mu_{j,J_\ell} - \mu_{J_i,J_\ell}\right) \bmod (q-1), \quad i = 0, \ldots, k-1 \tag{2.99}$$

  where all operations are for integers;
- Calculate the missing codeword symbols as

$$c_j = \sum_{i=0}^{k-1} c_{J_i}\alpha^{g_{i,j}} \quad j \notin \{J_0, J_1, \ldots, J_{k-1}\}. \tag{2.100}$$

Computing each missing symbol c_j requires k multiplications and $k-1$ sums in GF(q), plus $k(2k-3)$ sums of integers with reduction modulo the integer $q-1$. The maximum number of erased symbols to reconstruct is $n-k$, so the operations needed in GF(q) are in the worst case $\mathcal{O}(k(n-k))$, where $\mathcal{O}(\cdot)$ is the big-$\mathcal{O}$ Landau symbol. If the encoder is in systematic form, we just need to reconstruct the erased symbols in the systematic part, so often less than $n-k$.

Equation (2.97) may also be used for building a systematic RS encoder: In this case, the first k codeword symbols are the k message symbols, and the $n-k$ parity symbols are computed by the interpolation polynomial

$$c_j = u(\alpha^j) = \sum_{i=0}^{k-1} m_i\alpha^{g_{i,j}} \quad j = k, \ldots, n-1, \tag{2.101}$$

where m_i are the message symbols, the integers $g_{i,j}$ are

$$g_{i,j} = \left(\sum_{\substack{\ell=0 \\ \ell \neq i}}^{k-1} \mu_{j,\ell} - \mu_{i,\ell} \right) \bmod (q-1) \tag{2.102}$$

and $\mu_{i,j}$ is given in (2.98). The integers $g_{i,j}$ can be precomputed and stored. This systematic encoding method requires $k(n-k)$ multiplications in GF(q) and the storage of the $k(n-k)$ integers $g_{i,j}$. Note that (2.102) provides directly the parity submatrix of the generating matrix in systematic form for RS codes.

Example 2.7 Let us consider the $RS(7,5)$ code in Example 2.6 built on $GF(8)$ as defined by Table 2.2. From the table and the definition (2.98), we can compute all $\mu_{i,j}$. For example, since $\alpha^2 - \alpha^5 = \alpha^2 + \alpha^5 = \alpha^4$, we have $\mu_{2,5} = 4$. The complete description is

$$\{\mu_{i,j}\}_{0 \le i,j \le n-1} = \begin{bmatrix} - & 5 & 3 & 2 & 6 & 1 & 4 \\ 5 & - & 6 & 4 & 3 & 0 & 2 \\ 3 & 6 & - & 0 & 5 & 4 & 1 \\ 2 & 4 & 0 & - & 1 & 6 & 5 \\ 6 & 3 & 5 & 1 & - & 2 & 0 \\ 1 & 0 & 4 & 6 & 2 & - & 3 \\ 4 & 2 & 1 & 5 & 0 & 3 & - \end{bmatrix}. \tag{2.103}$$

Then, from (2.102) we get the elements of the RS generator matrix. For example $g_{0,5} = \mu_{5,1} + \mu_{5,2} + \mu_{5,3} + \mu_{5,4} - \mu_{0,1} - \mu_{0,2} - \mu_{0,3} - \mu_{0,4} \bmod 7 = 0 + 4 + 6 + 2 - 5 - 3 - 2 - 6 \bmod 7 = 3$, which means α^3 in the position $(1,6)$ of the generator matrix. Calculating all elements with (2.102) we have directly the generator matrix in systematic form as

$$\boldsymbol{G} = \left[\begin{array}{c|cc} & \alpha^3 & \alpha^6 \\ & \alpha^2 & \alpha^6 \\ \boldsymbol{I}_5 & \alpha^2 & \alpha^4 \\ & 1 & 1 \\ & \alpha^3 & \alpha^4 \end{array} \right]. \tag{2.104}$$

2.4.4 RS Codes for Errors and Erasures

The practical implementation of maximum-likelihood decoders is generally of prohibitive complexity when dealing with errors, except for very small codes. However, by exploiting their algebraic structure, efficient algorithms can be derived for RS codes, capable of recovering all configurations of up to t errors and e erasures, provided that $2t + e \le n - k$. In the presence of erasures only, all patterns of up to $n - k$ erasures are recovered, and the resulting decoder is thus also maximum likelihood (see (2.73)).

The following theorems describe some of these techniques. Among several possible variants, the algorithms, here derived in an original way, can be cast in the so-called frequency domain decoders.

Theorem 2.9 (RS decoding in the presence of errors) *Assume the message* $u(X) = u_0 + u_1 X + \cdots + u_{k-1}X^{k-1}$ *is encoded by an* $RS(n,k)$ *code in* $GF(q)$ *with* $n = q-1$, *yielding the codeword* $\mathbf{c} = [c_0\ c_1\ \cdots\ c_{n-1}] = [u(1)\ u(\alpha)\ \cdots\ u(\alpha^{n-1})]$ *which is transmitted through the channel. Denote the received vector as* $\mathbf{y} = [y_0\ y_1\ \cdots\ y_{n-1}]$ *with polynomial* $y(X) = y_0 + y_1 X + \cdots + y_{n-1}X^{n-1}$, *and assume* t *errors occurred at* t *unknown positions. Provided that* $2t \le n-k$, *the message symbols* $u_0, u_1, \ldots, u_{k-1}$ *are obtained with the following four steps, where all operations are in* $GF(q)$:[10]

1. *From the received vector, calculate*

$$Y_i = y(\alpha^i), \quad i = 0,\ldots,n-1; \tag{2.105}$$

2. *Find the solution* $\lambda_0, \lambda_1, \ldots, \lambda_{t-1}$ *of the* $(t \times t)$ *Toeplitz system*

$$\begin{bmatrix} Y_t & Y_{t+1} & \cdots & Y_{2t-1} \\ Y_{t-1} & Y_t & \cdots & Y_{2t-2} \\ \vdots & \vdots & & \vdots \\ Y_1 & Y_2 & \cdots & Y_t \end{bmatrix} \begin{bmatrix} \lambda_0 \\ \lambda_1 \\ \vdots \\ \lambda_{t-1} \end{bmatrix} = -\begin{bmatrix} Y_{2t} \\ Y_{2t-1} \\ \vdots \\ Y_{t+1} \end{bmatrix}; \tag{2.106}$$

3. *Initialize:*

$$a_i = \frac{1}{n} Y_{-i}, \quad i = 0,\ldots,k+t-1; \tag{2.107}$$

4. *For* $\ell = k-1, k-2, \ldots, 0$

$$u_\ell = \begin{bmatrix} a_{\ell+t} & \cdots & a_{\ell+1} & a_\ell \end{bmatrix} \begin{bmatrix} \lambda_0 & \cdots & \lambda_{t-1} & 1 \end{bmatrix}^{\mathsf{T}} \tag{2.108}$$

$$a_\ell \leftarrow a_\ell - u_\ell. \tag{2.109}$$

Note that if the field order is $q = 2^m$ *the factor* $1/n$ *is equal to* 1, *and the* + *and* − *operations are coincident.*

Proof The received polynomial can be written $y(X) = c(X) + e(X)$ where $e(X) = \sum_{i=1}^{t} e_{J_i} X^{J_i}$ is the error polynomial, and the t errors occurred at the unknown positions $J_1, J_2, \ldots, J_t$. Let us introduce the monic *error locator polynomial*

$$\lambda(X) = (X - \alpha^{J_1}) \cdots (X - \alpha^{J_t}) = \lambda_0 + \lambda_1 X + \cdots + \lambda_{t-1}X^{t-1} + \lambda_t X^t \tag{2.110}$$

of degree t, such that the zeros of the polynomial are in correspondence of the errors, that is, $\lambda(\alpha^{J_i}) = 0$. Note that $\lambda_t = 1$ by construction. According to (2.30) and (2.32), the transformed polynomials and theirs coefficients are indicated with capital letters,

[10] When not explicitly indicated the sums are from 0 to $n-1$, and the coefficient indexes are to be interpreted as mod$n$.

so $\Lambda(X) = \Lambda_0 + \Lambda_1 X + \cdots + \Lambda_{n-1}X^{n-1}$ with coefficients $\Lambda_i = \lambda(\alpha^i)$ is the DFT of $\lambda(X)$, the DFT of $y(X)$ is $Y(X)$ with coefficients $Y_i = y(\alpha^i)$, and so on. Next, define the polynomial

$$q(X) = u(X)\lambda(X). \tag{2.111}$$

Since $u(X)$ has maximum degree $k-1$ and $\lambda(X)$ has degree t, it follows that $q(X) = q_0 + q_1 X + \cdots q_{k+t-1}X^{k+t-1}$ has maximum degree $k+t-1$. Let us consider the coefficients Q_i of the DFT of $q(X)$. Observing that when there is an error the locator polynomial gives zero we have

$$\begin{aligned} Q_i = q(\alpha^i) &= u(\alpha^i)\lambda(\alpha^i) \\ &= (y_i - e_i)\Lambda_i = y_i\Lambda_i \quad i = 0, \ldots, n-1\,. \end{aligned} \tag{2.112}$$

By using (2.32), we rewrite y_i as inverse DFT from $Y(X)$, and then the q_ℓ as inverse DFT from the Q_i, obtaining

$$\begin{aligned} q_\ell &= \frac{1}{n}\sum_i y_i \Lambda_i \alpha^{-\ell i} = \frac{1}{n^2}\sum_i\sum_p Y_p \alpha^{-pi}\Lambda_i \alpha^{-\ell i} \\ &= \frac{1}{n^2}\sum_p Y_p \sum_i \Lambda_i \alpha^{-i(p+\ell)} = \frac{1}{n}\sum_p Y_p \lambda_{p+\ell} \\ &= \frac{1}{n}\sum_{p=0}^{t} \lambda_p Y_{p-\ell}\,. \end{aligned} \tag{2.113}$$

This equation is valid for any ℓ. Moreover, since $q(X)$ has maximum degree $k+t-1$, we know that the coefficients $q_{n-1}, q_{n-2}, \ldots, q_{k+t}$ of $q(X)$ are identically zero. These known coefficients can be used to derive the λ_i. In fact, since we assumed $n-k \geq 2t$, there are at least t leading zero coefficients, $q_{n-1} = q_{n-1} = \cdots = q_{n-t} = 0$, that substituted in (2.113) give

$$0 = \sum_{p=0}^{t} \lambda_p Y_{p+m} \quad m = 1, 2, \ldots, t\,. \tag{2.114}$$

This is a set of t equations in the t unknowns $\lambda_0, \lambda_1, \ldots, \lambda_{t-1}$. Rewriting it with $\lambda_t = 1$ gives the $t \times t$ linear system in Toeplitz form (2.106), whose solution gives $\lambda(X)$. Once we have $\lambda(X)$ we have to find the message symbols u_ℓ. To this aim, we observe that $q(X) = u(X)\lambda(X)$ implies

$$q_\ell = \sum_{m=0}^{t} \lambda_m u_{\ell-m} \qquad \ell = 0, \cdots, n-1, \tag{2.115}$$

where $u_i = 0$ for $i < 0$ and for $i > k-1$. Equating (2.115) to (2.113), we get

$$\lambda_0 u_\ell + \lambda_1 u_{\ell-1} + \cdots \lambda_t u_{\ell-t} = \frac{1}{n}(\lambda_0 Y_{-\ell} + \lambda_1 Y_{1-\ell} + \cdots \lambda_t Y_{t-\ell}), \tag{2.116}$$

where the right-hand side is known. Solving (2.116) for $\ell - t$ from $k-1$ to 0 gives the recursion of steps (3), (4). □

The linear system in (2.106) is in Toeplitz form (actually, in the form of Yule-Walker equations), and therefore it can be solved efficiently. For example, $\mathcal{O}(t^2)$ operations are needed using the extended Euclidean algorithm or the Berlekamp–Massey algorithm. For large systems, there are also algorithms requiring $\mathcal{O}(t \log t)$ operations. The recursion (4) in the theorem statement requires approximately $t\,k$ multiplications and $t\,k$ additions. Overall, the operations needed are a DFT (to obtain the $Y(\alpha^i)$), plus $\mathcal{O}(k\,t) + \mathcal{O}(t^2)$ operations, where $t \le (n-k)/2$.

In practice, since the number of errors is not a priori known, the algorithm is started by setting $t = t_{\max} = \lfloor (n-k)/2 \rfloor$ in (2.106): Then, it can be proved that the solution by means of the Euclidean algorithm or the Berlekamp–Massey algorithm of (2.106) provides a vector whose length is the actual number of errors, to be used in the subsequent steps (3) and (4). Also, it can be checked that, if the recursion (4) continues for $\ell = -1, \dots, -t$, and the $u_{-1}, \dots u_{-t}$ are not all zero, then we must have $2t > n-k$ (hence, the described one is a bounded distance decoder).

As an alternative to performing steps (3) and (4) in Theorem 2.9, after having derived $\lambda(X)$ we could explicitly find the position of the errors and correct them. To this aim we have to find the zero of the error locator polynomial. The simplest way is to compute it in all positions $\lambda(\alpha^0), \dots, \lambda(\alpha^{n-1})$, and check where it is zero. An efficient algorithm for this phase is due to Chien. Once we know the positions of the correct symbols, we could use (2.97) instead of steps (3) and (4) in Theorem 2.9, or a formula, due to Forney, requiring another polynomial called error evaluator. The advantage here is that the inverse matrix given by (2.97) can be reused for several received vectors if we know that the errors are in the same positions. This situation applies for example when parallel RS codes are employed to protect large packets.

Theorem 2.10 (RS decoding with errors and erasures) *Assume the message $u(X) = u_0 + u_1 X + \cdots + u_{k-1}X^{k-1}$ is encoded by an $RS(n,k)$ code in $GF(q)$ with $n = q-1$, yielding the codeword $\mathbf{c} = [c_0\ c_1\ \cdots\ c_{n-1}] = [u(1)\ u(\alpha)\ \cdots\ u(\alpha^{n-1})]$ which is transmitted through the channel. Denote the received vector as $\mathbf{y} = [y_0\ y_1\ \cdots\ y_{n-1}]$, assume t errors occurred at t unknown positions $J_1, J_2, ..., J_t$, and e erasures occurred at known positions $I_1, I_2, \dots, I_e$. If $2t + e \le n-k$, then the message symbols can be found as follows:*

1. *Define arbitrary values for y_i at the erasure locations, for example, $y_{I_i} = 0$, $i = 1, \dots, e$, the polynomial $y(X) = y_0 + y_1 X + \cdots + y_{n-1}X^{n-1}$, the erasure locator polynomial $\tilde{\lambda}(X) = (X - \alpha^{I_1}) \cdots (X - \alpha^{I_e}) = \tilde{\lambda}_0 + \tilde{\lambda}_1 X + \cdots + \tilde{\lambda}_{e-1}X^{e-1} + X^e$.*
2. *From the received vector and the erasure locator polynomial $\tilde{\lambda}(X)$, calculate*

$$Y_i = y(\alpha^i), \qquad i = 0, \dots, n-1 \tag{2.117}$$

$$\tilde{Y}_i = \sum_{m=0}^{e} \tilde{\lambda}_m Y_{m+i}, \qquad i = 1, \dots, 2t. \tag{2.118}$$

3. *Find the solution* $\lambda_0, \lambda_1, \ldots, \lambda_{t-1}$ *of the* $(t \times t)$ *Toeplitz system*

$$\begin{bmatrix} \tilde{Y}_t & \tilde{Y}_{t+1} & \cdots & \tilde{Y}_{2t-1} \\ \tilde{Y}_{t-1} & \tilde{Y}_t & \cdots & \tilde{Y}_{2t-2} \\ \vdots & \vdots & & \vdots \\ \tilde{Y}_1 & \tilde{Y}_2 & \cdots & \tilde{Y}_t \end{bmatrix} \begin{bmatrix} \lambda_0 \\ \lambda_1 \\ \vdots \\ \lambda_{t-1} \end{bmatrix} = - \begin{bmatrix} \tilde{Y}_{2t} \\ \tilde{Y}_{2t-1} \\ \vdots \\ \tilde{Y}_{t+1} \end{bmatrix}. \tag{2.119}$$

4. *Calculate the coefficients* $\overline{\lambda}_0, \overline{\lambda}_1, \ldots, \overline{\lambda}_{t+e}$ *of the polynomial* $\tilde{\lambda}(X)\lambda(X)$, *where* $\lambda(X) = \lambda_0 + \lambda_1 X + \cdots + \lambda_{t-1}X^{t-1} + X^t$.
5. *Initialize:*

$$a_i = \frac{1}{n} Y_{-i}, \quad i = 0, \ldots, k+t+e-1. \tag{2.120}$$

6. *For* $\ell = k-1, k-2, \ldots, 0$

$$u_\ell = \begin{bmatrix} a_{t+\ell} & \cdots & a_{\ell+1} & a_\ell \end{bmatrix} \begin{bmatrix} \overline{\lambda}_0 & \cdots & \overline{\lambda}_{t+e-1} & 1 \end{bmatrix}^{\mathsf{T}} \tag{2.121}$$

$$a_\ell = a_\ell - u_\ell. \tag{2.122}$$

Proof Define the error locator polynomial

$$\lambda(X) = (X - \alpha^{J_1}) \cdots (X - \alpha^{J_t}) = \lambda_0 + \lambda_1 X + \cdots + \lambda_{t-1}X^{t-1} + X^t \tag{2.123}$$

of degree t, with $\lambda(\alpha^{J_i}) = 0$, and the erasure locator polynomial

$$\tilde{\lambda}(X) = (X - \alpha^{I_1}) \cdots (X - \alpha^{I_e}) \tag{2.124}$$

of degree e, with $\tilde{\lambda}(\alpha^{I_i}) = 0$. Note that $\tilde{\lambda}(X)$ is completely known. Then, define the polynomial

$$q(X) = u(X)\tilde{\lambda}(X)\lambda(X). \tag{2.125}$$

Since $u(X)$ has maximum degree $k-1$, $\lambda(X)$ has degree t, and $\tilde{\lambda}(X)$ has degree e, it follows that $q(X) = q_0 + q_1 X + \cdots q_{k+t+e-1}X^{k+t+e-1}$ has maximum degree $k+t+e-1$. Evaluating this polynomial in α^i, and observing that when there is an error or an erasure the locator polynomials give zero, we have

$$\begin{aligned} Q_i = q(\alpha^i) &= u(\alpha^i)\tilde{\lambda}(\alpha^i)\lambda(\alpha^i) = (y_i - e_i)\tilde{\lambda}(\alpha^i)\lambda(\alpha^i) \\ &= y_i\tilde{\lambda}(\alpha^i)\lambda(\alpha^i) = \tilde{y}_i\lambda(\alpha^i) \quad i = 0, \ldots, n-1. \end{aligned} \tag{2.126}$$

The situation is similar to the previous case, and we can write

$$q_\ell = \frac{1}{n}\sum_{p=0}^{t} \lambda_p \tilde{Y}_{p-\ell} \tag{2.127}$$

where the transform of $\tilde{y}_i$ is

$$\tilde{Y}_\ell = \sum_i y_i \tilde{\Lambda}_i \alpha^{\ell i} = \frac{1}{n}\sum_m Y_m \sum_i \tilde{\Lambda}_i \alpha^{-i(m-\ell)} \tag{2.128}$$

$$= \sum_m Y_m \tilde{\lambda}_{m-\ell} = \sum_m \tilde{\lambda}_m Y_{m+\ell}. \tag{2.129}$$

Proceeding as in Theorem 2.9 we first obtain the coefficients λ_i, then the polynomial $\overline{\lambda}(X) = \tilde{\lambda}(X)\lambda(X)$. Going back to (2.126), we can also write $Q_i = y_i\overline{\lambda}(X)$, and therefore the final steps to recover the message are as in the previous Theorem 2.9, with the coefficients $\overline{\lambda}_i$ instead of the λ_i. □

Theorem 2.10, applied to the particular case in which we have erasures only, provides an alternative to the Lagrangian interpolation approach of Theorem 2.8, as follows.

Corollary 2.2 (RS decoding with erasures only) *Assume the message* $u(X) = u_0 + u_1X + \cdots + u_{k-1}X^{k-1}$ *is encoded by an* $RS(n,k)$ *code in* $GF(q)$ *with* $n = q-1$*, yielding the codeword* $\mathbf{c} = [c_0\ c_1\ \cdots\ c_{n-1}] = [u(1)\ u(\alpha)\ \cdots\ u(\alpha^{n-1})]$ *which is transmitted through the channel. Denote the received vector as* $\mathbf{y} = [y_0\ y_1\ \cdots\ y_{n-1}]$*, and assume* e *erasures occurred at known positions* $I_1, I_2, \ldots, I_e$.

Then, if $e \leq n-k$*, the message symbols can be simply found as follows:*

1. *Define arbitrary values for* y_i *at the erasure locations, for example,* $y_{I_i} = 0$, $i = 1, \ldots, e$*, the polynomial* $y(X) = y_0 + y_1X + \cdots + y_{n-1}X^{n-1}$*, the erasure locator polynomial* $\tilde{\lambda}(X) = (X - \alpha^{I_1}) \cdots (X - \alpha^{I_e}) = \tilde{\lambda}_0 + \tilde{\lambda}_1X + \cdots + \tilde{\lambda}_{e-1}X^{e-1} + X^e$.
2. *Initialize:*

$$a_i = \frac{1}{n} y(\alpha^{-i}), \quad i = 0, \ldots, k+e-1. \tag{2.130}$$

3. *For* $\ell = k-1, k-2, \ldots, 0$

$$u_\ell = \begin{bmatrix} a_{e+\ell} & \cdots & a_{\ell+1} & a_\ell \end{bmatrix} \begin{bmatrix} \tilde{\lambda}_0 & \cdots & \tilde{\lambda}_{e-1} & 1 \end{bmatrix}^{\mathsf{T}} \tag{2.131}$$

$$a_\ell = a_\ell - u_\ell. \tag{2.132}$$

2.4.5 An Alternative Decoding Algorithm for RS Codes

We here present an algorithm which can be used instead of Theorem 2.9 to directly find the message symbols in the presence of errors.

Below we indicate with $|\boldsymbol{A}|$ the determinant of a matrix $\boldsymbol{A}$.

Theorem 2.11 (RS decoding by zeroing determinants) *Assume the message* $u(X) = u_0 + u_1X + \cdots + u_{k-1}X^{k-1}$ *is encoded by an* $RS(n,k)$ *code in* $GF(q)$ *with* $n = q-1$*, yielding the codeword* $\mathbf{c} = [c_0\ c_1\ \cdots\ c_{n-1}] = [u(1)\ u(\alpha)\ \cdots\ u(\alpha^{n-1})]$ *which is transmitted through the channel. Denote the received vector as* $\mathbf{y} = [y_0\ y_1\ \cdots\ y_{n-1}]$*, with polynomial* $y(X) = y_0 + y_1X + \cdots + y_{n-1}X^{n-1}$*, and assume* t *errors occurred at* t *unknown positions. Provided that* $2t \leq n-k$*, the message symbols* $u_0, u_1, \ldots, u_{k-1}$ *are obtained with the following steps, where all operations are in* $GF(q)$*:*

1. *From the received vector, compute* $Y_i = \frac{1}{n}y(\alpha^i)$, $i = -k+1, \ldots, 2t$;
2. *Initialize* $a_i = Y_i$, $i = 1, \ldots, 2t$;

3. For $\ell = 0, -1, \dots, -k+1$

$$a_\ell = -\frac{\begin{vmatrix} a_{\ell+t} & a_{\ell+t+1} & \dots & a_{\ell+2t} \\ \vdots & \vdots & \vdots & \vdots \\ a_{\ell+1} & a_{\ell+2} & \dots & a_{\ell+t+1} \\ 0 & a_{\ell+1} & \dots & a_{\ell+t} \end{vmatrix}}{\begin{vmatrix} a_{\ell+t+1} & \dots & a_{\ell+2t} \\ \vdots & \vdots & \vdots \\ a_{\ell+3} & \dots & a_{\ell+t+2} \\ a_{\ell+2} & \dots & a_{\ell+t+1} \end{vmatrix}} \tag{2.133}$$

$$u_{-\ell} = Y_\ell - a_\ell. \tag{2.134}$$

Proof From (2.116), we see that it must hold (note that here the factor $1/n$ has been taken into account in step 1)

$$\begin{bmatrix} Y_t & Y_{t+1} & \dots & Y_{2t} \\ Y_{t-1} & Y_t & \dots & Y_{2t-1} \\ \vdots & \vdots & \vdots & \vdots \\ Y_1 & Y_2 & \dots & Y_{t+1} \\ Y_0 - u_0 & Y_1 & \dots & Y_t \\ Y_{-1} - u_1 & Y_0 - u_0 & \dots & Y_{t-1} \\ \vdots & \vdots & \vdots & \vdots \\ Y_{-k+1} - u_{k-1} & Y_{-k+2} - u_{k-2} & \dots & Y_{-k+1+t} \\ Y_{-k} & Y_{-k+1} - u_{k-1} & \dots & Y_{-k+t} \\ \vdots & \vdots & \vdots & \vdots \\ Y_{-k-t+1} & Y_{-k-t+2} & \dots & Y_{-k+1} - u_{k-1} \end{bmatrix} \begin{bmatrix} \lambda_0 \\ \lambda_1 \\ \vdots \\ 1 \end{bmatrix} = \begin{bmatrix} 0 \\ 0 \\ \vdots \\ \vdots \\ \vdots \\ 0 \end{bmatrix}. \tag{2.135}$$

In order for this system to have a nonzero solution, the matrix in the left-hand side must be rank-deficient. Hence, the determinant of each $(t+1) \times (t+1)$ submatrix must be zero. For the upper square submatrix, this implies

$$\begin{vmatrix} Y_t & Y_{t+1} & \dots & Y_{2t} \\ Y_{t-1} & Y_t & \dots & Y_{2t-1} \\ \vdots & \vdots & \vdots & \vdots \\ Y_1 & Y_2 & \dots & Y_{t+1} \\ Y_0 - u_0 & Y_1 & \dots & Y_t \end{vmatrix} = 0. \tag{2.136}$$

Thus, u_0 is the field element zeroing the determinant of this matrix. One possibility to find it is to check all q possible values. A better option is to write the determinant as Laplace expansion from the last row, from which it follows that it must be

$$(Y_0 - u_0)\begin{vmatrix} Y_{t+1} & Y_{t+2} & \dots & Y_{2t} \\ Y_t & Y_{t+1} & \dots & Y_{2t-1} \\ \vdots & \vdots & \vdots & \vdots \\ Y_2 & Y_3 & \dots & Y_{t+1} \end{vmatrix} + \begin{vmatrix} Y_t & Y_{t+1} & \dots & Y_{2t} \\ Y_{t-1} & Y_t & \dots & Y_{2t-1} \\ \vdots & \vdots & \vdots & \vdots \\ Y_1 & Y_2 & \dots & Y_{t+1} \\ 0 & Y_1 & \dots & Y_t \end{vmatrix} = 0 \tag{2.137}$$

that provides

$$Y_0 - u_0 = -\frac{\begin{vmatrix} Y_t & Y_{t+1} & \dots & Y_{2t} \\ Y_{t-1} & Y_t & \dots & Y_{2t-1} \\ \vdots & \vdots & \vdots & \vdots \\ Y_1 & Y_2 & \dots & Y_{t+1} \\ 0 & Y_1 & \dots & Y_t \end{vmatrix}}{\begin{vmatrix} Y_{t+1} & Y_{t+2} & \dots & Y_{2t} \\ Y_t & Y_{t+1} & \dots & Y_{2t-1} \\ \vdots & \vdots & \vdots & \vdots \\ Y_2 & Y_3 & \dots & Y_{t+1} \end{vmatrix}}. \tag{2.138}$$

Then, we can proceed similarly with the next submatrix to derive $Y_{-1} - u_1$, and then iteratively to find all u_ℓ. □

Finding the determinant of a $t \times t$ Toeplitz matrix requires no more than $\mathcal{O}(t^2)$ operations. Thus, the decoding algorithm from Theorem 2.11 requires $\mathcal{O}(kt^2)$ operations, plus the calculation of the DFT $y(\alpha^i)$.

2.5 Historical Notes

Shannon introduced the concept of channel capacity and demonstrated that good codes exist, therefore launching the challenge for finding specific families of codes that achieve capacity [1]. So, although some forms of coding had been developed before (see [2, 3]), it can be said that it was the Shannon's landmark paper to trigger the birth of the error-correcting codes research area, where specific mathematical tools have been developed over the years to design codes with efficient encoding and decoding algorithms. Among the first designs there is the Hamming code, already mentioned in the Shannon's 1948 paper as an example of efficient non trivial error correcting code and later fully described in [4], followed by the Golay code [5], and then many others. Among those inspired by algebraic constructions, the most important are the Reed–Muller codes [6–8], the BCH codes [9, 10], and the Reed–Solomon codes [11]. In their original 1960 paper [11], Reed and Solomon proposed to build the codewords as the input polynomials evaluated at all field elements: $(u(0), u(1), u(\alpha), \dots, u(\alpha^{q-2}))$. This would give a code with length $n = q = 2^m$, which is like that given by (2.82) and (2.86) with an additional parity symbol equal to $u(0) = u_0$. Since $0, 1, \alpha, \dots, \alpha^{q-2}$ are distinct and $u(X)$ has degree $k-1$, the maximum number of zeros in a codeword is $k-1$ and therefore $d_{\min} = q - (k-1) = n - k + 1$. This $RS(n = 2^m, k)$ code is still

MDS, but not cyclic. Indeed, since $u(1)+u(\alpha)+\cdots+u(\alpha^{q-2}) = -(q-1)u(0) = u(0)$, the original RS code is essentially like the single extended code in (2.92). RS codes are of vast interest and have several applications, and therefore they have been extensively studied [12–18]. The relation between Cauchy-based MDS codes and RS codes is discussed in [19–22].

2.6 Exercises

Exercise 2.1 (Fields) From the definition of field prove that, for any elements a and b in a field, $a \cdot 0 = 0 \cdot a = 0$, and $-(a \cdot b) = (-a) \cdot b = a \cdot (-b)$.

Exercise 2.2 (Euclidean algorithm for the greatest common divisor) Given two positive integers a and b, set $r_1 = a, r_0 = b$. Prove that the last nonzero remainder in the repeated remainders $r_{i+1} = r_{i-1} \bmod r_i$, $i = 1,2,\ldots$ is the greatest common divisor of a and b.

Exercise 2.3 (Finite fields) For p prime and arbitrary integers a, b, prove using the binomial expansion that $(a+b)^p \bmod p = (a^p + b^p) \bmod p$. Then, prove by induction that $a^p \bmod p = a \bmod p$ (which is known as the Fermat's little theorem).

Exercise 2.4 Prove Corollary 2.1.

Exercise 2.5 (Cyclic codes) Denote $c^{(1)}(X)$ the cyclic shift of one position of a codeword $c(X) = c_0 + c_1X + \cdots + c_{n-1}X^{n-1}$.

1. Show that we can write

$$c^{(1)}(X) = X\,c(X) + c_{n-1}(X^n - 1) = X\,c(X) \bmod (X^n - 1) \tag{2.139}$$

2. Assume a code with codewords $c(X) = q(X)g(X)$, where $g(X)$ is a fixed polynomial of degree $n-k$, and $q(X)$ are all polynomials with maximum degree $k-1$. Show that the code is cyclic if and only if $g(X)$ divides $X^n - 1$.

Exercise 2.6 Using the companion matrices, find a binary matrix representation of the parity-check matrix in (2.91).

Exercise 2.7 Derive the expressions of the weight enumerating functions (2.60), (2.61).

Exercise 2.8 (Hamming codes) For a given $m > 1$, build an $m \times n$ parity-check matrix whose columns are all possible nonzero binary m-tuples. Show that the corresponding Hamming code has parameters $(n,k) = (2^m - 1, 2^m - 1 - m)$, and that $d_{\min} = 3$. Which code is obtained for $m = 2$?

Exercise 2.9 (Dual of Hamming codes) The dual of the Hamming code $(n,k) = (2^m - 1, 2^m - 1 - m)$ is called simplex code or maximal-length feedback shift register code. It can be shown that it consists of the zero codeword and $2^{n-k} - 1$ codewords, all of weight 2^{n-k-1}. Verify this statement for $m = 2$ and for $m = 3$.

Exercise 2.10 (Weight enumerator of the dual code) Write the weight enumerating function for the $(7,3)$ simplex code, consisting of the zero codeword and other 7 codewords of weight 4. Then, using the MacWilliams identity, find the weight enumerator of its dual (which is the Hamming $(7,4)$ code).

Exercise 2.11 (Erasure correction with Hamming codes) Find the number of patterns of three erasures that can be corrected by the Hamming $(7,4)$ code. Find the number of erasures which can be corrected by the extended $(8,4)$ Hamming code. Generalize to $(2^m - 1, 2^m - 1 - m)$ Hamming codes. Find the pattern of erasures which can be corrected by a $(6,4)$ code obtained by puncturing (erasing) the last codeword bit of the Hamming $(7,4)$ code defined by (2.64).

Exercise 2.12 (Block error probability bound) Erasure patterns covering exactly all nonzero elements (support) of a codeword cannot be corrected. This in particular implies, for binary codes, that the number of correctable configurations of $d_{\min}$ erasures is $A_{d_{\min}}$.

Prove that the block error probability for binary codes is

$$P_B = 1 - P_c \tag{2.140}$$

with

$$P_c \geq \sum_{e=0}^{n-k} \left[\binom{n}{e} - \sum_{\ell=d_{\min}}^{e} \binom{n-\ell}{e-\ell} A_\ell \right] \epsilon^e (1-\epsilon)^{n-e}, \tag{2.141}$$

where for the case $d_{\min} > (n-k)/2$ the equality holds.

References

[1] C. Shannon, "A mathematical theory of communication," *Bell Syst. Tech. J.*, vol. 27, no. 3, pp. 379–423, 623–656, July 1948.

[2] A. Barg, "At the dawn of the theory of codes," *Math. Intelligencer*, vol. 15, no. 1, pp. 20–27, 1993.

[3] A. R. Calderbank, "The art of signaling: Fifty years of coding theory," *IEEE Trans. Inf. Theory*, vol. 44, no. 6, pp. 2561–2595, 1998.

[4] R. W. Hamming, "Error detecting and error correcting codes," *Bell Syst. Tech. J.*, vol. 29, no. 2, pp. 147–160, 1950.

[5] M. J. Golay, "Notes on digital coding," *Proc. IEEE*, vol. 37, p. 657, 1949.

[6] D. E. Muller, "Metric properties of Boolean algebra and their application to switching circuits, Univ. of Ill., Digital Computer Lab," Internal Report, Technical Report, 1953.

[7] D. E. Muller, "Application of boolean algebra to switching circuit design and to error detection," *Trans. I.R.E. Professional Group on Electronic Computers*, vol. EC-3, no. 3, pp. 6–12, 1954.

[8] I. Reed, "A class of multiple-error-correcting codes and the decoding scheme," *Trans. IRE Professional Group on Information Theory*, vol. 4, no. 4, pp. 38–49, 1954.

[9] A. Hocquenghem, "Codes correcteurs d'erreurs," *Chiffers*, vol. 2, pp. 147–156, 1959.

[10] R. Bose and D. Ray-Chaudhuri, "On a class of error correcting binary group codes," *Information and Control*, vol. 3, no. 1, pp. 68–79, 1960.

[11] I. Reed and G. Solomon, "Polynomial codes over certain finite fields," *J. Soc. Indust. Appl. Math.*, vol. 8, no. 2, pp. 300–304, June 1960.

[12] C. Chen, W. W. Peterson, and E. Weldon Jr, "Some results on quasi-cyclic codes," *Information and Control*, vol. 15, no. 5, pp. 407–423, 1969.

[13] E. Berlekamp, *Algebraic coding theory*. New York, NY, USA: McGraw-Hill, 1968.

[14] W. Peterson and E. Weldon, Jr., *Error-correcting codes*. Cambridge, MA, USA: MIT Press, 1972.

[15] F. MacWilliams and N. Sloane, *The theory of error-correcting codes*. Amsterdam, the Netherlands: North Holland Mathematical Libray, 1977, vol. 16.

[16] S. B. Wicker and V. K. Bhargava, *Reed-Solomon Codes and Their Applications*. New York, NY, USA: Wiley-IEEE Press, 1994.

[17] R. J. McEliece, *The theory of information and coding*. New York, NY, USA: Cambridge University Press, 2004.

[18] W. E. Ryan, S. Lin, and S. G. Wilson, *Channel Codes: Classical and Modern*. Cambridge University Press, 2024.

[19] R. Roth and G. Seroussi, "On generator matrices of MDS codes," *IEEE Trans. Inf. Theory*, vol. 31, no. 6, pp. 826–830, 1985.

[20] G. Seroussi, "A systolic Reed-Solomon encoder," *IEEE Trans. Inf. Theory*, vol. 37, no. 4, pp. 1217–1220, 1991.

[21] J. Bloemer, M. Kalfane, R. Karp, M. Karpinski, M. Luby, and D. Zuckerman, "An XOR-based erasure-resilient coding scheme," International Computer Science Institute, Berkeley, CA, Technical Report TR-95.048, Technical Report, 1995.

[22] J. Plank and L. Xu, "Optimizing Cauchy Reed-Solomon codes for fault-tolerant network storage applications," in *Proc. Fifth IEEE Int. Symp. Network Computing and Applications (NCA'06)*, 2006, pp. 173–180.

3 Erasure Channel Models and Bounds

In this chapter, we provide an overview of erasure channel models. We start with the simple case of memoryless erasure channels (Section 3.1). Section 3.2 discusses upper and lower bounds on the error probability over memoryless erasure channels. Finally, Section 3.3 illustrates various erasure channel models with memory.

3.1 Memoryless Erasure Channels

Following the introduction in Chapter 1, we first recall some basic definitions and facts about DMCs. We then address the specific case of memoryless erasure channels.

3.1.1 Discrete Memoryless Channels

A DMC accepts symbols from the alphabet $\mathcal{X}$ at its input, and outputs symbols from the alphabet $\mathcal{Y}$. The channel operates over n successive time instants, with y_i being the channel output corresponding to the input x_i at time i. Let us now denote by $P(\boldsymbol{y}|\boldsymbol{x})$ the probability of observing the n-tuple $\boldsymbol{y} = \begin{bmatrix} y_1 & y_2 & \cdots & y_n \end{bmatrix}$ at the output of the channel given that at the input $\boldsymbol{x} = \begin{bmatrix} x_1 & x_2 & \cdots & x_n \end{bmatrix}$. The channel behavior is fully characterized from a statistical viewpoint by the function $P(\boldsymbol{y}|\boldsymbol{x})$, with $\boldsymbol{x} \in \mathcal{X}^n$ and $\boldsymbol{y} \in \mathcal{Y}^n$.

In a DMC, the joint conditional probability

$$P(\boldsymbol{y}|\boldsymbol{x}) = P(y_1, y_2, \ldots, y_n | x_1, x_2, \ldots, x_n) \tag{3.1}$$

can be factored as

$$P(\boldsymbol{y}|\boldsymbol{x}) = \prod_{i=1}^{n} P(y_i|x_i). \tag{3.2}$$

At time i, the channel output given channel input is conditionally independent of all other inputs and outputs (*memoryless* channel). Thus, the channel is completely characterized by the transition probability $P(y|x)$, assumed to be constant over time.

Definition 3.1 (Channel transition matrix) Consider a DMC with input alphabet $\mathcal{X} = \{\mathsf{x}_1, \mathsf{x}_2, \ldots, \mathsf{x}_{|\mathcal{X}|}\}$ and output alphabet $\mathcal{Y} = \{\mathsf{y}_1, \mathsf{y}_2, \ldots, \mathsf{y}_{|\mathcal{Y}|}\}$. The channel transition matrix $\boldsymbol{\Pi}$ is a $|\mathcal{X}| \times |\mathcal{Y}|$ matrix, with the (i,j) entry being the channel transition probability

$$\pi_{i,j} = P(\mathsf{y}_j|\mathsf{x}_i). \tag{3.3}$$

Definition 3.2 (Symmetric channel) Consider a DMC with input alphabet $\mathcal{X}$ and output alphabet $\mathcal{Y}$ with $|\mathcal{X}| = |\mathcal{Y}|$. Let $\mathbf{\Pi}$ denote its transition matrix. The channel is said to be symmetric if each row of $\mathbf{\Pi}$ is a permutation of any other row, and each column of $\mathbf{\Pi}$ is a permutation of any other column.

Definition 3.3 (Weakly symmetric channel) Consider a DMC with input alphabet $\mathcal{X}$ and output alphabet $\mathcal{Y}$. Let $\mathbf{\Pi}$ denote its transition matrix. The channel is said to be weakly symmetric if $\mathbf{\Pi}$ can be decomposed in submatrices $\mathbf{\Pi}_i$ as

$$\mathbf{\Pi} = \left[\mathbf{\Pi}_1 \quad \mathbf{\Pi}_2 \quad \cdots \quad \mathbf{\Pi}_s\right], \tag{3.4}$$

where for every submatrix $\mathbf{\Pi}_i$ each row is a permutation of any other row and each column is a permutation of any other column.

We will briefly review some fundamental capacity results for DMCs. In the following, for discrete random variables (r.v.s) X and Y we denote by $P(x)$ and $P(y)$ the marginal distributions of X and Y. The joint distribution of X and Y is $P(x, y)$, and the conditional distribution of X given $Y = y$ is $P(x|y)$.

Definition 3.4 (Entropy) Consider a discrete r.v. X. The entropy of X is denoted by $\mathsf{H}(X)$ and it is defined as

$$\mathsf{H}(X) = \sum_{x} P(x) \log_2 \frac{1}{P(x)}, \tag{3.5}$$

where the sum is for x in the support of $P(x)$.

Definition 3.5 (Conditional entropy) Consider two discrete r.v.s X and Y. The conditional entropy of X given Y is defined as

$$\mathsf{H}(X|Y) = \sum_{x,y} P(x,y) \log_2 \frac{1}{P(x|y)} \tag{3.6}$$

$$= \sum_{y} P(y) \mathsf{H}(X|Y = y), \tag{3.7}$$

where the sum in (3.6) is for (x, y) in the support of $P(x|y)$, and where $\mathsf{H}(X|Y = y)$ in (3.7) denotes the entropy of X given that Y takes value y, that is,

$$\mathsf{H}(X|Y = y) = \sum_{x} P(x|y) \log_2 \frac{1}{P(x|y)}. \tag{3.8}$$

Definition 3.6 (Mutual information) Consider two discrete r.v.s X and Y. The mutual information between X and Y is denoted by $\mathsf{I}(X; Y)$ and it is defined as

$$\mathsf{I}(X;Y) = \mathsf{H}(X) - \mathsf{H}(X|Y) \tag{3.9}$$

$$= \sum_{x,y} P(x,y) \log_2 \frac{P(x,y)}{P(x)P(y)}, \tag{3.10}$$

where the sum is for (x, y) in the support of $P(x, y)$.

Definition 3.7 (Capacity of a DMC) Consider a DMC with input X and output Y. Its channel capacity is defined as

$$\mathsf{C} = \max_{P(x)} \mathsf{I}(X;Y). \tag{3.11}$$

The capacity of a channel provides an upper bound to the maximum code rate that can be used to transmit reliably (i.e., with arbitrarily low block error probability). It is important to remark that the rate definition assumed by (3.11) differs from Definition 2.14 by the unit of measurement: While the former is expressed in bits per channel use, the latter is expressed in information symbols per codeword symbol. To clarify the point, consider a code defined on GF(q) with blocklength n and $M = q^k$ codewords, with transmission that takes place over a DMC with input alphabet cardinality $|\mathcal{X}| = q$. The blocklength equals the number of channel uses, and the code rate in bits per channel use is

$$R = \frac{1}{n} \log_2 M \qquad \text{[bits/channel use]}. \tag{3.12}$$

Alternatively, we may express the rate according to Definition 2.14 as

$$R = \frac{k}{n} \qquad \text{[symbols/channel use]}. \tag{3.13}$$

When comparing rates and channel capacities, we will ensure to consistently use the same unit of measurement. In particular, unless otherwise stated, we will adopt the convention of using (3.13).

Remark 3.1. For (weakly) symmetric channels, the capacity is attained by choosing a uniform input distribution.

Theorem 3.1 (Channel Coding Theorem) *For all rates $R < \mathsf{C}$ and sufficiently large blocklength n, there exists an (n, nR) code $\mathcal{C}$ and a decoding rule such that the block error probability can be made arbitrarily small. Conversely, no such code exists for $R > \mathsf{C}$.*

The Channel Coding Theorem partitions the rate into two regions. All rates $R < \mathsf{C}$ are said to be *achievable*: For $R < \mathsf{C}$, reliable communication is possible. Conversely, the rates exceeding C are said to be *not achievable*, that is, for $R > \mathsf{C}$ reliable communication is not possible.

3.1.2 The Binary Erasure Channel

The memoryless binary erasure channel, commonly referred to as the BEC, is one of the simplest channels within the class of DMCs. The BEC accepts binary symbols (bits) at its input and outputs symbols from a ternary set. The channel input and output alphabets are $\mathcal{X} = \{0, 1\}$ and $\mathcal{Y} = \{0, 1, \mathsf{E}\}$, where the output symbol E denotes an erasure (see Figure 1.2). When the channel output is either 0 or 1, we know the bit value is correct. When the channel output is an erasure, we have complete uncertainty about the value of the corresponding bit. The channel transition probabilities of a BEC are

$$P(0|0) = P(1|1) = 1 - \epsilon \tag{3.14}$$

$$P(1|0) = P(0|1) = 0 \tag{3.15}$$

$$P(\mathsf{E}|0) = P(\mathsf{E}|1) = \epsilon, \tag{3.16}$$

where ϵ is the channel erasure probability. The channel transition matrix is

$$\mathbf{\Pi} = \begin{bmatrix} P(0|0) & P(1|0) & P(\mathsf{E}|0) \\ P(0|1) & P(1|1) & P(\mathsf{E}|1) \end{bmatrix} = \begin{bmatrix} 1-\epsilon & 0 & \epsilon \\ 0 & 1-\epsilon & \epsilon \end{bmatrix}. \tag{3.17}$$

From a direct inspection of (3.17), we can see that the BEC does not fulfill the symmetric condition of Definition 3.2. However, the BEC belongs to the broader class of weakly symmetric channels (Definition 3.3). In fact, $\mathbf{\Pi}$ can be decomposed into two matrices

$$\mathbf{\Pi}_1 = \begin{bmatrix} 1-\epsilon & 0 \\ 0 & 1-\epsilon \end{bmatrix} \quad \text{and} \quad \mathbf{\Pi}_2 = \begin{bmatrix} \epsilon \\ \epsilon \end{bmatrix}. \tag{3.18}$$

For both $\mathbf{\Pi}_1$ and $\mathbf{\Pi}_2$, each row is a permutation of the other row, and for $\mathbf{\Pi}_1$, each column is a permutation of the other column ($\mathbf{\Pi}_2$ has a single column).

A natural generalization of a BEC is obtained by increasing the input alphabet size to $q > 2$. In this case, the input alphabet is made of q elements, and the output alphabet is the input alphabet plus the erasure symbol. The channel is referred to as QEC, with q denoting the erasure channel model order. The transition diagram for a memoryless QEC is shown in Figure 3.1. The channel transition probabilities are $P(y|x) = 1 - \epsilon$ if $y = x$ and $P(y|x) = \epsilon$ if $y = \mathsf{E}$. The QEC is weakly symmetric.

From now on, we adopt the convention of matching the finite field of the code to the erasure channel model order. Since in the BEC case we choose $\mathcal{X}$ to be the binary finite field GF(2), we will continue along this path and focus on the case where $\mathcal{X} = \text{GF}(q)$, with q being a prime or a power of a prime. Although not strictly necessary, this choice is convenient from an algorithmic perspective, as the tools available to operate on the field elements will facilitate the development of practical encoding and decoding algorithms.

As the QEC is a weakly symmetric DMC, the maximum in (3.11) is attained by a uniformly distributed input, that is, by setting $P(x) = 1/q$ for all $x \in \mathcal{X}$. Thus, by noting that $\mathsf{H}(X) = \log_2 q$, that $\mathsf{H}(X|Y = \mathsf{E}) = \log_2 q$, and that $\mathsf{H}(X|Y = y) = 0$ for any $y \in \mathcal{Y} \setminus \{\mathsf{E}\}$, following (3.7), we find that the capacity of the QEC is

$$\mathsf{C} = (1 - \epsilon) \log_2 q \qquad \text{[bits/channel use]} \tag{3.19}$$

or, more often, provided as

$$\mathsf{C} = 1 - \epsilon \qquad \text{[symbols/channel use].} \tag{3.20}$$

Figure 3.2 depicts the capacity of the QEC.

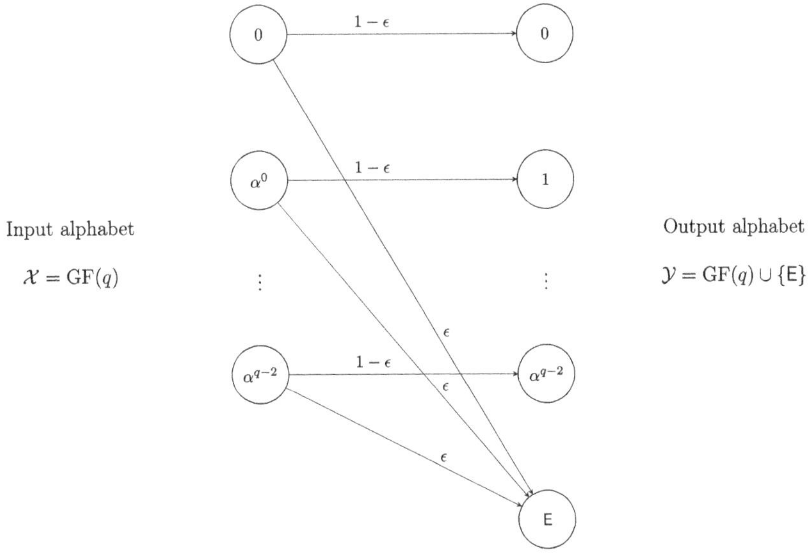

Figure 3.1 Transition diagram for a QEC.

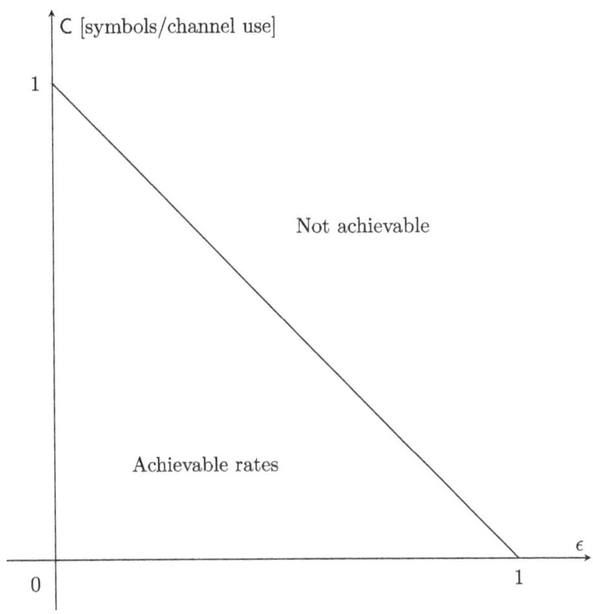

Figure 3.2 Capacity of the QEC in symbols per channel use.

3.2 Bounds on the Block Error Probability

Channel capacity provides the upper limit to the rate achievable with a vanishing small block error probability. The limit can be approached (in the absence of feedback) by codes with large blocklength n. However, it is often important to determine the ultimate performance bounds for erasure correcting codes when the blocklength is moderate or small. We will see that simple yet tight bounds on the error probability achievable by an (n,k) erasure code are available, and can serve as a benchmark to analyze the performance of a code. In the following derivations, we denote by e the number of erasures in a codeword, and by E the corresponding r.v.. Before delving into the finite-length bounds, we first introduce the code ensembles that will be analyzed in terms of error probability.

3.2.1 Code Ensembles

Definition 3.8 (Code ensemble) A code ensemble is a set of codes, endowed with a probability distribution.

Consider a code ensemble $\mathscr{C}$ and draw a random code $\mathcal{C}$ from it according to the probability distribution $P(\mathcal{C})$. We may be interested in determining some average properties of $\mathcal{C}$ (e.g., the average minimum distance, the average weight distribution, the average error probability). We will see how a proper selection of the distribution $P(\mathcal{C})$ allows deriving bounds to the performance of the best codes for a given block-length and code rate. Before doing so, we describe some ensembles that will be used for the purpose.

Definition 3.9 (Linear code ensemble, parity-check-based) We denote by $\mathscr{H}^q_{n,m}$ the ensemble of all (n,k) linear block codes over $\mathrm{GF}(q)$, with $k \geq n-m$, and with $P(\mathcal{C})$ being the *parity-check induced* distribution defined as follows. Consider a random matrix $\boldsymbol{H}$ from the set of all possible $m \times n$ matrices over $\mathrm{GF}(q)$, with $\boldsymbol{H}$ uniformly distributed,

$$P(\boldsymbol{H}) = q^{-nm}. \tag{3.21}$$

For a code $\mathcal{C} \in \mathscr{H}^q_{n,m}$, we denote by $\mathcal{S}_{\mathcal{C}}$ the set of all $m \times n$ matrices over $\mathrm{GF}(q)$ which are valid parity-check matrices for $\mathcal{C}$. We say that $\mathcal{C}$ follows the parity-check induced distribution if

$$P(\mathcal{C}) = \sum_{\boldsymbol{H} \in \mathcal{S}_{\mathcal{C}}} P(\boldsymbol{H}) = |\mathcal{S}_{\mathcal{C}}| q^{-nm}. \tag{3.22}$$

Sampling a code from $\mathscr{H}^q_{n,m}$ is particularly simple: It is sufficient to construct an $m \times n$ parity-check matrix $\boldsymbol{H}$ over $\mathrm{GF}(q)$ by selecting each element independently and uniformly at random over $\mathrm{GF}(q)$.

The dual of $\mathscr{H}^q_{n,m}$ is the ensemble $\mathscr{G}^q_{n,k}$.

Definition 3.10 (Linear code ensemble, generator based) We denote by $\mathscr{G}^q_{n,k}$ the ensemble of all (n,k') linear block codes over $\mathrm{GF}(q)$, with $k' \leq k$, and with $P(\mathcal{C})$ being the *generator induced* distribution defined as follows. Consider a random matrix $\boldsymbol{G}$ from the set of all possible $k \times n$ matrices over $\mathrm{GF}(q)$, with $\boldsymbol{G}$ uniformly distributed,

$$P(\boldsymbol{G}) = q^{-nk}. \tag{3.23}$$

For a code $\mathcal{C} \in \mathscr{G}^q_{n,k}$, we denote by $\mathcal{S}_{\mathcal{C}}$ the set of all $k \times n$ matrices over GF(q) which are generator matrices for $\mathcal{C}$. We say that $\mathcal{C}$ follows the generator induced distribution if

$$P(\mathcal{C}) = \sum_{\boldsymbol{G} \in \mathcal{S}_{\mathcal{C}}} P(\boldsymbol{G}) = |\mathcal{S}_{\mathcal{C}}| q^{-nk}. \tag{3.24}$$

Drawing a random $\mathcal{C}$ following the generator-induced distribution is also simple: It is sufficient to construct a $k \times n$ generator check matrix $\boldsymbol{G}$ over GF(q) by selecting each element independently and uniformly at random over GF(q).

For completeness, we introduce two additional code ensembles that are particularly important.

Definition 3.11 (Linear code ensemble) We denote by $\mathscr{L}^q_{n,k}$ the ensemble of all (n, k) linear block codes over GF(q), where

$$P(\mathcal{C}) = \left|\mathscr{L}^q_{n,k}\right|^{-1}. \tag{3.25}$$

Definition 3.12 We denote by $\mathscr{R}^q_{n,k}$ the ensemble of all block codes with blocklength n and q^k codewords in GF$(q)^n$, where

$$P(\mathcal{C}) = \left|\mathscr{R}^q_{n,k}\right|^{-1}. \tag{3.26}$$

When the field order is not specified, we assume the ensemble consisting of binary codes only. We thus have the binary ensembles $\mathscr{H}_{n,m}, \mathscr{G}_{n,k}, \mathscr{L}_{n,k}, \mathscr{R}_{n,k}$.

3.2.2 Singleton Bound

The simplest bound on the block error probability of any (n, k) linear erasure code is a lower bound that follows from Singleton's upper bound on the minimum distance of an (n, k) linear block code (Theorem 2.4). Singleton's minimum distance bound states that the minimum distance of an (n, k) linear block code satisfies the inequality $d_{\min} \leq n - k + 1$. The bound arises from the observation that any subset of $n - k + 1$ or more columns of the $(n - k) \times n$ code's parity-check matrix must contain linear dependencies. It follows that any erasure pattern with $n - k + 1$ or more erasures leads to an equation system whose rank is less than the number of unknowns. With reference to the incomplete decoding rule introduced in Section 2.2, the argument can be used to develop a simple lower bound on the block error probability achievable by any (n, k) linear block code $\mathcal{C}$ as the probability of observing more than $n - k$ erasures at the channel output, that is,

$$P_B(\mathcal{C}) \geq P_B^{(\mathrm{S})} \tag{3.27}$$

with

$$P_B^{(\mathrm{S})} = \sum_{e=n-k+1}^{n} \binom{n}{e} \epsilon^e (1 - \epsilon)^{n-e}. \tag{3.28}$$

We will refer next to the expression in (3.28) as the Singleton bound on the block error probability of (n,k) linear block codes. It may be observed that the bound is independent of the erasure channel order q, and that the bound is achieved over a QEC by any (n,k) MDS code over GF(q). The bound holds for linear codes, and it has a simple algebraic interpretation: Given that an (n,k) linear code is specified by $n-k$ parity equations, decoding (according to the incomplete decoding rule, that is, solving the equations with respect to the erased symbols) cannot succeed if the number of erasures exceeds the number of equations.

For large n and fixed ϵ, the computation of the right-hand side of (3.28) may be carried out efficiently by approximating the binomial distribution of the number of erasures via a normal distribution with mean $n\epsilon$ and variance $n\epsilon(1-\epsilon)$. The Singleton bound can be hence approximated as

$$P_B^{(\mathsf{S})} \approx \mathsf{Q}\left[(\mathsf{C}-R)\sqrt{\frac{n}{\epsilon(1-\epsilon)}}\right] \tag{3.29}$$

with $\mathsf{C} = 1-\epsilon$ and the Q-function defined as

$$\mathsf{Q}(x) = \frac{1}{\sqrt{2\pi}}\int_x^{\infty} \exp\left(-\frac{u^2}{2}\right)\mathrm{d}u. \tag{3.30}$$

Figure 3.3 depicts the Singleton bound for rate-1/2 codes and various blocklengths.

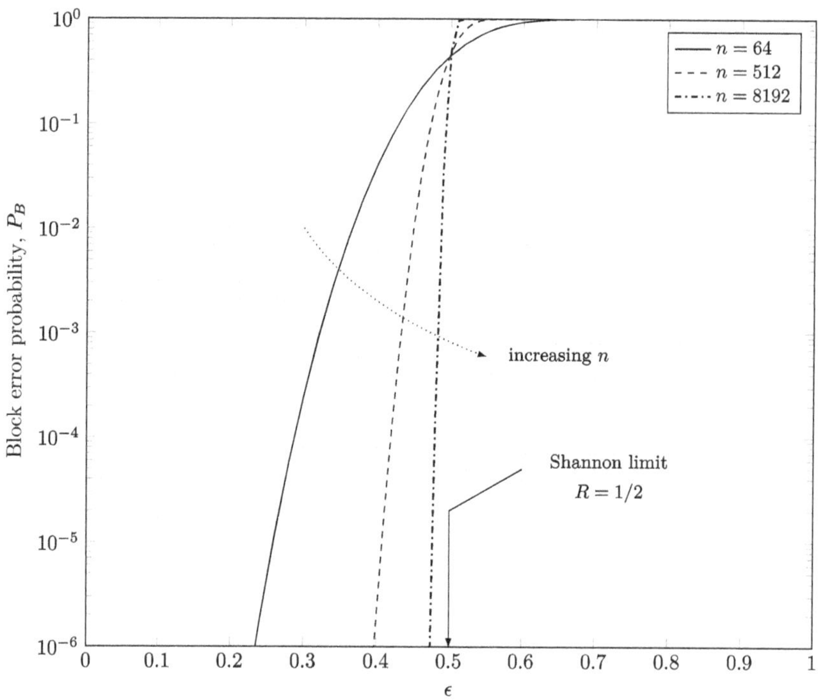

Figure 3.3 Singleton bound for rate-1/2 codes with various blocklengths n.

A similar bound, which holds for any (n, M) code $\mathcal{C}$ (codes with M codewords, not necessarily linear) is

$$P_B(\mathcal{C}) \geq \sum_{e=\lfloor n-\log_q M \rfloor+1}^{n} \binom{n}{e} \epsilon^e (1-\epsilon)^{n-e} \left(1 - \frac{q^{n-e}}{M}\right). \tag{3.31}$$

3.2.3 Random Coding Bounds

Random coding bounds provide upper bounds on the average error probability of a random code from a given ensemble. For an ensemble $\mathscr{C}$, the expected value of the block error probability is

$$\mathsf{E}[P_B(\mathcal{C})] = \sum_{\mathcal{C} \in \mathscr{C}} P_B(\mathcal{C}) P(\mathcal{C}). \tag{3.32}$$

Note that there must exist at least one code in the ensemble with a block error probability that is equal to or lower than the ensemble average error probability. Hence, the block error probability provided by (3.32) is achievable by at least one code in the ensemble. Sometimes, direct evaluation of (3.32) can be complex, so it is convenient to use tight upper bounds on the average block error probability that admit a simpler computation. Over erasure channels, we should distinguish between two types of bounds. The Gallager random coding bound captures the exponential decay of the block error probability with the blocklength n. The Berlekamp random coding bound is tighter and, in its original derivation, relies on the probability distribution of the rank of random matrices, discussed in Appendix B.

3.2.3.1 Gallager Random Coding Bound

Consider a random code $\mathcal{C}$ in $\mathscr{R}^q_{n,k}$. Denote by $P_B(\mathcal{C})$ the error probability when $\mathcal{C}$ is used to transmit over a QEC with erasure probability ϵ. Then

$$\mathsf{E}[P_B(\mathcal{C})] < P_B^{(\mathsf{G})} \tag{3.33}$$

with

$$P_B^{(\mathsf{G})} = \exp(-nE_r(R)), \tag{3.34}$$

where $E_r(R)$ is called *random coding exponent*. For the codes in GF(q) with rate $R = k/n$, it is given by

$$E_r(R) = \begin{cases} -\ln\left(\frac{1-\epsilon}{q} + \epsilon\right) - R \ln q & 0 < R < R_c \\ \mathsf{D}(1-R \| \epsilon) & R_c \leq R \leq 1-\epsilon \\ 0 & R > 1-\epsilon, \end{cases} . \tag{3.35}$$

In (3.35), $\mathsf{D}(1-R\|\epsilon)$ denotes the Kullback–Leibler divergence between two Bernoulli distributions with parameters ϵ and $1-R$

$$\mathsf{D}(1-R\|\epsilon) = (1-R)\ln\frac{1-R}{\epsilon} + R\ln\frac{R}{1-\epsilon}, \tag{3.36}$$

and $R_c = (1-\epsilon)/(\epsilon(q-1)+1)$ is called the *channel critical rate*. Remarkably, the bound (3.33) holds for the average error probability of linear random codes too, that is, for the average error probability of random codes in $\mathscr{L}^q_{n,k}$.

3.2.3.2 Berlekamp Random Coding Bound

Consider a random code $\mathcal{C}$ in $\mathscr{R}^q_{n,k}$. Denote by $P_B(\mathcal{C})$ the error probability when $\mathcal{C}$ is used to transmit over a QEC with erasure probability ϵ. Then

$$\mathsf{E}[P_B(\mathcal{C})] < P_B^{(\mathsf{B})} \tag{3.37}$$

with

$$P_B^{(\mathsf{B})} = \sum_{e=n-k+1}^{n} \binom{n}{e} \epsilon^e (1-\epsilon)^{n-e} + \sum_{e=0}^{n-k} \binom{n}{e} \epsilon^e (1-\epsilon)^{n-e} q^{-n+k+e}. \tag{3.38}$$

The bound follows from the union bound on the block error probability, conditioned on the transmission of a codeword $\boldsymbol{c}$ and on observing e erasures at the decoder input. In particular, consider two random codewords $\boldsymbol{c}$ and $\boldsymbol{c}'$, whose elements are independent and identically distributed (i.i.d.) with a uniform distribution in GF(q). Recalling Definition 2.10, we denote by $P(\boldsymbol{c} \to \boldsymbol{c}')$ the pairwise error probability. We assume to break ties always towards $\boldsymbol{c}'$. Hence, $P(\boldsymbol{c} \to \boldsymbol{c}')$ is the probability that $\boldsymbol{c}$ and $\boldsymbol{c}'$ are indistinguishable after passing through a QEC. If we condition on the number of erasures $E = e$ introduced by the channel, since $\boldsymbol{c}$ and $\boldsymbol{c}'$ cannot be distinguished at the channel output if they do not differ in any of the $n-e$ nonerased positions, we have

$$P(\boldsymbol{c} \to \boldsymbol{c}' | E = e) = q^{-(n-e)}. \tag{3.39}$$

Via the union bound, the block error probability conditioned on e erasures and on the transmission of $\boldsymbol{c}$ is upper bounded as

$$P(\text{error}|E = e, \boldsymbol{c}) \le q^{-(n-e)}(q^k - 1) < q^{-(n-k-e)}. \tag{3.40}$$

Observe that the union bound is independent of the choice of $\boldsymbol{c}$ as soon as $\boldsymbol{c}$ is randomly picked from a code in the ensemble. Finally, (3.37) follows by averaging over the number of erasures and observing that $q^{-(n-k-e)} > 1$ for $e > n-k$.

The bound also holds for the binary linear code ensemble $\mathscr{H}_{n,m}$ in the form

$$P_B^{(\mathsf{B})} = \sum_{e=m+1}^{n} \binom{n}{e} \epsilon^e (1-\epsilon)^{n-e} + \sum_{e=1}^{m} \binom{n}{e} \epsilon^e (1-\epsilon)^{n-e} 2^{-m+e}. \tag{3.41}$$

Over the QEC, the bound for the linear ensemble $\mathscr{H}^q_{n,m}$ can be tightened to

$$P_B^{(\mathsf{B})} = \sum_{e=m+1}^{n} \binom{n}{e} \epsilon^e (1-\epsilon)^{n-e} + \frac{1}{q-1} \sum_{e=1}^{m} \binom{n}{e} \epsilon^e (1-\epsilon)^{n-e} q^{-m+e}. \tag{3.42}$$

The proof for the linear case relies on the result in Appendix B which states that an $m \times e$ random matrix $\boldsymbol{A}$ with $m \ge e$ and with elements independently and uniformly in GF(q) is rank deficient with probability

$$P(\text{rank}(\boldsymbol{A}) < e) < \frac{1}{q-1} q^{-(m-e)}. \tag{3.43}$$

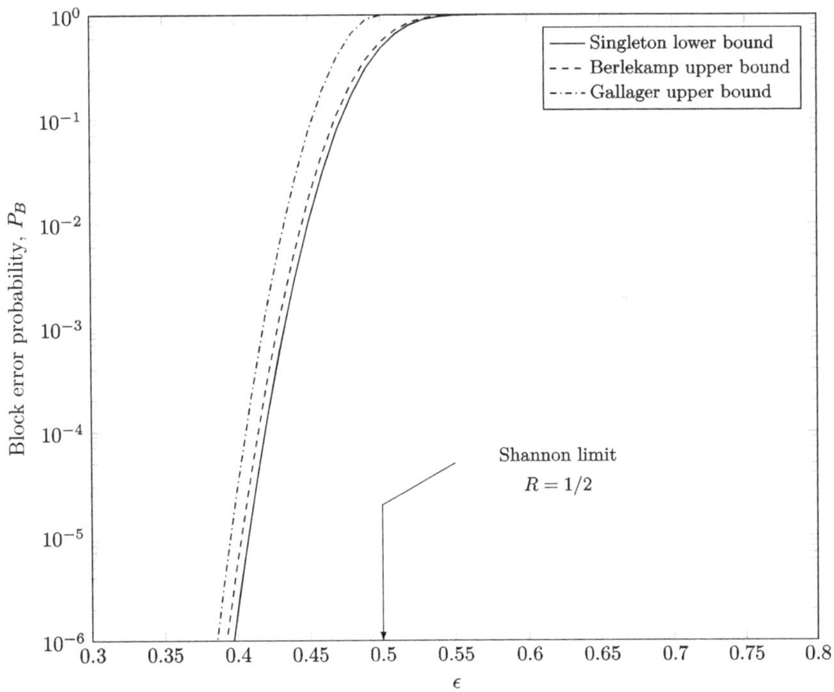

Figure 3.4 Singleton, Berlekamp, and Gallager bounds for rate-1/2 codes with blocklength $n = 512$, BEC.

In fact, decoding consists of solving a system of e unknowns in m equations. If $e > m$, no unique solution exists (see Section 2.4). If $e \leq m$, then a unique solution exists if and only if the rank of the system matrix equals e. It is easy to prove that (3.42) upper bounds the average block error probability of a linear random code $\mathcal{C}$ in $\mathscr{L}^q_{n,k}$.

The Singleton, Berlekamp, and Gallager bounds for rate-1/2 codes with blocklength $n = 512$ over the BEC are depicted in Figure 3.4. Remarkably, already for this relatively short blocklength, the two random coding bounds are very close to the Singleton lower bound, meaning that there exist codes tightly approaching the Singleton bound. The Berlekamp-type bound is in general tighter than the Gallager-type one.

Remark 3.2 (Bounds for the QEC with large q). When the order of the QEC is large, both Berlekamp and Gallager random coding bounds can be simplified. In particular, Berlekamp's bound can be tightly approximated by (3.28) and (3.29), whereas for Gallager's bound

$$P_B^{(\mathsf{G})} \approx \exp[-n\mathsf{D}(1 - R\|\epsilon)], \qquad R \leq 1 - \epsilon. \tag{3.44}$$

3.3 Erasure Channels with Memory

In the previous sections, we have considered memoryless erasure channels. However, in many practical cases, communication channels introduce erasures in a correlated manner. For example, when transmission takes place over a mobile wireless channel, and no interleaving is employed, packet losses are typically clustered in bursts of variable length, with average burst durations that scale with the coherence time of the channel. The use of a channel interleaver of sufficient depth may remove correlation, at the expense of an increased communication latency that may be, in some cases, unacceptable. We hence consider next performance limits for erasure channels with memory.

In the following, we discuss correlated erasure channel models without distinction regarding the underlying erasure channel model order: We simply refer to erasures affecting symbols, where symbols have to be intended as bits if the erasure channel model order is two.

3.3.1 Single Burst Erasure Channels

The single burst erasure channel (SBUEC) channel produces a single erasure burst within a codeword block. This simple model allows assessing the suitability of a code for correlated erasure channels. On the SBUEC, an n-symbols codeword is transmitted and the channel introduces a *solid* burst of L erasures within the n-symbols block. Two variants of SBUEC models can be obtained if the erasure burst is allowed to affect the block in a cyclic manner or not. In the first case, the erasure burst starts at a given symbol index i with i uniformly distributed in $(1,n)$. The symbols in position $i, i \bmod n+1, \ldots, (i+L-2) \bmod n+1$ are erased, while the other symbols are received unerased (see Figure 3.5), that is,

$$y_\ell = \begin{cases} \mathsf{E} & \text{if } \ell \in \{i, i \bmod n+1, \ldots, (i+L-2) \bmod n+1\} \\ x_\ell & \text{otherwise.} \end{cases} \quad (3.45)$$

In the second case, the erasure burst starts at a given symbol index i with i uniformly distributed in $(1, n-L+1)$ and the symbols in position $i, i+1, \ldots, i+L-1$ are erased, while the other symbols are received unerased. We refer to the former model as a *cyclic* SBUEC, and the latter is referred to as an *acyclic* SBUEC. The performance of an (n,k) linear code $\mathcal{C}$ with a given erasure decoding algorithm over a cyclic/acyclic

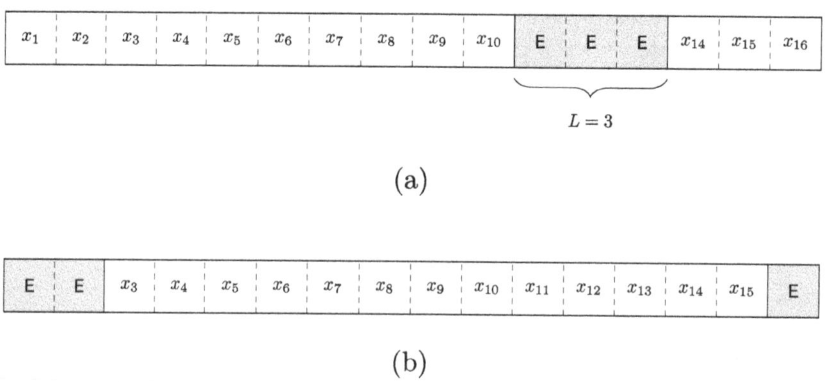

Figure 3.5 Single burst erasure channel, with $L = 3$.

SBUEC can be analyzed by finding, under the given erasure decoding, what is the maximum value of L for which decoding succeeds in resolving the erasures for all the possible shifts of the erasure burst. This value of L is usually referred as maximum tolerable burst length (MTBL), and it is denoted by $L_{\max}$. We stress that the MTBL for a given code depends on the decoding algorithm used to recover the erasures. Codes with a practical erasure decoding algorithm that yields a large MTBL are typically referred to as burst erasure correcting codes. A simple upper bound on the maximum achievable MTBL by any (n, k) linear block code is given by

$$L_{\max} \leq n - k. \tag{3.46}$$

In fact, any erasure burst with length exceeding $n - k$ cannot be recovered since the number of erasures would exceed the number of parity-check equations.

3.3.2 Block Erasure Channels

The SBUEC introduces a single burst of erasures in each transmission block. The SBUEC has a limited capability to analyze the behavior of a given code/decoder pair when dealing with correlated erasures since multiple erasure bursts are not taken into account. A simple channel model which addresses this point is the block erasure channel (BLEC). In the BLEC, each codeword is partitioned into B blocks of equal length $L = n/B$ symbols (we implicitly assume here that B divides n)

$$\boldsymbol{x} = \begin{bmatrix} \boldsymbol{x}_1 & \boldsymbol{x}_2 & \cdots & \boldsymbol{x}_B \end{bmatrix}. \tag{3.47}$$

Each block is either correctly received (with probability $1 - \epsilon$) or erased (with probability ϵ), as illustrated in Figure 3.6. Erasures affect each block independently. By introducing on $\boldsymbol{y}$ the same partitioning used for $\boldsymbol{x}$, we have

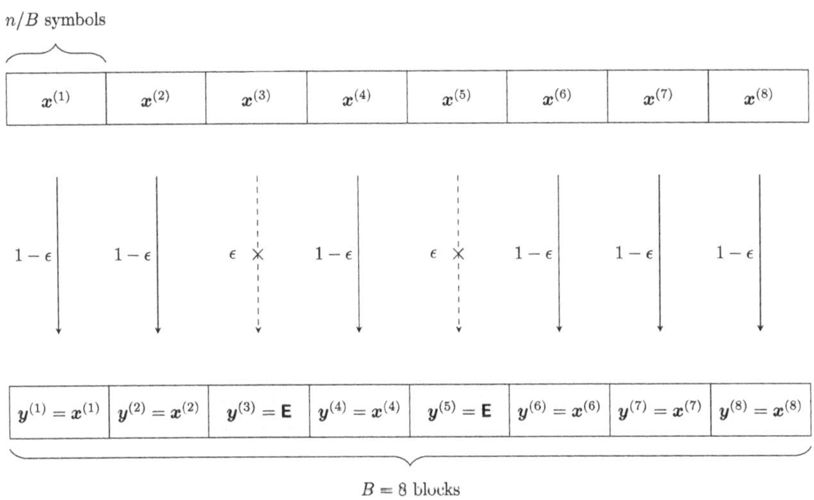

Figure 3.6 Block erasure channel, with $B = 8$.

$$\boldsymbol{y} = \begin{bmatrix} y_1 & y_2 & \cdots & y_B \end{bmatrix}. \tag{3.48}$$

The probability of observing $\boldsymbol{y}$ given that $\boldsymbol{x}$ has been transmitted is

$$P(\boldsymbol{y}|\boldsymbol{x}) = \prod_{\ell=1}^{B} P(\boldsymbol{y}_\ell|\boldsymbol{x}_\ell) \tag{3.49}$$

with

$$P(\boldsymbol{y}_\ell|\boldsymbol{x}_\ell) = \begin{cases} 1-\epsilon & \text{if } \boldsymbol{y}_\ell = \boldsymbol{x}_\ell \\ \epsilon & \text{if } \boldsymbol{y}_\ell = \mathsf{E} \end{cases}, \tag{3.50}$$

where E denotes a block erasure. For a fixed B, the BLEC is among the simplest examples of *nonergodic* channels: Each codeword experiences only a finite number (B) of independent channel realizations. As a result, the BLEC has zero capacity, that is, for any code rate, no matter how large n is, there is a nonzero block error probability. To prove this it is sufficient to observe that the probability that the entire codeword is erased is ϵ^B, which is bounded away from zero for any finite B and $\epsilon > 0$. The performance limits of a nonergodic channel are hence frequently characterized in terms of *outage probability*. To do so for the case of a BLEC, let us introduce a parallel erasure channel model that mimics the behavior of the BLEC. Within this model, transmission takes place over B parallel independent QECs. Each QEC is characterized by an erasure probability that can be either zero (good channel) or one (bad channel). We introduce the QEC state vector

$$\boldsymbol{\sigma} = \begin{bmatrix} \sigma_1 & \sigma_2 & \cdots & \sigma_B \end{bmatrix} \tag{3.51}$$

and the corresponding random vector

$$\boldsymbol{\Sigma} = \begin{bmatrix} \Sigma_1 & \Sigma_2 & \cdots & \Sigma_B \end{bmatrix}. \tag{3.52}$$

In particular, we define the state of the generic ℓth subchannel to be bad by associating a value 0 to the state σ_ℓ, whereas when the subchannel state is good, we set σ_ℓ to 1. For a given QEC state vector $\boldsymbol{\sigma}$, we denote the *instantaneous channel capacity* by $\mathsf{C}(\boldsymbol{\sigma})$. The corresponding random variable is given by $\mathsf{C}(\boldsymbol{\Sigma})$. The definition of outage probability follows.

Definition 3.13 For a given code rate R, the BLEC outage probability is

$$P_{\text{out}} = P(\mathsf{C}(\boldsymbol{\Sigma}) < R). \tag{3.53}$$

It is easy to show that the outage probability is given here by

$$P_{\text{out}} = \sum_{\ell=\lfloor B(1-R)\rfloor+1}^{B} \binom{B}{\ell} \epsilon^{\ell}(1-\epsilon)^{B-\ell}. \tag{3.54}$$

In fact, we have that

$$\mathsf{C}(\boldsymbol{\Sigma}) = \frac{1}{B}\sum_{\ell=1}^{B} \mathsf{C}(\Sigma_\ell), \tag{3.55}$$

where $\mathsf{C}(\Sigma_\ell) = 1$ [symbols/channel use] if $\Sigma_\ell = 1$, whereas, for the case where $\Sigma_\ell = 0$, we have $\mathsf{C}(\Sigma_\ell) = 0$ [symbols/channel use].

Theorem 3.2 *All (n, nR) linear MDS codes achieve the outage probability* (3.54).

Proof Denote by ℓ the number of block erasures introduced by a BLEC. Since the BLEC introduces a number of symbol erasures in $\{0, L, 2L, \ldots, BL\}$, we have that an MDS code with blocklength $n = BL$ and rate R succeeds in decoding whenever $\ell L \leq n(1-R)$, that is, whenever the number of erasures does not exceed the number of parity-check equations defining the code. It follows that decoding succeeds whenever $\ell \leq B(1-R)$. Being ℓ integer, we can re-state the successful decoding condition as

$$\ell \leq \lfloor B(1-R) \rfloor \tag{3.56}$$

which leads to the Singleton bound on the block error probability

$$P_B^{(\mathsf{S})} = \sum_{\ell=\lfloor B(1-R)\rfloor+1}^{B} \binom{B}{\ell} \epsilon^{\ell} (1-\epsilon)^{B-\ell}. \tag{3.57}$$

□

Theorem 3.2 introduces the Singleton bound for the BLEC, that is lower bound on the block error probability achievable by any (n, k) linear code. An upper bound on the average performance of random linear codes is given by Berlekamp's random coding bound for the BLEC. Consider a random code $\mathcal{C} \in \mathscr{L}_{n,k}^{q}$. Denote by $P_B(\mathcal{C})$ the block error probability when $\mathcal{C}$ is used to transmit over a BLEC with erasure probability ϵ. Then

$$\mathsf{E}[P_B(\mathcal{C})] < P_B^{(\mathsf{B})} \tag{3.58}$$

with

$$\begin{aligned} P_B^{(\mathsf{B})} = & \sum_{\ell=\lfloor B(1-R)\rfloor+1}^{B} \binom{B}{\ell} \epsilon^{\ell} (1-\epsilon)^{B-\ell} \\ & + \frac{1}{q-1} \sum_{\ell=0}^{\lfloor B(1-R)\rfloor} \binom{B}{\ell} \epsilon^{\ell} (1-\epsilon)^{B-\ell} q^{-n(1-R-\ell/B)}. \end{aligned} \tag{3.59}$$

The Singleton and Berlekamp bounds for (256, 128) codes over BLECs with $B = 8$ and $B = 16$ are shown in Figure 3.7.

Remark 3.3. The BLEC allows establishing a bridge with the analysis of the wider class of block fading channels. For example, we can introduce the notion of *diversity* of a code when used to communicate over a BLEC. We start by observing that, if we send a symbol directly through an erasure channel, the receiver will lose the symbol with probability ϵ. If instead we send the same symbol through two parallel independent erasure channels, the receiver will lose the symbol only if both channels erase it, thus with probability ϵ^2. We say in this case that we have diversity 2. In general, diversity denotes the presence of parallel independent channels over which replicas of the transmitted signal is made available to the receiver. Diversity is very important in communication systems, for example to deal with random fluctuations of the received signal due to multipath propagation. Parallel channels can be produced by exploiting space through multiple antennas (space diversity) or frequency

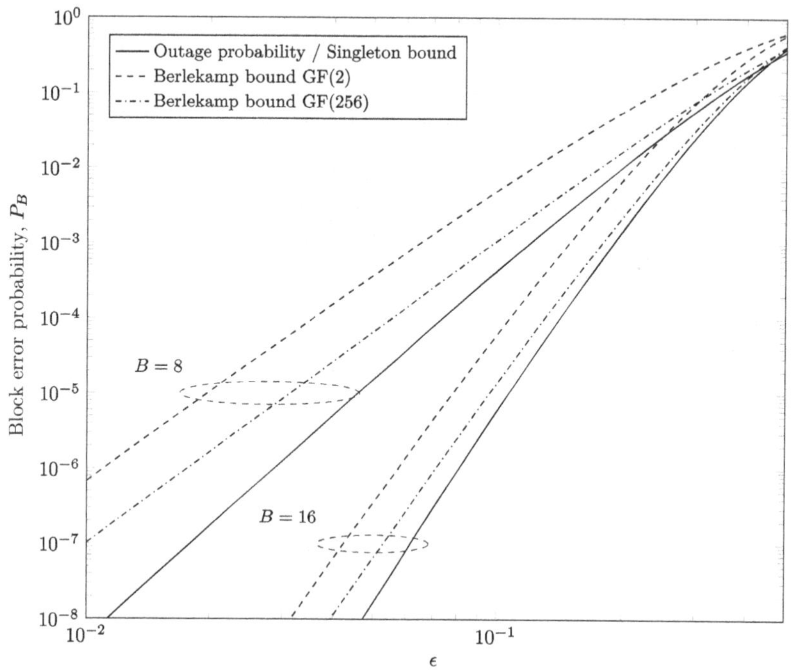

Figure 3.7 Singleton bound (outage probability) compared with the Berlekamp bound for rate-$1/2$ codes with blocklength $n = 256$ symbols and dimension $k = 128$ symbols.

(frequency diversity) or time (time diversity) or other degrees of freedom offered by the transmission system such as polarization. In its simplest form, sending replicas of the transmitted signal over D parallel channels can be seen as a repetition D code. Higher rates can be obtained by the joint use of codes and diversity, with codewords spread over the available parallel channels. For example, coding and interleaving are commonly adopted to exploit time diversity in channels with memory. For the BLEC with coding, if for small ϵ, the block error probability has the form $P_B \approx a\,\epsilon^D$ (where a is a suitably chosen constant), we say that we have a *code diversity* D, which can therefore be defined in general as

$$D(\mathcal{C}) = \lim_{\epsilon \to 0} \frac{\log P_B(\mathcal{C}, \epsilon)}{\log \epsilon}. \tag{3.60}$$

In a log–log plot, a diversity D means that for small ϵ, the error probability follows a straight line with slope D. Note that for a given number B of parallel independent channels, the diversity achieved by a repetition code is $D = B$.

3.3.3 Gilbert Erasure Channel

The Gilbert erasure channel (GEC) introduces a further level of sophistication in modeling correlated erasure channels. The GEC model relies on a two-state Markov chain, where the two states are denoted as G ("good") and B ("bad"). The state diagram is represented in Figure 3.8. When the channel is in the good state, symbols are correctly

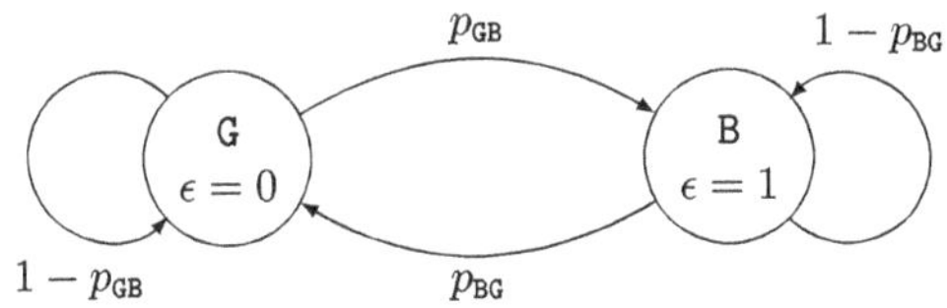

Figure 3.8 Gilbert erasure channel.

received ($\epsilon = 0$). On the contrary, when the channel is in the bad state, symbols are erased with probability one ($\epsilon = 1$). If we denote by S_t, the state of the channel at time t, we define the transition probabilities as

$$P(S_t = \mathsf{G}|S_{t-1} = \mathsf{G}) = 1 - p_{\mathrm{GB}} \qquad P(S_t = \mathsf{B}|S_{t-1} = \mathsf{G}) = p_{\mathrm{GB}} \tag{3.61}$$

and

$$P(S_t = \mathsf{B}|S_{t-1} = \mathsf{B}) = 1 - p_{\mathrm{BG}} \qquad P(S_t = \mathsf{G}|S_{t-1} = \mathsf{B}) = p_{\mathrm{BG}} \tag{3.62}$$

where we assume all transition probabilities to be strictly positive ($0 < p_{\mathrm{GB}} < 1$ and $0 < p_{\mathrm{BG}} < 1$, that is, there are no absorbing states). The steady state probabilities of being is state G and of being in state B are denoted as π_{G} and π_{B}, respectively, and they are given by

$$\pi_{\mathrm{G}} = \frac{p_{\mathrm{BG}}}{p_{\mathrm{GB}} + p_{\mathrm{BG}}} \quad \text{and} \quad \pi_{\mathrm{B}} = \frac{p_{\mathrm{GB}}}{p_{\mathrm{GB}} + p_{\mathrm{BG}}}. \tag{3.63}$$

Being erasure only possible in the bad state, and being the erasure probability in the bad state equal to one, we have that the channel average erasure probability is

$$\bar{\epsilon} = \pi_{\mathrm{B}}. \tag{3.64}$$

A reasonable measure of the channel correlation is given by the average erasure burst length, which equals the average permanence in the bad state, denoted as

$$\delta = \frac{1}{p_{\mathrm{BG}}}. \tag{3.65}$$

The GEC can be fully specified by either the couple $(p_{\mathrm{GB}}, p_{\mathrm{BG}})$ or by the couple $(\bar{\epsilon}, \delta)$. The second option has the advantage of giving immediate evidence of the erasure channel characteristics in terms of erasure probability and correlation. The GEC is ergodic. Moreover, the channel memory does not help in increasing capacity with respect to the memoryless erasure channel with erasure probability $\bar{\epsilon}$. It follows that the capacity of the $(\bar{\epsilon}, \delta)$ GEC is

$$\mathsf{C} = 1 - \bar{\epsilon} \qquad \text{[symbols/channel use]}. \tag{3.66}$$

The Singleton bound for (n, k) codes over the $(\bar{\epsilon}, \delta)$ GEC is

$$P_B^{(\mathsf{S})} = \sum_{e=n-k+1}^{n} Q(n, e), \tag{3.67}$$

where $Q(n, e)$ is the probability of observing e erasures in n channel uses and can be obtained as

$$Q(n, e) = Q(n, e, \mathsf{G}) + Q(n, e, \mathsf{B}). \tag{3.68}$$

Here, $Q(t, e, \mathtt{S})$ is the probability of observing e erasures after t channel uses and of being in state S at time t, and it can be computed recursively as

$$Q(t, e, \mathtt{G}) = Q(t-1, e, \mathtt{G})(1 - p_{\mathtt{GB}}) + Q(t-1, e, \mathtt{B})p_{\mathtt{BG}} \tag{3.69}$$

and

$$Q(t, e, \mathtt{B}) = Q(t-1, e-1, \mathtt{B})(1 - p_{\mathtt{BG}}) + Q(t-1, e-1, \mathtt{G})p_{\mathtt{GB}} \tag{3.70}$$

with initial conditions

$$Q(1,0,\mathtt{G}) = \pi_{\mathtt{G}} \qquad Q(1,1,\mathtt{B}) = \pi_{\mathtt{B}} \tag{3.71}$$

$$Q(1,1,\mathtt{G}) = 0 \qquad Q(1,0,\mathtt{B}) = 0. \tag{3.72}$$

The knowledge of the distribution $Q(n, e)$ allows computing in a straightforward manner the Berlekamp bound for the linear ensemble $\mathscr{L}^q_{n,k}$, which is given by

$$P_B^{(\mathsf{B})} = \sum_{e=n-k+1}^{n} Q(n,e) + \frac{1}{q-1}\sum_{e=0}^{n-k} Q(n,e)q^{-m+e}. \tag{3.73}$$

The Singleton and Berlekamp bounds for (256, 128) codes and various average burst lengths δ are shown in Figure 3.9.

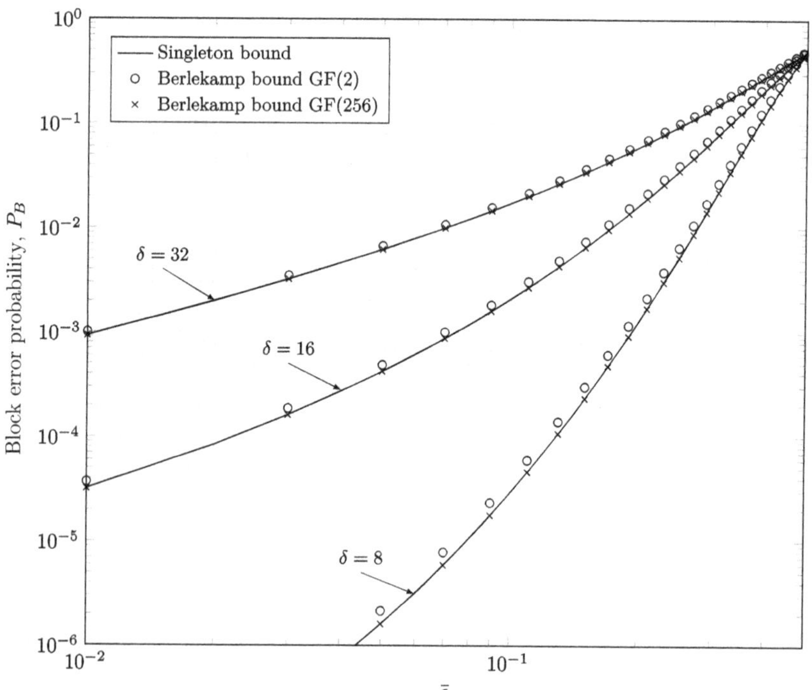

Figure 3.9 Singleton bound compared with the Berlekamp bounds for rate-1/2 codes with blocklength $n = 256$ symbols and dimension $k = 128$ symbols.

3.4 Historical Notes

The BEC was originally introduced in [1]. The generalization provided by the QEC found a remarkable application in direct-detection (photon counting) optical communications as a surrogate model for analysis [2] and code design. The foundations of information theory were laid by Shannon in his seminal paper [3], where the concepts of entropy and mutual information as information measures, as well as the notion of channel capacity, were introduced. The bound provided by (3.31) can be found in [4, Theorem 38]. Bounds in the form of (3.29) were introduced in [4] for general DMCs (see also [5, Part IV]). We refer in particular to the normal approximation [4, 6] $\log_q M \approx n\mathsf{C} - \sqrt{n\mathsf{V}}\mathsf{Q}^{-1}(P_B)$ with $\mathsf{V} = \epsilon(1-\epsilon)$ being the dispersion of the QEC. The Gallager random coding bound dates back to R. Gallager [7], whereas the Berlekamp random coding bound is due to E. Berlekamp [8]. For a survey of Gallager and Berlekamp random coding bounds and their application to the QEC, we refer the reader to [9]. The average error probability of a binary linear random code $\mathcal{C}$ with uniform distribution in $\mathscr{L}_{n,k}$ was derived by Ashikhmin [4, Theorem 6], and it is tightly upper bounded by Berlekamp's random coding bound (3.41) down to blocklengths as short as a few tens of bits. Burst erasure channels and the relevant LDPC code designs are discussed in [10, 11]. The class of codes that achieve the bound (3.46) under (optimum) ML decoding is surprisingly large. In fact, besides MDS codes (which obviously attain the bound), it is possible to show that all cyclic codes do so, as well as many linear block codes whose parity-check matrix satisfies mild conditions [12]. An information theoretic treatment of the BLEC can be found in [13], while a characterization of the performance for certain block codes over general block fading channels can be found in [14]. The GEC is named after the E. Gilbert, who introduced a two-states channel model where the good state is error-free, and the bad state corresponds to a BSC with a given error probability [15]. In this context, the GEC can be seen as a Gilbert channel in the original sense of [15], where the error probability in the bad state is set to $1/2$, and where the decoder is equipped with complete channel state information (i.e., the decoder is informed on the state of the channel for each bit transmission.) A generalization of the GEC model was introduced in [16], where an arbitrary number of states having different erasure probabilities was considered.

3.5 Exercises

Exercise 3.1 The BSC is a DMC with 2 input symbols and 2 output symbols. For simplicity, we may denote the input and output alphabets as $\mathcal{X} = \mathcal{Y} = \{0, 1\}$. The channel cross-over probability is $p = P(1|0) = P(0|1)$. What is the capacity of the channel? Plot it and compare it with the capacity of a BEC with erasure probability p.

Exercise 3.2 Consider now the binary error and erasure channel, which is a DMC with 2 input symbols and 3 output symbols. The input and output alphabets are $\mathcal{X} = \{0, 1\}$ and $\mathcal{Y} = \{0, 1, \mathsf{E}\}$, respectively, where the E denotes an erasure. The channel transition probabilities are $P(1|0) = P(0|1) = p$, $P(\mathsf{E}|0) = P(\mathsf{E}|1) = \epsilon$ and $P(0|0) = P(1|1) = 1 - p - \epsilon$. Is the channel symmetric? What is the capacity of the channel?

Exercise 3.3 The Z-channel is a DMC with 2 input symbols and 2 output symbols, which we denote as $\mathcal{X} = \mathcal{Y} = \{0, 1\}$. The channel transition probabilities are $P(0|1) = p$, $P(1|1) = 1 - p$ and $P(0|0) = 1$. Note that the channel is nonsymmetric. Derive its capacity, and write a computer program for its numerical evaluation. Compare the result with the mutual information achievable by using a uniform input distribution.

Exercise 3.4 A q-ary symmetric channel is a DMC with q input symbols $\mathcal{X} = \{\mathsf{x}_1, \mathsf{x}_2, \ldots, \mathsf{x}_q\}$ and q output symbols $\mathcal{Y} = \{\mathsf{y}_1, \mathsf{y}_2, \ldots, \mathsf{y}_q\}$, and with channel transition probabilities $\pi_{i,j} = p/(q-1)$ for $j \neq i$, and $\pi_{i,j} = 1 - p$ for $i = j$. Note that the channel is fully defined by its error probability p. Derive the capacity of the q-ary symmetric channel as a function of the error probability. Show that, as q grows large, the capacity tends to that of a QEC with erasure probability p.

Exercise 3.5 Consider the parity-check based ensemble $\mathscr{H}^q_{n,m}$. All codes belonging to the ensemble have the same blocklength n. Some of the codes, however, are characterized by idle symbols and some others by an excess in dimension. The first case occurs when at least one row of $\boldsymbol{H}$ has Hamming weight 1. Since the corresponding parity-check equation involves one encoded symbol only, this symbol must be permanently set to zero in all codewords. The second case occurs when $\text{rank}(\boldsymbol{H}) < m$, in which case the dimension of the corresponding code is $k > n - m$. Let $\mathtt{I}$ be the event that a code drawn in the linear random block code ensemble $\mathscr{H}^q_{n,m}$ has at least one idle symbol. Prove that

$$P(\mathtt{I}) = 1 - \left(1 - \frac{n(q-1)}{q^n}\right)^m \tag{3.74}$$

and that

$$P(\mathtt{I}) < (n - m + 1)q^{-(n-m)}. \tag{3.75}$$

Exercise 3.6 Compute the Berlekamp random coding bound for binary codes with parameters (7,4), over a BEC with erasure probability $\epsilon = 0.01$. Compare the value with the block error probability of a (7,4) Hamming code from (1.5). Is the Hamming code performing better or worse than the upper bound on the average performance of random codes?

Exercise 3.7 Complete the formal proof of the tighter version of Berlekamp's random coding bound given in (3.42). *Hint:* Write $P(\text{error}) = \sum_{e=0}^{n} P(\text{error}|E = e)P(E = e)$ and use the result on the probability of singularity for random matrices provided in Appendix B.

Exercise 3.8 An exponential form of the Berlekamp bound for large q can be obtained by applying the Chernoff bound to the right-hand side of (3.29) yielding

$$P_B^{(\mathsf{B})} \approx \exp\left[-n\frac{(\mathsf{C} - R)^2}{2\epsilon(1-\epsilon)}\right]. \tag{3.76}$$

Show that the right-hand side of (3.76) holds for Gallager's bound with large q, that is,

$$P_B^{(\mathrm{G})} \approx \exp\left[-n\frac{(\mathsf{C}-R)^2}{2\epsilon(1-\epsilon)}\right]. \tag{3.77}$$

The exponent in (3.76) and (3.77) is known to be an approximation to the channel reliability function

$$E(R) = \lim_{n\to\infty} \sup_{\mathcal{C}} \frac{-\log_2 P_B(\mathcal{C})}{n} \tag{3.78}$$

for rates near capacity [4], with $\mathcal{C} \in \mathscr{R}^q_{n,nR}$ [17]. *Hint:* Consider Taylor's expansion in R around $R = 1-\epsilon$ of Gallager's error exponent for large q.

Exercise 3.9 Consider the (7,4) Hamming code defined by the parity-check matrix

$$\boldsymbol{H} = \begin{bmatrix} 1 & 1 & 1 & 0 & 1 & 0 & 0 \\ 0 & 1 & 1 & 1 & 0 & 1 & 0 \\ 1 & 1 & 0 & 1 & 0 & 0 & 1 \end{bmatrix}. \tag{3.79}$$

Assuming transmission over a cyclic SBUEC, what is the code MTBL? Consider now an equivalent code, obtained by permuting the last and the second-to-last columns, yielding the parity-check matrix

$$\boldsymbol{H} = \begin{bmatrix} 1 & 1 & 1 & 0 & 1 & 0 & 0 \\ 0 & 1 & 1 & 1 & 0 & 0 & 1 \\ 1 & 1 & 0 & 1 & 0 & 1 & 0 \end{bmatrix}. \tag{3.80}$$

What is the MTBL of the equivalent code?

Exercise 3.10 Consider a modified GEC, where the two states are characterized by the erasure probabilities $\epsilon_\mathrm{G}, \epsilon_\mathrm{B}$ (with $\epsilon_\mathrm{G} < \epsilon_\mathrm{B}$). What is the channel capacity?

References

[1] P. Elias, "Coding for two noisy channels," in *Proc. Third London Symp. Inf. Theory*, London, UK, Sep. 1955, pp. 61–74.

[2] J. Massey, "Capacity, cut-off rate, and coding for a direct-detection optical channel," *IEEE Trans. Commun.*, vol. 29, no. 11, pp. 1615–1621, Nov. 1981.

[3] C. Shannon, "A mathematical theory of communication," *Bell Syst. Tech. J.*, vol. 27, no. 3, pp. 379–423, 623–656, July 1948.

[4] Y. Polyanskiy, H. V. Poor, and S. Verdú, "Channel coding rate in the finite blocklength regime," *IEEE Trans. Inf. Theory*, vol. 56, no. 5, pp. 2307–2359, May 2010.

[5] Y. Polyanskiy and Y. Wu, *Information theory: From coding to learning*. Cambridge, England: Cambridge University Press, Jan. 2025.

[6] V. Strassen, "Asymptotische Abschätzungen in Shannons Informationstheorie," in *Trans. Third Prague Conf. Information Theory*, Prague, Czech Republic, June 1962, pp. 689–723.

[7] R. Gallager, *Information theory and reliable communication*. New York, NY, USA: Wiley, 1968.

[8] E. Berlekamp, "The technology of error-correcting codes," *Proc. IEEE*, vol. 68, no. 5, pp. 564–593, May 1980.

[9] G. Liva, E. Paolini, and M. Chiani, "Bounds on the error probability of block codes over the q-ary erasure channel," *IEEE Trans. Commun.*, vol. 61, no. 6, pp. 2156–2165, June 2013.

[10] M. Yang and W. Ryan, "Design of LDPC codes for two-state fading channel models," in *Proc. 5th Int. Symp. Wireless Personal Multimedia Communications*, Honolulu, HI, USA, Oct. 2002, pp. 986–990.

[11] E. Paolini and M. Chiani, "Construction of near-optimum burst erasure correcting low-density parity-check codes," *IEEE Trans. Commun.*, vol. 57, no. 5, pp. 1320–1328, May 2009.

[12] M. Fossorier, "Universal burst error correction," in *Proc. IEEE Int. Symp. Inf. Theory*, 2006, pp. 1969–1973.

[13] A. Guillen i Fabregas, "Coding in the block-erasure channel," *IEEE Trans. Inf. Theory*, vol. 52, no. 11, pp. 5116–5121, Nov. 2006.

[14] M. Chiani, "Error probability for block codes over channels with block interference," *IEEE Trans. Inf. Theory*, vol. 44, no. 7, pp. 2998–3008, Nov. 1998.

[15] E. Gilbert, "Capacity of a burst-noise channel," *Bell Syst. Tech. J.*, vol. 39, no. 5, pp. 1253–1265, Sep. 1960.

[16] S. Kumar, J.-F. Chamberland, and H. D. Pfister, "First-passage time and large-deviation analysis for erasure channels with memory," *IEEE Trans. Inf. Theory*, vol. 59, no. 9, pp. 5547–5565, Sep. 2013.

[17] C. Shannon, R. Gallager, and E. Berlekamp, "Lower bounds to error probability for coding on discrete memoryless channels. I," *Information and Control*, vol. 10, no. 1, pp. 65–103, Jan. 1967.

4 Low-Density Parity-Check Codes

This chapter introduces low-density parity-check (LDPC) codes, a class of powerful codes whose structure can be represented on sparse graphs. These codes are important in erasure correction applications because of their very good tradeoff between performance and decoding complexity. Using LDPC codes, erasure correction can be performed via iterative decoding, with a low complexity scaling as a linear function of the blocklength, or via ML decoding, with a complexity that is polynomial in the codeword length and affordable in several applications. This chapter provides the main definitions in LDPC coding theory and describes some notable LDPC code ensembles. Additional properties of LDPC codes, related to their weight distributions, are described in Chapter 5, while iterative and ML erasure decoding algorithms are presented in Chapter 6.

4.1 Basics

An LDPC code is a linear block code which may be represented by at least one parity-check matrix $\boldsymbol{H} = \{H_{i,j}\}$ in *sparse* form, that is, with relatively few nonzero entries. In a binary LDPC code, both the encoded symbols forming a codeword and the entries of the parity-check matrix belong to GF(2). Given a binary n-tuple $\boldsymbol{y} = \begin{bmatrix} y_1 & y_2 & \dots & y_n \end{bmatrix} \in \mathrm{GF}(2)^n$ with length equal to the blocklength n, all operations necessary to calculate its *syndrome* $\boldsymbol{s} = \boldsymbol{y}\,\boldsymbol{H}^{\mathsf{T}}$ are in GF(2), where $\boldsymbol{s} = \begin{bmatrix} s_1 & s_2 & \dots & s_m \end{bmatrix} \in \mathrm{GF}(2)^m$, $m \geq n-k$ is the number of parity-check equations, and k is the LDPC code dimension. The syndrome $\boldsymbol{s}$ is null if and only if $\boldsymbol{y}$ is a codeword. Besides binary ones, LDPC codes may also be constructed on nonbinary finite fields. In this case, all codeword symbols and all entries of the parity-check matrix belong to $\mathrm{GF}(q)$. Given an n-tuple $\boldsymbol{y} \in \mathrm{GF}(q)^n$, its syndrome is still given by $\boldsymbol{s} = \boldsymbol{y}\,\boldsymbol{H}^{\mathsf{T}}$, where, however, additions and multiplications are performed in $\mathrm{GF}(q)$. Again, the syndrome $\boldsymbol{s}$ is null if and only if $\boldsymbol{y}$ is a codeword.

4.1.1 Tanner Graph

The *Tanner graph* of an LDPC code with parity-check matrix $\boldsymbol{H}$ is a bipartite graph consisting of a set $\{\mathsf{v}_1, \mathsf{v}_2, \dots, \mathsf{v}_n\}$ of n variable nodes (VNs), and a set $\{\mathsf{c}_1, \mathsf{c}_2, \dots, \mathsf{c}_m\}$ of m check nodes (CNs). Each VN is bijectively associated with one symbol in the codeword and each CN is bijectively associated with one linear constraint (corresponding to one row of $\boldsymbol{H}$). The jth VN v_j, $j \in \{1,2,\ldots,n\}$, and the ith CN c_i, $i \in \{1,2,\ldots,m\}$, are connected by an edge if and only if the entry $H_{i,j}$ of $\boldsymbol{H}$ is nonzero. In this case, the VN v_j and the CN c_i are said to be neighboring nodes. We use the symbol $\mathcal{E}$ to denote the set of all Tanner graph edges, and $E = |\mathcal{E}|$ to denote its cardinality. A Tanner graph may in principle be drawn for any parity-check matrix representing any linear block code. However, if the linear block code is an LDPC code and a sparse parity-check matrix is considered, then the Tanner graph is *sparse*, that is, it is characterized by a relatively low number of edges (compared with the total number $n \cdot m$ of possible edges connecting VNs to CNs). In the Tanner graph of an LDPC code, the *degree* of a node is the number of edges connected to the node. Each VN and each CN should be thought as equipped with *sockets* to which edges are connected. Obviously, the number of sockets of a VN or CN is equal to its degree.

If the LDPC code is constructed over GF(q), then the edges in its Tanner graph are labeled. In particular, the label of an edge connecting VN v_j to CN c_i is equal to the corresponding entry $H_{i,j} \in \text{GF}(q) \setminus \{0\}$ of the parity-check matrix. In the Tanner graph of a binary LDPC code, edges are unlabeled.

Example 4.1 Consider a binary linear block code with blocklength $n = 10$, dimension $k = 5$, and represented by the parity-check matrix

$$\boldsymbol{H} = \begin{bmatrix} 1 & 0 & 0 & 0 & 1 & 0 & 1 & 0 & 1 & 0 \\ 0 & 0 & 1 & 1 & 1 & 1 & 0 & 0 & 0 & 0 \\ 0 & 1 & 0 & 0 & 0 & 1 & 0 & 0 & 1 & 1 \\ 1 & 1 & 0 & 1 & 0 & 0 & 0 & 1 & 0 & 1 \\ 0 & 0 & 1 & 0 & 0 & 0 & 1 & 1 & 1 & 1 \end{bmatrix}. \tag{4.1}$$

The Tanner graph corresponding to $\boldsymbol{H}$ is depicted in Figure 4.1, where VNs are represented by circles, CNs are represented as squares, VN v_j, $j \in \{1,2,\ldots,10\}$, is associated with the codeword bit corresponding to the jth column of $\boldsymbol{H}$, and CN c_i, $i \in \{1,2,\ldots,5\}$, is associated with the parity-check equation corresponding to the ith row of $\boldsymbol{H}$.

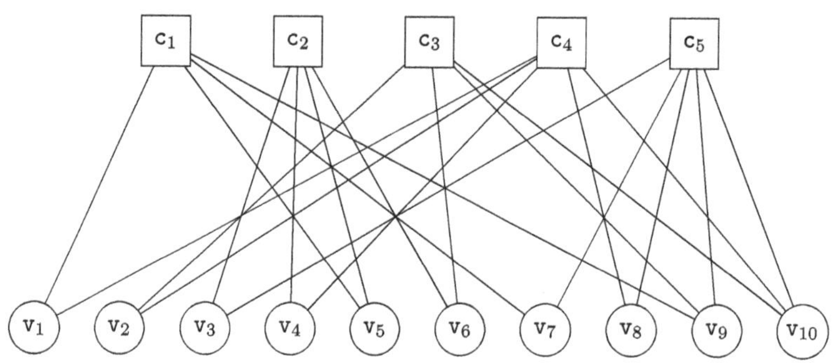

Figure 4.1 Tanner graph corresponding to the parity-check matrix (4.1) of a binary linear block code with $n = 10$ and $k = 5$. The number of edges is $E = 22$.

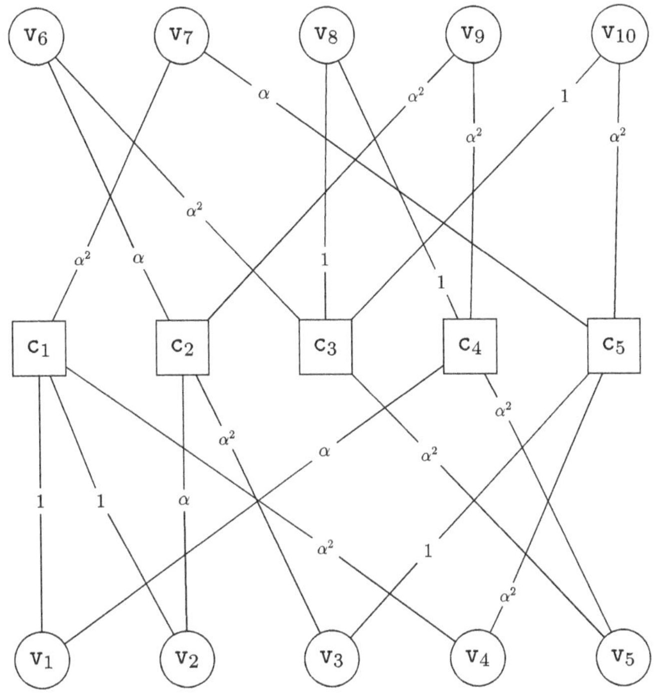

Figure 4.2 Tanner graph corresponding to the parity-check matrix (4.2) of a nonbinary linear block code with $n = 10$ and $k = 5$. The number of edges is $E = 20$, and edge labels are indicated on the edges.

Example 4.2 Consider a linear block code constructed on GF(4) with blocklength $n = 10$, dimension $k = 5$, and represented by the parity-check matrix

$$\boldsymbol{H} = \begin{bmatrix} 1 & 1 & 0 & \alpha^2 & 0 & 0 & \alpha^2 & 0 & 0 & 0 \\ 0 & \alpha & \alpha^2 & 0 & 0 & \alpha & 0 & 0 & \alpha^2 & 0 \\ 0 & 0 & 0 & 0 & \alpha^2 & \alpha^2 & 0 & 1 & 0 & 1 \\ \alpha & 0 & 0 & 0 & \alpha^2 & 0 & 0 & 1 & \alpha^2 & 0 \\ 0 & 0 & 1 & \alpha^2 & 0 & 0 & \alpha & 0 & 0 & \alpha^2 \end{bmatrix}, \qquad (4.2)$$

where α is a primitive element of the field. The Tanner graph corresponding to $\boldsymbol{H}$ is depicted in Figure 4.2, where VNs are represented by circles, CNs are represented as squares, VN $\mathtt{v}_j$, $j \in \{1,2,\ldots,10\}$, is associated with the codeword symbol corresponding to the jth column of $\boldsymbol{H}$, and CN $\mathtt{c}_i$, $i \in \{1,2,\ldots,5\}$, is associated with the linear constraint corresponding to the ith row of $\boldsymbol{H}$.

Important graphical structures in the Tanner graph of an LDPC code are cycles. As we will see, in fact, cycles have relationships with the most harmful structures for LDPC decoders. Formally, for nonnegative integer l, a cycle of length $2l$ is defined as follows.

Definition 4.1 (Cycle) A cycle of length $2l$ is a subset of $2l$ edges forming a path such that the first and the last node of the path coincide. The path touches exactly l VNs and l CNs.

Example 4.3 Consider the Tanner graph depicted in Figure 4.2. The set $\{\mathsf{v}_1, \mathsf{c}_1, \mathsf{v}_2, \mathsf{c}_2, \mathsf{v}_9, \mathsf{c}_4\}$ of $l = 3$ VNs and $l = 3$ CNs represents a cycle of length $2l = 6$.

4.1.2 Regular and Irregular LDPC Codes

An LDPC code is said to be *irregular* when either the VN degrees or the CN degrees in its Tanner graph are allowed to vary, as opposed to *regular* LDPC codes in which all VNs have the same degree as well as all CNs. For an irregular LDPC code, it is a common practice to describe the VN and CN degree profiles in its Tanner graph by means of a degree distribution pair. Letting λ_d be the fraction of edges which are connected to VNs of degree d and ρ_d be the fraction of edges connected to CNs of degree d, we define the two polynomials

$$\lambda(x) = \sum_{d=1}^{\mathsf{d}_v} \lambda_d x^{d-1} \tag{4.3}$$

and

$$\rho(x) = \sum_{d=2}^{\mathsf{d}_c} \rho_d x^{d-1} \tag{4.4}$$

called the *edge-oriented* VN and CN degree distribution polynomials, respectively. In (4.3) and (4.4), d_v and d_c denote the maximum VN and CN degrees, respectively. Some care should be taken to the fact that λ_d represents the fraction of the E edges connected to VNs of degree d, and not the fraction of VNs of degree d, and similarly for ρ_d. For this reason, the two polynomials $\lambda(x)$ and $\rho(x)$ are dubbed "edge-oriented" distributions. Their pair is compactly denoted by (λ, ρ), which represents the edge-oriented degree distribution pair for the considered LDPC code. Clearly, we must have $\sum_{d=1}^{\mathsf{d}_v} \lambda_d = \sum_{d=2}^{\mathsf{d}_c} \rho_d = 1$. In (4.4) we impose a minimum CN degree equal to 2 since degree-1 CNs would introduce "dummy" symbols always equal to zero.

A node-oriented degree distribution pair for an LDPC code may also be defined. By observing that the number of VNs of degree d, hereafter denoted by n_d, is equal to $n_d = E\lambda_d/d$ and that the total number of VNs (equal to the blocklength n) may be expressed as $n = \sum_{d=1}^{\mathsf{d}_v} n_d$, the fraction of VNs of degree d, denoted by Λ_d, is given by

$$\Lambda_d = \frac{\lambda_d/d}{\sum_{t=1}^{\mathsf{d}_v} \lambda_t/t} = \frac{\lambda_d/d}{\int \lambda}, \tag{4.5}$$

where we have used the conventional notation $\int \lambda = \int_0^1 \lambda(x)\mathrm{d}x$. Similarly, the number m_d of CNs of degree d may be expressed as $m_d = E\rho_d/d$ and we have $m = \sum_{t=2}^{\mathsf{d}_c} m_t$. Therefore, the fraction of CNs of degree d, denoted by P_d, is given by

$$P_d = \frac{\rho_d/d}{\sum_{t=1}^{\mathsf{d}_c} \rho_t/t} = \frac{\rho_d/d}{\int \rho}, \tag{4.6}$$

where again $\int \rho = \int_0^1 \rho(x)\mathrm{d}x$. The node-oriented variable and check node degree distribution polynomials are defined as

$$\Lambda(x) = \sum_{d=1}^{\mathsf{d}_v} \Lambda_d \, x^d \tag{4.7}$$

and

$$P(x) = \sum_{d=2}^{\mathsf{d}_v} P_d \, x^d, \tag{4.8}$$

respectively. The pair (Λ, P) represents the LDPC code node-oriented degree distribution pair. Since the number of edges may be expressed both as $E = n/\int \lambda$ and as $E = m \int \rho$, the number m of CNs may be expressed as $m = n \int \rho / \int \lambda$. The code rate of the LDPC code then fulfills

$$R \geq 1 - \frac{\int \rho}{\int \lambda}, \tag{4.9}$$

where equality holds if and only if all m parity-check equations are linearly independent.

In the special case of a regular LDPC code, the edge-oriented degree distribution pair is represented by the two polynomials $\lambda(x) = x^{\mathsf{d}_v - 1}$ and $\rho(x) = x^{\mathsf{d}_c - 1}$, while the node-oriented one by the two polynomials $\Lambda(x) = x^{\mathsf{d}_v}$ and $P(x) = x^{\mathsf{d}_c}$. The code is referred to as a $(\mathsf{d}_v, \mathsf{d}_c)$-regular LDPC code.

Example 4.4 The Tanner graph in Figure 4.1 is characterized by an edge-oriented degree distribution pair

$$\lambda(x) = \frac{8}{11}x + \frac{3}{11}x^2 \quad \text{and} \quad \rho(x) = \frac{6}{11}x^3 + \frac{5}{11}x^4 \tag{4.10}$$

and by a node-oriented degree distribution pair

$$\Lambda(x) = \frac{8}{10}x^2 + \frac{2}{10}x^3 \quad \text{and} \quad P(x) = \frac{3}{5}x^4 + \frac{2}{5}x^5. \tag{4.11}$$

4.1.3 Encoding of LDPC Codes

As for any other linear block code, encoding of an LDPC code $\mathcal{C}(n,k)$ may be performed by computing the product between the source message $\boldsymbol{u}$ of k symbols and a $k \times n$ code generator matrix $\boldsymbol{G}$, yielding the codeword $\boldsymbol{c} = \boldsymbol{u}\,\boldsymbol{G}$. Starting from the (sparse) parity-check matrix $\boldsymbol{H}$ defining the LDPC code $\mathcal{C}(n,k)$, a generator matrix in systematic form can be obtained following the procedure addressed in Section 2.3. It should be remarked, however, that the generator matrix of an LDPC code is generally a *dense* matrix (in case of $\boldsymbol{G}$ in systematic form, $\boldsymbol{G} = \left[\boldsymbol{I}_k|\boldsymbol{P}\right]$, the submatrix $\boldsymbol{P}$ is dense) which yields an overall encoding complexity that scales as the square of the blocklength n for a constant code rate. For this reason, the problem of encoding of LDPC codes is more commonly tackled by exploiting the parity-check matrix and its sparseness.

Systematic encoding, where the k information symbols appear in the first k positions of the codeword, may be recast as a *decoding* problem over the erasure channel; under this perspective, a codeword is transmitted over an erasure channel that always cancels the last $n - k$ codeword symbols, letting the first k symbols be correctly received, and the decoder is tasked with reconstructing the missing symbols. Since the first k columns of the generator matrix in systematic form are linearly independent by construction, the set of indices $\{1,2,\ldots,k\}$ is an information set (see Definition 2.23) and the last $n - k$ codeword symbols can be obtained my running an ML decoder on $\boldsymbol{y} = \left[y_1 \;\; y_2 \;\; \cdots \;\; y_n\right] = \left[u_1 \;\; \ldots \;\; u_k \;\; \mathsf{E} \;\; \ldots \;\; \mathsf{E}\right]$. Any efficient ML erasure decoder able to exploit the $\boldsymbol{H}$ matrix sparseness (see Section 6.2) may be used to this purpose. As an instance of this approach to encoding, it can be shown that the parity-check matrix of any LDPC code can be reduced to an approximate lower triangular form, as depicted in Figure 4.3, by row and column permutations only. The parameter g, representing the "gap" to an exact triangular form, defines the encoding complexity. In fact, while the calculation of the $n - k - g$ redundant symbols corresponding to the triangular square matrix $\boldsymbol{T}$ can be performed with linear complexity via back substitutions, the computation of remaining g redundant symbols involves a complexity in the order of $n + g^2$. Thus, if the initial $\boldsymbol{H}$ matrix yields a gap g that is small compared with n, encoding can be performed in a relatively efficient way.

The above encoding strategies are general-purpose, in that they may be in principle applied to any LDPC code, regardless of its $\boldsymbol{H}$ matrix structure. A different approach consists of specifically designing the $\boldsymbol{H}$ matrix structure in order to favor

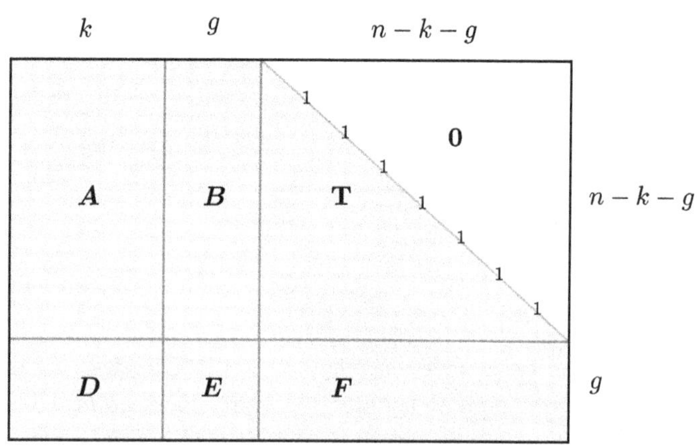

Figure 4.3 Parity-check matrix $\boldsymbol{H}$ in approximate triangular form. The white part of the matrix, denoted by $\mathbf{0}$, is null; the other parts, in light gray, are nonzero but sparse.

efficient encoding (while still achieving a good performance). The IRA codes presented in Section 4.2.2 are an example of such an approach. In this respect, it is worth mentioning that some $\boldsymbol{H}$ matrix structures are designed to facilitate LDPC encoding targeting a *hardware* implementation. In contrast, since erasure correcting LDPC codes are typically implemented in *software*, such approaches are not addressed in this book.

4.2 LDPC Code Ensembles

In LDPC coding theory, it is a common practice to reason in terms of code ensembles rather than of single codes. The design and analysis of LDPC codes, both in the finite-length regime and in the asymptotic regime where their blocklength n is allowed to tend to infinity, mainly relies on expected ensemble parameters. In this section, some notable LDPC code ensembles are described. The analysis starts from *unstructured* ensembles, described in Section 4.2.1. Then, in Section 4.2.2, two examples of *partially structured* ensembles are addressed. They differ from the unstructured ones in that some of the edges connecting the VN sockets to the CN sockets are drawn fulfilling some constraints that depend on the ensemble under consideration. Finally, Section 4.2.3 introduces the *protograph* ensemble, characterized by a higher level of structure and by a finer control in the connectivity between VNs and CNs.

4.2.1 Unstructured Irregular and Regular LDPC Code Ensembles

The Tanner graph in Figure 4.4 represents a different way of illustrating the one in Figure 4.1. Here, the VN sockets are connected to the CN sockets by edges which are shuffled by an edge interleaver (represented by the rectangle labeled as $\boldsymbol{\Pi}$). If both the VN sockets and the CN sockets are indexed from 1 to E, then the edge interleaver may be represented as an $E \times E$ permutation matrix $\boldsymbol{\Pi}$ bijectively mapping each VN socket

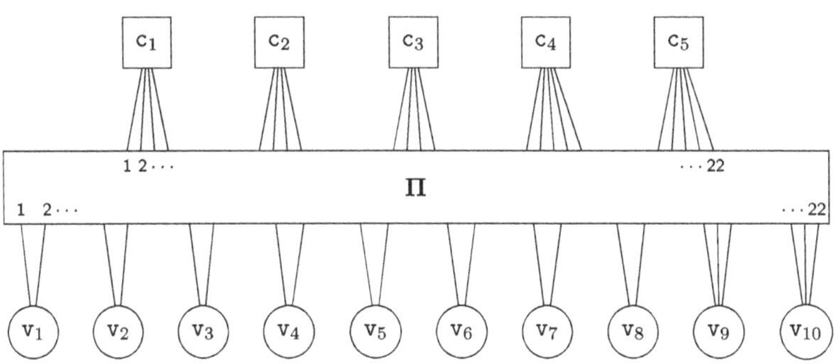

Figure 4.4 Tanner graph representation for an unstructured binary LDPC code ensemble of length $n = 10$ and edge- and node-degree distribution pairs in Example 4.4.

index into one CN socket index. An edge is drawn in the Tanner graph to connect each pair of sockets corresponding to a 1 entry in $\mathbf{\Pi}$.

Example 4.5 Both the VN sockets and the CN sockets in Figure 4.4 are indexed from 1 to 22, since there are $E = 22$ edges in the Tanner graph. In this example, indexes for both the VN and the CN sockets increase progressively from left to right. The permutation matrix $\mathbf{\Pi}$ yielding the specific Tanner graph in Figure 4.1 has nonzero entries in positions $(1,1)$, $(2,9)$, $(3,13)$, $(4,17)$, $(5,5)$, $(6,7)$, $(7,10)$, $(8,11)$, $(9,3)$, $(10,12)$, $(11,18)$, $(12,20)$, $(13,4)$, $(14,8)$, $(15,15)$, $(16,21)$, $(17,2)$, $(18,6)$, $(19,14)$, $(20,16)$, $(21,19)$, and $(22,22)$.

The Tanner graph representation in Figure 4.4 is helpful to introduce the concept of binary unstructured irregular LDPC code ensemble with length n and degree distribution pair (λ, ρ), denoted by $\mathscr{C}^n_{\lambda,\rho}$. The procedure to construct the ensemble $\mathscr{C}^n_{\lambda,\rho}$ starts from n VNs and

$$m = n\frac{\int \rho}{\int \lambda} \tag{4.12}$$

CNs, each having a predefined number of sockets so that the desired degree distribution pair (λ, ρ) is obtained. As in the example in Figure 4.4, we can imagine that the VN degrees and the CN degrees increase from left to right and that the indexes for both the VN and the CN sockets increase progressively from left to right. Constructing the $\mathscr{C}^n_{\lambda,\rho}$ ensemble consists of generating all possible $E!$ permutation matrices $\mathbf{\Pi}$ and of connecting the VN sockets to the CN sockets according to each such matrix to obtain a parity-check matrix $\boldsymbol{H}$ representing a binary linear block code. Unless explicitly stated, the probability distribution of a random code $\mathcal{C} \in \mathscr{C}^n_{\lambda,\rho}$ is the one induced by a random choice of the permutation matrix $\mathbf{\Pi}$ according to a uniform probability measure. With reference to the compact representation used in Example 4.5, the random generation of $\mathbf{\Pi}$ may be implemented by drawing E integer numbers, without replacement, starting from the set $\{1,2,\ldots,E\}$. The resulting probability of each permutation matrix to be generated is obviously $1/E!$.

An unstructured LDPC code ensemble may also be constructed on a finite field $\mathrm{GF}(q)$ with $q > 2$, in which case we denote it by $\mathscr{C}^{n,q}_{\lambda,\rho}$. To construct the $\mathscr{C}^{n,q}_{\lambda,\rho}$ ensemble we start again from n VNs and a number of CNs given by (4.12), each node having a predefined number of sockets corresponding to an overall degree distribution pair (λ, ρ), and also in this case, we take all possible permutation matrices $\mathbf{\Pi}$, of size $E \times E$, to connect VN sockets to CN ones. In addition, for each such $\mathbf{\Pi}$ matrix, we perform all possible labelings of the E Tanner graph edges using the elements of $\mathrm{GF}(q) \setminus \{0\}$, the total number of such labelings being $(q-1)^E$. Each choice of $\mathbf{\Pi}$ and of the edge labeling yields a Tanner graph that is associated with a sparse parity-check matrix representing an LDPC code constructed on $\mathrm{GF}(q)$. Unless explicitly stated, the probability distribution of a random code $\mathcal{C} \in \mathscr{C}^{n,q}_{\lambda,\rho}$ is the one induced by a random choice of the permutation matrix $\mathbf{\Pi}$, according to a uniform probability measure, and a random choice of the edge labels, again according to a uniform probability measure. The random labels may be generated by drawing E i.i.d. random variables, each characterized by a uniform probability mass function (p.m.f.) on $\mathrm{GF}(q) \setminus \{0\}$.

In case all VNs are characterized by the same number d_v of sockets and all CNs are characterized by the same number d_c of sockets, the unstructured LDPC code ensemble is said to be regular. In the binary case, this ensemble is denoted by $\mathscr{C}^n_{\mathsf{d}_v,\mathsf{d}_c}$. If the ensemble is constructed over $\mathrm{GF}(q)$, it is denoted by $\mathscr{C}^{n,q}_{\mathsf{d}_v,\mathsf{d}_c}$.

By definition, the design rate of an ensemble $\mathscr{C}^{n,q}_{\lambda,\rho}$, denoted by R, is

$$\mathrm{R} = 1 - \frac{\int \rho}{\int \lambda} \tag{4.13}$$

which reduces to

$$\mathrm{R} = 1 - \frac{\mathsf{d}_v}{\mathsf{d}_c} \tag{4.14}$$

for regular $\mathscr{C}^{n,q}_{\mathsf{d}_v,\mathsf{d}_c}$ ensembles.

Remark 4.1. Let the random variable $R(\mathcal{C})$ represent the code rate of a random code $\mathcal{C} \in \mathscr{C}^{n,q}_{\mathsf{d}_v,\mathsf{d}_c}$. We have $R(\mathcal{C}) \geq \mathrm{R}$ for all realizations of $\mathcal{C}$ with equality if and only if all m linear constraints are linearly independent. Moreover, for any constant $\epsilon > 0$, we have $P(R(\mathcal{C}) - \mathrm{R} > \epsilon) \to 0$ as $n \to \infty$. A proof of this latter property is proposed in Exercise 5.1 in Chapter 5. This proof makes use of some results developed throughout the proof of Theorem 5.11.

Remark 4.2. A stronger rate concentration result for the binary ensemble $\mathscr{C}^n_{\lambda,\rho}$ is the following. Let

$$v(u) = \left(\sum_d \frac{\lambda_d}{1 + u^d} \right)^{-1} \left(\sum_d \frac{\lambda_d u^{d-1}}{1 + u^d} \right). \tag{4.15}$$

If the global maximum in $[0, \infty)$ of the function

$$\Psi(u) = -\Lambda'(1) \log_2\left(\frac{1 + u\, v(u)}{1 + v(u)} \right) + \sum_d \Lambda_d \log_2(1 + u^d)$$
$$+ \frac{\Lambda'(1)}{P'(1)} \sum_d P_d \log_2\left(1 + \left(\frac{1 - v(u)}{1 + v(u)} \right)^d \right) - \Lambda(1) \tag{4.16}$$

is attained at $u = 1$, then there exists $B > 0$ such that, for any $\xi > 0$, we have $P(R(\mathcal{C}) - \mathrm{R} > \epsilon) \leq \exp(-Bn\xi)$ for all $n > n_0$, where n_0 depends on ξ and on the degree distribution pair (λ, ρ).

Remark 4.3. In the above-described ensemble generation procedure, some choices of the $\mathbf{\Pi}$ matrix yield multiple edges between a VN and a CN. When that happens, the corresponding entry of the parity-check matrix is set equal to the sum (in the finite field) of the labels of the involved edges.

Remark 4.4. For any choice of the $E \times E$ permutation matrix $\mathbf{\Pi}$ and for any labeling of the E edges, the number of nonzero entries of the obtained parity-check matrix is upper bounded by the quantity $E = n / \int \lambda$, while its total number of entries is equal to R n^2. Hence, for a constant degree distribution pair (λ, ρ), the parity-check matrix sparsity (defined as the ratio of the number of zero entries to the total number of entries) converges to 1 as n grows.

The above remarks remain valid also for the partially structured ensembles that will be next introduced in Section 4.2.2. Also for these ensembles, the probability that the sample code rate deviates from the design rate by any constant vanishes as n grows; parallel edges between any two nodes correspond, in the parity-check matrix, to the sum of their labels; the parity-check matrix sparsity converges to 1 as n grows for fixed degree distributions.

4.2.2 Partially Structured Ensembles

4.2.2.1 Irregular Repeat-Accumulate Ensemble

Figure 4.5 depicts the Tanner graph of an LDPC code belonging to a binary irregular repeat-accumulate (IRA) ensemble, where **Π** represents again an edge interleaver. The ensemble is denoted by $\mathscr{I}^{n}_{\psi,\mathsf{d}_c}$, where n is the blocklength and d_c is the degree of all CNs but the first one (denoted by c_1 in the figure) whose degree is $\mathsf{d}_c - 1$. In the example Tanner graph in Figure 4.5, we have $n = 18$ and $\mathsf{d}_c = 7$. The total number of edges in the Tanner graph is $E = m\,\mathsf{d}_c - 1$, where m is the number of CNs.

The Tanner graph is composed of an unstructured part and of a structured one. The unstructured part corresponds to the edges permuted by the edge interleaver **Π** and to the associated VNs ($\mathsf{v}_1, \mathsf{v}_2, \ldots, \mathsf{v}_{11}$ in the figure). These VNs are said to be of type V_1, and the edges connected to them are said to be of type E_1. The structured part of the Tanner graph is composed of m VNs of which all have degree 2 and one has degree 1. These VNs ($\mathsf{v}_{12}, \mathsf{v}_{13}, \ldots, \mathsf{v}_{18}$ in the figure) are said to be of type V_2 and their edges, connected to the CNs in an open loop, are said to be of type E_2. The number of VNs of type V_1 is denoted by n_1.

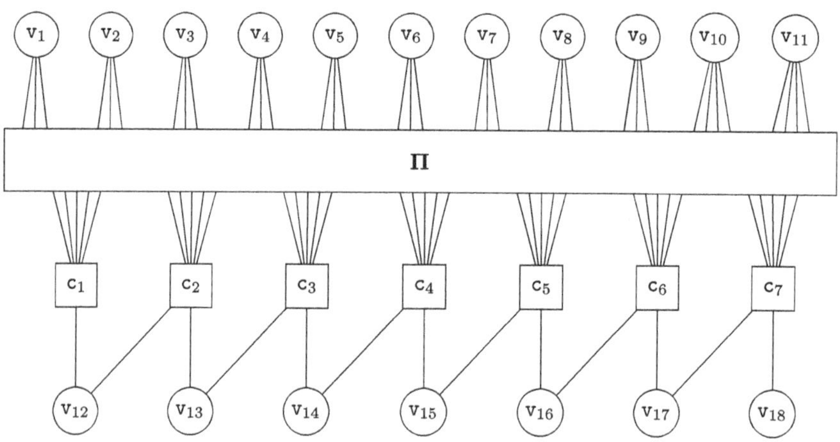

Figure 4.5 Tanner graph representation of a binary IRA LDPC code ensemble of length $n = 18$.

The polynomial $\psi(x) = \sum_{j=2}^{\mathsf{d}_v} \psi_j x^{j-1}$ represents the edge-perspective degree distribution for edges of type E_1. Accordingly, the parameter ψ_j denotes the fraction of edges of type E_1 connected to VNs of type V_1 of degree j. The corresponding node-oriented degree distribution is $\Psi(x) = \sum_{j=2}^{\mathsf{d}_v} \Psi_j x^j$ where Ψ_j is the fraction of VNs of type V_1 having degree j. A nonzero fraction of VNs of type V_1 having degree 2 may in principle exist. It is however a common practice to design IRA ensembles characterized by $\psi_2 = 0$.

Similar to previously presented ensembles, the construction of the IRA ensemble consists of generating all possible $(m(\mathsf{d}_c - 2))!$ permutation matrices $\mathbf{\Pi}$ and of connecting the sockets of the type-V_1 VNs to the CN sockets of type E_1 according to each such matrix, to obtain a parity-check matrix $\boldsymbol{H}$ representing a binary linear block code. The probability distribution of a random code $\mathcal{C} \in \mathscr{I}^n_{\psi,\mathsf{d}_c}$ is the one induced by a random choice of the permutation matrix $\mathbf{\Pi}$ according to a uniform probability measure.

The design rate of an IRA ensemble $\mathscr{I}^n_{\psi,\mathsf{d}_c}$ is

$$\mathrm{R} = \frac{(\mathsf{d}_c - 2) \int \psi}{1 + (\mathsf{d}_c - 2) \int \psi}, \tag{4.17}$$

where as usual $\int \psi = \int_0^1 \psi(x) \mathrm{d}x$. Moreover, the overall VN degree distribution from an edge perspective is given by

$$\lambda(x) = \frac{2}{\mathsf{d}_c} x + \sum_{j=2}^{\mathsf{d}_v} \frac{(\mathsf{d}_c - 2)\psi_j}{\mathsf{d}_c} x^{j-1} \tag{4.18}$$

and its CN degree distribution from an edge perspective by $\rho(x) = x^{\mathsf{d}_c - 1}$. Both degree distributions apply to the limiting case where the number of VNs and the number of CNs tend to infinity, for fixed ψ and d_c (and therefore for fixed R). In this case, in fact, the ratio of the number of type-V_2 VNs of degree 2 to the total number of type-V_2 VNs tends to 1, as well as the ratio of the number of CNs of degree d_c to the total number of CNs.

Example 4.6 Consider the IRA LDPC code ensemble whose Tanner graph representation is depicted in Figure 4.5. We have $\psi(x) = \frac{27}{35}x^2 + \frac{8}{35}x^3$ and $\mathsf{d}_c = 7$. Applying (4.17), since $\int \psi = \frac{11}{35}$, we obtain $\mathrm{R} = \frac{11}{18}$ as expected. Moreover, applying (4.18) we obtain $\lambda(x) = \frac{2}{7}x + \frac{27}{49}x^2 + \frac{8}{49}x^3$.

Systematic encoding of a binary IRA code may be performed efficiently by regarding the VNs of type V_1 and the VNs of type V_2 as associated with the bits composing the message to be encoded and with the redundant bits generated by the encoder, respectively. For example, with reference to Figure 4.5, given the message bits corresponding to $\mathsf{v}_1, \mathsf{v}_2, \ldots, \mathsf{v}_{11}$, the value of the bit corresponding to v_{12} is calculated as the sum in GF(2) of the bits corresponding to the type-V_1 VNs connected to c_1. Next, the value of the bit associated with v_{13} is calculated as the sum in GF(2) of the bits corresponding to the type-V_1 VNs connected to c_2 and of the bit corresponding

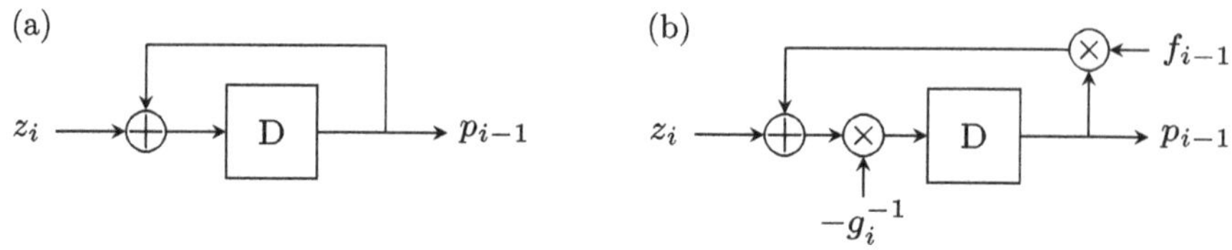

Figure 4.6 (a) An accumulator. (b) A time-variant accumulator. All operations are in the finite field to which z_i, p_{i-1}, g_i, and f_{i-1} belong.

to v_{12}. The process is iterated until all bits associated with $\mathsf{v}_{12}, \mathsf{v}_{13}, \ldots, \mathsf{v}_{18}$ have been computed.

The described systematic encoding algorithm may be implemented as follows. Each message bit is first repeated a number of times equal to the degree of the corresponding type-V_1 VN, yielding a sequence of $(\mathsf{d}_c - 2)m = \frac{n_1}{\int \psi}$ bits. The so-obtained bits, corresponding to the sockets of the type-V_1 VNs, are permuted according to the edge interleaver $\mathbf{\Pi}$. The permuted sequence is then partitioned into m segments, where each segment is composed of $\mathsf{d}_c - 2$ contiguous bits; for each segment, the sum in GF(2) of the bits in the segment is calculated, leading to a sequence of m bits. Finally, a binary accumulator with initial state 0 is fed with the m sums in GF(2). The m bits output by the accumulator are appended to the n_1 message bits, yielding the codeword of length $n = n_1 + m$.

A binary accumulator, depicted in Figure 4.6(a), is a linear encoder over GF(2) with code rate equal to 1. At every clock cycle, it generates one output bit by processing one input bit. The accumulator structure comprises one latch, represented as a square labeled by D in the figure, and one adder in GF(2). The bit stored in the latch is called the *state* of the accumulator and constitutes the accumulator output. Letting $\boldsymbol{z} = \begin{bmatrix} z_1 & z_2 & \ldots & z_m \end{bmatrix} \in \mathrm{GF}(2)^m$ and $\boldsymbol{p} = \begin{bmatrix} p_1 & p_2 & \ldots & p_m \end{bmatrix} \in \mathrm{GF}(2)^m$ be the accumulator input and output binary sequences, respectively, denoting the initial state by p_0, and letting $p_0 = 0 \in \mathrm{GF}(2)$, we have $p_1 = z_1$, $p_2 = p_1 + z_2 = z_1 + z_2$, and so on up to $p_m = z_1 + z_2 + \cdots + z_m$, all sums being in GF(2). Each output bit is then the sum in GF(2) of all previous input bits, which justifies the name "accumulator."

Remark 4.5. Every bit equal to 1 at the input of the accumulator enforces a state transition (from 0 to 1 or from 1 to 0), while every input bit equal to 0 lets the next state be equal to the current state. Since the accumulator state coincides with its output, a burst of output bits all equal to each other is observed when the accumulator is fed with a run of bits all equal to 0. Whenever a 1 bit occurs at the accumulator input, the output is inverted and a new output burst is started until the next 1 at the input is encountered.

Example 4.7 Consider the Tanner graph in Figure 4.5. Denote by $\boldsymbol{A}$ the 7×11 matrix with elements in GF(2), whose row of index $i \in \{1,2,\ldots,7\}$ is associated with CN c_i and whose column of index $j \in \{1,2,\ldots,11\}$ is associated with VN v_j. The element of $\boldsymbol{A}$ in position (i,j) is equal to 1 if c_i is connected to v_j and is equal to 0 otherwise. A specific matrix $\boldsymbol{A}$ is associated with each realization of the edge interleaver $\boldsymbol{\Pi}$. Let

$$\boldsymbol{A} = \begin{bmatrix} 1 & 0 & 0 & 0 & 0 & 1 & 1 & 1 & 0 & 0 & 1 \\ 0 & 0 & 0 & 0 & 1 & 1 & 1 & 0 & 0 & 1 & 1 \\ 1 & 1 & 1 & 0 & 0 & 0 & 0 & 1 & 1 & 0 & 0 \\ 0 & 1 & 1 & 1 & 0 & 0 & 0 & 0 & 1 & 1 & 0 \\ 0 & 0 & 1 & 1 & 1 & 0 & 0 & 0 & 0 & 1 & 1 \\ 0 & 0 & 0 & 1 & 1 & 1 & 0 & 0 & 0 & 1 & 1 \\ 1 & 1 & 0 & 0 & 0 & 0 & 1 & 1 & 1 & 0 & 0 \end{bmatrix} \tag{4.19}$$

and let the binary message to be encoded be $\boldsymbol{u} = [0\ 1\ 0\ 0\ 1\ 1\ 0\ 1\ 0\ 1\ 0]$, where the left-most bit is associated with VN v_1. The vector $\boldsymbol{z}$ which represents the accumulator input may simply be calculated $\boldsymbol{z} = \boldsymbol{u}\,\boldsymbol{A}^{\mathsf{T}}$, obtaining $\boldsymbol{z} = \begin{bmatrix} 0 & 1 & 0 & 0 & 0 & 1 & 0 \end{bmatrix}$. Letting the initial state of the accumulator be zero and applying the reasoning in Remark 4.5, we have $\boldsymbol{p} = \begin{bmatrix} 0 & 1 & 1 & 1 & 1 & 0 & 0 \end{bmatrix}$. Note that the parity-check matrix of the code is

$$\boldsymbol{H} = \left[\begin{array}{c|ccccccc} & 1 & 0 & 0 & 0 & 0 & 0 & 0 \\ & 1 & 1 & 0 & 0 & 0 & 0 & 0 \\ & 0 & 1 & 1 & 0 & 0 & 0 & 0 \\ \boldsymbol{A} & 0 & 0 & 1 & 1 & 0 & 0 & 0 \\ & 0 & 0 & 0 & 1 & 1 & 0 & 0 \\ & 0 & 0 & 0 & 0 & 1 & 1 & 0 \\ & 0 & 0 & 0 & 0 & 0 & 1 & 1 \end{array}\right]. \tag{4.20}$$

To construct an IRA code ensemble over a finite field GF(q) with $q > 2$, besides generating all possible $(m(\mathsf{d}_c - 2))!$ permutation matrices $\boldsymbol{\Pi}$ we need to label the edges of the Tanner graph. A first approach consists of assigning a label equal to $1 \in \text{GF}(q)$ to each edge of type E_2 and, for each $\boldsymbol{\Pi}$, of performing all possible labelings of the $m(\mathsf{d}_c - 2)$ edges of type E_1 using the elements of GF$(q) \setminus \{0\}$. A second approach consists of performing, for each $\boldsymbol{\Pi}$, all possible labelings of the $E = m\,\mathsf{d}_c - 1$ edges in the Tanner graph using the elements of GF$(q) \setminus \{0\}$. In either case, each choice of $\boldsymbol{\Pi}$ and of the edge labeling yields a sparse parity-check matrix representing an IRA code. As usual, the probability distribution of a random code in either of the two ensembles is the one induced by a random choice of $\boldsymbol{\Pi}$, based on a uniform probability measure, and by drawing independently, each based on a uniform probability distribution over GF$(q) \setminus \{0\}$, those edge labels that have not been set to $1 \in \text{GF}(q)$.

Encoding of nonbinary IRA codes may be performed efficiently, similarly to the binary case. Each message symbol is repeated a number of times equal to the degree of the corresponding type-V_1 VN; the so-obtained $(\mathsf{d}_c - 2)m$ symbols are permuted according to the edge interleaver $\boldsymbol{\Pi}$ and each of them is multiplied by the label of the corresponding type-E_1 edge; the obtained symbols are partitioned into m segments, each composed of $\mathsf{d}_c - 2$ contiguous symbols; finally, the sum in GF(q) of the symbols in each segment is calculated. In case all edges of type E_2 are labeled with $1 \in \text{GF}(q)$, an accumulator as the one shown in Figure 4.6(a) is fed with the m sums in GF(q). As opposed to the binary case, the adder must perform addition in $\mathrm{GF}(q)$ and the latch must store one symbol in $\mathrm{GF}(q)$. In case also the type-E_2 edge labels may be different from $1 \in \mathrm{GF}(q)$, a time-variant accumulator as the one shown in Figure 4.6(b) is fed with the m sums in $\mathrm{GF}(q)$. In the figure, g_i represents the label of the edge connecting VN of index $n_1 + i$ with CN of index i, for $i \in \{1,\ldots,m\}$. Moreover, f_i represents the label of the edge connecting VN of index $n_1 + i$ with CN of index $i + 1$, for $i \in \{2,\ldots,m\}$ (conventionally, $f_0 = 0$).

Remark 4.6. Although the IRA code ensemble has been addressed in the case where all CNs but one have the same degree d_c and one CN has degree $\mathsf{d}_c - 1$, IRA code ensembles characterized by an irregular degree profile may also be constructed, in which case a CN degree distribution must be used. In this case, we can denote by ϕ_i the fraction of edges of type E_1 connected to CNs of degree $i + 2$. We obtain the edge-oriented and node-oriented degree distribution polynomials $\phi(x) = \sum_{d=1}^{\mathsf{d}_c-2} \phi_d x^{d-1}$ and $\Phi(x) = \sum_{d=1}^{\mathsf{d}_c-2} \Phi_d x^d$, with $\Phi_d = (\phi_d/d)/(\sum_j \phi_j/j)$. The design rate of the corresponding IRA ensemble, that we denote by $\mathscr{I}^n_{\psi,\phi}$, is

$$\mathsf{R} = \left(1 + \frac{\sum_{d=1}^{\mathsf{d}_c-2} \phi_d/d}{\sum_{d=1}^{\mathsf{d}_v} \psi_d/d}\right)^{-1}. \tag{4.21}$$

IRA ensembles with an irregular CN degree profile are considered in Section 7.3 in the framework of density evolution analysis.

4.2.2.2 LDPC Code Ensemble with Separated VNs

The Tanner graph of an LDPC code belonging to a binary separated variable nodes (SVN) ensemble is illustrated in Figure 4.7, where $\mathbf{\Pi}$ represents an edge interleaver. The ensemble is denoted by $\mathscr{V}^n_{\psi,\mathsf{d}_c}(\nu_2)$, where n is the blocklength and d_c is the constant CN degree (we have $\mathsf{d}_c = 6$ in the example Tanner graph in Figure 4.7). The function $\psi(x)$ is a polynomial representing an edge-perspective degree distribution. Its meaning, together with that of the parameter ν_2, will be clarified in the remainder of this section.

There are three different edge types, denoted by E_1, E_2, E_3, and three different VN types, denoted by V_1, V_2, V_3. Since all m CNs have the same degree d_c, the total number of edges in the Tanner graph is $E = m\,\mathsf{d}_c$. In Figure 4.7, the set of VNs of type V_1 is $\{\mathsf{v}_1,\mathsf{v}_2,\ldots,\mathsf{v}_{11}\}$, while the sets of VNs of types V_2 and V_3 are $\{\mathsf{v}_{12},\mathsf{v}_{13}\}$ and $\{\mathsf{v}_{14}\}$, respectively. The VNs of type V_1 may have irregular degrees, but none of them is allowed to have degree 2. All type-V_1 VNs are connected only to edges of type E_1, depicted with solid lines in the example in Figure 4.7. Each VN of type V_2 has degree 2 and is connected to two edges of type E_2, depicted with dotted lines, while each VN of type V_3 has degree 3 and is connected to three edges of type E_3, depicted with dashed lines. Moreover, each CN is connected to $\mathsf{d}_c - 1$ edges of type E_1 and to one edge either of type E_2 or of type E_3. The $\mathsf{d}_c - 1$ sockets of a CN that are connected to edges of type E_1 will be referred to as type-E_1 sockets.

Constructing the $\mathscr{V}^n_{\psi,\mathsf{d}_c}(\nu_2)$ ensemble consists of generating all possible $(m(\mathsf{d}_c - 1))!$ permutation matrices $\mathbf{\Pi}$ and of connecting the sockets of the type-V_1 VNs to the

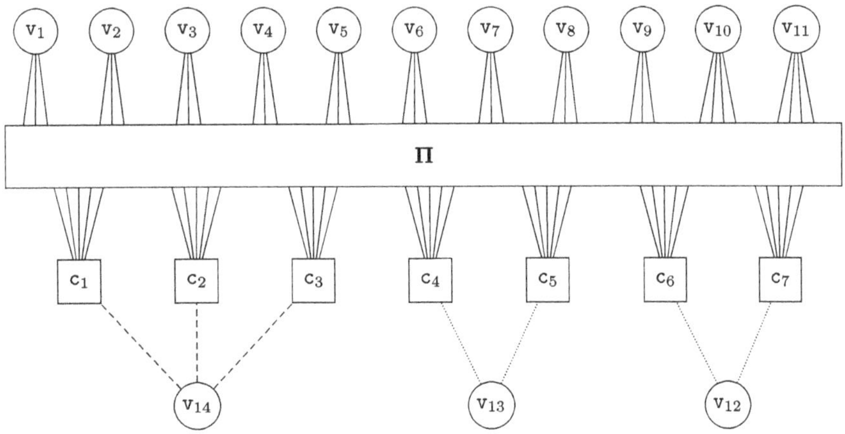

Figure 4.7 Tanner graph representation of a binary SVN LDPC code ensemble of length $n = 14$.

CN sockets of type $\mathtt{E}_1$ according to each such matrix, to obtain a parity-check matrix $\boldsymbol{H}$ representing a binary linear block code. Unless explicitly stated, the probability distribution of a random code $\mathcal{C} \in \mathscr{V}^n_{\psi,\mathsf{d}_c}(\nu_2)$ is the one induced by a random choice of the permutation matrix $\boldsymbol{\Pi}$ according to a uniform probability measure.

The main feature of this ensemble is that VNs of type $\mathtt{V}_2$ and $\mathtt{V}_3$ are *separated*, meaning that they do not share any neighboring CN with any other VN of type $\mathtt{V}_2$ or $\mathtt{V}_3$. The number m of CNs in an SVN ensemble has always to fulfill the identity $m = 2n_2 + 3n_3$, where $0 < n_2 \leq \lfloor m/2 \rfloor$ and $0 \leq n_3 \leq \lfloor m/3 \rfloor$ are the numbers of VNs of type $\mathtt{V}_2$ and $\mathtt{V}_3$, respectively. We also define n_1 as the number of VNs of type $\mathtt{V}_1$. Note that in an SVN ensemble, all VNs of degree 2 (all of which are of type $\mathtt{V}_2$) are separated, while in general, some degree-3 VNs are separated and some others are not.[1] Moreover, we write $n_1 = \sum_{j=3}^{\mathsf{d}_v} n_{1,j}$, where $n_{1,j}$ is the number of degree-j VNs of type $\mathtt{V}_1$.

The fraction of edges of type $\mathtt{E}_1$ connected to degree-j VNs of type $\mathtt{V}_1$ is denoted by ψ_j, while the fraction of type-$\mathtt{V}_1$ VNs of degree j by $\Psi_j = n_{1,j}/n_1$. In polynomial form, we write

$$\psi(x) = \sum_{j=3}^{\mathsf{d}_v} \psi_j x^{j-1} \qquad \text{and} \qquad \Psi(x) = \sum_{j=3}^{\mathsf{d}_v} \Psi_j x^j. \tag{4.22}$$

Moreover, for $i \in \{1,2,3\}$ we define $\delta_i = n_i/n$ and $\delta_{1,j} = n_{1,j}/n$. Finally, we define the parameter

$$\nu_2 = \frac{2n_2}{m} = \frac{2n_2}{2n_2 + 3n_3} \tag{4.23}$$

[1] The definition of SVN ensemble here provided may in principle be extended to the case in which the smallest VN degree larger than two is larger than three. The reason we stick to the described case is that the SVN code ensembles with good erasure correction properties are characterized by a number of degree-3 VNs that is sufficient to saturate the number of separated nodes, whenever $n_2 < \lfloor m/2 \rfloor$.

as the ratio of the number of edges connected to type-$\mathtt{V}_2$ VNs to the total number of edges connected to separated VNs. The parameter ν_2 fulfills $0 < \nu_2 \leq 1$, where $\nu_2 = 1$ indicates that all separated VNs have degree 2.

It is easy to show that the design rate of the ensemble $\mathscr{V}^n_{\psi,\mathsf{d}_c}(\nu_2)$ is

$$\mathsf{R} = 1 - \frac{6}{6(\mathsf{d}_c - 1)\int\psi + \nu_2 + 2}, \tag{4.24}$$

where $\int\psi = \int_0^1 \psi(x)\mathrm{d}x$. The overall VN degree distribution from an edge perspective of the ensemble $\mathscr{V}^n_{\psi,\mathsf{d}_c}(\nu_2)$ is given by

$$\lambda(x) = \frac{\nu_2}{\mathsf{d}_c}x + \frac{1 - \nu_2 + (\mathsf{d}_c - 1)\psi_3}{\mathsf{d}_c}x^2 + \sum_{j=4}^{\mathsf{d}_v}\frac{(\mathsf{d}_c - 1)\psi_j}{\mathsf{d}_c}x^{j-1}, \tag{4.25}$$

while its CN degree distribution from an edge perspective is $\rho(x) = x^{\mathsf{d}_c - 1}$. Conversely, given the edge-perspective degree distribution pair $\lambda(x)$, $\rho(x) = x^{\mathsf{d}_c-1}$ of an SVN ensemble, its parameters $\nu_2, \psi_3, \ldots, \psi_{\mathsf{d}_v}$ may be calculated as

$$\begin{aligned}
\nu_2 &= \lambda_2\,\mathsf{d}_c \\
\psi_3 &= \frac{\lambda_3\,\mathsf{d}_c + \lambda_2\,\mathsf{d}_c - 1}{\mathsf{d}_c - 1} \\
\psi_j &= \frac{\lambda_j\,\mathsf{d}_c}{\mathsf{d}_c - 1} \qquad j \in \{4, \ldots, \mathsf{d}_v\}.
\end{aligned} \tag{4.26}$$

Moreover, its parameters $\delta_2, \delta_3, \delta_{1,3}, \ldots, \delta_{1,\mathsf{d}_v}$ may be calculated as

$$\begin{aligned}
\delta_2 &= \frac{\nu_2}{2\,\mathsf{d}_c \int\lambda} \\
\delta_3 &= \frac{1 - \nu_2}{3\,\mathsf{d}_v \int\lambda} \\
\delta_{1,j} &= \frac{(\mathsf{d}_c - 1)\psi_j}{j\,\mathsf{d}_c \int\lambda} \qquad j \in \{3, 4, \ldots, \mathsf{d}_v\}.
\end{aligned} \tag{4.27}$$

Example 4.8 For the SVN Tanner graph in Figure 4.7, the number of VNs of type $\mathtt{V}_1$ is $n_1 = 11$, the number of VNs of type $\mathtt{V}_2$ is $n_2 = 2$, and the number of VNs of type $\mathtt{V}_3$ is $n_3 = 1$. Moreover we have $\mathsf{d}_v = 4$, $n_{1,3} = 9$, and $n_{1,4} = 2$. The design rate is $\mathsf{R} = 1/2$. There are 35 edges of type $\mathtt{E}_1$, so the edge-perspective degree distribution $\psi(x)$ is

$$\psi(x) = \frac{27}{35}x^2 + \frac{8}{35}x^3. \tag{4.28}$$

Moreover, the node-perspective degree distribution $\Psi(x)$ is

$$\Psi(x) = \frac{9}{11}x^3 + \frac{2}{11}x^4. \tag{4.29}$$

The value of ν_2 is $\nu_2 = 4/7$ and the overall degree distribution pair from an edge perspective is

$$\lambda(x) = \frac{2}{21}x + \frac{5}{7}x^2 + \frac{4}{21}x^3 . \tag{4.30}$$

We also have $\delta_2 = \frac{2}{14}$, $\delta_3 = \frac{1}{14}$, $\delta_{1,3} = \frac{9}{14}$, $\delta_{1,4} = \frac{2}{14}$.

As in the unstructured case, also an SVN ensemble may be constructed over a finite field GF(q) with $q > 2$. In this case, we denote the ensemble by $\mathscr{V}^{n,q}_{\psi,\mathsf{d}_c}(\nu_2)$. To construct the $\mathscr{V}^{n,q}_{\psi,\mathsf{d}_c}(\nu_2)$ ensemble, we generate again all possible $(m(\mathsf{d}_c - 1))!$ permutation matrices $\mathbf{\Pi}$. In addition, for each such $\mathbf{\Pi}$ matrix, we perform all possible labelings of the $E = m\,\mathsf{d}_c$ Tanner graph edges (regardless their type) using the elements of GF$(q) \setminus \{0\}$. Each choice of $\mathbf{\Pi}$ and of the edge labeling yields, a Tanner graph that is associated with a sparse parity-check matrix representing an LDPC code constructed on GF(q). The probability distribution of a random code $\mathcal{C} \in \mathscr{V}^{n,q}_{\psi,\mathsf{d}_c}(\nu_2)$ is the one induced by a random choice of the permutation matrix $\mathbf{\Pi}$, according to a uniform probability measure, and by drawing the edge labels independently, each according to a uniform probability distribution over GF$(q) \setminus \{0\}$.

4.2.3 Protograph LDPC Code Ensemble

In a protograph LDPC code ensemble, the connections between the VNs and the CNs are defined by a small bipartite graph called the *protograph*. The number of VNs in a protograph is denoted by n_v and the number of CNs in it by n_c. Working with protograph-based LDPC code ensembles, it is relatively easy to include in the analysis also *punctured* symbols, that is, symbols which are not transmitted and which are therefore characterized by minimum reliability from the decoder perspective. Of the n_v VNs in the protograph, n_u VNs are transmitted and n_p VNs are punctured, where n_u and n_p fulfill $n_v = n_u + n_p$. The sets of indices of the n_v VNs and n_c CNs in the protograph are denoted by $\mathcal{J}$ and $\mathcal{I}$, respectively, while the sets of indexes of the unpunctured and punctured VNs are denoted by $\mathcal{J}_u$ and $\mathcal{J}_p$ respectively.

The number of edges in the protograph is E_p. Protograph edges are indexed using two different conventions. The first one consists of using the set $\Phi = \{1,2,\ldots,E_\mathrm{p}\}$. In this case, for all $i \in \mathcal{I}$, $\Gamma_i \subseteq \Phi$ denotes the set of indices of the edges connected to the CN of index i in the protograph. The cardinality of Γ_i is equal to the degree of the corresponding CN, denoted by s_i. Similarly, for all $j \in \mathcal{J}$, $\Xi_j \subseteq \Phi$ denotes the set of indices of the edges connected to the VN of index j in the protograph. The cardinality of Ξ_j is equal to the degree of the corresponding VN, denoted by t_j. This indexing convention is used in Chapter 5 to develop the weight distribution of protograph ensembles, as well as in Chapter 9 to describe generalized LDPC protograph ensembles. The second indexing convention consists of indexing a protograph edge by a pair (i,j), where $i \in \mathcal{I}$ and $j \in \mathcal{J}$ are the indices of the protograph CN and VN to which the edge is connected. As opposed to the previous convention, in which each of the E_p edges possesses its own index, in this case, two protograph edges connected

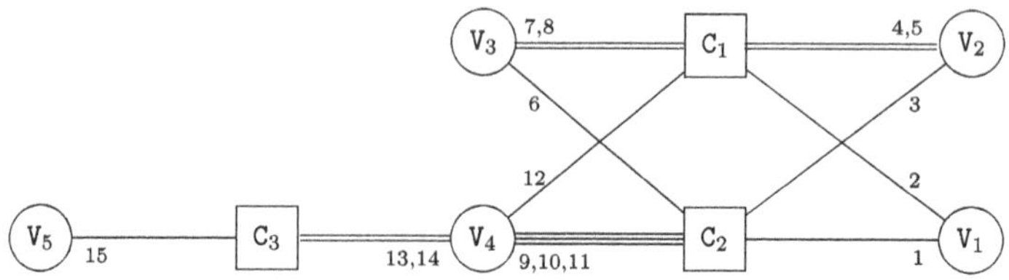

Figure 4.8 Protograph defining an ARJA LDPC code ensemble with design rate R = 1/2.

to the same pair of nodes have the same index. This indexing rule is used in Chapter 7 to describe density evolution of protograph LDPC code ensembles.

The node connectivity of the protograph is described by an $n_c \times n_v$ base matrix $\boldsymbol{B} = \{B_{ij}\}$, where B_{ij} denotes the number of edges connecting CN $i \in \mathcal{I}$ to VN $j \in \mathcal{J}$.

Example 4.9 Figure 4.8 shows a protograph known as the accumulate-repeat jagged-accumulate (ARJA) protograph with rate 1/2. The number of VNs in the protograph is $n_v = 5$ and the number of CNs in it is $n_c = 3$. The five VNs are indexed with the elements of the set $\mathcal{J} = \{1,2,3,4,5\}$ and the three CNs by the elements of the set $\mathcal{I} = \{1,2,3\}$. Of the five protograph VNs, v_1, v_2, v_3, and v_5 are transmitted while v_4 is punctured, so $\mathcal{J}_u = \{1,2,3,5\}$, $n_u = 4$, $\mathcal{J}_p = \{4\}$, and $n_p = 1$. The set of edge indexes is $\Phi = \{1,2,\ldots,15\}$ and we have, for example, $\Gamma_1 = \{2,4,5,7,8,12\}$ and $\Xi_4 = \{9,10,11,12,13,14\}$. Adopting the second indexing rule for the edges, edges 9, 10, and 11 are all indexed as $(2,4)$. The base matrix for this protograph is

$$\boldsymbol{B} = \begin{bmatrix} 1 & 2 & 2 & 1 & 0 \\ 1 & 1 & 1 & 3 & 0 \\ 0 & 0 & 0 & 2 & 1 \end{bmatrix}. \tag{4.31}$$

The protograph defines an ensemble of LDPC codes over the finite field GF(q). To draw an LDPC code over GF(q) from the protograph ensemble, a *copy-and-permute* operation with copy (or "lifting") factor Q is first performed on the protograph as follows. Each VN v_j, $j \in \mathcal{J}$, is copied Q times, producing Q VNs which are said to be of *type* j. Each CN c_i, $i \in \mathcal{I}$ is also copied Q times, producing Q CNs which are said to be of *type* i. Each edge connecting VN v_j and CN c_i in the protograph is copied Q times, and these edges connect the *copies* of VN v_j and the *copies* of CN c_i in a one-to-one fashion. The Q edges being copies of the protograph edge of index $g \in \Phi$ are said to be of *type* g. Edges of different types are permuted separately and the edge permutation procedure is equivalent to drawing E_{p} permutation matrices of size $Q \times Q$ independently of each other, each uniformly at random. Next, $E = QE_{\mathrm{p}}$ i.i.d. random variables are drawn by sampling a uniform p.m.f. on GF$(q) \setminus \{0\}$, and the E edges in the Tanner graph are labeled with the so-obtained E nonzero symbols.

The described protograph LDPC code ensemble over GF(q) is sometimes referred to in the literature as *unconstrained* protograph ensemble. In the *constrained* version of it, all edges of the same type are characterized by the same label in GF$(q) \setminus \{0\}$. Therefore, to draw an LDPC code from the constrained protograph ensemble over

GF(q), the E_p edges in the protograph are first labeled with elements of GF$(q) \setminus \{0\}$ and then the copy-and-permute operation is performed, also copying the edge labels.

Regardless the protograph ensemble from which the LDPC code is drawn (unconstrained or constrained), the obtained Tanner graph has Qn_v VNs, of which Qn_u are transmitted and Qn_p are punctured, and Qn_c CNs. The blocklength of the LDPC code is $n = Qn_u$. The dimension of the LDPC code is $k = Q(n_v - n_c)$ if all parity-check equations are linearly independent. The design rate is therefore

$$\mathrm{R} = \frac{n_v - n_c}{n_u}. \tag{4.32}$$

Example 4.10 The design rate R of the ARJA protograph depicted in Figure 4.8 is equal to 1/2, due to $n_v = 5$, $n_u = 4$, and $n_c = 3$. If VN v_5 and CN c_3 are removed from the protograph together with all of the edges connected to them, we obtain a protograph known as repeat jagged-accumulate (RJA) protograph. The base matrix for this protograph is

$$\boldsymbol{B} = \begin{bmatrix} 1 & 2 & 2 & 1 \\ 1 & 1 & 1 & 3 \end{bmatrix}. \tag{4.33}$$

Letting all VNs being transmitted ($n_v = n_u = 4$) and no VN being punctured ($n_p = 0$) the design rate of the corresponding ensemble is again $\mathrm{R} = 1/2$.

Unless explicitly stated, the set of transmitted VNs contains an information set of the unpunctured code, that is, of the length-Qn_v code before puncturing. The concept of information set has been introduced in Definition 2.23.

4.3 Historical Notes

Low-density parity-check codes were developed by R. Gallager in his doctoral dissertation completed in 1960. A revised version of this dissertation was published as a monograph in 1963 [1], while a short version of it had been published one year before [2]. With the exception of works, for example, by M. Taylor [3], V. Zyablov and M. Pinsker [4], and M. Tanner [5] (this latter work introducing the graphical description later named "Tanner graph"), LDPC codes did not receive a significant attention until the mid-nineties when, directly after the invention of turbo codes by C. Berrou and A. Glavieux [6], the scientific community started to realize their potential. Binary LDPC codes in [1] are regular, since each code in the ensemble is defined by a parity-check matrix having a constant Hamming weight d_v and d_c per column and per row, respectively. The ensemble is generated by dividing the parity-check matrix into d_v submatrices, where each submatrix is a block of m/d_v contiguous rows and where the Hamming weight of each column in each submatrix is exactly one. While the first submatrix has a fixed position of the 1 entry, for the other submatrices all possible column permutations of the first submatrix are considered. Nonbinary regular LDPC codes constructed on finite additive groups were also introduced in [1] and the corresponding ensemble was defined in the same way as for the binary case. As such, Gallager's regular ensembles differ from the unstructured regular ensembles described in this chapter in two aspects: Parallel edges between a VN and a CN are forbidden in Gallager's construction, while they are allowed in the unstructured case; Gallager's nonbinary codes were originally constructed on finite additive groups while the described unstructured codes are constructed on finite fields.

One of the main architects of LDPC codes rediscovery was D. MacKay, whose papers [7–9] published between 1995 and 1999, two of which co-authored by R. Neal, provided experimental evidence about the capability of Gallager's LDPC codes to perform close to turbo codes over binary symmetric channels and additive white Gaussian noise (AWGN) channels. The paper [9] also showed that sequences of Gallager's LDPC codes may be constructed with rate arbitrarily close to the Shannon limit and capable to achieve, over symmetric stationary ergodic noise, an arbitrarily small block error probability under a hard-decision typical set decoder while retaining parity-check matrix sparsity. In 1998 D. MacKay also presented, with M. Davey, the first construction of nonbinary LDPC codes on finite fields [10]. (More details on the code constructions and numerical results [10] are available in [11].) It is worth mentioning that, in parallel, the work of M. Sipser and D. Spielman on expander codes [12] also contributed significantly to the rediscovery of LDPC codes.

In parallel, a number of works investigating the use of irregular Tanner graphs provided a breakthrough in LDPC coding theory and practice. Early works in this area, by M. Luby *et al.*, date back to 1997 and 1998 [13, 14] (extended versions of these works, appearing in 2001, are [15, 16]). They investigated the asymptotic behavior of iterative erasure decoders and error hard-decision decoders with binary messages on irregular graphs, developed the first examples of degree distribution pair design, and provided experimental results illustrating the performance achieved by irregular LDPC codes. Fundamental steps forward were then represented by the discovery by A. Shokrollahi of capacity achieving sequences for the erasure channel [17], the development by T. Richardson *et al.* of density evolution analysis for soft- or hard-decision belief propagation LDPC decoding with exchange of soft messages [18, 19] and of efficient encoding strategies [20].

The IRA ensemble was introduced by H. Jin *et al.* in [21] as a generalization of the repeat-accumulate ensemble described in [22] and [23]. In its binary form, the LDPC code ensemble with separated variable nodes (SVN ensemble) appears in Section 5.3 of C. Di's Ph.D. Thesis [24]. There, an ensemble of irregular LDPC codes in which αn $(0 \leq \alpha \leq 1)$ degree-2 VNs exhibit the separation property was analyzed. The SVN ensemble described in this chapter is obtained by letting $\alpha = 1$ in the ensemble of [24], by generalizing the code construction to $\mathrm{GF}(q)$ for $q \geq 2$, and by letting some degree-3 VNs to be separated in case there are CNs that are not connected to any separated VN of degree 2. Interestingly, even before [24], the separation property of degree-2 VNs was proposed in [9], where the "Construction 2A" yields binary LDPC codes with $m/2$ separated degree-2 VNs and all the other VNs having degree 3. The nonbinary SVN ensemble, as presented in this chapter, was analyzed in [25] along with its weight distribution. LDPC codes constructed from protographs were introduced under this name in the 2003 Technical Report [26] by J. Thorpe. The idea of constructing an LDPC code through a copy and permute procedure appeared almost in parallel in the US patent [27] by T. Richardson and V. Novichkov, where the nomenclature "projective graph" was used instead of protograph. A related construction, multi-edge type LDPC codes, developed by T. Richardson and R. Urbanke, appeared in [28].

A very important topic in LDPC coding theory, not addressed by this chapter, is represented by algebraic constructions of LDPC codes, the most famous example of which is the construction based on finite geometries proposed by Kou *et al.* [29]. For a thorough treatment of algebraic LDPC code constructions we refer the reader to W. Ryan and S. Lin [30]. Finally, we point out that the rate concentration result for unstructured ensembles mentioned in Remark 4.2 appears in [31, Lemma 7].

4.4 Exercises

Exercise 4.1 Consider the edge-oriented degree distribution polynomials

$$\lambda(x) = \frac{2}{8} + \frac{6}{8}x \qquad\qquad \rho(x) = \frac{2}{8}x + \frac{6}{8}x^2 . \tag{4.34}$$

Determine the design rate R of the corresponding ensemble. Next, consider a binary linear block code with parity-check matrix

$$\boldsymbol{H} = \begin{bmatrix} 1 & 1 & 0 & 1 & 0 \\ 1 & 0 & 1 & 0 & 1 \\ 0 & 0 & 1 & 1 & 0 \end{bmatrix} . \tag{4.35}$$

Draw the code Tanner graph and establish if the code is a member of the unstructured ensemble specified by the above degree distribution pair.

Exercise 4.2 Consider a binary IRA ensemble with blocklength $n = 12$, node-oriented degree distribution of the type-$\mathtt{E}_1$ edges given by $\Psi(x) = x^3$, and check node degree $\mathsf{d}_c = 5$. Find the edge-perspective distribution $\psi(x)$, the ensemble design rate, and the number of check nodes in any Tanner graph in the ensemble. Write explicitly the parity-check matrix of one code belonging to such an IRA ensemble and such that the bits associated with the $k = n\mathrm{R}$ type-$\mathtt{V}_1$ VNs form an information set.

Exercise 4.3 Consider the binary IRA code corresponding to the parity-check matrix found in Exercise 4.2. Letting the type-$\mathtt{V}_1$ VNs and the type-$\mathtt{V}_2$ VNs be associated with the message bits and the redundant bits, respectively, find the codeword corresponding to the message $\boldsymbol{u} = \begin{bmatrix} 1 & 1 & \cdots & 1 \end{bmatrix}$, using the efficient systematic encoding procedure of IRA codes based on the $\boldsymbol{H}$ matrix, and therefore without using the generator matrix $\boldsymbol{G}$.

Exercise 4.4 Consider the binary linear block code $\mathcal{C}$ characterized by the parity-check matrix

$$\boldsymbol{H} = \begin{bmatrix} 0 & 1 & 0 & 0 & 1 & 0 & 1 & 1 \\ 1 & 0 & 0 & 0 & 0 & 1 & 1 & 1 \\ 0 & 0 & 1 & 1 & 0 & 1 & 1 & 0 \\ 0 & 0 & 1 & 1 & 1 & 0 & 0 & 1 \end{bmatrix}. \tag{4.36}$$

Derive the edge-oriented degree distribution polynomials, $\lambda(x)$ and $\rho(x)$ and draw the code Tanner graph. Next, assuming that $\boldsymbol{H}$ has been obtained by lifting a protograph base matrix $\boldsymbol{B}$, write the expression of $\boldsymbol{B}$ and determine the lifting factor Q.

Exercise 4.5 Consider the parity-check matrix

$$\boldsymbol{H} = \begin{bmatrix} 1 & 1 & 0 & 0 & 0 & 0 & 0 & 1 & 1 \\ 0 & 1 & 1 & 0 & 0 & 0 & 1 & 0 & 1 \\ 1 & 0 & 1 & 0 & 0 & 0 & 1 & 1 & 0 \\ 0 & 1 & 0 & 1 & 0 & 1 & 0 & 0 & 0 \\ 0 & 0 & 1 & 1 & 1 & 0 & 0 & 0 & 0 \\ 1 & 0 & 0 & 0 & 1 & 1 & 0 & 0 & 0 \end{bmatrix}. \tag{4.37}$$

Assume that the given code is a realization of a protograph ensemble. Write the expression of the base matrix $\boldsymbol{B}$ and find the lifting factor Q. Write the degree distribution polynomials $\lambda(x)$ and $\rho(x)$ corresponding to $\boldsymbol{H}$, both from an edge perspective and from a node perspective. Next, let A be the protograph ensemble defined by $\boldsymbol{H}$ and B be the unstructured ensemble defined by $\lambda(x)$ and $\rho(x)$. Is the claim $B \subset A$ true or false? Justify your answer.

Exercise 4.6 Consider a binary IRA LDPC code ensemble with $m = 10$ CNs. Write the structured submatrix of the parity-check matrix of any code in the ensemble, corresponding to the VNs of type $\mathtt{V}_2$. (The structure of this submatrix clarifies why IRA codes are sometimes called "LDPC staircase" codes.) Next, with reference to the systematic encoder, imagine to replace the binary accumulator of Figure 4.6(a) with the linear feedback shift register circuit depicted in Figure 4.9. What is the new submatrix of $\boldsymbol{H}$ corresponding to type-$\mathtt{V}_2$ VNs?

Exercise 4.7 Write a computer program that implements efficient systematic encoding of IRA codes. The program should read the parity-check matrix from a text file. Since the $m \times m$ submatrix of $\boldsymbol{H}$ associated with type-$\mathtt{V}_2$ VNs has a fixed structure, only the submatrix of $\boldsymbol{H}$ corresponding to VNs of type $\mathtt{V}_1$ can be stored in the file and read by the program; it is convenient to store only the nonzero entries of this submatrix

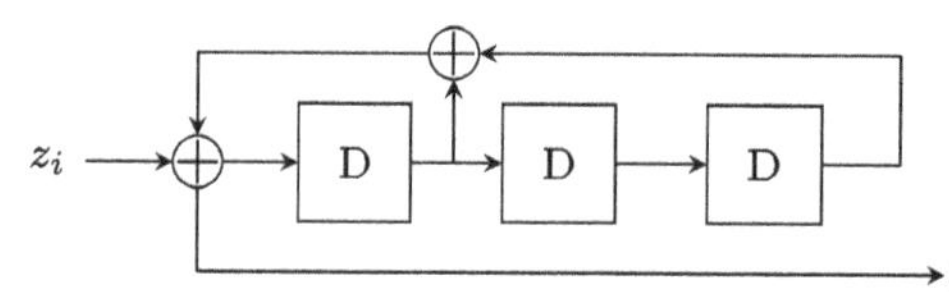

Figure 4.9 Linear feedback shift register circuit generalizing the binary accumulator.

since $\boldsymbol{H}$ is a low-density matrix. The program may, for example, encode a randomly generated message, check that the generated codeword fulfills (2.45), and output both the message and the codeword.

Exercise 4.8 Extend the computer program developed in Exercise 4.7 to the case where the accumulator in the efficient systematic IRA encoder is replaced by a linear feedback shift register circuit, an example of which is depicted in Figure 4.9. The specific structure of this circuit (meaning the bits that are fed back) should be accepted by the program as an input.

Exercise 4.9 Consider the following parity-check matrix, defining a linear block code $\mathcal{C} \subset \mathrm{GF}(2)^{12}$:

$$\boldsymbol{H} = \begin{bmatrix} 0 & 1 & 0 & 1 & 1 & 1 & 0 & 0 & 0 & 0 & 1 & 1 \\ 0 & 1 & 0 & 0 & 1 & 1 & 0 & 0 & 1 & 1 & 0 & 1 \\ 0 & 1 & 1 & 1 & 0 & 0 & 1 & 0 & 0 & 1 & 1 & 0 \\ 0 & 0 & 1 & 0 & 0 & 0 & 1 & 1 & 1 & 1 & 1 & 0 \\ 1 & 1 & 1 & 1 & 0 & 0 & 0 & 1 & 0 & 0 & 0 & 1 \\ 1 & 0 & 0 & 0 & 0 & 1 & 0 & 1 & 1 & 1 & 0 & 1 \end{bmatrix}. \tag{4.38}$$

Explain why $\mathcal{C}$ belongs to the ensemble $\mathscr{V}^{12}_{\psi,\mathrm{d}_c}(\nu_2)$ (separated VNs), find the value of the parameter ν_2 and the edge-perspective degree distribution polynomial $\psi(x)$ of type-$\mathtt{V}_1$ edges. Finally, verify that the actual rate of $\mathcal{C}$ equals the ensemble design rate R.

References

[1] R. Gallager, *Low-density parity-check codes.* Cambridge, MA, USA: MIT Press, 1963.

[2] R. Gallager, "Low-density parity-check codes," *IRE Trans. Inf. Theory*, vol. 8, no. 1, pp. 21–28, Jan. 1962.

[3] M. G. Taylor, "Reliable information storage in memories designed from unreliable components," *Bell Syst. Tech. J.*, vol. 47, no. 10, pp. 2299–2337, 1968.

[4] V. V. Zyablov and M. S. Pinsker, "Decoding complexity of low-density codes for transmission in a channel with erasures," *Probl. Peredachi Inf.*, vol. 10, no. 1, pp. 15–28, 1974.

[5] M. Tanner, "A recursive approach to low complexity codes," *IEEE Trans. Inf. Theory*, vol. 27, no. 5, pp. 533–547, Sep. 1981.

[6] C. Berrou, A. Glavieux, and P. Thitimajshima, "Near Shannon limit error-correcting coding and decoding: Turbo-codes," in *Proc. IEEE Int. Conf. Commun.*, Geneva, Switzerland, May 1993.

[7] D. MacKay and R. Neal, "Good codes based on very sparse matrices," in *Proc. 5th IMA Conference Cirencester*, ser. Lecture Notes in Computer Science, C. Boyd, Ed., vol. 1025. Springer Berlin Heidelberg, Dec. 1995, pp. 100–111.

[8] D. MacKay and R. Neal, "Near Shannon limit performance of low density parity check codes," *IET Electron. Lett.*, vol. 33, no. 6, pp. 457–458, Mar. 1997.

[9] D. MacKay, "Good error-correcting codes based on very sparse matrices," *IEEE Trans. Inf. Theory*, vol. 45, no. 2, pp. 399–431, Mar. 1999.

[10] M. Davey and D. MacKay, "Low density parity check codes over GF(q)," *IEEE Commun. Lett.*, vol. 2, no. 6, pp. 70–71, June 1998.

[11] M. Davey, "Error-correction using low-density parity-check codes," Ph.D. dissertation, University of Cambridge, Cambridge, UK, Dec. 1999.

[12] M. Sipser and D. Spielman, "Expander codes," *IEEE Trans. Inf. Theory*, vol. 42, no. 6, pp. 1710–1722, Nov. 1996.

[13] M. Luby, M. Mitzenmacher, A. Shokrollahi, D. Spielman, and V. Stemann, "Practical loss-resilient codes," in *Proc. 29th Symp. Theory Computing*, El Paso, TX, USA, May 1997, pp. 150–159.

[14] M. G. Luby, M. Mitzenmacher, M. A. Shokrollahi, and D. A. Spielman, "Analysis of low density codes and improved designs using irregular graphs," in *Proc. Annual ACM Symp. Theory Comput.*, Dallas, TX, USA, May 1998, pp. 249–258.

[15] M. Luby, M. Mitzenmacher, A. Shokrollahi, and D. Spielman, "Efficient erasure correcting codes," *IEEE Trans. Inf. Theory*, vol. 47, no. 2, pp. 569–584, Feb. 2001.

[16] M. Luby, M. Mitzenmacher, A. Shokrollahi, and D. Spielman, "Improved low-density parity-check codes using irregular graphs," *IEEE Trans. Inf. Theory*, vol. 47, no. 2, pp. 585–598, Feb. 2001.

[17] M. A. Shokrollahi, "New sequences of linear time erasure codes approaching the channel capacity," in *Proc. Int. Symp. Applied Algebra, Algebraic Algorithms and Error-Correcting Codes*, Honolulu, HI, USA, Nov. 1999, pp. 65–76.

[18] T. Richardson, A. Shokrollahi, and R. Urbanke, "Design of capacity-approaching irregular low-density parity-check codes," *IEEE Trans. Inf. Theory*, vol. 47, no. 2, pp. 619–637, Feb. 2001.

[19] T. Richardson and R. Urbanke, "The capacity of low-density parity-check codes under message-passing decoding," *IEEE Trans. Inf. Theory*, vol. 47, no. 2, pp. 599–618, Feb. 2001.

[20] T. Richardson and R. Urbanke, "Efficient encoding of low-density parity-check codes," *IEEE Trans. Inf. Theory*, vol. 47, no. 2, pp. 638–656, Feb. 2001.

[21] H. Jin, A. Khandekar, and R. McEliece, "Irregular repeat-accumulate codes," in *Proc. IEEE Int. Symp. Turbo Codes and Rel. Topics*, Brest, France, Sep. 2000, pp. 1–8.

[22] L. Ping, W. Leung, and N. Phamdo, "Low density parity check codes with semi-random parity check matrix," *Electronics Letters*, vol. 35, no. 1, p. 38–39, Jan. 1999.

[23] D. Divsalar, H. Jin, and R. J. McEliece, "Coding theorems for turbo-like codes," in *Proc. Ann. Allerton Conf. on Commun., Control, and Comput.*, Monticello, IL, USA, Sep. 1998, pp. 201–210.

[24] C. Di, "Asymptotic and finite-length analysis of low-density parity-check codes," Ph.D. dissertation, École Polytechnique Fédérale de Lausanne, Lausanne, Switzerland, 2004.

[25] G. Garrammone, E. Paolini, G. Liva, and B. Matuz, "Non-binary LDPC erasure codes with separated low-degree variable nodes," *IEEE Trans. Commun.*, vol. 63, no. 11, pp. 3937–3949, Nov. 2015.

[26] J. Thorpe, "Low-density parity-check (LDPC) codes constructed from protographs," NASA JPL, Pasadena, CA, USA, IPN Progress Report 42-154, Aug. 2003.

[27] T. Richardson and V. Novichkov, "Methods and apparatus for decoding LDPC codes," Oct. 14, 2003, US Patent 6,633,856.

[28] T. Richardson and R. Urbanke, "Multi-edge type LDPC codes," Presented at the Workshop honoring Prof. Bob McEliece on his 60th birthday, Pasadena, CA, USA, May 2002.

[29] Y. Kou, S. Lin, and M. Fossorier, "Low-density parity-check codes based on finite geometries: A rediscovery and new results," *IEEE Trans. Inf. Theory*, vol. 47, no. 7, pp. 2711–2736, Nov. 2001.

[30] W. Ryan, S. Lin, and S. Wilson, *Channel codes - Classical and modern.* New York, NY, USA: Cambridge University Press, 2024.

[31] C. Measson, A. Montanari, and R. Urbanke, "Maxwell construction: The hidden bridge between iterative and maximum a posteriori decoding," *IEEE Trans. Inf. Theory*, vol. 54, no. 12, pp. 5277–5307, Dec. 2008.

5 Weight Distribution and Minimum Distance of LDPC Codes

This chapter addresses structural properties of LDPC codes, focusing in particular on their weight distributions and minimum distance. Given an LDPC code ensemble and assuming to draw randomly a code of length n from it, the goal is to express the expectation of the random variable representing the number of codewords of some weight $0 \le w \le n$ of the drawn code. To this aim, we follow an approach based on the concept of assignments in Tanner graphs and on generating functions. Asymptotic exponents of the weight distributions are then addressed. This chapter delves into substantial mathematical details. Readers not interested in these more advanced topics can skip this chapter and proceed directly to Chapter 6 without losing the overall context of the book.

5.1 Assignments in Tanner Graphs

We start by formally introducing the concepts of CN socket assignment and of VN assignment, useful to give a graphical interpretation to the syndrome checking operation $\boldsymbol{s} = \boldsymbol{y}\,\boldsymbol{H}^{\mathsf{T}}$.

Definition 5.1 (CN socket assignment) A CN socket assignment is a subset of the CN sockets along with a symbol in $\mathrm{GF}(q) \setminus \{0\}$ associated with each such socket. It is said to have weight l if it comprises l sockets.

Definition 5.2 (VN assignment) A VN assignment is a subset of the VNs along with a symbol in $\mathrm{GF}(q) \setminus \{0\}$ associated with each such VN. It is said to have weight w if it comprises w VNs.

Definition 5.3 (Active edges) All edges connected to a VN in a VN assignment are called active edges.

A CN socket assignment is said to be *valid* if the following condition holds: Assuming that each CN socket not in the assignment is associated with $0 \in \mathrm{GF}(q)$, all linear constraints imposed by the CNs are satisfied. In the particular case of a binary linear block code, a CN socket assignment is valid whenever for each CN, either the subset of CN sockets belonging to the assignment has even cardinality or no CN socket belongs to it.

Importantly, every VN assignment induces a CN socket assignment, as explained next. The symbol in $\mathrm{GF}(q) \setminus \{0\}$ associated with the generic VN in the VN assignment is propagated toward its neighboring CNs along the edges connected to the VN (all of which are active edges). In each such edge, the propagated symbol is multiplied by the edge label, generating a symbol in $\mathrm{GF}(q) \setminus \{0\}$ that reaches the CN socket to which the edge is connected. The set of all such CN sockets, along with their so-generated nonzero symbols, represents the CN socket assignment induced by the VN assignment.

Given an LDPC code represented by a sparse parity-check matrix $\boldsymbol{H}$ and considering its corresponding Tanner graph, there clearly exists a bijective mapping between its nonzero codewords and the VN assignments in the Tanner graph, whose induced CN socket assignment is valid. Given a codeword $\boldsymbol{c}$, the corresponding VN assignment is the subset of the VNs that correspond to the nonzero symbols in $\boldsymbol{c}$. Moreover, each such VN is associated with the corresponding symbol of $\boldsymbol{c}$. A zero syndrome is obtained if and only if the CN socket assignment induced by this VN assignment is valid.

5.2 Basics of Generating Functions

Generating functions play an important role in the solution of enumeration problems. In this section, we review some basic results that are especially useful in dealing with weight enumerators.

Definition 5.4 (Univariate generating function) Consider a sequence $\{A_w\}_{w \geq 0}$ of complex or real numbers. The generating function of this sequence is the formal power series $A(W) = \sum_{w=0}^{\infty} A_w W^w$ in the indeterminate W. We use the notation

$$A_w = \mathrm{coeff}(A(W), W^w) \tag{5.1}$$

and we refer to A_w as the coefficient of W^w in $A(W)$.

Definition 5.5 (Multivariate generating function) Consider a d-dimensional sequence $\{A_{w_1,w_2,\ldots,w_d}\}_{w_1 \geq 0, w_2 \geq 0, \ldots, w_d \geq 0}$ of complex or real numbers. The multivariate generating function of this sequence is the formal multivariate power series

$$A(W_1, W_2, \ldots, W_d) = \sum_{w_1=0}^{\infty} \sum_{w_2=0}^{\infty} \cdots \sum_{w_d=0}^{\infty} A_{w_1,w_2,\ldots,w_d} W_1^{w_1} W_2^{w_2} \cdots W_d^{w_d} \tag{5.2}$$

in the indeterminates $W_1, W_2, \ldots, W_d$. Letting $\boldsymbol{W} = \begin{bmatrix} W_1 & W_2 & \ldots & W_d \end{bmatrix}$ and $\boldsymbol{w} = \begin{bmatrix} w_1 & w_2 & \ldots & w_d \end{bmatrix}$, we also use the compact notation $A(\boldsymbol{W}) = \sum_{\boldsymbol{w}} A_{\boldsymbol{w}} \boldsymbol{W}^{\boldsymbol{w}}$ where the sum is over all $\boldsymbol{w}$ such that $w_i \geq 0$ for all $i \in \{1,2,\ldots,d\}$. Similarly to the univariate case, we use the notation

$$A_{\boldsymbol{w}} = \mathrm{coeff}(A(\boldsymbol{W}), \boldsymbol{W}^{\boldsymbol{w}}) \tag{5.3}$$

and we refer to $A_{\boldsymbol{w}}$ as the coefficient of $\boldsymbol{W}^{\boldsymbol{w}}$ in $A(\boldsymbol{W})$.

Consider two univariate generating functions $A_1(W) = \sum_{w=0}^{\infty} A_{1,w} W^w$ and $A_2(W) = \sum_{w=0}^{\infty} A_{2,w} W^w$ in the same indeterminate W and take their product

$A_1(W)A_2(W)$. As it is immediate to recognize, this product is itself a generating function in the indeterminate W, and for all integers $w \geq 0$, we have

$$\mathrm{coeff}(A_1(W)A_2(W), W^w) = \sum_{\boldsymbol{w}} \mathrm{coeff}(A_1(W), W^{w_1})\mathrm{coeff}(A_2(W), W^{w_2}), \tag{5.4}$$

where the sum is over all $\boldsymbol{w} = \begin{bmatrix} w_1 & w_2 \end{bmatrix}$ such that $w_1 \geq 0$, $w_2 \geq 0$, and $w_1 + w_2 = w$. The previous result still holds in the more general case of $n \geq 2$ generating functions $A_1(W)$, $A_2(W)$, ..., $A_n(W)$. This is stated by the following lemma.

Lemma 5.1 *Let $A_1(W)$, $A_2(W)$, ..., $A_m(W)$ be univariate generating functions. Moreover, let $\boldsymbol{w} = \begin{bmatrix} w_1 & w_2 & \dots & w_m \end{bmatrix}$. For all integers $w \geq 0$, the coefficient of W^w in the generating function $\prod_{t=1}^m A_t(W)$ may be expressed as*

$$\mathrm{coeff}\left(\prod_{t=1}^{m} A_t(W), W^w\right) = \sum_{\boldsymbol{w}} \prod_{t=1}^{m} \mathrm{coeff}(A_t(W), W^{w_t}), \tag{5.5}$$

where the sum is over all m-tuples $\boldsymbol{w}$ such that $w_t \geq 0$ for all $t \in \{1,2,\dots,m\}$ and such that $\sum_{t=1}^m w_t = w$.

Next, we introduce an asymptotic result which, as we will se in Section 5.6, is particularly useful for the development of the *weight spectral shape* of LDPC code ensemble sequences. This result, expressed by Lemma 5.2, makes use of the notion of exponential equivalence, that is formally defined as follows.

Definition 5.6 (Exponential equivalence) Let a_n and b_n be two real-valued sequences, where $b_n \neq 0$ for all n. We say that a_n is *exponentially equivalent* to b_n as $n \to \infty$, writing $a_n \doteq b_n$, if and only if

$$\lim_{n\to\infty} \frac{1}{n} \ln \frac{a_n}{b_n} = 0. \tag{5.6}$$

Lemma 5.2 *Let $\mathbf{W} = \begin{bmatrix} W_1 & W_2 & \dots & W_d \end{bmatrix}$ and $A(\mathbf{W})$ be a multivariate generating function expressed by (5.2) such that $A_{\boldsymbol{w}} \geq 0$ for all $\boldsymbol{w}$ and $A(\mathbf{0}) > 0$. Let n be a positive integer and $\boldsymbol{\alpha} = \begin{bmatrix} \alpha_1 & \alpha_2 & \dots & \alpha_d \end{bmatrix}$ be such that $\alpha_t n$ is a positive integer for all $t \in \{1,2,\dots,d\}$. We have*

$$\mathrm{coeff}((A(\mathbf{W}))^n, \mathbf{W}^{n\boldsymbol{\alpha}}) \doteq \exp\left\{ n \ln\left(\frac{A(\mathbf{z}_0)}{\prod_{t=1}^d z_{0,t}^{\alpha_t}} \right) \right\}, \tag{5.7}$$

where $\mathbf{z}_0 = \begin{bmatrix} z_{0,1} & z_{0,2} & \dots, z_{0,d} \end{bmatrix}$ and $z_{0,1}, z_{0,2}, \dots, z_{0,d}$ are the unique positive solutions to the d simultaneous equations

$$z_{0,t} \left.\frac{\partial \ln A}{\partial W_t}\right|_{\mathbf{z}_0} = \frac{z_{0,t} \frac{\partial A}{\partial W_t}(\mathbf{z}_0)}{A(\mathbf{z}_0)} = \alpha_t \qquad \forall t \in \{1,2,\dots,d\}. \tag{5.8}$$

5.3 Weight Enumerating Functions of Replicated Codes

Let $\mathcal{C}(n,k)$ be a linear block code over GF(q) with length n, dimension k, and weight enumerating function $A(W) = 1 + \sum_{t=1}^{n} A_t W^t$. Moreover, let $\boldsymbol{H}$ be an $m \times n$ parity-check matrix for $\mathcal{C}(n,k)$. We call replicated code with base code $\mathcal{C}(n,k)$ and replica factor Q, denoting it by $\mathcal{C}_{\text{com}}(nQ,kQ)$, the linear block code of length nQ and dimension kQ defined by the $mQ \times nQ$ parity-check matrix

$$\boldsymbol{H}_{\text{com}} = \begin{bmatrix} \boldsymbol{H} & \boldsymbol{0} & \boldsymbol{0} & \dots & \boldsymbol{0} \\ \boldsymbol{0} & \boldsymbol{H} & \boldsymbol{0} & \dots & \boldsymbol{0} \\ \boldsymbol{0} & \boldsymbol{0} & \boldsymbol{H} & \dots & \boldsymbol{0} \\ \vdots & \vdots & \vdots & \ddots & \vdots \\ \boldsymbol{0} & \boldsymbol{0} & \boldsymbol{0} & \dots & \boldsymbol{H} \end{bmatrix}. \tag{5.9}$$

Note that every codeword of the replicated code $\mathcal{C}_{\text{com}}(nQ,kQ)$ is a concatenation of Q codewords of $\mathcal{C}(n,k)$, which implies that the replicated code and its base code have the same minimum distance. If we represent the linear block code $\mathcal{C}(n,k)$ as a square with n sockets (similar to the CNs of the Tanner graphs depicted in Figures 4.1 and 4.2), then the replicated code $\mathcal{C}_{\text{com}}(nQ,kQ)$ may be graphically represented as a collection of Q such squares. This is illustrated in Figure 5.1 assuming $n = 7$ and $Q = 5$.

The weight enumerating function of the replicated code $\mathcal{C}_{\text{com}}(nQ,kQ)$, denoted by $A_{\text{com}}(W)$, depends on the one of its base codes $\mathcal{C}(n,k)$. It is expressed by the following as well as the subsequent lemma.

Lemma 5.3 (Weight distribution of replicated codes – First formulation) *Let $\mathcal{C}_{\text{com}}(nQ,kQ)$ be a replicated code with base code $\mathcal{C}(n,k)$ and replica factor Q. Let $A(W)$ be the weight distribution function of the base code. Then, the weight enumerating function of $\mathcal{C}_{\text{com}}(nQ,kQ)$ is $A_{\text{com}}(W) = (A(W))^Q$, and therefore the multiplicity of replicated codewords of weight w may be expressed as*

$$A_w^{\text{com}} = \text{coeff}\Big((A(W))^Q, W^w\Big). \tag{5.10}$$

Proof This is a simple consequence of Lemma 5.1. Let us recall that every codeword of the replicated code is a concatenation of Q codewords of its base code. For $t \in \{1,2,\dots,Q\}$, let $0 \le w_t \le n$ be the weight of the tth such base codeword and let $\boldsymbol{w} = \begin{bmatrix} w_1 & w_2 & \dots & w_Q \end{bmatrix}$. The number of codewords of the replicated code that

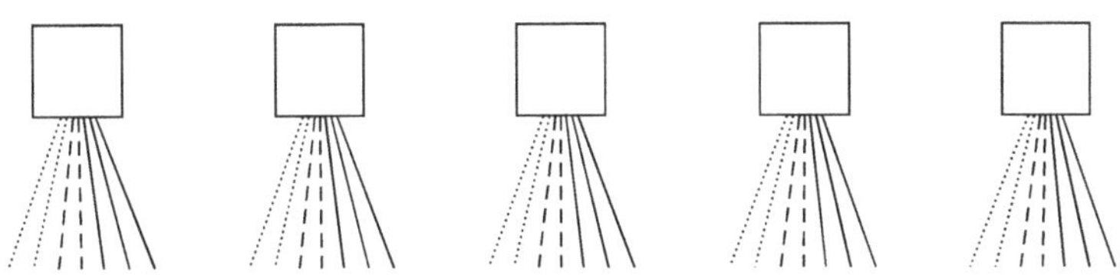

Figure 5.1 A replicated code of length $nQ = 35$ obtained from a base code of length $n = 7$ and replica factor $Q = 5$.

correspond to a given $\boldsymbol{w}$ is $\prod_{t=1}^{Q} \text{coeff}(A(W), W^{w_t})$, so the multiplicity of codewords of weight w of the replicated code is

$$A_w^{\text{com}} = \sum_{\boldsymbol{w}} \prod_{t=1}^{Q} \text{coeff}(A(W), W^{w_t}), \tag{5.11}$$

where the sum is over all $\boldsymbol{w}$ such that $\sum_{t=1}^{Q} w_t = w$. Equation (5.10) now directly follows from (5.5). □

Example 5.1 With reference to Figure 5.1, assume that $k = 4$ and that the base code $\mathcal{C}(7,4)$ is the binary Hamming code of length 7. The minimum distance of the replicated code is 3 and, due to Lemma 5.3, we have $A_{\text{com}}(W) = \left(1 + 7\,W^3 + 7\,W^4 + W^7\right)^5 = 1 + 35\,W^3 + 35\,W^4 + 490\,W^6 + 985\,W^7 + 490\,W^8 + 3430\,W^9 + 10430\,W^{10} + \ldots + W^{35}$.

Lemma 5.4 (Weight distribution of replicated codes – Second formulation) *Let $\mathcal{C}_{\text{com}}(nQ,kQ)$ be a replicated code with base code $\mathcal{C}(n,k)$ and replica factor Q. Let D be the minimum distance of the base code. Let $\boldsymbol{d} = \begin{bmatrix} d_0 & d_D & d_{D+1} & \ldots & d_n \end{bmatrix}$ be a vector whose tth element d_t is a nonnegative integer representing the number of times a base codeword of weight t, $t \in \{0, D, D+1, \ldots, n\}$, is present in a codeword of $\mathcal{C}_{\text{com}}(nQ,kQ)$. This means that, to form the codeword of $\mathcal{C}_{\text{com}}(nQ,kQ)$, d_t base codewords of Hamming weight t are used. Let A_t be the number of base codewords of weight t. Then, the multiplicity of replicated codewords of weight w may be expressed as*

$$A_w^{\text{com}} = \sum_{\boldsymbol{d} \in \mathcal{N}_w} \binom{Q}{d_0, d_D, d_{D+1}, \ldots, d_n} \prod_{i=D}^{n} A_t^{d_t}, \tag{5.12}$$

where $\mathcal{N}_w$ is the set of all vectors $\boldsymbol{d}$ that are solutions of the system of simultaneous Diophantine equations

$$d_0 + \sum_{t=D}^{n} d_t = Q \tag{5.13}$$

$$\sum_{t=D}^{n} t\, d_t = w\,. \tag{5.14}$$

Proof For a given $\boldsymbol{d}$, the multinomial coefficient in (5.12) represents the number of ways in which d_0 weights equal to 0, d_D weights equal to D, d_{D+1} weights equal to $D+1$, up to d_n weights equal to n may be assigned to the Q concatenated base codewords. This number must be multiplied by the number of possible choices of base codewords of weight t, for all $t \in \{D, D+1, \ldots, n\}$. The multiplicity A_w^{com} is the sum of the obtained number over all vectors $\boldsymbol{d}$ fulfilling the two constraints. The first constraint on $\boldsymbol{d}$ follows from the fact that the sum of its elements represents the number of base codewords whose concatenation forms the replicated codeword. This number must be equal to the replica factor Q. The second constraint is due to the fact that the sum of the weights of each concatenated base codeword must equal the weight w of the replicated codeword; overall, d_t base codewords of weight t are present. □

For reasons that will be clear when addressing the average weight distribution of protograph LDPC code ensembles, in addition to its ordinary weight enumerating function, it is useful to define a *vector* weight enumerating function of a replicated code. To this purpose, we start by introducing this concept for a generic linear block code.

Definition 5.7 (Vector weight enumerating function) Consider a linear block code $\mathcal{C}(n,k)$, let $\boldsymbol{c} = \begin{bmatrix} c_1 & c_2 & \dots & c_n \end{bmatrix}$ be a generic codeword of $\mathcal{C}(n,k)$, and let $I = \{1,2,\dots,n\}$. Let $I_1, I_2, \dots, I_T$ be a partition of the set I into $1 \le T \le n$ nonempty subsets. For all $t \in \{1,2,\dots,T\}$ let $\mathbf{c}_t$ be the vector containing all codeword symbols c_i such that $i \in I_t$, and let $0 \le w_t \le |I_t|$ be the Hamming weight of $\boldsymbol{c}_t$. Then, the vector weight of the codeword $\boldsymbol{c} \in \mathcal{C}(n,k)$, associated with the partition I_1, I_2, $\dots$, I_T, is defined as the vector $\boldsymbol{w} = \begin{bmatrix} w_1 & w_2 & \dots & w_T \end{bmatrix}$. Moreover, letting $\boldsymbol{W} = \begin{bmatrix} W_1 & W_2 & \dots & W_T \end{bmatrix}$ and $A_{\boldsymbol{w}}$ be the multiplicity of codewords $\boldsymbol{c} \in \mathcal{C}(n,k)$ of vector weight $\boldsymbol{w}$, the vector weight-enumerating function of $\mathcal{C}(n,k)$ is defined as

$$A(\boldsymbol{W}) = \sum_{\boldsymbol{w}} A_{\boldsymbol{w}} \boldsymbol{W}^{\boldsymbol{w}} . \tag{5.15}$$

Importantly, as explicitly remarked in its definition, the vector weight enumerating function (5.15) depends on the considered partition of the set of coordinates $I = \{1,2,\dots,n\}$. As a special case, for $T = 1$ (no partition) the vector weight enumerating function coincides with the ordinary weight enumerating function. As another special case, for $T = n$ (meaning $I_1 = \{1\}, I_2 = \{2\}, \dots, I_n = \{n\}$) we have $w_t \in \{0,1\}$ for all $t \in \{1,2,\dots,n\}$, and each vector $\boldsymbol{w} = \begin{bmatrix} w_1 & w_2 & \dots & w_n \end{bmatrix}$ is bijectively associated with a codeword support (equivalently, with the subset of all codewords sharing that support).

In analogy with Lemma 5.3, we have the following result.

Lemma 5.5 (Vector weight distribution of replicated codes) *Let $\mathcal{C}_{\mathsf{com}}(nQ,kQ)$ be a replicated code with with base code $\mathcal{C}(n,k)$ and replica factor Q. Let $A(\boldsymbol{W})$ be the weight enumerating function of the base code. Then, the vector weight enumerating function of $\mathcal{C}_{\mathsf{com}}(nQ,kQ)$ is $A_{\mathsf{com}}(\boldsymbol{W}) = (A(\boldsymbol{W}))^Q$, and therefore the multiplicity of replicated codewords of weight w may be expressed as*

$$A_{\boldsymbol{w}}^{\mathsf{com}} = \mathrm{coeff}\Big((A(\boldsymbol{W}))^Q, \boldsymbol{W}^{\boldsymbol{w}}\Big) . \tag{5.16}$$

Example 5.2 Assume again that for the replicated code in Figure 5.1, $k = 4$ and the base code $\mathcal{C}(7,4)$ is the binary Hamming code of length 7 represented by the parity-check matrix (2.64). Let $T = 3$ and $I_1 = \{1,2\}$ (corresponding to the dotted edges), $I_2 = \{3,4\}$ (dashed edges), and $I_3 = \{5,6,7\}$ (solid edges). Letting $\boldsymbol{W} = \begin{bmatrix} W_1 & W_2 & W_3 \end{bmatrix}$, it is easy to verify that the vector weight enumerating function of the base code is

$$A(\boldsymbol{W}) = 1 + 2W_1W_2W_3 + 2W_1W_3^2 + W_2W_3^3 + W_2W_3^2 + W_1^2W_3^2 + 2W_1W_2W_3^2 \\ + W_2^2W_3 + W_1^2W_2W_3 + W_1^2W_2 + 2W_1W_2^2W_3 + W_1^2W_2^2W_3^3 . \quad (5.17)$$

Due to Lemma 5.5, we also have

$$A_{\mathsf{com}}(\boldsymbol{W}) = (A(\boldsymbol{W}))^5 \\ = 1 + 5W_1^2W_2 + 10W_1^4W_2^2 + 10W_1^6W_2^3 + 5W_1^8W_2^4 + W_1^{10}W_2^5 \\ + 10W_1W_2W_3 + 5W_1^2W_2W_3 + 5W_2^2W_3 + \cdots + W_1^{10}W_2^{10}W_3^{15} . \quad (5.18)$$

Example 5.3 Let the base code of a replicated code be a (7,6) linear block code over GF(4) with parity-check matrix $\boldsymbol{H} = \begin{bmatrix} 1 & 1 & 1 & 1 & 1 & 1 & 1 \end{bmatrix}$. Let $T = 7$ and $\boldsymbol{W} = \begin{bmatrix} W_1 & W_2 & \ldots & W_7 \end{bmatrix}$. The vector weight enumerating function of the code relevant to the partition $I_1 = \{1\}, I_2 = \{2\}, \ldots, I_7 = \{7\}$ is a multivariate polynomial in the indeterminate $\boldsymbol{W}$ with 121 terms whose initial and final terms are

$$A(\boldsymbol{W}) = 1 + 3W_1W_2 + 3W_1W_3 + 3W_2W_3 + 6W_1W_2W_3 + 3W_1W_4 + 3W_2W_4 \\ + 6W_1W_2W_4 + 3W_3W_4 + 6W_1W_3W_4 + 6W_2W_3W_4 + \cdots \\ + 183W_1W_3W_4W_5W_6W_7 + 183W_2W_3W_4W_5W_6W_7 \\ + 546W_1W_2W_3W_4W_5W_6W_7 . \quad (5.19)$$

For a replica factor $Q = 3$, applying Lemma 5.5, we find that the vector weight enumerating function of the replicated code, $A_{\mathsf{com}}(\boldsymbol{W}) = (A(\boldsymbol{W}))^3$, is a multivariate polynomial in the indeterminate $\boldsymbol{W}$ with 16132 terms whose initial and final terms are

$$A_{\mathsf{com}}(\boldsymbol{W}) = 1 + 9W_1W_2 + 27W_1^2W_2^2 + 27W_1^3W_2^3 + 9W_1W_3 + 9W_2W_3 + \cdots \\ + 54854982\, W_1W_2^3W_3^3W_4^3W_5^3W_6^3W_7^3 \\ + 163665684\, W_1^2W_2^3W_3^3W_4^3W_5^3W_6^3W_7^3 \\ + 162771336\, W_1^3W_2^3W_3^3W_4^3W_5^3W_6^3W_7^3 . \quad (5.20)$$

5.4 Weight Distributions of Unstructured LDPC Code Ensembles

In this section, weight distributions of unstructured LDPC code ensembles are discussed, starting with regular ensembles, and then generalizing the results to irregular ones.

5.4.1 Regular Ensembles

A convenient way to obtain an expression for the expected number $\bar{A}_w$ of codewords of weight w of a random LDPC code whose realizations belong to the unstructured regular LDPC code ensemble $\mathscr{C}^{n,q}_{\mathsf{d}_v,\mathsf{d}_c}$, is by recalling the existence of a bijective mapping between codewords and VN assignments whose induced CN socket assignment is valid. For every integer $0 < w \leq n$, all VN assignments of weight w have the same probability to induce a valid CN socket assignment when drawing the $E \times E$ permutation matrix $\mathbf{\Pi}$ and the E nonzero edge labels. This is because both $\mathbf{\Pi}$ and the edge labels are drawn according to a uniform probability measure as described in Section 4.2.1.

Since the total number of VN assignments of weight w is equal to $\binom{n}{w}(q-1)^w$, we may write

$$\bar{A}_w = \binom{n}{w}(q-1)^w P(\text{valid}|w), \tag{5.21}$$

where "valid$|w$" is the conditional event that, upon drawing the permutation matrix $\mathbf{\Pi}$ and the edge labels, the CN socket assignment induced by the generic VN assignment of weight w is valid. Since all valid CN socket assignments have the same probability to be induced by a VN assignment, we may further develop (5.21) as

$$\bar{A}_w = \binom{n}{w}(q-1)^w \frac{C_w^{\text{valid}}}{\binom{n\mathsf{d}_v}{w\mathsf{d}_v}(q-1)^{w\mathsf{d}_v}}, \tag{5.22}$$

where we have denoted by C_w^{valid} the total number of valid CN socket assignments that are induced by a VN assignment of weight w, and where $\binom{n\mathsf{d}_v}{w\mathsf{d}_v}(q-1)^{w\mathsf{d}_v}$ is the total number of possible induced CN socket assignments. The number C_w^{valid} is clearly equal to the number of codewords of weight $w\mathsf{d}_v$ of the replicated code whose base code is the $(\mathsf{d}_c, \mathsf{d}_c - 1)$ linear block code over GF(q) with parity-check matrix $\boldsymbol{H} = \begin{bmatrix} 1 & 1 & 1 & \cdots & 1 \end{bmatrix}$ and with replica factor m. Due to Lemma 5.3 and recalling (2.61), we have

$$\bar{A}_w = \binom{n}{w} \frac{\text{coeff}\left(\left(P_{q,\mathsf{d}_c}(Z)\right)^m, Z^{w\mathsf{d}_v}\right)}{\binom{n\mathsf{d}_v}{w\mathsf{d}_v}(q-1)^{w(\mathsf{d}_v-1)}}. \tag{5.23}$$

In the special case of a binary unstructured LDPC code ensemble $\mathscr{C}^{n}_{\mathsf{d}_v,\mathsf{d}_c}$, the expression (5.23) becomes

$$\bar{A}_w = \binom{n}{w} \frac{\text{coeff}\left((P_{2,\mathsf{d}_c}(Z))^m, Z^{w\mathsf{d}_v}\right)}{\binom{n\mathsf{d}_v}{w\mathsf{d}_v}}, \tag{5.24}$$

where $P_{2,\mathsf{d}_c}(Z)$ is given by (2.60).

5.4.2 Irregular Ensembles

The next step consists of extending the expression (5.23) to an irregular ensemble $\mathscr{C}^{n,q}_{\lambda,\rho}$. To achieve this goal we start by letting $\boldsymbol{w} = \begin{bmatrix} w_2 & w_3 & \ldots & w_{\mathsf{d}_v} \end{bmatrix}$ be the vector weight of a length-n sequence with elements in GF(q). The generic coordinate w_j of $\boldsymbol{w}$ is an integer representing the portion of the total weight w of the length-n sequence that is apportioned to VNs of degree $2 \leq j \leq \mathsf{d}_v$. Therefore we have

$0 \leq w_j \leq \Lambda_j n$ for all $j \in \{2,3,\ldots,\mathsf{d}_v\}$. A simple way to develop an expression of the expected number of codewords of weight w of an LDPC code drawn in $\mathscr{C}_{\lambda,\rho}^{n,q}$ consists of approaching the problem similarly to the regular case, distinguishing however the different contributions of different vector weights $\boldsymbol{w}$.

Let $\mathcal{W}_w$ be the set of all vector weights $\boldsymbol{w} = \begin{bmatrix} w_2 & w_3 & \ldots & w_{\mathsf{d}_v} \end{bmatrix}$ such that $0 \leq w_j \leq \Lambda_j n$ for all $j \in \{2,3,\ldots,\mathsf{d}_v\}$ and such that $\sum_{j=2}^{\mathsf{d}_v} w_j = w$. The number of VN assignments of weight w corresponding to a vector weight $\boldsymbol{w} \in \mathcal{W}_w$ is given by $\binom{\Lambda_2 n}{w_2}(q-1)^{w_2} \cdots \binom{\Lambda_{\mathsf{d}_v} n}{w_{\mathsf{d}_v}}(q-1)^{w_{\mathsf{d}_v}}$. Thus we may write

$$\begin{aligned}\bar{A}_w &= \sum_{\boldsymbol{w} \in \mathcal{W}_w} \binom{\Lambda_2 n}{w_2}(q-1)^{w_2} \cdots \binom{\Lambda_{\mathsf{d}_v} n}{w_{\mathsf{d}_v}}(q-1)^{w_{\mathsf{d}_v}} P(\text{valid}|\boldsymbol{w}) \\ &= (q-1)^w \sum_{\boldsymbol{w} \in \mathcal{W}_w} P(\text{valid}|\boldsymbol{w}) \prod_{j=2}^{\mathsf{d}_v} \binom{\Lambda_j n}{w_j},\end{aligned} \tag{5.25}$$

where "valid$|\boldsymbol{w}$" is the conditional event that, upon drawing the permutation matrix $\boldsymbol{\Pi}$ and the edge labels, the CN socket assignment induced by the generic VN assignment of vector weight $\boldsymbol{w}$ is valid. Letting $C_{\boldsymbol{w}}^{\text{valid}}$ be the number of valid CN socket assignments induced by a VN assignment of vector weight $\boldsymbol{w}$, the expression of $\bar{A}_w$ in (5.25) may be developed by expressing $P(\text{valid}|\boldsymbol{w})$ as the ratio of $C_{\boldsymbol{w}}^{\text{valid}}$ to the total number of CN socket assignments induced by a VN assignment of vector weight $\boldsymbol{w}$, yielding

$$\bar{A}_w = (q-1)^w \sum_{\boldsymbol{w} \in \mathcal{W}_w} \frac{C_{\boldsymbol{w}}^{\text{valid}}}{\binom{\sum_{j=2}^{\mathsf{d}_v} \Lambda_j j n}{\sum_{j=2}^{\mathsf{d}_v} w_j j}(q-1)^{\sum_{j=2}^{\mathsf{d}_v} w_j j}} \prod_{j=2}^{\mathsf{d}_v} \binom{\Lambda_j n}{w_j}. \tag{5.26}$$

The final step consists of finding an expression for $C_{\boldsymbol{w}}^{\text{valid}}$, which may be accomplished in the following way. Let $\boldsymbol{l} = \begin{bmatrix} l_2 & l_3 & \ldots & l_{\mathsf{d}_c} \end{bmatrix}$ be a vector weight for the CN socket assignments. The ith coordinate $0 \leq l_i \leq P_i m$ of $\boldsymbol{l}$ is an integer denoting the portion of the total CN socket assignment weight that is apportioned to the CNs of degree $2 \leq i \leq \mathsf{d}_c$. Since the weight of a CN socket assignment induced by a VN assignment of vector weight $\boldsymbol{w}$ is $\sum_{j=2}^{\mathsf{d}_v} w_j j$ we have $\sum_{i=2}^{\mathsf{d}_c} l_i = \sum_{j=2}^{\mathsf{d}_v} w_j j$. Next, let $C_{l_i}^{\text{valid}}$ be the number of valid CN socket assignments of weight l_i over the $P_i m$ CNs of degree i. We have

$$C_{\boldsymbol{w}}^{\text{valid}} = \sum_{\boldsymbol{l}} C_{l_2}^{\text{valid}} C_{l_3}^{\text{valid}} \cdots C_{l_{\mathsf{d}_c}}^{\text{valid}}, \tag{5.27}$$

where the sum is over all $\boldsymbol{l}$ such that $0 \leq l_i \leq P_i m$ for all $i \in \{2,3,\ldots,\mathsf{d}_c\}$ and such that $\sum_{i=2}^{\mathsf{d}_c} l_i = \sum_{j=2}^{\mathsf{d}_v} w_j j$. For all $i \in \{2,3,\ldots,\mathsf{d}_c\}$, the number $C_{l_i}^{\text{valid}}$ is equal to the number of codewords of weight l_i of a $(P_i\, m i, P_i\, m\,(i-1))$ replicated code whose base code is a $(i,(i-1))$ linear block code over GF(q) with parity-check matrix $\boldsymbol{H} = \begin{bmatrix} 1 & 1 & 1 & \cdots & 1 \end{bmatrix}$ and whose replica factor is $P_i m$. Hence we obtain

$$C_{\boldsymbol{w}}^{\text{valid}} = \sum_{\boldsymbol{l}} \prod_{i=2}^{\mathsf{d}_c} \operatorname{coeff}\left(\left(P_{q,i}(Z)\right)^{P_i m}, Z^{l_i}\right) \tag{5.28}$$

$$= \operatorname{coeff}\left(\prod_{i=2}^{\mathsf{d}_c} \left(P_{q,i}(Z)\right)^{P_i m}, Z^{\sum_{j=2}^{\mathsf{d}_v} w_j j}\right), \tag{5.29}$$

where (5.28) and (5.29) follow from Lemma 5.3 and Lemma 5.1, respectively. Incorporating (5.29) in (5.26) we finally obtain

$$\bar{A}_w = \sum_{\boldsymbol{w} \in \mathcal{W}_w} \frac{\operatorname{coeff}\left(\prod_{i=2}^{\mathsf{d}_c} \left(P_{q,i}(Z)\right)^{P_i m}, Z^{\sum_{j=2}^{\mathsf{d}_v} w_j j}\right)}{\binom{\sum_{j=2}^{\mathsf{d}_v} \Lambda_j j n}{\sum_{j=2}^{\mathsf{d}_v} w_j j} (q-1)^{\sum_{j=2}^{\mathsf{d}_v} w_j (j-1)}} \prod_{j=2}^{\mathsf{d}_v} \binom{\Lambda_j n}{w_j}. \tag{5.30}$$

An equivalent expression of (5.30) is proposed in Exercise 5.5.

5.5 Weight Distributions of Protograph LDPC Code Ensembles

Consider an unconstrained protograph LDPC code ensemble over GF(q), and a length-n LDPC code drawn from it. Assume that we assign an element of GF(q) to each VN (either transmitted or punctured) in the lifted Tanner graph with $n = Qn_v$ VNs. The subset of the VNs with nonzero value represents a VN assignment, which induces a CN socket assignment. This VN assignment has a vector weight $\boldsymbol{w} = \begin{bmatrix} w_1 & w_2 & \dots & w_{n_v} \end{bmatrix}$, where w_j is an integer representing the portion of the total weight of the obtained length-n vector in GF$(q)^n$ that is apportioned to VNs of type $j \in \mathcal{J}$. The VN assignment induces a CN socket assignment with vector weight $\boldsymbol{l} = \boldsymbol{l}(\boldsymbol{w}) = (l_g)_{g \in \Phi}$ where $l_g = w_j$ whenever $g \in \Lambda_j$ (i.e., whenever the edge of index g in the protograph is connected to the VN of index j in the protograph). Since, for every $j \in \mathcal{J}$, t_j elements of $\boldsymbol{l}(\boldsymbol{w})$ have the value w_j, the total weight of $\boldsymbol{l}(\boldsymbol{w})$ is equal to $l = \sum_{j=1}^{n_v} w_j t_j$. The vector in GF$(q)^n$ corresponding to the transmitted VNs is a codeword of the protograph LDPC code if and only if the induced CN socket assignment is valid. The codeword weight w is the number of transmitted VNs in the VN assignment.

Let $\bar{A}_w$ be the expected number of codewords of weight w in an LDPC code drawn randomly, according to the procedure described in Section 4.2.2, from the protograph ensemble. Let

$$\mathcal{W}_w = \left\{\boldsymbol{w} : \sum_{j \in \mathcal{J}_u} w_j = w;\ 0 \le w_j \le Q\ \forall j \in \mathcal{J}\right\}. \tag{5.31}$$

We observe that all VN assignments with the same vector weight $\boldsymbol{w}$ have the same probability to induce a valid CN socket assignment after the copy-and-permute procedure is performed, and the edge labels are drawn. Therefore, $\bar{A}_w$ may again be expressed as the sum over all admissible vectors $\boldsymbol{w}$, associated with a weight w, of the number of VN assignments of vector weight $\boldsymbol{w}$ each multiplied by the probability that its induced CN socket assignment is valid. Denoting by $C_{\boldsymbol{w}}^{\text{valid}}$ the number of valid CN socket assignments induced by a VN assignment of vector weight $\boldsymbol{w}$, and observing that the number of VN assignments of vector weight $\boldsymbol{w}$ is given by $\prod_{j\in\mathcal{J}}\binom{Q}{w_j}(q-1)^{w_j}$ and the total number of CN socket assignments induced by a VN assignment of vector weight $\boldsymbol{w}$ is $\prod_{j\in\mathcal{J}}\binom{Q}{w_j}^{t_j-1}(q-1)^{(t_j-1)w_j}$, we obtain

$$\begin{aligned}\bar{A}_w &= \sum_{\boldsymbol{w}\in\mathcal{W}_w} \frac{C_{\boldsymbol{w}}^{\text{valid}}}{\prod_{j\in\mathcal{J}}\binom{Q}{w_j}^{t_j}(q-1)^{w_j t_j}} \prod_{j\in\mathcal{J}}\binom{Q}{w_j}(q-1)^{w_j} \\ &= \sum_{\boldsymbol{w}\in\mathcal{W}_w} \frac{C_{\boldsymbol{w}}^{\text{valid}}}{\prod_{j\in\mathcal{J}}\binom{Q}{w_j}^{t_j-1}(q-1)^{(t_j-1)w_j}}. \end{aligned} \tag{5.32}$$

Note that in the first equality of (5.32), the fraction multiplying the number $\prod_{j\in\mathcal{J}}\binom{Q}{w_j}(q-1)^{w_j}$ of VN assignments of vector weight $\boldsymbol{w}$ is the conditional probability $P(\text{valid}|\boldsymbol{w})$ that, upon performing the copy-and-permute operation and drawing the edge labels, the CN socket assignment induced by the VN assignment is valid.

Next, we develop (5.32) by finding an expression for the quantity $C_{\boldsymbol{w}}^{\text{valid}}$. Recall that s_i is the length of CNs of type $i \in \mathcal{I}$ and that the induced CN socket assignment has vector weight $\boldsymbol{l} = \boldsymbol{l}(\boldsymbol{w}) = (l_g)_{g\in\Phi}$ where $l_g = w_j$ whenever $g \in \Lambda_j$. We denote by $\boldsymbol{l}_i(\boldsymbol{w}) = \left(l_g\right)_{g\in\Gamma_i}$ the subvector of $\boldsymbol{l}(\boldsymbol{w})$ of length s_i whose elements are those indexed by Γ_i. We can write

$$C_{\boldsymbol{w}}^{\text{valid}} = \prod_{i\in\mathcal{I}} C_{\boldsymbol{l}_i(\boldsymbol{w})}^{\text{valid}}, \tag{5.33}$$

where $C_{\boldsymbol{l}_i(\boldsymbol{w})}^{\text{valid}}$ is the number of valid CN socket assignments of vector weight $\boldsymbol{l}_i(\boldsymbol{w})$ over the Q CNs of type i. Obviously, $C_{\boldsymbol{l}_i(\boldsymbol{w})}^{\text{valid}}$ coincides with the number of codewords of vector weight $\boldsymbol{l}_i(\boldsymbol{w})$ of the replicated code formed by the Q CNs of type i. Letting $A_i(\boldsymbol{W})$ be the vector weight enumerating function (for $T = s_i$, see Definition 5.7) of the $(s_i, s_i - 1)$ linear block code associated with a CN of type i, from Lemma 5.5, we have

$$\bar{A}_w = \sum_{\boldsymbol{w}\in\mathcal{W}_w} \frac{\prod_{i\in\mathcal{I}} \operatorname{coeff}\Big((A_i(\boldsymbol{W}))^Q, \boldsymbol{W}^{\boldsymbol{l}_i(\boldsymbol{w})}\Big)}{\prod_{j\in\mathcal{J}}\binom{Q}{w_j}^{t_j-1}(q-1)^{(t_j-1)w_j}}. \tag{5.34}$$

Remark 5.1. In the binary case, sometimes we know in advance that some vector weights $\boldsymbol{w}$ give no contribution to the right-hand side of (5.34). In particular, assume that there exists at least one CN type $i \in \mathcal{I}$ such that $\sum_{j\in\mathcal{J}} B_{i,j} w_j$ is an odd integer. In this case, we know in advance that

$$\operatorname{coeff}\Big((A_i(\boldsymbol{W}))^Q, \boldsymbol{W}^{\boldsymbol{l}_i(\boldsymbol{w})}\Big) = 0 \tag{5.35}$$

and therefore we can avoid to calculate the corresponding term in the summation. This is because the quantity $\sum_{j\in\mathcal{J}} B_{i,j} w_j$ represents the weight of the CN socket assignment over the Q CNs of type i that is induced by the VN assignment of vector weight $\boldsymbol{w}$. If this weight is odd, there is no way to let this CN socket assignment be valid.

5.6 Weight Spectral Shapes

This section is devoted to the analysis of an important function associated with the average weight distribution of code ensembles, called the *weight spectral shape* or the *growth rate of the weight distribution* and representing the exponent of the weight enumerator. Although this function can be defined for code ensembles beyond LDPC ones, the section is mostly focused on the weight spectral shape of LDPC codes, along with its efficient computation. The main application of the weight spectral shape of LDPC codes, an upper bound on the cumulative distribution function (c.d.f.) of the minimum distance of a random LDPC code, will be illustrated. Weight spectral shapes are especially important, since they allow to characterize (from a code ensemble perspective) how the minimum distance of LDPC codes scales as the blocklength grows.

5.6.1 Basics

Next, we define the weight spectral shape function for a sequence $\{\mathscr{C}_n\}$ of block code ensembles over $\mathrm{GF}(q)$. This means that every element of the sequence is an ensemble of block codes constructed over $\mathrm{GF}(q)$ and that all codes in $\mathscr{C}_n$ have the same blocklength n.

Example 5.4 Sequences of LDPC code ensembles may be constructed with all the ensembles described in Section 4.2. For example, by $\{\mathscr{C}^{n,q}_{\mathsf{d}_v,\mathsf{d}_c}\}$, we denote a sequence of unstructured regular LDPC code ensembles constructed over $\mathrm{GF}(q)$, with variable node degree d_v and check node degree d_c.

Definition 5.8 (Weight spectral shape) Let $\{\mathscr{C}_n\}$ be a sequence of block code ensembles over $\mathrm{GF}(q)$. For any rational ω such that $0 \le \omega \le 1$, the weight spectral shape of the ensemble sequence is defined as

$$G(\omega) = \lim_{n\to\infty} \frac{1}{n} \log_q \bar{A}_{\omega n} = \frac{1}{\ln q} \lim_{n\to\infty} \frac{1}{n} \ln \bar{A}_{\omega n} \tag{5.36}$$

when the limit exists. The limit assumes the inclusion of only those positive integers n for which ωn is an integer and $\bar{A}_{\omega n}$ is positive.

The next two lemmas will be extensively used throughout this paragraph, together with Lemma 5.2 and Lemma 5.3, to calculate the weight spectral shape functions of sequences of the LDPC code ensembles introduced in Section 4.2.

Lemma 5.6 *Let n be a positive integer and $\boldsymbol{\beta} = \begin{bmatrix}\beta_1 & \beta_2 & \dots & \beta_d\end{bmatrix}$, where $\beta_1, \beta_2, \dots, \beta_d$ are nonnegative rational numbers such that $\sum_{i=1}^{d} \beta_i = 1$ and such that $\beta_1 n$,*

$\beta_2 n, \ldots, \beta_d n$ are integer numbers. Let $\mathsf{H_e}(\boldsymbol{\beta}) = -\sum_{i=1}^{d} \beta_i \ln \beta_i$ be the Shannon's entropy of the distribution $\boldsymbol{\beta}$ in nats. Then we have

$$\frac{\exp\{n\mathsf{H_e}(\boldsymbol{\beta})\}}{(n+1)^d} \le \binom{n}{\beta_1 n, \beta_2 n, \cdots, \beta_d n} \le \exp\{n\mathsf{H_e}(\boldsymbol{\beta})\} \tag{5.37}$$

and therefore

$$\binom{n}{\beta_1 n, \beta_2 n, \cdots, \beta_d n} \doteq \exp\{n\mathsf{H_e}(\boldsymbol{\beta})\}. \tag{5.38}$$

In the special case of $d = 2$, we have the tighter bounds

$$\frac{\exp\left\{n\mathsf{H_e}(\beta) - \frac{1}{12n\beta(1-\beta)}\right\}}{\sqrt{2\pi n\beta(1-\beta)}} < \binom{n}{\beta n} < \frac{\exp\{n\mathsf{H_e}(\beta)\}}{\sqrt{2\pi n\beta(1-\beta)}} \tag{5.39}$$

and therefore

$$\binom{n}{\beta n} \doteq \exp\{n\,\mathsf{H_e}(\beta)\}, \tag{5.40}$$

where $\mathsf{H_e}(\beta) = -\beta \ln \beta - (1-\beta)\ln(1-\beta)$ is the entropy of a Bernoulli r.v. with parameter β.

Lemma 5.7 *Let $x_1, x_2, \ldots, x_d$ be real numbers and n be an integer. Then we have*

$$\sum_{t=1}^{d} \exp\{n\, x_t\} \doteq \exp\left\{n \max_{t \in \{1,2,\ldots,d\}} \{x_t\}\right\}. \tag{5.41}$$

5.6.2 Weight Spectral Shape of Linear Random Ensembles

Before addressing the weight spectral shapes of LDPC codes, in this section, we develop the weight spectral shape of the (parity-check-based) linear random ensemble $\mathscr{H}_{n,m}^{q}$. This spectral shape function will be used as reference for the LDPC code case.

Recall that a Tanner graph representation may be drawn for any parity-check matrix of any linear block code constructed over $\mathrm{GF}(q)$. We can then calculate the expected multiplicity of codewords of weight w for a code drawn randomly in $\mathscr{H}_{n,m}^{q}$ by reasoning on the Tanner graph associated with the generic realization of the random parity-check matrix defining the ensemble. We have

$$\bar{A}_w = \binom{n}{w}(q-1)^w P(\text{valid}|w), \tag{5.42}$$

where $P(\text{valid}|w)$ is the probability that the CN socket assignment induced by a VN assignment of weight w is valid. In the equivalent Tanner graph perspective, the procedure to draw a code $\mathcal{C} \in \mathscr{H}_{n,m}^{q}$ consists of connecting each VN to all CNs, of assigning to each edge a label picked uniformly at random in $\mathrm{GF}(q)$, independently for each edge, and of removing all edges whose label is equal to $0 \in \mathrm{GF}(q)$. Therefore, we have

$$P(\text{valid}|w) = [P(X_1 + X_2 + \cdots + X_w = 0)]^m, \tag{5.43}$$

where $X_1, X_2, \ldots, X_w$ are discrete i.i.d. random variables uniform in GF(q). Since $P(X_1 + X_2 + \cdots + X_w = 0) = \frac{1}{q}$ (irrespective of w), expressing the number m of CNs as $m = (1 - \mathrm{R})n$, we obtain

$$\bar{A}_w = \binom{n}{w}(q-1)^w q^{-(1-\mathrm{R})n} . \tag{5.44}$$

The weight spectral shape function may now be obtained by letting $w = \omega n$ and by applying Lemma 5.6 to the binomial coefficient $\binom{n}{\omega n}$. This immediately yields

$$G(\omega) = \frac{1}{\ln q}\left(\mathsf{H_e}(\omega) + \omega \ln(q-1) - (1-\mathrm{R}) \ln q\right) . \tag{5.45}$$

5.6.3 Weight Spectral Shape of Unstructured LDPC Code Ensembles

5.6.3.1 Regular Ensembles

We start the analysis of the weight spectral shape of LDPC code ensembles by the simplest case, an unstructured regular $\mathscr{C}^{n,q}_{\mathsf{d}_v,\mathsf{d}_c}$ ensemble, whose expected number of weight-w codewords is expressed by (5.23). Application of Lemma 5.6 yields

$$\binom{n\mathsf{d}_v}{w\mathsf{d}_v} = \binom{n\mathsf{d}_v}{n\omega\mathsf{d}_v} \doteq \exp\{n\mathsf{d}_v \mathsf{H_e}(\omega)\} \tag{5.46}$$

and

$$\binom{n}{w} = \binom{n}{n\omega} \doteq \exp\{n\mathsf{H_e}(\omega)\} . \tag{5.47}$$

Moreover, applying Lemma 5.2, we obtain

$$\begin{aligned} \mathrm{coeff}\left(\left(P_{q,\mathsf{d}_c}(Z)\right)^m, Z^{w\mathsf{d}_v}\right) &= \mathrm{coeff}\left(\left(P_{q,\mathsf{d}_c}(Z)\right)^m, Z^{m\omega\mathsf{d}_c}\right) \\ &\doteq \exp\left\{m\left[\ln P_{q,\mathsf{d}_c}(z_0) - \omega\mathsf{d}_c \ln z_0\right]\right\} \\ &= \exp\left\{n\left[(1-\mathrm{R}) \ln P_{q,\mathsf{d}_c}(z_0) - \omega\, \mathsf{d}_v \ln z_0\right]\right\}, \end{aligned} \tag{5.48}$$

where z_0 is the unique positive root to the equation

$$z_0 \left.\frac{\mathrm{d} \ln P_{q,\mathsf{d}_c}}{\mathrm{d}Z}\right|_{z_0} = \mathsf{d}_c\, \omega . \tag{5.49}$$

To obtain (5.48), we have used the relationships $n\mathsf{d}_v = m\mathsf{d}_c$ in the first equality and $m/n = \mathsf{d}_v/\mathsf{d}_c = 1 - \mathrm{R}$ in the third one, R being the ensemble design rate. From the above exponential equivalences, we obtain the expression of the weight spectral shape of the $\mathscr{C}^{n,q}_{\mathsf{d}_v,\mathsf{d}_c}$ ensemble as

$$\begin{aligned} G(\omega) = \frac{1}{\ln q}&\left[(1-\mathrm{R}) \ln P_{q,\mathsf{d}_c}(z_0) - \omega\, \mathsf{d}_v \ln z_0 - (\mathsf{d}_v - 1)\mathsf{H_e}(\omega)\right. \\ &\left. - \omega(\mathsf{d}_v - 1) \ln(q-1)\right], \end{aligned} \tag{5.50}$$

where $z_0 = z_0(\omega)$ is the unique positive real root to (5.49). A few remarks on the weight spectral shape of $\mathscr{C}^{n,q}_{\mathsf{d}_v,\mathsf{d}_c}$ ensembles follow.

Remark 5.2. According to Definition 5.8, the weight spectral shape function is defined for those rational numbers $0 \le \omega \le 1$ such that the limit (5.36) exists. From the conducted analysis it turns out, however, that $G(\omega)$ given by (5.50) together with the constraint equation (5.49) is well-defined for any real $0 \le \omega \le 1$ (besides being well-defined for all rational $0 \le \omega \le 1$) and that this function is continuous in $[0,1] \subset \mathbb{R}$. Unless explicitly stated, we will refer to the extension of the weight spectral shape function to the real domain.

Remark 5.3. It is easy to verify that $z_0 = 1$ when $\omega = \frac{q-1}{q}$ and that

$$G\left(\frac{q-1}{q}\right) = \mathsf{R}, \tag{5.51}$$

this value represents the maximum of the weight spectral shape function (see Exercise 5.2).

Remark 5.4. In a few cases, the $G(\omega)$ function expressed by (5.50) admits a closed form. This happens, in particular, when $\mathsf{d}_c \in \{6,7,8,9\}$ and $q = 2$. In fact, for $q = 2$ and $\mathsf{d}_c \in \{6,7\}$, making the substitution $y_0 = z_0^2$ in (5.49), we obtain a cubic equation from which we can explicitly calculate $z_0 = z_0(\omega)$. Similarly, for $q = 2$ and $\mathsf{d}_c \in \{8,9\}$ the substitution $y_0 = z_0^2$ yields a quadratic equation from which $z_0 = z_0(\omega)$ may be explicitated.

Remark 5.5. The value $\omega^* = \inf\{\omega \in (0,1) : G(\omega) > 0\}$ is often referred to as the critical fractional codeword weight. Regardless of the field order q, for a $\mathscr{C}^{n,q}_{\mathsf{d}_v,\mathsf{d}_c}$ ensemble we have $\omega^* > 0$ if $\mathsf{d}_v \ge 3$ and $\omega^* = 0$ if $\mathsf{d}_v = 2$. This fact may be better understood by developing an expression of the weight spectral shape for small values of ω. Recalling (2.61), it follows from (5.49) that $z_0 \to 0$ as $\omega \to 0$. We may therefore recast (5.49) as

$$2\binom{\mathsf{d}_c}{2}(q-1)\, z_0^2\,(1 + \mathsf{o}(1)) = \mathsf{d}_c\, \omega \tag{5.52}$$

or, equivalently, as

$$z_0 = \left[\frac{\omega}{(\mathsf{d}_c - 1)(q-1)}\right]^{\frac{1}{2}} (1 + \mathsf{o}(1))\,. \tag{5.53}$$

Here, $\mathsf{o}(\cdot)$ stands for the little-o Landau symbol. Substituting the above expression of x_0 into (5.50), after simple algebraic manipulation, we obtain

$$G(\omega) = \frac{\omega \ln \omega}{\ln q}\left(\frac{\mathsf{d}_v}{2} - 1\right) + \frac{\omega}{\ln q}\left[\frac{\mathsf{d}_v}{2}\ln(\mathsf{d}_c - 1) - \left(\frac{\mathsf{d}_v}{2} - 1\right)\ln(q-1)\right] + \mathsf{o}(\omega) \tag{5.54}$$

which simplifies to

$$G(\omega) = \omega\, \frac{\ln(\mathsf{d}_c - 1)}{\ln q} + \mathsf{o}(\omega) \tag{5.55}$$

when $\mathsf{d}_v = 2$.

Remark 5.6. The function $G(\omega)$ is everywhere differentiable in $(0,1)$. As from the previous two expressions developed for small ω, we have $G(0) = 0$ (by continuous extension of $G(\omega)$ in $\omega = 0$) and $\lim_{\omega \to 0^+} G'(\omega) = -\infty$ when $\mathsf{d}_v \geq 3$, while $G(0) = 0$ and $G'(0) = \ln(\mathsf{d}_c - 1)/\ln(q-1) > 0$ when $\mathsf{d}_v = 2$. In both cases the function $G(\omega)$ starts from 0 when $\omega = 0$. However, while in the $\mathsf{d}_v \geq 3$ case it has a unique extremal point for $\omega \in (0, \frac{q-1}{q})$ (as it decreases, reaches a negative minimum and then increases up to its maximum value R for $\omega = \frac{q-1}{q}$), in the $\mathsf{d}_v = 2$ case it increases monotonically for all $\omega \in (0, \frac{q-1}{q})$.

Remark 5.7. For $\mathsf{d}_v \geq 3$, the critical fractional codeword weight $\omega^* > 0$ of a regular unstructured ensemble $\mathscr{C}^{n,q}_{\mathsf{d}_v,\mathsf{d}_c}$ may be calculated by simply solving the system of simultaneous equations

$$z_0 \left. \frac{\mathrm{d} \ln P_{q,\mathsf{d}_c}}{\mathrm{d}Z} \right|_{z_0} = \mathsf{d}_c\, \omega^* \tag{5.56}$$

and

$$(1 - \mathrm{R}) \ln P_{q,\mathsf{d}_c}(z_0) - \omega^* \mathsf{d}_v \ln z_0 - (\mathsf{d}_v - 1)\mathsf{H}_\mathsf{e}(\omega^*) - \omega^*(\mathsf{d}_v - 1)\ln(q-1) = 0 \tag{5.57}$$

in the two unknowns ω^* and z_0.

Example 5.5 The weight spectral shape functions of unstructured regular LDPC code ensembles $\mathscr{C}^{n,q}_{3,6}$ with $q \in \{2,4,8,16\}$ are illustrated in Figure 5.2, where a zoom of the plot for $\omega \in (0,0.1)$ is also shown. All of these ensembles are characterized by a design rate $\mathrm{R} = 1/2$. For each value of $q \in \{2,4,8,16\}$, the weight spectral shape is plotted in the range $\omega \in (0, \frac{q-1}{q})$ and the weight spectral shape of a parity-check based random linear block ensemble, given by (5.45), is also depicted as a solid curve labeled as "random". The values of the critical fractional codeword weights ω^* for regular ensembles with $\mathsf{d}_v = 3$ and $\mathsf{d}_c = 6$, calculated for a broader set of q with the method explained in Remark 5.7, are shown in Table 5.1 along with the corresponding value of z_0. Moreover, the values of ω^* for some regular ensembles with $\mathsf{d}_v = 4$ and $\mathsf{d}_c = 6$, corresponding to design rate $\mathrm{R} = 1/3$, are shown in Table 5.2 along with the corresponding values of z_0.

5.6.3.2 Irregular Ensembles

To develop the weight spectral shape of unstructured irregular LDPC code ensembles, we proceed similarly to the irregular case, starting from (5.30). Let us define $\omega_j = w_j/n$ for all $j \in \{2,3,\ldots,\mathsf{d}_v\}$. Applying Lemma 5.6, we may write

$$\prod_{j=2}^{\mathsf{d}_v} \binom{\Lambda_j n}{\omega_j n} \doteq \exp\left\{ n \sum_{j=2}^{\mathsf{d}_v} \Lambda_j \mathsf{H}_\mathsf{e}\left(\frac{\omega_j}{\Lambda_j}\right) \right\} \tag{5.58}$$

and

$$\binom{\sum_{j=2}^{\mathsf{d}_v} j\Lambda_j n}{\sum_{j=2}^{\mathsf{d}_v} j\omega_j n} \doteq \exp\left\{ n \frac{1}{\int \lambda} \mathsf{H}_\mathsf{e}\left(\beta \textstyle\int \lambda\right) \right\}, \tag{5.59}$$

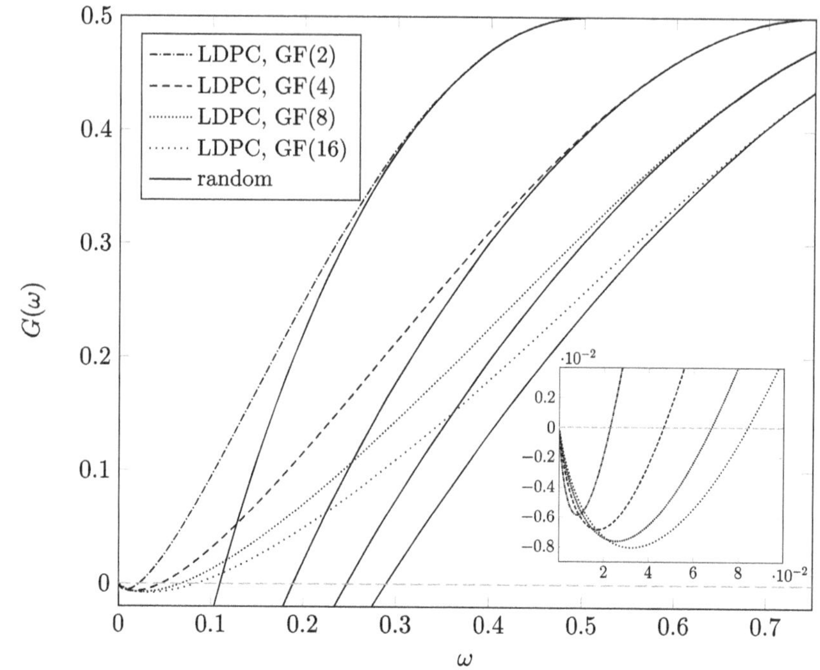

Figure 5.2 Weight spectral shape functions for $R = 1/2$ unstructured regular LDPC ensembles $\mathscr{C}^q_{3,6}$, for $q \in \{2,4,8,16\}$. The solid curves represent the weight spectral shape functions of $R = 1/2$ random linear parity-check ensembles for the same values of the Galois field order q.

Table 5.1 Critical fractional codeword weights ω^* of unstructured regular LDPC code ensembles with $\mathsf{d}_v = 3$ and $\mathsf{d}_c = 6$ constructed over GF(q), for $q = 2^p$ and $p \in \{1,2,\ldots,10\}$

q	ω^*	z_0
2	0.022733	0.069506
4	0.046531	0.053262
8	0.067787	0.038262
16	0.084057	0.026007
32	0.094456	0.016904
64	0.099262	0.010609
128	0.099373	0.006482
256	0.095898	0.003879
512	0.089924	0.002284
1024	0.082398	0.001328

where we have defined β as

$$\beta = \sum_{j=2}^{\mathsf{d}_v} j\omega_j \,. \tag{5.60}$$

Table 5.2 Critical fractional codeword weights ω^* of unstructured regular LDPC code ensembles with $\mathsf{d}_v = 4$ and $\mathsf{d}_c = 6$ constructed over GF(q), for $q = 2^p$ and $p \in \{1,2,\ldots,6\}$

q	ω^*	z_0
2	0.128836	0.195336
4	0.226959	0.135504
8	0.298591	0.089287
16	0.349098	0.056421
32	0.383351	0.034527
64	0.405214	0.020637

Moreover, we have

$$(q-1)^{\sum_{j=2}^{\mathsf{d}_v}(j-1)\omega_j n} = \exp\left\{ n \sum_{j=2}^{\mathsf{d}_v} (j-1)\omega_j \ln(q-1) \right\}. \tag{5.61}$$

Next, we need to develop an exponentially equivalent expression for $C_{\boldsymbol{w}}^{\text{valid}}$ given by (5.29). To do so, we consider the right-hand side of equality (a) in (5.29), and we observe that, by Lemma 5.2 and by defining $l_i = \epsilon_i m$, we have

$$\text{coeff}\left(\left(P_{q,i}(Z)\right)^{P_i m}, Z^{\epsilon_i m} \right) \doteq \exp\left\{ m \left[P_i \ln P_{q,i}(z_{0,i}) - \epsilon_i \ln z_{0,i} \right] \right\}, \tag{5.62}$$

where $z_{0,i}$ is the unique positive solution to

$$P_i\, z_{0,i} \left. \frac{\partial \ln P_{q,i}}{\partial Z} \right|_{z_{0,i}} = \epsilon_i . \tag{5.63}$$

Therefore, letting $\boldsymbol{\epsilon} = \left[\epsilon_2 \quad \epsilon_3 \quad \ldots \quad \epsilon_{\mathsf{d}_c} \right]$, we may write

$$\begin{aligned} C_{\boldsymbol{w}}^{\text{valid}} &\doteq \sum_{\boldsymbol{\epsilon}} \exp\left\{ m \sum_{i=2}^{\mathsf{d}_c} \left[P_i \ln P_{q,i}(z_{0,i}) - \epsilon_i \ln z_{0,i} \right] \right\} \\ &= \exp\left\{ n(1-R) \max_{\boldsymbol{\epsilon}} \sum_{i=2}^{\mathsf{d}_c} \left[P_i \ln P_{q,i}(z_{0,i}) - \epsilon_i \ln z_{0,i} \right] \right\}, \end{aligned} \tag{5.64}$$

where the equality follows from $m = (1-\mathrm{R})n$, from Lemma 5.7, and from $\exp x$ being monotonically increasing with x. Since the vector $\boldsymbol{l}$ is subject to the constraint $\sum_{i=2}^{\mathsf{d}_c} l_i = \sum_{j=2}^{\mathsf{d}_v} w_j j$, the maximization in (5.64) is subject to

$$\sum_{i=2}^{\mathsf{d}_c} \epsilon_i = \frac{\sum_{j=2}^{\mathsf{d}_v} j w_j}{m} = \frac{\beta}{(1-\mathrm{R})}. \tag{5.65}$$

Applying Lagrange multipliers, it is immediate to obtain

$$\ln z_{0,i} = \mu \qquad \forall i \in \{2,3,\ldots,\mathsf{d}_c\}, \tag{5.66}$$

where μ is the (unique) Lagrange multiplier. We can therefore conclude that all $z_{0,i}$ are equal to each other and we can let $z_0 = z_{0,i}$ for all $i \in \{2, 3, \ldots, \mathsf{d}_c\}$. Substituting in (5.64) yields

$$C_{\boldsymbol{w}}^{\text{valid}} \doteq \exp\left\{n(1-\mathrm{R})\left[\sum_{i=2}^{\mathsf{d}_c} P_i \ln P_{q,i}(z_0) - \frac{\beta}{1-\mathrm{R}} \ln z_0\right]\right\}$$
$$= \exp\left\{n\left[(1-\mathrm{R}) \sum_{i=2}^{\mathsf{d}_c} P_i \ln P_{q,i}(z_0) - \beta \ln z_0\right]\right\}. \tag{5.67}$$

Summing both sides of (5.63) over i, we may conclude that z_0 is the unique positive real root to

$$(1-\mathrm{R})\, z_0 \sum_{i=2}^{\mathsf{d}_c} P_i \left.\frac{\partial \ln P_{q,i}}{\partial Z}\right|_{z_0} = (1-\mathrm{R})\, z_0 \sum_{i=2}^{\mathsf{d}_c} P_i \frac{\frac{\partial P_{q,i}(Z)}{\partial Z}(z_0)}{P_{q,i}(z_0)} = \beta\,. \tag{5.68}$$

Incorporating (5.58), (5.59), (5.60), and (5.67) into (5.30), we obtain

$$\bar{A}_w \doteq \sum_{\boldsymbol{w}} \exp\left\{n\, S\left(\frac{1}{n}\boldsymbol{w}\right)\right\}, \tag{5.69}$$

where, letting $\boldsymbol{\omega} = \frac{1}{n}\boldsymbol{w}$,

$$S(\boldsymbol{\omega}) = (1-R) \sum_{i=2}^{\mathsf{d}_c} P_i \ln P_{q,i}(z_0) - \beta \ln z_0 + \sum_{j=2}^{\mathsf{d}_v} \Lambda_j \mathsf{H_e}\left(\frac{\omega_j}{\Lambda_j}\right)$$
$$- \frac{1}{\int \lambda} \mathsf{H_e}\left(\beta \textstyle\int \lambda\right) - \sum_{j=2}^{\mathsf{d}_v} (j-1)\omega_j \ln(q-1)\,. \tag{5.70}$$

Application of Lemma 5.7 to (5.69) allows expressing the weight spectral shape of a $\mathscr{C}^q_{\lambda,\rho}$ ensemble as

$$G(\omega) = \frac{1}{\ln q} \max_{\boldsymbol{\omega}} S(\boldsymbol{\omega}), \tag{5.71}$$

where the maximization is subject to

$$\sum_{j=2}^{\mathsf{d}_v} \omega_j = \omega\,. \tag{5.72}$$

Denoting by μ the Lagrange multiplier, for all $j \in \{2, 3, \ldots, \mathsf{d}_v\}$ at the maximum, we must have

$$\frac{\partial S(\boldsymbol{\omega})}{\partial \omega_j} = \mu\,. \tag{5.73}$$

After simple algebraic manipulation, this latter equality may be recast as

$$-j \ln z_0 + \ln \frac{\Lambda_j - \omega_j}{\omega_j} + j \ln \frac{\beta \int \lambda}{1 - \beta \int \lambda} - \ln (q-1)^{j-1} = \mu \tag{5.74}$$

or, equivalently, as

$$\ln\left(\frac{1}{z_0}\frac{\beta f\lambda}{(1-\beta f\lambda)}\frac{1}{(q-1)}\right)^j + \ln\frac{\Lambda_j-\omega_j}{\omega_j} + \ln(q-1) = \mu\,. \tag{5.75}$$

We may therefore conclude that, for any ω, the value $G(\omega)$ of the weight spectral shape function is obtained by solving a system of $\mathsf{d}_v + 2$ simultaneous equations, represented by (5.60), (5.68), (5.72), and (5.75) ($\mathsf{d}_v - 1$ equations), in the $\mathsf{d}_v + 2$ unknowns $\omega_2, \ldots, \omega_{\mathsf{d}_v}, z_0, \beta$, and μ.

We will now show that the obtained $(\mathsf{d}_v+2)\times(\mathsf{d}_v+2)$ system may be reduced to a 4×4 system, irrespective of the value of d_v. To this purpose, let us define two new variables x_0 and y_0 as

$$x_0 = e^{-\mu}\,. \tag{5.76}$$

and

$$y_0 = \frac{1}{z_0}\frac{\beta f\lambda}{(1-\beta f\lambda)}\frac{1}{(q-1)}, \tag{5.77}$$

respectively. Using these two definitions (5.75) may be recast in the form

$$(q-1)x_0y_0^j = \frac{\omega_j}{\Lambda_j-\omega_j} \tag{5.78}$$

from which we have

$$1+(q-1)x_0y_0^j = \frac{\Lambda_j}{\Lambda_j-\omega_j}\,. \tag{5.79}$$

Note that the quantity $1+(q-1)XY^j$ is the input–output weight enumerating function of a $(j,1)$ repetition code, that will be denoted in this chapter by $B_j(X,Y)$. The two variables x_0 and y_0 are subject to two constraints that involve only ω and β, as follows. Dividing each side of (5.78) by the corresponding side of (5.79), we obtain

$$\frac{(q-1)x_0y_0^j}{1+(q-1)x_0y_0^j} = \frac{\omega_j}{\Lambda_j}. \tag{5.80}$$

Multiplying both sides of (5.80) by Λ_j and summing over j, we obtain the first constraint, that can be written in the form

$$x_0\sum_{j=2}^{\mathsf{d}_v}\Lambda_j\left.\frac{\partial\ln B_j}{\partial X}\right|_{x_0,y_0} = \omega\,. \tag{5.81}$$

Moreover, multiplying both sides of (5.80) by $j\Lambda_j$ and summing over j, we obtain the second constraint, that can be written in the form

$$y_0\sum_{j=2}^{\mathsf{d}_v}\Lambda_j\left.\frac{\partial\ln B_j}{\partial Y}\right|_{x_0,y_0} = \beta\,. \tag{5.82}$$

Next, we show that the weight spectral shape can be expressed as a function of the four variables x_0, y_0, z_0, and β. Substituting $\mu = \ln x_0$ (which follows from (5.76)) into (5.74), multiplying both sides by ω_j, and summing over j, we obtain

$$-\beta \ln z_0 = -\omega \ln x_0 - \sum_{j=2}^{\mathsf{d}_v} \omega_j \ln \frac{\Lambda_j - \omega_j}{\omega_j} + \beta \ln \frac{1 - \beta \int \lambda}{\beta \int \lambda}. \tag{5.83}$$

Incorporating in (5.70) and using (5.79) and performing simple algebraic manipulation leads to

$$G(\omega) = \frac{1}{\ln q}\left[(1-\mathrm{R}) \sum_{i=2}^{\mathsf{d}_c} P_i \ln P_{q,i}(z_0) - \omega \ln x_0 \right.$$
$$\left. + \sum_{j=2}^{\mathsf{d}_v} \Lambda_j \ln B_j(x_0, y_0) + \frac{1}{\int \lambda} \ln(1 - \beta \textstyle\int \lambda)\right]. \tag{5.84}$$

We can therefore conclude that the weight spectral shape function of an unstructured irregular LDPC code ensemble sequence $\{\mathscr{C}_{\lambda,\rho}^{n,q}\}$ is given by (5.84), where x_0, y_0, z_0, and β are the unique positive real roots of the four simultaneous equations (5.68), (5.77), (5.81), and (5.82).

Remark 5.8. It is easy to verify that for $\omega = \frac{q-1}{q}$, the solution to the 4×4 system is $x_0 = 1$, $y_0 = 1$, $z_0 = 1$, $\beta = \frac{1}{\int \lambda} \frac{q-1}{q}$, and that

$$G\left(\frac{q-1}{q}\right) = \mathrm{R}. \tag{5.85}$$

The above value represents the maximum of the weight spectral shape function.

Remark 5.9. Let $\mathsf{p} = \min\{j \in \{2,3,\ldots,\mathsf{d}_v\} : \lambda_j = 0\}$ be the minimum VN degree. The critical fractional codeword weight ω^* of an unstructured irregular ensemble sequence $\{\mathscr{C}_{\lambda,\rho}^{n,q}\}$ is positive whenever $\mathsf{p} > 0$, regardless the value of the field order q. It is also positive whenever $\mathsf{p} = 2$ and $0 < \lambda_2 \rho'(1) < 1$, where $\rho(x)$ is defined in (4.4) and $\rho'(1) = \sum_{i=2}^{\mathsf{d}_c} (i-1)\rho_i$ (note that we have $\rho'(1) = \mathsf{d}_c - 1$ if all CNs have the same degree d_c), regardless q. The critical fractional codeword weight ω^* is equal to zero for $\mathsf{p} = 2$, regardless q. This fact may be better understood by observing that for small values of ω, the weight spectral shape function may be written as

$$G(\omega) = \frac{\omega}{\ln q}\left(1 - \frac{\mathsf{p}}{2}\right) + \frac{\omega}{\ln q} \ln\left(\lambda_{\mathsf{p}} (\rho'(1))^{\frac{\mathsf{p}}{2}} \left(\frac{\mathsf{p}\, \omega \int \lambda}{q-1}\right)^{\frac{\mathsf{p}}{2}-1}\right) + \mathsf{o}(\omega) \tag{5.86}$$

which simplifies to

$$G(\omega) = \omega \ln_q (\lambda_2 \rho'(1)) + \mathsf{o}(\omega) \tag{5.87}$$

for $\mathsf{p} = 2$. A proof of (5.86) is proposed in Exercise 5.3 at the end of this chapter. As from the previous expressions we have $G(0) = 0$ (by continuous extension of $G(\omega)$ in $\omega = 0$) for all $\mathsf{p} \geq 2$. Moreover, we have $\lim_{\omega \to 0^+} G'(\omega) = -\infty$ when $\mathsf{p} \geq 3$ and $G'(0) = \ln_q(\lambda_2 \rho'(1))$ when $\mathsf{p} = 2$. Both in the case $\mathsf{p} \geq 3$ and in the case $\mathsf{p} = 2$ and $0 < \lambda_2 \rho'(1) < 1$ the function $G(\omega)$ decreases as ω increases for values of ω close to zero, it reaches a negative minimum and then increases up to its maximum value R for $\omega = \frac{q-1}{q}$. Conversely, in the $\mathsf{p} = 2$ and $\lambda_2 \rho'(1) > 1$ case, it increases monotonically for all $\omega \in (0, \frac{q-1}{q})$.

Remark 5.10. For both the case $\mathsf{p} \geq 3$ and the case $\mathsf{p} = 2$ and $0 < \lambda_2 \rho'(1) < 1$, the critical fractional codeword weight $\omega^* > 0$ of an irregular unstructured ensemble sequence $\{\mathscr{C}^{n,q}_{\lambda,\rho}\}$ may be calculated by simply solving the system of five simultaneous equations (5.68), (5.77), (5.81), (5.82), and

$$(1-\mathsf{R}) \sum_{i=\mathsf{r}}^{\mathsf{d}_c} P_i \ln P_{q,i}(z_0) - \omega^* \ln x_0 \quad + \sum_{j=\mathsf{p}}^{\mathsf{d}_v} \Lambda_j \ln B_j(x_0, y_0) + \frac{1}{\int \lambda} \ln(1 - \beta \textstyle\int \lambda) = 0 \tag{5.88}$$

in the five unknowns ω^*, x_0, y_0, z_0, and β.

Example 5.6 Consider an irregular unstructured ensemble sequence with the same edge-oriented degree distribution pair $\lambda(x) = \frac{8}{11}x + \frac{3}{11}x^2$ and $\rho(x) = \frac{6}{11}x^3 + \frac{5}{11}x^4$, here denoted by $(\lambda_{\mathsf{a}}, \rho_{\mathsf{a}})$, that was considered in Example 4.4. The design rate is $\mathsf{R} = 1/2$. Since for the degree distribution pair $(\lambda_{\mathsf{a}}, \rho_{\mathsf{a}})$ we have $\mathsf{p} = 2$ and $\lambda_2 \rho'(1) > 1$, we expect $\omega^* = 0$ for all q. A plot of the weight spectral shape curves for $q \in \{2,4,8,16\}$ is illustrated in Figure 5.3, where each weight spectral shape curve is plotted in the range $(0, \frac{q-1}{q})$ of ω. A zoom for $\omega \in (0, 0.03)$ is also shown. For each value of $q \in \{2,4,8,16\}$ the weight spectral shape of a parity-check-based random linear block ensemble is also depicted as a solid curve labeled as "random". As it is possible to observe, we have $G(0) = 0$ and $\lim_{\omega \to 0^+} G(\omega) > 0$ for all considered values of q, as expected.

Example 5.7 Consider an irregular unstructured ensemble sequence with edge-oriented degree distribution pair

$$\lambda_{\mathsf{b}}(x) = 0.062498\, x + 0.479743\, x^2 + 0.049808\, x^5 + 0.117758\, x^8 + 0.290192\, x^{29}$$
$$\rho_{\mathsf{b}}(x) = x^8$$

corresponding to a design rate $\mathsf{R} = 1/2$. Since in this case $\mathsf{p} = 2$ and $\lambda_2 \rho'(1) = 0.5$, we expect $\omega^* > 0$ for every Galois field order q. Figure 5.4 depicts the weight spectral shape curves corresponding to this degree distribution pair for $q \in \{2,4,8,16\}$. For each q, the weight spectral shape is plotted in the range $\omega \in (0, \frac{q-1}{q})$ and a zoom is shown for $\omega \in (0, 0.03)$. Similarly to Figures 5.2 and 5.3, the weight spectral shape curves of parity-check based random linear block ensembles are also depicted as solid curves labeled as "random." The values of the critical fractional codeword weights ω^*, calculated with the augmented 5×5 system discussed in Remark 5.10, are shown in Table 5.3 along with the corresponding values of x_0, y_0, z_0, and β.

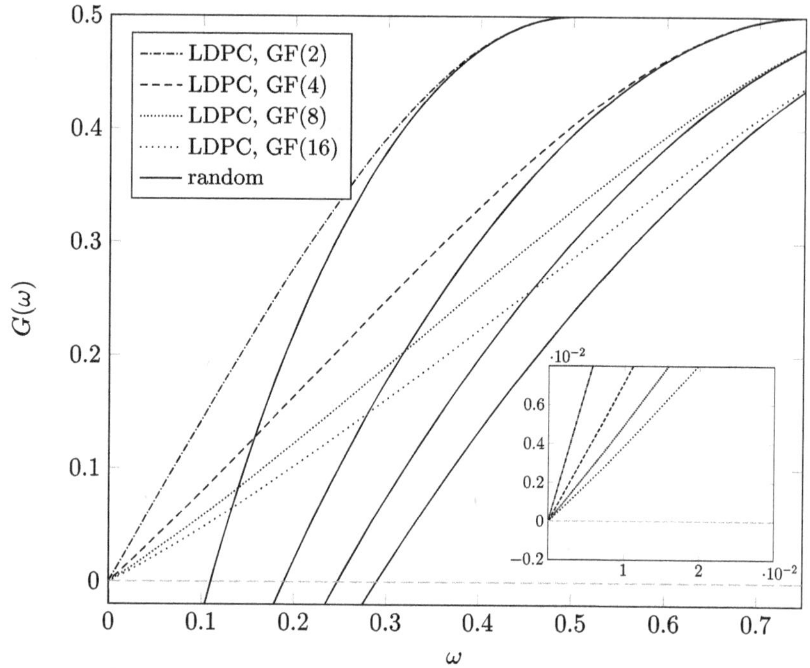

Figure 5.3 Weight spectral shape functions for $R = 1/2$ unstructured irregular LDPC code ensemble sequences with degree distribution pair $(\lambda, \rho) = (\lambda_{\mathsf{a}}, \rho_{\mathsf{a}})$.

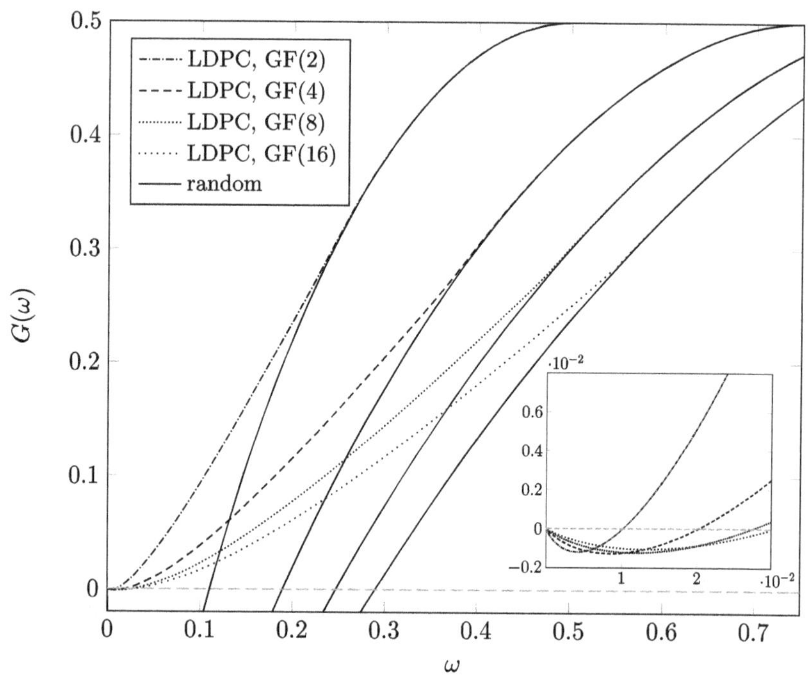

Figure 5.4 Weight spectral shape functions for $R = 1/2$ unstructured irregular LDPC code ensemble sequences with degree distribution pair $(\lambda, \rho) = (\lambda_{\mathsf{b}}, \rho_{\mathsf{b}})$.

Table 5.3 Critical fractional codeword weights ω^* of unstructured irregular LDPC code ensembles with degree distribution pair $(\lambda, \rho) = (\lambda_b, \rho_b)$ constructed over GF(q), for $q = 2^p$ and $p \in \{1,2,\ldots,6\}$

q	ω^*	x_0	y_0	z_0	β
2		0.780164	0.212668	0.026989	0.025681
4	0.020134	0.758704	0.184033	0.020494	0.050347
8	0.027427	0.739869	0.150919	0.014404	0.067450
16	0.030323	0.026007	0.116368	0.009425	0.072837
32	0.028510	0.728998	0.083533	0.005782	0.066376
64	0.023057	0.739127	0.055190	0.003343	0.051703

5.6.4 Weight Spectral Shape of Protograph LDPC Code Ensembles

The expected multiplicity of codewords of weight w for a code drawn randomly from a protograph ensemble is given by (5.34). Let $\omega = w/n$ and $\boldsymbol{\omega} = (\omega_j)_{j\in\mathcal{J}} = (1/n)\boldsymbol{w}$. Moreover, let $\boldsymbol{\lambda}_i(\boldsymbol{\omega}) = (1/n)\,\boldsymbol{l}_i(n\boldsymbol{\omega})$. With a slight abuse of notation, relabel ω_j as ω_g whenever VN $j \in \mathcal{J}$ is the one connected to the edge $g \in \Phi$ in the protograph. From Lemma 5.2, we have

$$\begin{aligned}\operatorname{coeff}\left((A_i(\boldsymbol{W}))^Q, \boldsymbol{W}^{\boldsymbol{l}_i(\boldsymbol{w})}\right) &= \operatorname{coeff}\left((A_i(\boldsymbol{W}))^Q, \boldsymbol{W}^{n_u \boldsymbol{\lambda}_i(\boldsymbol{\omega})Q}\right)\\ &\doteq \exp\left\{n\left[\frac{1}{n_u}\ln A_i(\boldsymbol{z}_i) - \sum_{g\in\Gamma_i}\omega_g \ln z_g\right]\right\},\end{aligned} \tag{5.89}$$

where the vector $\boldsymbol{z}_i = \left(z_g\right)_{g\in\Gamma_i}$ contains the positive solutions to the equations

$$z_g \left.\frac{\partial \ln A_i}{\partial W_g}\right|_{\boldsymbol{z}_i} = n_u \omega_g \qquad g \in \Gamma_i\,. \tag{5.90}$$

Lemma 5.6 applied to the denominator of (5.34) leads to

$$\begin{aligned}&\prod_{j\in\mathcal{J}}\binom{Q}{w_j}^{t_j-1}(q-1)^{(t_j-1)w_j}\\ &\doteq \exp\left\{n\left[\frac{1}{n_u}\sum_{j\in\mathcal{J}}(t_j-1)\mathsf{H}_\mathsf{e}(n_u\omega_j) + \omega_j(t_j-1)\ln(q-1)\right]\right\}.\end{aligned} \tag{5.91}$$

Application of the above asymptotic results and of Lemma 5.7 to (5.34) yields

$$G(\omega) = \frac{1}{\ln q}\max_{\boldsymbol{\omega}} S(\boldsymbol{\omega}), \tag{5.92}$$

where

$$S(\omega) = \sum_{i \in \mathcal{I}} \left(\frac{1}{n_u} \ln A_i(z_i) - \sum_{g \in \Gamma_i} \omega_g \ln z_g \right) \tag{5.93}$$

$$- \sum_{j \in \mathcal{J}} \left[\left(\frac{q_j - 1}{n_u} \right) \mathsf{H}_{\mathsf{e}}(n_u \omega_j) + \omega_j (q_j - 1) \ln(q-1) \right] \tag{5.94}$$

and where the maximization is subject to the constraints $\sum_{j \in \mathcal{J}_u} \omega_j = \omega$ and $0 \leq \omega_j \leq 1/n_u \; \forall j \in \mathcal{J}$. Using Lagrange multipliers, we obtain that for every $j \in \mathcal{J}_u$, we must have

$$\frac{\partial S(\omega)}{\partial \omega_j} = \mu \tag{5.95}$$

and for every $j \in \mathcal{J}_p$, we must have

$$\frac{\partial S(\omega)}{\partial \omega_j} = 0\,. \tag{5.96}$$

First, we consider the term on the left-hand side of both equations. This may be developed as

$$\frac{\partial S(\omega)}{\partial \omega_j} = \sum_{i \in \mathcal{I}} \sum_{g \in \Gamma_i} \left[\frac{\frac{\partial A_i}{\partial W_g}(z_i)}{n_u A_i(z_i)} - \frac{\omega_g}{z_g} \right] \frac{\partial z_g}{\partial \omega_j} - \sum_{g \in \Lambda_j} \ln z_g$$

$$+ (t_j - 1) \ln \left(\frac{n_u \omega_j}{1 - n_u \omega_j} \right) - (t_j - 1) \ln(q-1)\,. \tag{5.97}$$

The term in square brackets is equal to zero for any $i \in \mathcal{I}$, $g \in \Gamma_i$, from (5.90). Therefore, for all $j \in \mathcal{J}$, we must have

$$(t_j - 1) \ln \left(\frac{n_u \omega_j}{(q-1)(1 - n_u \omega_j)} \right) - \sum_{g \in \Lambda_j} \ln z_g = \begin{cases} \mu & \text{if } j \in \mathcal{J}_u \\ 0 & \text{if } j \in \mathcal{J}_p\,. \end{cases} \tag{5.98}$$

We conclude that for each ω, the evaluation of the weight spectral shape for a protograph LDPC code ensemble constructed over GF(q) requires solution of a system of $n_v + E_p + 1$ simultaneous equations in the same number of unknowns. The weight spectral shape function is given by

$$G(\omega) = \frac{1}{n_u \ln q} \left[\sum_{i \in \mathcal{I}} \left(\ln A_i(z_i) - n_u \sum_{g \in \Gamma_i} \omega_g \ln z_g \right) \right.$$

$$\left. - \sum_{j \in \mathcal{J}} \left((t_j - 1) \mathsf{H}_{\mathsf{e}}(n_u \omega_j) + n_u \omega_j (t_j - 1) \ln(q-1) \right) \right]. \tag{5.99}$$

The equations are: (5.90) for each $i \in \mathcal{I}$, $g \in \Gamma_i$ (together giving E_p equations); (5.98) for each $j \in \mathcal{J}$ (together giving n_v equations); and $\sum_{j \in \mathcal{J}_u} \omega_j = \omega$. The unknowns are: $\{\omega_j\}$ (n_v variables); $\{z_g\}$ (E_p variables); and μ (one variable).

The following lemma can, in general, reduce the dimension of the system of equations to be solved, when parallel edges are present in the protograph.

Lemma 5.8 *If two edges $a,b \in \Phi$ are connected to the same pair of VN and CN in the protograph, then $z_a = z_b$.*

Proof Consider two parallel edges $a,b \in \Phi$ which connect VN $j \in \mathcal{J}$ to CN $i \in \mathcal{I}$. Since the local weight enumerator $A_i(\boldsymbol{W})$ is symmetric in the variables $\{z_g\}$, $g \in \Gamma_i$, it follows that if a solution exists to the system of equations with $(z_a, z_b) = (\alpha, \beta)$, then another solution must exist with $(z_a, z_b) = (\beta, \alpha)$, all other variables being unchanged. Then, the hypothesis that $\beta \neq \alpha$ leads to a contradiction (specifically, it contradicts the uniqueness of the positive solution z_i to (5.90) for CN type i). □

Remark 5.11. The solution to the $(n_v + E_p + 1) \times (n_v + E_p + 1)$ system for $\omega = \frac{q-1}{q}$ is $\omega_j = \frac{1}{n_u}\frac{q-1}{q}$ for all $j \in \mathcal{J}_u$ and for all $j \in \mathcal{J}_u$, $z_g = 1$ for all $g \in \Phi$, and $\mu = 0$ and we have

$$G\left(\frac{q-1}{q}\right) = \mathrm{R}. \tag{5.100}$$

The above value represents the maximum of the weight spectral shape function.

Remark 5.12. The critical fractional codeword weight ω^* of a protograph ensemble sequence may be directly calculated by solving an augmented system of $n_v + E_p + 2$ simultaneous equations in the same number of unknowns. The equations are again (5.90) for each $i \in \mathcal{I}$, $g \in \Gamma_i$, (5.98) for each $j \in \mathcal{J}$, and $\sum_{j \in \mathcal{J}_u} \omega_j = \omega^*$, together with $S(\omega) = 0$ where $S(\omega)$ is given by (5.93). The unknowns are again $\{w_j\}$, $\{z_g\}$, and μ, together with ω^*.

Example 5.8 Consider the protograph of an ARJA LDPC code ensemble of design rate $\mathrm{R} = 1/2$, depicted in Figure 4.8. The Q VNs being copies of v_4 are punctured. As discussed in Example 4.9 we have $n_v = 5$ and $E_p = 15$ and so, to calculate one point on the weight spectral shape curve, we must solve a 21×21 system. Use of Lemma 5.8 reduces this to a 16×16 system. As described in Example 4.10, an RJA protograph ensemble with design rate $\mathrm{R} = 1/2$ is obtained by removing v_5, c_3, and all edges connected to them (in the RJA case, all VNs are transmitted). Since $n_v = 4$ and $E_p = 12$, evaluation in this case requires solution of a 17×17 system, which is reduced to a 13×13 system on invoking again Lemma 5.8. The growth rate curves of both ARJA and RJA ensembles over GF(q), $q \in \{2,4\}$, are shown in Figure 5.5. The solid curves in Figure 5.5 labeled as "random" correspond to the weight spectral shapes of (parity-check based) linear random ensembles with rate-$1/2$ over GF(q), $q \in \{2,4\}$, and are used as benchmarks. For each q the plot is shown for $\omega \leq (q-1)/q$. The critical fractional codeword weight values, calculated using the augmented system of equations described in Remark 5.12, are reported in Table 5.4 for $q \in \{2,4,8,16\}$.

Interestingly, the weight spectral shape of protograph LDPC code ensembles may be recast in a form similar to the one obtained for unstructured ensembles. For all $j \in \mathcal{J}$, let $B_j(X,Y) = 1 + (q-1)XY^{t_j}$ be the input–output weight enumerating

Table 5.4 Critical fractional codeword weight values for rate-1/2 ARJA and RJA protograph ensembles over GF(q), for $q \in \{2,4,8,16\}$

q	ω^* (ARJA)	ω^* (RJA)
2	0.014401	0.013316
4	0.030871	0.028576
8	0.046681	0.043253
16	0.059427	0.055104

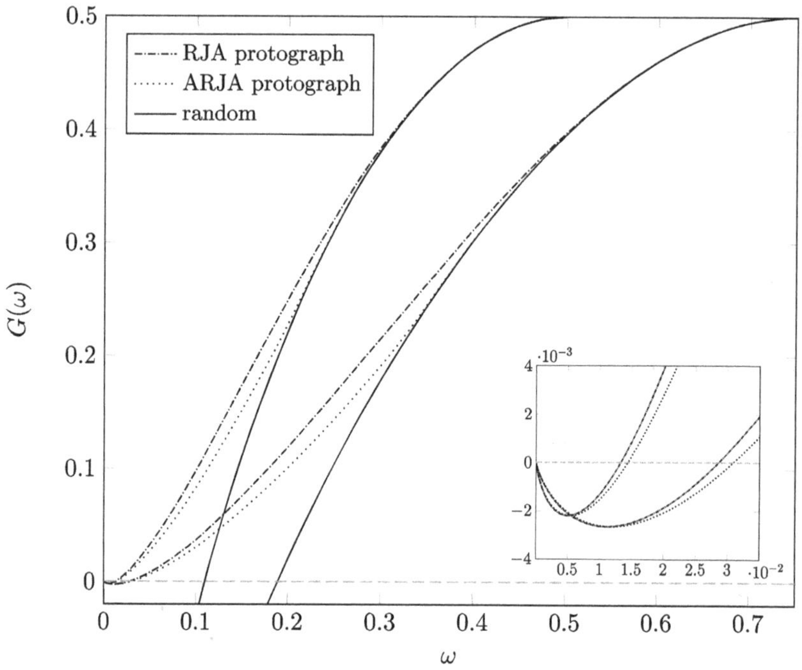

Figure 5.5 Weight spectral shape functions for $R = 1/2$ ARJA (dotted) and RJA (dot-dashed) protograph LDPC code ensemble sequences.

function of a $(t_j, 1)$ repetition code. Then it is possible to show (see Exercise 5.8) that the weight spectral shape of any protograph LDPC code ensemble constructed over GF(q) is given by

$$G(\omega) = \frac{1}{\ln q}\left[\frac{1}{n_u}\sum_{i\in\mathcal{I}} \ln A_i(z_i) - \omega \ln x_0 \right. \\ \left. + \frac{1}{n_u}\sum_{j\in\mathcal{J}_u} \ln B_j(x_0, y_{0,j}) + \frac{1}{n_u}\sum_{j\in\mathcal{J}_p} \ln B_j(1, y_{0,j})\right]. \tag{5.101}$$

The $n_v + E_p + 1$ variables x_0, $y_{0,j}$ (for all $j \in \mathcal{J}$), and z_g (for all $g \in \Phi$) appearing in (5.101) are the unique positive roots of the $n_v + E_p + 1$ simultaneous equations

$$\frac{1}{n_u} x_0 \sum_{j \in \mathcal{J}_u} \frac{\frac{\partial B_j}{\partial X}(x_0, y_{0,j})}{B_j(x_0, y_{0,j})} = \omega \tag{5.102}$$

$$\frac{z_g \frac{\partial A_i}{\partial W_g}(z_i)}{A_i(z_i)} = \begin{cases} x_0 \frac{\partial B_g}{\partial X}(x_0, y_{0,g})/B_g(x_0, y_{0,g}) & \forall\, i \in \mathcal{I}, j(g) \in \mathcal{J}_u \\ \frac{\partial B_g}{\partial X}(1, y_{0,g})/B_g(1, y_{0,g}) & \forall\, i \in \mathcal{I}, j(g) \in \mathcal{J}_p \end{cases} \tag{5.103}$$

and

$$\left(\prod_{g \in \Lambda_j} z_g\right)^{1/t_j} = \begin{cases} x_0 y_{0,j}^{t_j - 1} & \forall\, j \in \mathcal{J}_u \\ y_{0,j}^{t_j - 1} & \forall\, j \in \mathcal{J}_p \,. \end{cases} \tag{5.104}$$

Note that in (5.103), $j(g) \in \mathcal{J}$ is the index of the VN in the protograph connected to edge $g \in \Phi$. Moreover, the variable $y_{0,j}$ and the function $B_j(X,Y)$ have been relabeled as $y_{0,g}$ and $B_g(X,Y)$, respectively, whenever the VN $j \in \mathcal{J}$ is the one connected to the edge $g \in \Phi$ in the protograph.

5.7 Minimum Distance Analysis

In Section 5.6, the weight spectral shape of an ensemble of codes has been defined and efficient techniques to calculate the weight spectral shapes of some LDPC code ensembles have been presented. Next, we discuss an important application of the weight spectral shape function of LDPC code ensembles, namely, the possibility to upper bound the c.d.f. of the minimum distance of random LDPC codes and to prove a linear minimum distance property for some LDPC ensembles. We start by developing an upper bound on the c.d.f. of the minimum distance of a random code drawn from an ensemble (Lemma 5.9). This lemma is exploited to prove an upper bound on the average weight enumerating function of expurgated ensembles (Theorem 5.10), a result that will be useful in later chapters, and to prove the above-mentioned linear minimum distance growth property for regular LDPC codes (Theorem 5.11).

Consider an ensemble of block codes and let $\mathcal{C}$ be a random code drawn from it according to some probability distribution. Let the random variable $D_{\min} \in \{1,2,\ldots,n\}$ be the minimum distance of $\mathcal{C}$, where n is the length of all codes in the ensemble. The following lemma establishes a simple upper bound on the probability $P(D_{\min} \le d)$ for any $d \in \{1,2,\ldots,n\}$.

Lemma 5.9 *Let $\mathcal{C}$ be a random code drawn from an ensemble of block codes all of length n. Let the random variable $D_{\min}$ be the minimum distance of $\mathcal{C}$. Moreover, let $\bar{A}_w$ be the expected multiplicity of codewords of weight w in $\mathcal{C}$. Then*

$$P(D_{\min} \le d) \le \min\left\{1, \sum_{w=1}^{d} \bar{A}_w\right\}. \tag{5.105}$$

Proof Let the random variable A_w be the multiplicity of codewords of weight w in $\mathcal{C}$. Since $A_w \geq 0$, by Markov's inequality, we have

$$P(A_w \geq 1) \leq \bar{A}_w \,. \tag{5.106}$$

Hence, we can write

$$\begin{aligned} P(D_{\min} \leq d) &= P\left(\bigcup_{w=1}^{d} \{A_w \geq 1\}\right) \\ &\leq \sum_{w=1}^{d} P(A_w \geq 1) \\ &\leq \sum_{w=1}^{d} \bar{A}_w \,. \end{aligned} \tag{5.107}$$

The proof is completed by observing that we also have $P(D_{\min} \leq d) \leq 1$. □

Definition 5.9 (Ensemble expurgation) Consider an ensemble of block codes $\mathscr{C}$, all of length n, and let d^* be a positive integer. By ensemble expurgation with respect to d^*, we mean the operation consisting of removing from $\mathscr{C}$ all codes whose minimum distance is less than or equal to d^*.

The ensemble expurgation yields a new ensemble of smaller cardinality, $\mathscr{C}^{\text{ex}}$, that is referred to as *expurgated ensemble*. If a probability distribution $Q(\mathcal{C})$ exists on the codes in the original ensemble, the probability distribution on the codes in the expurgated ensemble is obtained by re-normalizing all probabilities.

Theorem 5.10 *Let $\mathscr{C}$ be an ensemble of block codes, all of length n, $\bar{A}_w$ be the expected multiplicity of codewords of weight w in a random code drawn from $\mathscr{C}$, and $F(d) = \min\{1, \sum_{w=1}^{d} \bar{A}_w\}$. For any $0 < p^* < 1$, let $d^* = \max\{d > 0 : F(d) \leq p^*\}$ and $\mathscr{C}^{\text{ex}}$ be the expurgated ensemble with respect to d^*. Then, the expected multiplicity of codewords of weight w of a random code drawn from $\mathscr{C}^{\text{ex}}$, $\bar{A}_w^{\text{ex}}$, fulfills*

$$\bar{A}_w^{\text{ex}} = 0 \quad \text{if } 1 \leq w \leq d^* \tag{5.108}$$

and

$$\bar{A}_w^{\text{ex}} \leq \frac{1}{1-p^*} \bar{A}_w \quad \text{if } d^* < w \leq n. \tag{5.109}$$

Proof Let $D_{\min}$ be the minimum distance of a random code drawn in the ensemble $\mathscr{C}$ and define $p = P(D_{\min} \leq d^*)$. On invoking Lemma 5.9, we have $p \leq p^*$. We can partition the original ensemble into the ensemble of codes with minimum distance at least d^* (the expurgated ensemble) and the ensemble of codes with minimum distance smaller than d^* (the ones removed during the expurgation). Letting the expected multiplicity of codewords of weight w over this latter ensemble be $\bar{A}'_w$, we can write $\bar{A}_w = (1-p)\bar{A}_w^{\text{ex}} + p\bar{A}'_w$ from which

$$\bar{A}_w^{\text{ex}} = \frac{1}{1-p}\bar{A}_w - \frac{p}{1-p}\bar{A}'_w \leq \frac{1}{1-p}\bar{A}_w \leq \frac{1}{1-p^*}\bar{A}_w, \tag{5.110}$$

where the last inequality is due to $p \leq p^*$. Although the obtained upper bound on $\bar{A}_w^{\text{ex}}$ is valid for any $1 \leq w \leq n$, for $1 \leq w \leq d^*$ we have $\bar{A}_w^{\text{ex}} = 0$ by definition of expurgation. □

As we will show in the remainder of the section, application of Lemma 5.9 to LDPC code ensembles, in conjunction with the weight spectral shape function and its properties, allows to prove rigorously that some LDPC code ensemble sequences are characterized by a linear growth of the minimum distance. We start by providing a formal definition for this concept.

Definition 5.10 [Linear minimum distance growth] Let $\{\mathscr{C}_n\}$ be a sequence of block code ensembles with weight spectral shape function $G(\cdot)$. For any n, let the random variable $D_{\min}^{(n)}$ represent the minimum distance of a random code $\mathcal{C}$ drawn in $\mathscr{C}_n$ according to some probability distribution. We say that the ensemble sequence has a linear minimum distance growth when the critical fractional codeword weight ω^* is strictly positive and when

$$\lim_{n\to\infty} P\left(D_{\min}^{(n)} \leq \omega n\right) = 0 \tag{5.111}$$

for any constant rational ω such that $0 < \omega < \omega^*$. The limit assumes the inclusion of only those positive integers n for which ωn is an integer.

In the following theorem, we discuss fulfillment of Definition 5.10 for the simplest LDPC code ensembles introduced in this chapter, namely, the unstructured regular ensemble described in Section 5.4.

Theorem 5.11 *For every* $\mathsf{d}_v \geq 3$, *a sequence* $\{\mathscr{C}_{\mathsf{d}_v,\mathsf{d}_c}^{n,q}\}$ *of unstructured regular LDPC code ensembles has a linear minimum distance growth.*

Proof First of all, observe from Remark 5.5 that for $\mathscr{C}_{\mathsf{d}_v,\mathsf{d}_c}^{n,q}$ ensembles, $\omega^* = 0$ if $\mathsf{d}_v = 2$ and $\omega^* > 0$ if $\mathsf{d}_v \geq 3$. Hence, fulfillment of condition (5.111) is possible only in the $\mathsf{d}_v \geq 3$ case. Let us recall that the expected multiplicity of codewords of weight w for a random code $\mathcal{C} \in \mathscr{C}_{\mathsf{d}_v,\mathsf{d}_c}^{n,q}$ is given by (5.23). Combining Lemma 5.4 and Lemma 5.6, and using (2.61), we may write

$$\begin{aligned}
&\operatorname{coeff}\left(\left(P_{q,\mathsf{d}_c}(Z)\right)^m, Z^{w\mathsf{d}_v}\right)\\
&\quad\leq \sum_{\boldsymbol{d}\in\mathcal{N}_{w\mathsf{d}_v}} \exp\left\{m\,\mathsf{H}_{\mathsf{e}}\left(\frac{1}{m}\boldsymbol{d}\right)\right\} \prod_{t=2}^{\mathsf{d}_c}\left(\binom{\mathsf{d}_c}{t}p_{q,t}\right)^{d_t}\\
&\quad= \sum_{\boldsymbol{d}\in\mathcal{N}_{w\mathsf{d}_v}} \exp\left\{m\left[\mathsf{H}_{\mathsf{e}}\left(\frac{1}{m}\boldsymbol{d}\right) + \sum_{t=2}^{\mathsf{d}_c}\frac{d_t}{m}\ln\left(\binom{\mathsf{d}_c}{t}p_{q,t}\right)\right]\right\},
\end{aligned} \tag{5.112}$$

where $\mathcal{N}_{w\mathsf{d}_v}$ is the set of all vectors $\boldsymbol{d} = \begin{bmatrix} d_0 & d_2 & \dots & d_{\mathsf{d}_c}\end{bmatrix}$ that are nonnegative integer solutions to $d_0 + \sum_{t=2}^{\mathsf{d}_c} d_t = m$ and $\sum_{t=2}^{\mathsf{d}_c} t\,d_t = w\,\mathsf{d}_v$. The last expression may be further bounded above as

$$\begin{aligned}
&\mathrm{coeff}\left(\left(P_{q,\mathsf{d}_c}(Z)\right)^m, Z^{w\mathsf{d}_v}\right)\\
&\quad \le |\mathcal{N}_{w\mathsf{d}_v}| \exp\left\{ m \max_{\boldsymbol{d} \in \mathcal{N}_{w\mathsf{d}_v}} \left\{ \mathsf{H}_\mathsf{e}\left(\frac{1}{m}\boldsymbol{d}\right) + \sum_{t=2}^{\mathsf{d}_c} \frac{d_t}{m} \ln\left(\binom{\mathsf{d}_c}{t} p_{q,t}\right)\right\}\right\}\\
&\quad \le |\mathcal{N}_{w\mathsf{d}_v}| \exp\left\{ m \max_{\boldsymbol{\nu}} \left\{ \mathsf{H}_\mathsf{e}(\boldsymbol{\nu}) + \sum_{t=2}^{\mathsf{d}_c} \nu_t \ln\left(\binom{\mathsf{d}_c}{t} p_{q,t}\right)\right\}\right\}, \qquad (5.113)
\end{aligned}$$

where $\boldsymbol{\nu} = \begin{bmatrix}\nu_0 & \nu_2 & \nu_3 & \dots & \nu_{\mathsf{d}_c}\end{bmatrix} \in \mathbb{R}^{\mathsf{d}_c}$ and where the maximization is subject to $\nu_i \ge 0$ for all $i \in \{0, 2, 3, \dots, \mathsf{d}_c\}$,

$$\sum_{t=2}^{\mathsf{d}_c} t\,\nu_t = \frac{w}{n}\mathsf{d}_c \qquad (5.114)$$

$$\nu_0 + \sum_{t=2}^{\mathsf{d}_c} \nu_t = 1\,. \qquad (5.115)$$

The constrained maximization problem may be solved by Lagrange multipliers. Denoting the Lagrange multipliers by λ and μ, we obtain

$$-\ln \nu_t - 1 + \ln\left(\binom{\mathsf{d}_c}{t} p_{q,t}\right) + \lambda t + \mu = 0 \qquad (5.116)$$

for all $t \in \{0, 2, 3, \dots, \mathsf{d}_c\}$; due to $p_{q,0} = 1$, this may be recast as

$$\nu_t = \nu_0 \binom{\mathsf{d}_c}{t} p_{q,t}\, z_0^t \qquad (5.117)$$

for all $t \in \{0, 2, 3, \dots, \mathsf{d}_c\}$, where $z_0 = e^\lambda > 0$. Substituting (5.117) in (5.114) and in (5.115), and dividing term by term, yields

$$\frac{z_0 \frac{\mathrm{d}P_{q,\mathsf{d}_c}}{\mathrm{d}Z}(z_0)}{P_{q,\mathsf{d}_c}(z_0)} = \frac{w}{n}\mathsf{d}_c \qquad (5.118)$$

which coincides with (5.49). Moreover, substituting (5.117) into the objective function, after simple algebraic manipulation, we find that at the maximum we have

$$\begin{aligned}
\mathsf{H}_\mathsf{e}(\boldsymbol{\nu}) + \sum_{t=2}^{\mathsf{d}_c} \nu_t \ln\left(\binom{\mathsf{d}_c}{t} p_{q,t}\right) &= -\ln \nu_0 - \frac{w}{n}\mathsf{d}_c \ln z_0\\
&= \ln P_{q,\mathsf{d}_c}(z_0) - \frac{w}{n}\mathsf{d}_c \ln z_0, \qquad (5.119)
\end{aligned}$$

where the second equality easily follows from (5.115) and (5.117). Based on the solution of the maximization problem, we can then write

$$\mathrm{coeff}\left(\left(P_{q,\mathsf{d}_c}(Z)\right)^m, Z^{w\mathsf{d}_v}\right) \le |\mathcal{N}_{w\mathsf{d}_v}| \exp\left\{ n\left[(1-\mathsf{R}) \ln P_{q,\mathsf{d}_c}(z_0) - \frac{w}{n}\mathsf{d}_v \ln z_0\right]\right\}, \qquad (5.120)$$

where $z_0 > 0$ solves (5.49).

Next, we upper bound the cardinality of the set $\mathcal{N}_{w\mathsf{d}_v}$. To this aim, observe that every local CN codeword has weight at least 2. Hence, for all $t \in \{2, 3, \dots, \mathsf{d}_c\}$ we have

$$2\,d_j \le \sum_{t=2}^{\mathsf{d}_c} 2\,d_t \le \sum_{t=2}^{\mathsf{d}_c} t\,d_t = w\,\mathsf{d}_v \tag{5.121}$$

from which we conclude that $0 \le d_j \le \frac{w\mathsf{d}_v}{2}$, $j \in \{2,3,\ldots,\mathsf{d}_c\}$. Since d_0 is constrained to fulfill $d_0 = m - \sum_{t=2}^{\mathsf{d}_c} d_t$, the number of vectors $\boldsymbol{d} \in \mathcal{N}_{w\mathsf{d}_v}$ is bounded as $|\mathcal{N}_{w\mathsf{d}_v}| \le \left(\frac{w\mathsf{d}_v}{2}\right)^{\mathsf{d}_c-1}$ and we write

$$\begin{aligned}
&\mathrm{coeff}\Big(\big(P_{q,\mathsf{d}_c}(Z)\big)^m, Z^{w\mathsf{d}_v}\Big)\\
&\quad \le \left(\frac{w\,\mathsf{d}_v}{2}\right)^{\mathsf{d}_c-1} \exp\Big\{n\Big[(1-\mathrm{R})\ln P_{q,\mathsf{d}_c}(z_0) - \frac{w}{n}\,\mathsf{d}_v \ln z_0\Big]\Big\}.
\end{aligned} \tag{5.122}$$

To find an upper bound to $\bar{A}_w$ we use again Lemma 5.6 (upper and lower bounds for the binomial coefficient), obtaining

$$\begin{aligned}
\bar{A}_w &< (\mathsf{d}_v)^{1/2}\left(\frac{w\,\mathsf{d}_v}{2}\right)^{\mathsf{d}_c-1} \exp\left\{\frac{1}{12\,\mathsf{d}_v\,w(1-\frac{w}{n})}\right\}\\
&\quad \times \exp\Big\{n\Big[(1-\mathrm{R})\ln P_{q,\mathsf{d}_c}(z_0) - \frac{w}{n}\,\mathsf{d}_v \ln z_0\Big]\Big\}\\
&\quad \times \exp\Big\{n\Big[-(\mathsf{d}_v-1)\mathsf{H}_{\mathsf{e}}\Big(\frac{w}{n}\Big) - \frac{w}{n}(\mathsf{d}_v-1)\ln(q-1)\Big]\Big\}\\
&= (\mathsf{d}_v)^{1/2}\left(\frac{w\,\mathsf{d}_v}{2}\right)^{\mathsf{d}_c-1} \exp\left\{\frac{1}{12\,\mathsf{d}_v\,w\,(1-\frac{w}{n})}\right\} e^{n\,G(\frac{w}{n})\ln q}\\
&\le (\mathsf{d}_v)^{1/2}\left(\frac{w\,\mathsf{d}_v}{2}\right)^{\mathsf{d}_c-1} \exp\left\{\frac{1}{6\,\mathsf{d}_v\,w}\right\} e^{n\,G(\frac{w}{n})\ln q},
\end{aligned} \tag{5.123}$$

where $G(\cdot)$ is the weight spectral shape function given by (5.50) and where the last inequality follows from $w(1-\frac{w}{n}) = n\,\omega(1-\omega)$ and from the function $\omega(1-\omega)$ having a maximum at $\omega = 1/2$.

Let the random variable $D_{\min}$ be the minimum Hamming distance of a random code $\mathcal{C} \in \mathscr{C}^{n,q}_{\mathsf{d}_v,\mathsf{d}_c}$. Moreover, let

$$f(w) = (\mathsf{d}_v)^{1/2}\left(\frac{w\,\mathsf{d}_v}{2}\right)^{\mathsf{d}_c-1} \exp\left\{\frac{1}{6\,\mathsf{d}_v\,w}\right\}. \tag{5.124}$$

For any constant and rational $0 < \delta < 1$ such that δn is an integer, the probability $P(D_{\min} \le \delta n)$ may be upper bounded as

$$P\Big(D^{(n)}_{\min} \le \delta n\Big) \le \min\left\{1, \sum_{w=1}^{\delta n} f(w) e^{n\,G(\frac{w}{n})\ln q}\right\}. \tag{5.125}$$

To develop the right-hand side of (5.125), let us analyze the generic term of the sum corresponding to some w independent of n. As n becomes large, we have $\frac{w}{n} \to 0$ and therefore we can resort on the expression of the weight spectral shape function in Remark 5.5, valid for small $\omega = \frac{w}{n}$.

Using the expression for $\mathsf{d}_v \geq 3$, after simple algebraic manipulation, we obtain

$$f(w)e^{n\,G(\frac{w}{n})\ln q} = f(w)(\mathsf{d}_c-1)\left(\frac{(\mathsf{d}_c-1)w}{q}\right)^{w\left(\frac{\mathsf{d}_v}{2}-1\right)}\frac{1}{n^{w\left(\frac{\mathsf{d}_v}{2}-1\right)}} + \mathsf{o}\left(\frac{1}{n^{w\left(\frac{\mathsf{d}_v}{2}-1\right)}}\right)$$
$$= c(w)\frac{1}{n^{w\left(\frac{\mathsf{d}_v}{2}-1\right)}} + \mathsf{o}\left(\frac{1}{n^{\frac{\mathsf{d}_v}{2}-1}}\right), \tag{5.126}$$

where, by definition, $c(w) = f(w)(\mathsf{d}_c-1)\left(\frac{(\mathsf{d}_c-1)w}{q}\right)^{w\left(\frac{\mathsf{d}_v}{2}-1\right)}$. The previous equation reveals that for $w = 1$, we have

$$f(1)e^{n\,G(\frac{1}{n})\ln q} = c(1)\frac{1}{n^{\frac{\mathsf{d}_v}{2}-1}} + \mathsf{o}\left(\frac{1}{n^{\frac{\mathsf{d}_v}{2}-1}}\right), \tag{5.127}$$

while for any $w \geq 2$ independent of n, we have

$$f(w)e^{n\,G(\frac{w}{n})\ln q} = \mathsf{o}\left(\frac{1}{n^{\frac{\mathsf{d}_v}{2}-1}}\right). \tag{5.128}$$

Due to the above observation, defining $\bar{w} = \lceil\frac{2\mathsf{d}_c}{\mathsf{d}_v-2}\rceil + 2$, for $\frac{1}{n} < \delta < \frac{q-1}{q}$ we can write

$$\sum_{w=1}^{\delta n} f(w)e^{n\,G(\frac{w}{n})\ln q}$$
$$= c(1)\frac{1}{n^{\frac{\mathsf{d}_v}{2}-1}} + \mathsf{o}\left(\frac{1}{n^{\frac{\mathsf{d}_v}{2}-1}}\right) + \sum_{w=\bar{w}}^{\delta n} f(w)e^{n\,G(\frac{w}{n})\ln q} \tag{5.129}$$
$$< c(1)\frac{1}{n^{\frac{\mathsf{d}_v}{2}-1}} + \mathsf{o}\left(\frac{1}{n^{\frac{\mathsf{d}_v}{2}-1}}\right) + \sum_{w=\bar{w}}^{\delta n} f(w)\left(e^{n\,G(\frac{\bar{w}}{n})\ln q} + e^{n\,G(\delta)\ln q}\right) \tag{5.130}$$
$$< c(1)\frac{1}{n^{\frac{\mathsf{d}_v}{2}-1}} + \mathsf{o}\left(\frac{1}{n^{\frac{\mathsf{d}_v}{2}-1}}\right) + nf(n)\left(e^{n\,G(\frac{\bar{w}}{n})\ln q} + e^{n\,G(\delta)\ln q}\right) \tag{5.131}$$
$$= c(1)\frac{1}{n^{\frac{\mathsf{d}_v}{2}-1}} + \mathsf{o}\left(\frac{1}{n^{\frac{\mathsf{d}_v}{2}-1}}\right) + nf(n)e^{n\,G(\delta)\ln q}. \tag{5.132}$$

The inequality (5.130) is due to the properties of the weight spectral shape function of $\mathscr{C}^{n,q}_{\mathsf{d}_v,\mathsf{d}_c}$ ensembles summarized in Remark 5.6. Accordingly, for any $\bar{w} < w < \delta n$ and $\delta < \frac{q-1}{q}$ we have either $\exp\{nG(\frac{w}{n})\ln q\} < \exp\{nG(\frac{\bar{w}}{n})\ln q\}$ or $\exp\{nG(\frac{w}{n})\ln q\} < \exp\{nG(\delta)\ln q\}$ and therefore $\exp\{nG(\frac{w}{n})\ln q\}$ is also bounded by the sum of the two. Moreover, the inequality (5.131) is due to $f(w) < f(n)$ for all $\bar{w} \leq w \leq \delta n$. The last equality follows from the fact that, with the above choice of $\bar{w}$, we have

$$nf(n)e^{nG(\frac{\bar{w}}{n})\ln q} = \mathsf{o}\left(\frac{1}{n^{\frac{\mathsf{d}_v}{2}-1}}\right). \tag{5.133}$$

From (5.125) and (5.132), it is immediate to recognize that $P(D^{(n)}_{\min} \leq \delta n) \to 0$ in the limit where $n \to \infty$, for any constant rational $0 < \delta < \omega^*$. Notably, since the bound (5.132) is tight for large enough n, $P(D^{(n)}_{\min} \leq \delta n)$ converges to zero as fast as $n^{-\frac{\mathsf{d}_v}{2}+1}$. □

5.8 Historical Notes

Gallager [1] developed the average weight distribution and the weight spectral shape of both his binary and nonbinary regular LDPC code ensembles. Actually, the $B_{jkq}(\omega)$ function in [1, eq. (5.13)] coincides with (5.50).

The weight distribution of unstructured LDPC code ensembles, including asymptotic exponent analysis and minimum distance linear growth properties, was addressed by Burshtein and Miller in [2], Orlitsky *et al.* in [3], and Di *et al.* in [4]. Moreover, a combinatorial expression of the average weight distribution of LDPC code ensembles with separated low-degree variable nodes appears in Di [5], together a solution for its weight spectral shape.

A first approach to calculate the average weight distribution and the weight spectral shape of binary protograph-based LDPC codes was proposed in 2005 by S. Fogal *et al.* [6]. The approach in [6] follows the method of types and yields a formulation of each point of the weight spectral shape function as the solution, to be found via numerical methods, of a constrained optimization problem. One year later, another approach was proposed by D. Divsalar [7]. The combinatorial expression of the average weight distribution in [7] is identical to the one obtained in Section 5.5 for $q = 2$; the weight spectral shape is still expressed as the solution of a constrained optimization problem to be solved numerically. The approach of [7] was inherited by [8], in which Abu Surra *et al.* presented methods to calculate several enumerators for binary protograph LDPC code ensembles. More recently, the approach was also extended to protograph LDPC code ensembles constructed on GF(q) by Dolecek *et al.* [9]. The method to calculate the weight spectral shape of protograph LDPC code ensembles described in this chapter differs from the above approaches in that it expresses each point of the weight spectral shape as the solution of a system of polynomial equations. Moreover, it allows to solve directly for the critical fractional codeword weight. This method was proposed in [10].

5.9 Exercises

Exercise 5.1 Consider a sequence $\{\mathscr{C}^{n,q}_{\mathsf{d}_v,\mathsf{d}_c}\}$ of unstructured regular LDPC code ensembles with design rate R, and let C be a random code in $\mathscr{C}^{n,q}_{\mathsf{d}_v,\mathsf{d}_c}$. Let the random variable $R(C)$ be the code rate of C. Show that for any constant $\epsilon > 0$, we have $P(R(C) - \mathrm{R} > \epsilon) \to 0$ as $n \to \infty$. *Hint:* Take $Z = R(C) - \mathrm{R} \geq 0$. Bound $\mathsf{E}[Z]$ from above as $\mathsf{E}[Z] \leq \frac{1}{n}\ln_q(n+1) + \frac{1}{n}\ln_q f(n)$, where $f(w)$ is defined in (5.124) (use Jensen's inequality and $\bar{A}_w \leq f(w)\exp\{n\,G(\frac{w}{n})\ln q\}$). Conclude by Markov's inequality.

Exercise 5.2 With reference to Remark 5.3, prove that the weight spectral shape function for unstructured regular $\mathscr{C}^{n,q}_{\mathsf{d}_v,\mathsf{d}_c}$ ensembles, given by (5.50), takes its maximum for $\omega = \frac{q-1}{q}$. *Hint:* Regard the weight spectral shape as a function of ω and z_0 and find its global maximum in the domain defined by $0 \leq \omega \leq 1$ and $z_0 > 0$. Show that the maximizing (ω, z_0) fulfills the constraint (5.49).

Exercise 5.3 Prove that for small values of ω, the weight spectral shape function of a sequence of unstructured irregular LDPC code ensembles $\mathscr{C}^{n,q}_{\lambda,\rho}$ assumes the form (5.86). *Hint*:

(1) Using (5.60) and (5.72), show that $\beta \to 0^+$ as $\omega \to 0^+$.
(2) From step (1) and (5.68), show that $z_0^2 = \mathsf{o}(1)$ and $z_0^i = \mathsf{o}(\beta)$ for all $i > 2$. Consequently, show that

$$z_0(\beta) = \left(\frac{\beta \int \lambda}{(q-1)\rho'(1)}\right)^{\frac{1}{2}} (1 + \mathsf{o}(1)).$$

(3) From steps (1) and (2), and (5.77), show that

$$y_0(\beta) = \left(\frac{\rho'(1)\beta \int \lambda}{q-1}\right)^{\frac{1}{2}} (1 + \mathsf{o}(1)).$$

(4) From step (1) and (5.82), show that $x_0 y_0^{\mathsf{p}} = \mathsf{o}(1)$ and $x_0 y_0^j = \mathsf{o}(\beta)$ for all $j > \mathsf{p}$. Using step (3) show that

$$x_0(\beta) = \frac{1}{\mathsf{p}\Lambda_{\mathsf{p}}}(q-1)^{\frac{\mathsf{p}-2}{2}} (\rho'(1) \textstyle\int \lambda)^{-\frac{\mathsf{p}}{2}} \beta^{\frac{2-\mathsf{p}}{2}} (1 + \mathsf{o}(1)).$$

(5) Incorporating in (5.81) the previous expressions of $y_0(\beta)$ and $x_0(\beta)$, we get

$$\beta(\omega) = \mathsf{p}\omega(1 + \mathsf{o}(1)).$$

(6) Conclude the proof using (5.84).

Exercise 5.4 Elaborating on Exercise 5.3 shows that, if the minimum VN degree p fulfills $\mathsf{p} \geq 3$ and if the critical fractional codeword weight ω^* is small, we can approximate ω^* as

$$\omega^* \approx \exp(1) \cdot \left(\lambda_{\mathsf{p}}^{\frac{2}{\mathsf{p}-2}} (\rho'(1))^{\frac{\mathsf{p}}{\mathsf{p}-2}} \frac{\mathsf{p}\int \lambda}{q-1}\right)^{-1}.$$

Verify numerically the quality of the approximation for regular LDPC code ensembles with $\mathsf{p} = 3$.

Exercise 5.5 Prove that the expected number of codewords of Hamming weight w over an unstructured irregular LDPC code ensemble, given by (5.30), can be equivalently expressed as

$$\bar{A}_w = \sum_{l=1}^{n} \frac{\mathrm{coeff}\left(B(W,Z), W^w Z^l\right) \mathrm{coeff}\left(\prod_{i=2}^{\mathsf{d}_c} \left(P_{q,i}(Z)\right)^{P_i m}, Z^l\right)}{\binom{\sum_{j=2}^{\mathsf{d}_v} \Lambda_j j n}{l} (q-1)^l},$$

where $B(W,Z) = \prod_{j=2}^{\mathsf{d}_v} \left(1 + (q-1)WZ^j\right)^{\Lambda_j n}$.

Exercise 5.6 Consider an ensemble of regular $(\mathsf{d}_v, \mathsf{d}_c)$ LDPC codes of length n over GF(q) in which all nonzero elements of the parity-check matrix are constrained to be equal to $1 \in \mathrm{GF}(q)$. The ensemble is constructed as explained in Section 4.2.1 but with unlabeled edges. Let $\boldsymbol{f} = \begin{bmatrix} f_0 & f_1 & \dots & f_{q-2} \end{bmatrix}$ be the symbol multiplicity vector of the generic codeword, that is, f_i is the number of symbols equal to $\alpha^i \in \mathrm{GF}(q)$ in the codeword. Let $\Upsilon_{\boldsymbol{f}}$ be the multiplicity of codewords with symbol multiplicity vector $\boldsymbol{f}$ in the linear block code of length d_c and parity-check matrix $\begin{bmatrix} 1 & 1 & \cdots & 1 \end{bmatrix}$, and $\Upsilon(\boldsymbol{F}) = \sum_{\boldsymbol{f}} \Upsilon_{\boldsymbol{f}} \boldsymbol{F}^{\boldsymbol{f}}$ be the corresponding multivariate enumerating function. Show that the expected number of codewords of weight w of a random code in the ensemble is

$$\bar{A}_w = \sum_{\boldsymbol{f} \in \mathcal{F}_w} \binom{n}{n-w, f_0, \dots, f_{q-2}} \frac{\operatorname{coeff}\left((\Upsilon(\boldsymbol{F}))^m, \boldsymbol{F}^{\mathsf{d}_v \boldsymbol{f}}\right)}{\binom{\mathsf{d}_v n}{\mathsf{d}_v (n-w), \mathsf{d}_v f_0, \dots, \mathsf{d}_v f_{q-2}}},$$

where $\mathcal{F}_w = \{\boldsymbol{f} : \sum_{i=0}^{q-2} f_i = w\}$.

Exercise 5.7 Develop an expression of the weight spectral shape for the regular ensemble considered in Exercise 5.6.

Exercise 5.8 Consider the protograph shown in Figure 4.8. In it, remove one of the two edges labeled as 4 and 5, thus making the degree of VN v_2 equal to 2. For $q \in \{2, 4, 8, 16\}$ numerically compute the critical fractional codeword weight ω^* for the new protograph, showing that the obtained values are lower than the ones reported in Table 5.4 for the original protograph.

Exercise 5.9 Consider the binary protograph LDPC code ensemble defined by the base matrix

$$\boldsymbol{B} = \begin{bmatrix} 1 & 2 & 0 \\ 0 & 3 & 3 \end{bmatrix}. \tag{5.134}$$

The second protograph VN (with degree 5) is punctured, so the ensemble parameters are $n_v = 3$, $n_u = 2$, and $n_c = 2$. Calculate the ensemble design rate. Compute numerically and represent graphically the weight spectral shape function of the protograph ensemble. Compare its critical fractional codeword weight ω^* with that of a binary, unstructured regular LDPC code ensemble with $\mathsf{d}_v = 3$ and $\mathsf{d}_c = 6$, $\omega^* = 0.0227$.

Exercise 5.10 Consider a sequence of unstructured regular LDPC code ensembles $\{\mathscr{C}^{n,q}_{\mathsf{d}_v,\mathsf{d}_c}\}$ with $\mathsf{d}_c \geq 3$, let $\bar{A}_w^{(n)}$ be the multiplicity of codewords of weight w of a random code in the ensemble of length n and $G(\omega)$ be the weight spectral shape function of the ensemble sequence. Show that there exists a sequence of positive integers $d_n \to \infty$ and a sequence of real numbers $\theta_n \geq 0$ such that $\lim_{n\to\infty} \sum_{h=1}^{d_n} \bar{A}_h^{(n)} = 0$, $\lim_{n\to\infty} n\theta_n / d_n = 0$, and $\frac{1}{n} \log_q \bar{A}_{\omega n} \leq G(\omega) + \theta_n$. *Hint*: Use (5.125) and subsequent development as well as (5.123).

Exercise 5.11 Theorem 5.11 shows that a sequence of unstructured regular LDPC code ensembles exhibits a linear minimum distance growth for $\mathsf{d}_v \geq 3$. Following steps similar to the ones in the proof of Theorem 5.11 show that also an SVN ensemble $\mathscr{V}^n_{\psi,\mathsf{d}_c}(\nu_2)$ holds a linear minimum distance growth.

References

[1] R. Gallager, *Low-density parity-check codes*. Cambridge, MA, USA: MIT Press, 1963.

[2] D. Burshtein and G. Miller, "Asymptotic Enumeration Methods for Analyzing LDPC Codes," *IEEE Trans. Inf. Theory*, vol. 50, no. 6, pp. 1115–1131, June 2004.

[3] A. Orlitsky, K. Viswanathan, and J. Zhang, "Stopping set distribution of LDPC code ensembles," *IEEE Trans. Inf. Theory*, vol. 51, no. 3, pp. 929–953, Mar. 2005.

[4] C. Di, R. Urbanke, and T. Richardson, "Weight distribution of low-density parity-check codes," *IEEE Trans. Inf. Theory*, vol. 52, no. 11, pp. 4839–4855, Nov. 2006.

[5] C. Di, "Asymptotic and finite-length analysis of low-density parity-check codes," Ph.D. dissertation, École Polytechnique Fédérale de Lausanne, Lausanne, Switzerland, 2004.

[6] S. Fogal, R. McEliece, and J. Thorpe, "Enumerators for protograph ensembles of LDPC codes," in *Proc. IEEE Int. Symp. Inf. Theory*, Adelaide, Australia, Sep. 2005, pp. 2156–2160.

[7] D. Divsalar, "Ensemble weight enumerators for protograph LDPC codes," in *Proc. IEEE Int. Symp. Inf. Theory*, Seattle, WA, USA, Jul. 2006, pp. 1554–1558.

[8] S. Abu-Surra, D. Divsalar, and W. E. Ryan, "Enumerators for protograph-based ensembles of LDPC and generalized LDPC codes," *IEEE Trans. Inf. Theory*, vol. 57, no. 2, pp. 858–886, Feb. 2011.

[9] L. Dolecek, D. Divsalar, Y. Sun, and B. Amiri, "Non-binary protograph-based LDPC codes: Enumerators, analysis, and designs," *IEEE Trans. Inf. Theory*, vol. 60, no. 7, pp. 3913–3940, July 2014.

[10] E. Paolini and M. Flanagan, "Efficient and exact evaluation of the weight spectral shape and typical minimum distance of protograph LDPC codes," *IEEE Commun. Lett.*, vol. 20, no. 11, pp. 2141–2144, Nov. 2016.

6 Iterative and Maximum-Likelihood Decoding of LDPC Codes

This chapter discusses different erasure decoding algorithms for LDPC codes. We first introduce iterative decoding of LDPC codes in Section 6.1, describing different types of iterative decoders, their equivalence in terms of erasure correction performance, and their relation to ML decoders. Efficient ML erasure decoding of LDPC codes is then introduced in Section 6.2.

6.1 Iterative Decoding

Iterative erasure decoding of LDPC codes can be performed according to two approaches, namely, using the belief propagation (BP) algorithm or via peeling decoding. The BP decoder can be described as a message-passing algorithm, where messages are exchanged by VNs and CNs along the edges of the code Tanner graph. Messages carry estimates of the bit values, which are updated iteratively. BP decoding is general in that it can be applied to any communication channel. When applied to erasure channels, it takes a simplified form. Peeling decoding, on the contrary, is specific to erasure channels. In this section, we introduce BP decoding of LDPC codes in its general formulation (Section 6.1.1). We then particularize its message update equations to erasure channels (Section 6.1.2). Peeling decoding is outlined in Section 6.1.3. The equivalence between the two algorithms is discussed by analyzing their erasure correction capability, introducing the notion of stopping set (Section 6.1.4). While both algorithms are described for binary LDPC codes, they can be easily extended to the nonbinary case, as shown in Section 6.1.5. Section 6.1.6 discusses the application to recover packet erasures.

6.1.1 Belief Propagation Decoding of Binary LDPC Codes

Consider the MAP decoding rule (see Definition 2.12),

$$\hat{x}_i = \arg\max_{x_i \in \mathrm{GF}(2)} P(x_i | \boldsymbol{y}). \tag{6.1}$$

BP decoding of LDPC codes approximates the rule by performing an iterative message-passing algorithm over the Tanner graph of the code. The messages exchanged between a VN v_i, and its neighboring CNs carry estimates on the value of the codeword bit x_i. The estimates are expressed in terms of bit probabilities (*beliefs*) or functions of them, for example, logarithmic probability ratios (log-probability ratios, or *L-values*)[1]. The messages are updated at each iteration, yielding refined estimates of the codeword bit values. When the BP algorithm is described in the domain of log-probability ratios, it is convenient to rewrite the bit-wise MAP decoding rule in terms of the logarithmic a posteriori probability (APP) ratio

$$L_i^{\mathsf{APP}} = \log \frac{P(X_i = 0|\mathbf{y})}{P(X_i = 1|\mathbf{y})} \tag{6.2}$$

as

$$\hat{x}_i = \begin{cases} 0 & \text{if } L_i^{\mathsf{APP}} > 0 \\ 1 & \text{if } L_i^{\mathsf{APP}} < 0. \end{cases} \tag{6.3}$$

The case where $L_i^{\mathsf{APP}} = 0$ (*tie*) can be addressed either by choosing arbitrarily between 0 and 1 or by declaring $\hat{x}_i$ erased. Note that, while on several DMCs the occurrence of $L_i^{\mathsf{APP}} = 0$ is typically infrequent, the probability of a tie over the BEC is often nonnegligible. In fact, when erasure decoding fails, the value of some codeword bits remains undetermined, which is equivalent to stating that their log-APP ratio is zero.

To describe how the BP algorithm can be used to estimate log-APP ratios of the codeword bits, we focus on the messages exchanged by VNs and CNs. Let us denote by

$$L_i = \log \frac{P(y_i|X_i = 0)}{P(y_i|X_i = 1)} \tag{6.4}$$

the log-likelihood ratio (LLR) for the ith codeword bit. The message sent by a VN v_i to a neighboring CN c_j is denoted as $L_{i\to j}$, whereas the message received by v_i from c_j is denoted as $L_{i\leftarrow j}$. The L-values are defined as

$$L_{i\to j} = \log \frac{P\big(X_i = 0|L_i, \{L_{i\leftarrow z}\}_{z\neq j}\big)}{P\big(X_i = 1|L_i, \{L_{i\leftarrow z}\}_{z\neq j}\big)} \tag{6.5}$$

and

$$L_{i\leftarrow j} = \log \frac{P\big(X_i = 0|\{L_{z\to j}\}_{z\neq i}\big)}{P\big(X_i = 1|\{L_{z\to j}\}_{z\neq i}\big)}. \tag{6.6}$$

In (6.5), we used the shorthand $\{L_{i\leftarrow z}\}_{z\neq j}$ to denote the set of messages sent to v_i by all its neighbors with the exclusion of c_j, that is,

$$\{L_{i\leftarrow z}\}_{z\neq j} = \{L_{i\leftarrow z} : \mathsf{c}_z \in \mathcal{N}(\mathsf{v}_i) \setminus \{\mathsf{c}_j\}\}. \tag{6.7}$$

Similarly, $\{L_{z\to j}\}_{z\neq i}$ denotes the set of messages sent to c_j by all its neighbors, with the exclusion of v_i,

$$\{L_{z\to j}\}_{z\neq i} = \{L_{z\to j} : \mathsf{v}_z \in \mathcal{N}(\mathsf{c}_j) \setminus \{\mathsf{v}_i\}\}. \tag{6.8}$$

The computation of messages at VNs and CNs is illustrated in Figure 6.1.

[1] In the following, the terms "message" and "beliefs" will be used interchangeably, also for the case where the messages are defined in the log-probability ratio domain.

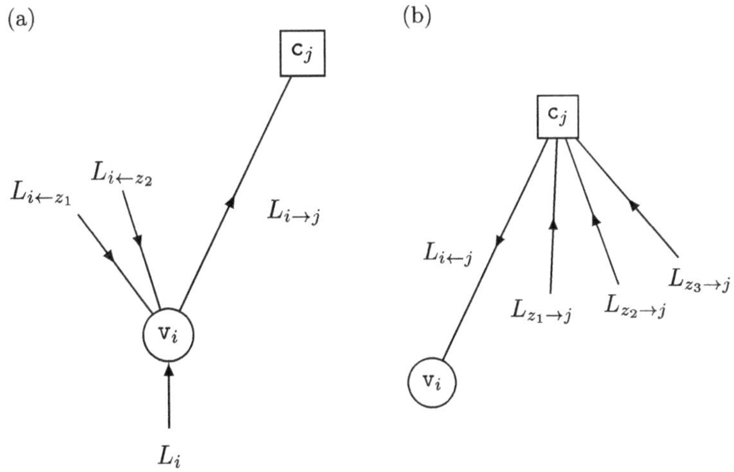

Figure 6.1 (a): Message update at a degree-3 VN and (b): at a degree-4 CN.

Remark 6.1 (Extrinsic information). The message sent by v_i to c_j is computed by explicitly excluding the message previously sent by c_j to v_i. Similarly, the calculation of the message sent by c_j to v_i excludes the belief provided by v_i to c_j. This principle, which lies at the core of the BP algorithm, can be intuitively explained by the need to provide *extrinsic* beliefs to a node. That is, the message a node sends to a neighboring node must not contain information already available at the target node. Owing to this observation, the messages computed at the output of VNs and CNs are often referred to as *extrinsic information*. Instead, the messages a node receives from its neighbors are referred to as *priors* or *a priori information*.

It is easy to observe that the sign of a message (L-value) defines which bit value is more probable. The magnitude of the message quantifies the degree of belief in the corresponding decision. In a limiting case, a message may take value $+\infty$ or $-\infty$, indicating full certainty about the bit having value 0 or 1, respectively. In contrast, an L-value equal to zero indicates total uncertainty about the value of the associated bit.

A crucial, simplifying assumption in the evaluation of (6.5) and (6.6) is to consider the messages at the input of a node as *independent*. By developing (6.5) and (6.6) under this assumption, the node update rules reduce to

$$L_{i\rightarrow j} = L_i + \sum_{z:\mathsf{c}_z \in \mathcal{N}(\mathsf{v}_i)\setminus\{\mathsf{c}_j\}} L_{i\leftarrow z} \tag{6.9}$$

for the VN case, and

$$L_{i\leftarrow j} = 2\tanh^{-1}\left[\prod_{z:\mathsf{v}_z \in \mathcal{N}(\mathsf{c}_j)\setminus\{\mathsf{v}_i\}} \tanh\left(\frac{L_{z\rightarrow j}}{2}\right)\right] \tag{6.10}$$

for the CN case, with

$$\tanh(x) = \frac{e^x - e^{-x}}{e^x + e^{-x}} \tag{6.11}$$

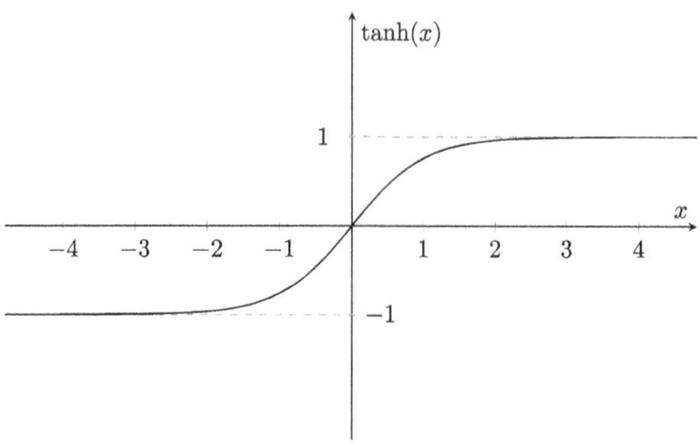

Figure 6.2 Hyperbolic tangent function.

denoting the hyperbolic tangent, depicted in Figure 6.2 (see Exercise 6.3 and Exercise 6.4). Note the odd symmetry of the hyperbolic tangent function: It implies that the sign of a CN output message is given by the product of the signs of the CN input messages.

BP decoding proceeds as follows. The decoder is initialized by feeding each VN with the corresponding channel LLR (6.4). Moreover, all outbound CN messages are initially set to zero. Each iteration is divided into two steps: a first half-iteration, referred to as *VN step*, and a second half-iteration, referred to as *CN step*. During the VN step, (6.5) is evaluated at all VNs, with VNs sending a message to each of their neighbors. The messages are computed based on the channel LLRs and the messages sent by the CNs in the previous half-iteration. In the CN step, (6.6) is evaluated at all the VNs. The message that a CN sends to each neighbor is computed based on the messages sent by the VNs in the previous half-iteration. After completing a pre-defined number of iterations $I_{\max}$, the log-APP ratios are produced at each VN as

$$L_i^{\mathsf{APP}} \approx \log \frac{P\big(X_i = 0 | L_i, \{L_{i\leftarrow j}\}\big)}{P\big(X_i = 1 | L_i, \{L_{i\leftarrow j}\}\big)} \tag{6.12}$$

$$= L_i + \sum_{z:\mathsf{c}_z \in \mathcal{N}(\mathsf{v}_i)} L_{i\leftarrow z}, \tag{6.13}$$

where $\{L_{i\leftarrow j}\}$ is the set of messages received by v_i from all its neighbors. The bit-wise decisions are finally taken according to (6.3). Crucially, (6.12) and (6.13) provide the exact value of the log-APP ratio only under strict conditions, whereas they yield only an approximation for most cases of interest. In particular, it is possible to show that the BP algorithm produces the exact log-APP ratios when the underlying Tanner graph is cycle-free (see Exercise 6.5). It is hardly possible to leverage this result since binary linear block codes with a cycle-free Tanner graph necessarily have a low minimum distance, impairing their error correction capability. On graphs with cycles, the algorithm yields only an approximation of the true log-APP ratios. Hence, BP decoding is suboptimal for codes whose Tanner graph possesses cycles. The reason for the sub-optimality can be understood by observing that the update rules at

VN and CN treat the belief provided by the neighboring nodes as independent. This assumption is correct on graphs without cycles or during the first $\lceil g/4 - 1 \rceil$ iterations (where g is the girth of the graph). In the latter case, when the number of iterations grows beyond $\lceil g/4 - 1 \rceil$, statistical dependencies among messages emerge, reducing the beliefs exchanged between the nodes to mere approximations of the correct terms (6.5) and (6.6).

Remark 6.2 (Early stopping). The BP algorithm can be stopped before reaching the final iteration. For example, by attempting a decision after each iteration according to (6.13) and (6.3), the decoder can output the decision $\hat{\boldsymbol{x}} = \begin{bmatrix} \hat{x}_1 & \hat{x}_2 & \dots & \hat{x}_n \end{bmatrix}$ whenever $\hat{\boldsymbol{x}}\boldsymbol{H}^{\mathsf{T}} = \boldsymbol{0}$, that is, when all check equations are satisfied.

Remark 6.3 (Decoding complexity). The number of operations required for a VN/CN update depends only on the node degree. Each iteration of the BP algorithm has a linear complexity in the blocklength n for a fixed degree distribution pair. Hence, for fixed $I_{\max}$, the complexity of BP decoding if $\mathcal{O}(n)$.

6.1.2 Belief Propagation Algorithm over the BEC

The calculation of the messages exchanged by VNs and CNs can be largely simplified on the BEC. Observe first that the channel LLR takes values in the ternary alphabet $\{-\infty, 0, +\infty\}$. In particular,

$$L_i = \begin{cases} +\infty & \text{if } y_i = 0 \\ 0 & \text{if } y_i = \mathsf{E} \\ -\infty & \text{if } y_i = 1. \end{cases} \tag{6.14}$$

By inspection of the node update rules (6.9) and (6.10), we can observe that the messages exchanged between VNs and CNs belong to the same alphabet. More importantly, a message either carries the correct sign (hence, communicating the correct bit value) or it is an erasure. These observations allow us to cast the BP algorithm by forcing messages to take values in the ternary set $\{0, 1, \mathsf{E}\}$. Having this alternative representation of the message values in mind, we introduce a new notation for the messages, with $m_{i\to j}$ and $m_{i\leftarrow j}$ denoting the message from v_i to c_j and the message from c_j to v_i, respectively. The VN update rule becomes

$$m_{i\to j} = \begin{cases} x_i & \text{if } y_i = x_i \text{ or } m_{i\leftarrow z} = x_i \text{ for any } z \in \mathcal{N}(\mathsf{v}_i) \setminus \mathsf{c}_j \\ \mathsf{E} & \text{otherwise} \end{cases} \tag{6.15}$$

that is, a VN communicates its correct bit value to a neighboring CN whenever the bit is not erased at the channel output, or any of the messages sent by the neighboring CNs – with the exclusion of the target CN – carries the correct bit value. If all messages at the VN input are erasures (including the channel output), then the VN forwards an erasure to the target CN. Examples of the VN updates are shown in Figure 6.3.

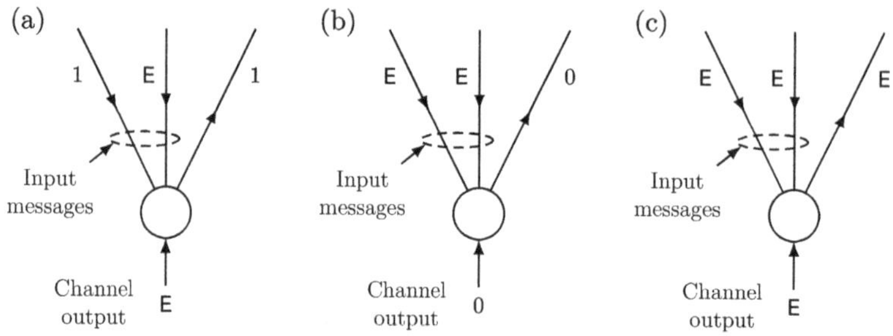

Figure 6.3 Examples of message updates at a degree-3 VN, computed according to (6.15). In (a) and (b), the output message carries the correct bit value. In (c), the channel output and the incoming messages (with the exclusion of the one from the target CN) are erasures, resulting in an extrinsic erasure message.

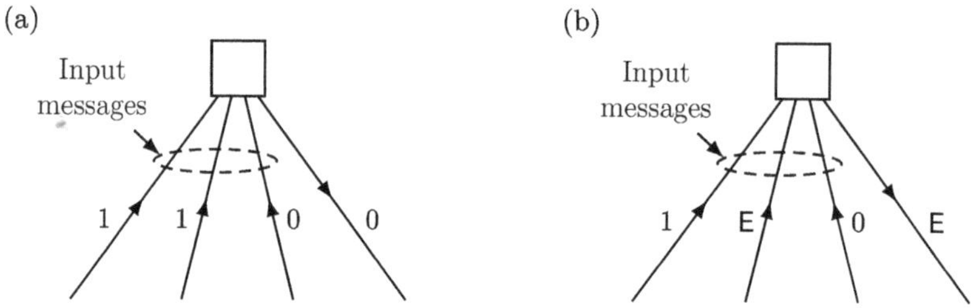

Figure 6.4 Examples of message updates at a degree-4 CN, computed according to (6.16). In (a), all input messages (excluding the one from the target VN) carry the correct value of the codeword bits involved in the corresponding parity-check equation. This allows the computation of the correct bit value for the target VN. In (b), an erasure at the CN input hinders resolving the bit value for the target VN.

The CN update rule is

$$m_{i \leftarrow j} = \begin{cases} \sum_{z:\mathsf{v}_z \in \mathcal{N}(\mathsf{c}_j)\setminus\{\mathsf{v}_i\}} x_z & \text{if } m_{z \to j} = x_z \text{ for all } z : \mathsf{v}_z \in \mathcal{N}(\mathsf{c}_j) \setminus \{\mathsf{v}_i\}, \\ \mathsf{E} & \text{otherwise} \end{cases} \tag{6.16}$$

where the sum in (6.16) is over GF(2). Here, a CN reveals the correct bit value to a neighboring VN if and only if all the messages received on edges connecting the CN to the other neighboring VN are not erasures. Examples of the CN updates are shown in Figure 6.4. The complete BP decoding algorithm is outlined in Algorithm 1.

Remark 6.4 (Stopping rule). In alternative to setting a maximum number of iterations, a sensible approach is to stop the message-passing iterations when all erasures have been solved or when the messages do not change between two consecutive iterations (signaling that the BP algorithm cannot resolve the residual erasures). Note that BP decoding over erasure channels requires at most e iterations, where e is the number of codeword bits erased at the channel output. Recalling that an (n, k) binary linear block code can solve up to $n - k$ erasures, it follows that the maximum number of iterations is upper bounded by $n - k$. In the following, we will always assume that BP decoding can iterate until no residual erasure can be resolved.

Algorithm 1: BP decoding on the BEC

for $j = 1 : m$ **do**
    for $\mathsf{v}_z \in \mathcal{N}(\mathsf{c}_j)$ **do**
        $m_{z \leftarrow j} = \mathsf{E}$;   `/* CN initialization */`
    end
end
while *maximum number of iterations not reached* **do**
    `/* VN update */`
    for $i = 1, \ldots, n$ **do**
        for $\mathsf{c}_j \in \mathcal{N}(\mathsf{v}_i)$ **do**
            if $y_i = x_i$ *or* $m_{i \leftarrow z} = x_i$ *for any* $z : \mathsf{c}_z \in \mathcal{N}(\mathsf{v}_i) \setminus \{\mathsf{c}_j\}$ **then**
                $m_{i \rightarrow j} = x_i$
            else
                $m_{i \rightarrow j} = \mathsf{E}$
            end
        end
    end
    `/* CN update */`
    for $i = 1, \ldots, m$ **do**
        for $\mathsf{v}_i \in \mathcal{N}(\mathsf{c}_j)$ **do**
            if $m_{z \rightarrow j} = x_z$ *for all* $z : \mathsf{v}_z \in \mathcal{N}(\mathsf{c}_j) \setminus \{\mathsf{v}_i\}$ **then**
                $m_{i \leftarrow j} = \sum_{z : \mathsf{v}_z \in \mathcal{N}(\mathsf{c}_j) \setminus \{\mathsf{v}_i\}} x_z$
            else
                $m_{i \leftarrow j} = \mathsf{E}$
            end
        end
    end
end
`/* Final decision */`
for $i = 1, \ldots, n$ **do**
    if $y_i = x_i$ *or* $m_{i \leftarrow z} = x_i$ *for any* $z : \mathsf{c}_z \in \mathcal{N}(\mathsf{v}_i)$ **then**
        $\hat{x}_i = x_i$
    else
        $\hat{x}_i = \mathsf{E}$
    end
end

Observe that Algorithm 1 performs message updates according to a *flooding scheduling*, that is, all messages are computed in parallel at each iteration. Over erasure channels, the behavior of the BP decoding algorithm can be simplified since only a message carrying an erasure may be modified (revealing the value of the corresponding codeword bit) over two consecutive iterations. In particular, at a given iteration, only CNs that receive a single erasure message update their outgoing messages, resolving the value of the bit value associated with the erasure. This observation lies at the core of a serial iterative erasure decoding algorithm, referred to as *peeling decoding*, discussed in Section 6.1.3.

6.1.3 Peeling Decoding

Peeling decoding is an iterative erasure decoding algorithm that resolves a single erasure at each iteration. At a given iteration, each VN can be in two possible states:

1. *Erased:* A VN is referred to as erased if the value of the codeword bit associated to it is unknown; that is, if the corresponding channel observation is an erasure, and no CN connected to the VN revealed the bit value in previous iterations.
2. *Known:* A VN is referred to as known if the corresponding channel observation equals the codeword bit value, or if the bit value has been *revealed* by a CN connected to the VN in previous iterations.

The peeling decoding algorithm works as follows. Each CN keeps a count of the number of erased VNs connected to it. The count is referred to as *residual degree* of the CN. During the ℓth iteration, the decoder

1. Picks a CN with residual degree equal to one;
2. Computes the value of the single erased VN bit, revealing its value;
3. Subtracts 1 to the residual degree of all the CNs connected to the revealed VN.

The decoder stops when no more CNs with residual degree one can be found.

Peeling decoding can be visualized as an edge-and-node removal process over the code Tanner graph. To proceed with the description, it is convenient to introduce the notion of *induced subgraph.*

Definition 6.1 (Induced subgraph) A subgraph $\mathcal{G}(\mathcal{S})$ of a Tanner graph $\mathcal{G}$ induced by the subset $\mathcal{S}$ of VNs is the graph composed of the VNs in $\mathcal{S}$, all CNs neighboring to the VNs in $\mathcal{S}$, and the edges connecting the VNs in $\mathcal{S}$ to their neighbors.

Let us denote by $\mathcal{S}_\ell$ the set of erased VNs ay iteration ℓ. Before the first iteration, the set $\mathcal{S}_0$ is initialized to the erasure pattern at the channel output. At the beginning, each CN stores the GF(2) sum of its neighboring known VNs bit values (*partial parity*). During the generic ℓth iteration, the decoder looks for a CN with residual degree one in $\mathcal{G}(\mathcal{S}_{\ell-1})$. The bit value associated with the single VN in $\mathcal{G}(\mathcal{S}_{\ell-1})$ connected to the selected CN is revealed by assigning to it the partial parity stored at the selected CN. The parities stored by all CNs connected to the revealed VN are updated, and their residual degree is reduced by 1. The revealed VN, its adjacent edges, and all its residual degree-0 neighbors are removed from the graph, resulting in a new induced subgraph $\mathcal{G}(\mathcal{S}_\ell)$, where $\mathcal{S}_\ell$ is obtained by removing the revealed VN from $\mathcal{S}_{\ell-1}$. Denoting by $e = |\mathcal{S}_0|$ the number of erasures introduced by the channel, decoding fails if, for any $\ell < e$, $\mathcal{G}(\mathcal{S}_\ell)$ does not contain CNs with residual degree one. In contrast, decoding succeeds if and only if, at the end of the peeling process, the induced subgraph is empty.

Example 6.1 Consider the binary linear block code defined by the Tanner graph in Figure 6.5a. The codeword $\boldsymbol{x} = \begin{bmatrix}1 & 1 & 1 & 0\end{bmatrix}$ is transmitted over a BEC. The channel output is $\boldsymbol{x} = \begin{bmatrix}1 & 1 & \mathsf{E} & \mathsf{E}\end{bmatrix}$. At the beginning of the peeling decoding process, the set of erased VNs is $\mathcal{S}_0 = \{\mathsf{v}_3, \mathsf{v}_4\}$. The induced subgraph $\mathcal{G}(\mathcal{S}_0)$ is depicted in Figure 6.5b. The partial parity values are depicted above each CN. At the first iteration, the only residual degree-1 CN is c_2. The partial parity stored in c_2 is assigned to its only neighbor v_3, revealing the value of the corresponding codeword bit. The partial parity at c_3 is updated accordingly. The VN v_3 and c_2 are removed from the graph, yielding the induced subgraph $\mathcal{G}(\mathcal{S}_1)$. During the second iteration (Figure 6.5c), the only residual degree-1 CN (c_3) is used to reveal the value of the bit associated with the only remaining erased VN (v_4). Note that the peeling decoding process succeeds; accordingly, the final induced subgraph is empty.

Remark 6.5 (Complexity of the peeling decoder). The peeling decoder runs for at most $e \leq n-k$ iterations, where e is the number of erasures introduced by the channel. At each iteration, a number of operations that depend only on the variable node degree is required to update the partial parities and the residual CN degrees. Thus, for a given LDPC code ensemble, the complexity of the peeling decoder is linear in the blocklength n.

It is essential to observe that BP and peeling decoding are equivalent in performance when the former is operated according to the stopping criterion of Remark 6.4. More precisely, both algorithms yield the same final decision for a given erasure pattern. This equivalence can be explained by analyzing the conditions that make BP and peeling decoding fail, as will be discussed further.

6.1.4 Stopping Sets

To explain the equivalence of BP and peeling decoding over erasure channels, it is essential to define which erasure patterns prevent both algorithms from successfully recovering all erasures. With this respect, a crucial role is played by *stopping sets*.

Definition 6.2 (Stopping set) A stopping set of a Tanner graph $\mathcal{G}$ is a subset $\mathcal{S}$ of VNs of $\mathcal{G}$ such that all neighboring CNs are connected to the VNs in $\mathcal{S}$ at least twice.

Let us consider the peeling decoder and assume that, at some iteration $\ell < e$ (e is the number of erasures at the channel output), the subset of erased VN $\mathcal{S}_\ell$ is a stopping

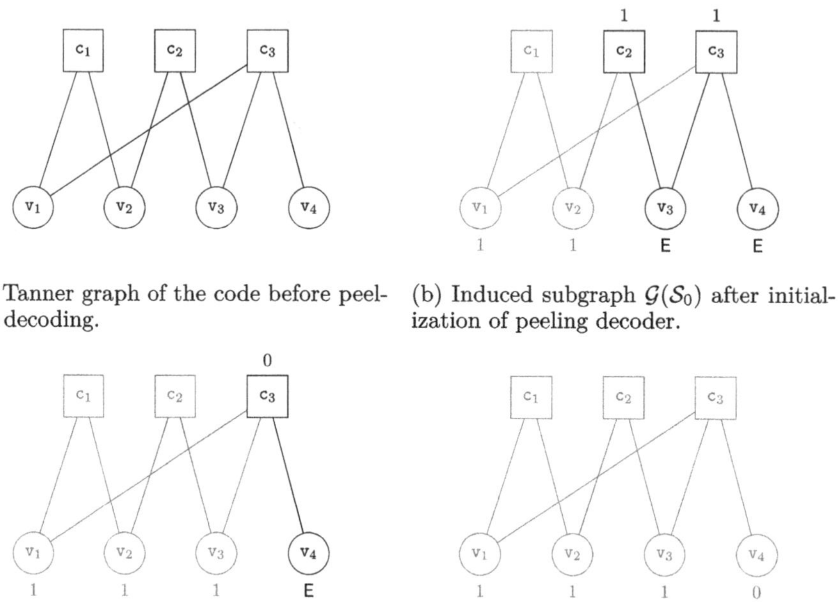

(a) Tanner graph of the code before peeling decoding.

(b) Induced subgraph $\mathcal{G}(\mathcal{S}_0)$ after initialization of peeling decoder.

(c) Induced subgraph $\mathcal{G}(\mathcal{S}_1)$ after first decoding iteration.

(d) Induced subgraph $\mathcal{G}(\mathcal{S}_1)$ after second decoding iteration.

Figure 6.5 Two iterations of peeling decoding. Even/odd parities are indicated by CN labels.

set. It is easy to recognize that this condition would prevent the peeling decoder from proceeding to any further erasure recovery. In fact, the induced subgraph $\mathcal{G}(\mathcal{S}_\ell)$ does not contain any CN with residual degree one.

In the BP decoding case, let us assume that the VNs forming a stopping set $\mathcal{S}$ are erased. Then, concerning Algorithm 1, we can observe that all edges connected to the VNs in $\mathcal{S}$ will carry an erasure message across all iterations. It follows that the values of the bits associated with the VNs in $\mathcal{S}$ will remain unknown, no matter how many BP decoding iterations will be performed.

Remark 6.6 (Necessary and sufficient conditions for decoding failure). The presence of an erasure pattern involving a stopping set is *sufficient* to hinder the success of both BP and peeling decoding. Conversely, suppose that the channel outputs an erasure pattern that does not involve any stopping set. Then, both BP and peeling decoding – with the former allowed to iterate long enough – will recover all the erasures. Thus, the occurrence of an erasure pattern involving a stopping set in the channel output is also *necessary* to prevent both BP and peeling decoding from succeeding.

Trivially, the union of stopping sets forms a stopping set. We define the *maximal stopping set* involved in an erasure pattern as the unique largest stopping set within the subgraph induced by the erasure pattern. In case of failure, BP and peeling decoding will generally resolve all erasures introduced by the channel, except those associated with the maximal stopping set.

Example 6.2 (Stopping set) Consider the Tanner graph of an (8,5) binary linear block code defined by the parity-check matrix

$$\boldsymbol{H} = \begin{bmatrix} 1 & 0 & 1 & 0 & 0 & 1 & 0 & 1 \\ 1 & 1 & 0 & 1 & 0 & 0 & 1 & 1 \\ 0 & 1 & 1 & 0 & 1 & 0 & 1 & 1 \\ 0 & 0 & 1 & 1 & 0 & 1 & 1 & 0 \\ 1 & 0 & 0 & 1 & 1 & 1 & 0 & 0 \end{bmatrix} \tag{6.17}$$

depicted in Figure 6.6(a). At the channel output, the peeling decoder observes

$$\boldsymbol{y} = \begin{bmatrix} 0 & 0 & \mathsf{E} & 0 & \mathsf{E} & 0 & \mathsf{E} & \mathsf{E} \end{bmatrix}. \tag{6.18}$$

The subgraph induced by the erasure pattern $\mathcal{S}_0 = \{\mathsf{v}_3, \mathsf{v}_5, \mathsf{v}_7, \mathsf{v}_8\}$ is depicted in Figure 6.6(b). During the first iteration, the residual degree-1 CN c_5 reveals the value of v_5. The peeling decoder stops, since the residual set of erased VN $\mathcal{S}_0 = \{\mathsf{v}_3, \mathsf{v}_7, \mathsf{v}_8\}$ forms a stopping set (Figure 6.6(c)).

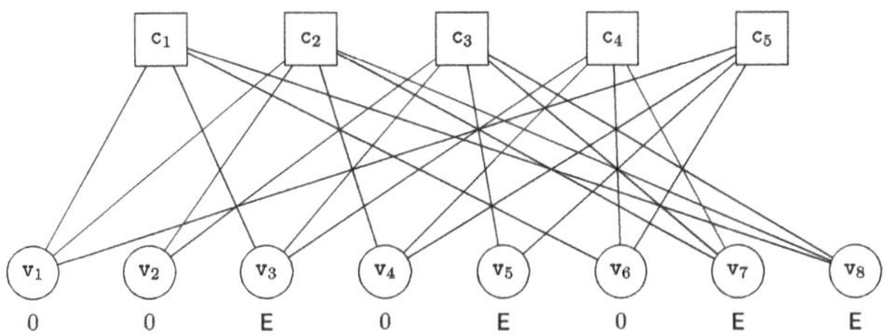

(a) Tanner graph of the code before peeling decoding.

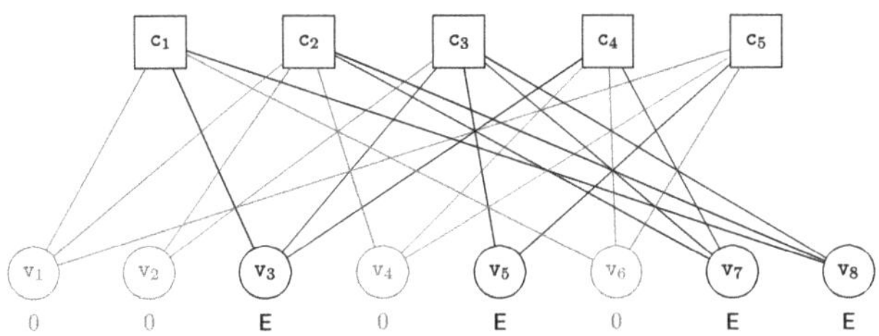

(b) Induced subgraph $\mathcal{G}(\mathcal{S}_0)$ after initialization of peeling decoder.

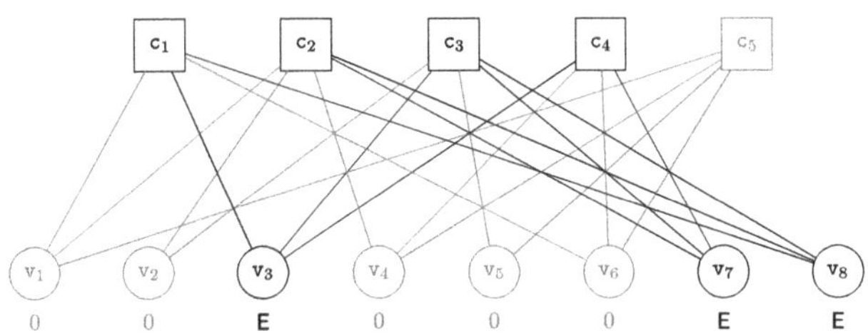

(c) Induced subgraph $\mathcal{G}(\mathcal{S}_1)$ after first decoding iteration.

Figure 6.6 Peeling decoding over the Tanner graph of Example 6.2.

Remark 6.7 (Stopping sets and codeword supports). The VNs associated with the support of a codeword always form a stopping set. In contrast, not all stopping sets correspond to the support of a codeword.

Remark 6.7 implies that an erasure pattern that cannot be resolved by BP and peeling decoding may be resolved by an ML decoder (this is the case for the erasure pattern of Example 6.2). In particular, denote by $t_{\min}$ the size of the smallest stopping set in the Tanner graph $\mathcal{G}$ of a binary linear block code $\mathcal{C}$ with minimum distance $d_{\min}$. We have that $t_{\min} \leq d_{\min}$. BP and peeling decoding applied to $\mathcal{G}$ will resolve all erasure patterns with less than $t_{\min}$ erasures, whereas ML decoding will resolve all erasure patterns with less than $d_{\min}$ erasures. We will further elaborate on this point in Section 6.2 and in Chapter 7, where the block error probability under iterative erasure decoding (either BP or peeling) will be put in relation with *stopping set enumerators*, counting the number of stopping sets of size $i = 0, \ldots, n$ in a Tanner graph.

Small-size stopping sets are typically associated with clusters of short cycles in the subgraph induced by the stopping set. Thus, using Tanner graphs with large girth is especially important to prevent the emergence of small-size stopping sets. In the case of LDPC codes, given a degree distribution pair $\lambda(x)$, $\rho(x)$ and a blocklength n, girth optimization techniques can be used to design the Tanner graph. For instance, one may use the progressive edge growth (PEG) algorithm, which is a greedy optimization algorithm. It constructs the Tanner graph by successively placing the edges so that the girth is as high as possible in every step. Another popular algorithm is approximate cycle extrinsic message degree (ACE), which tries to avoid cycles involving VNs that have a low number of connections to CNs outside the cycle. This way, even if all VNs involved in the cycle are erased, there is a nonnegligible chance that the VN will be recovered by a CN which is not part of the cycle.

The Tanner graph representation of a code is not unique, as it depends on the parity-check matrix form used to define the code. Hence, stopping sets and the performance of BP and peeling decoding inevitably depend on the Tanner graph used to describe the code.

6.1.5 Iterative Decoding of Nonbinary LDPC Codes

Iterative decoding of nonbinary LDPC over GF(q), $q > 2$ requires simple modifications of binary BP and peeling decoding when the underlying communication channel is a QEC with inputs in GF(q). Focusing on BP decoding, the idea is to either propagate messages that carry the correct value of a code symbol or an erasure. Let again m_i be the channel message, $m_{i\to j}$ be the outgoing message from v_i to c_j, and $m_{i\leftarrow j}$ the outgoing message from c_j to v_i. All three messages take values from GF$(q) \cup \{\mathsf{E}\}$. The following modifications of the iterative decoding algorithm must be considered. As detailed in Chapter 4, the code's parity-check matrix contains elements $H_{j,i}$ from GF(q), and the corresponding Tanner graph labels an edge connecting v_i to c_j with the nonzero element $H_{j,i}$ from GF(q). A generic message $m_{i\to j}$ from v_i to c_j has to be multiplied with the edge label before being processed in the CN. More specifically, in the nonbinary setting (6.16) should be modified as

$$m_{i \leftarrow j} = \begin{cases} h_{ji}^{-1} \sum_{z: \mathsf{v}_z \in \mathcal{N}(\mathsf{c}_j) \setminus \{\mathsf{v}_i\}} h_{jz} x_z & \text{if } m_{z \to j} = x_z \text{ for all } z : \mathsf{v}_z \in \mathcal{N}(\mathsf{c}_j) \setminus \{\mathsf{v}_i\} \\ \mathsf{E} & \text{otherwise.} \end{cases} \tag{6.19}$$

The modifications applied to peeling decoding fall along the same line: When computing the value of a generic erased symbol x_i, the peeling decoder solves a nonbinary equation associated with a residual degree-1 CN by setting

$$x_i = h_{ji}^{-1} \sum_{z: \mathsf{v}_z \in \mathcal{N}(\mathsf{c}_j) \setminus \{\mathsf{v}_i\}} h_{jz} x_z, \tag{6.20}$$

where symbol values on the right-hand side of (6.20) are known.

Remark 6.8. As for its binary counterpart, iterative (BP/peeling) erasure decoding of a nonbinary LDPC code over GF(q), $q > 2$ fails whenever the channel erases the symbols associated with a stopping set in the code Tanner graph. It follows that iterative erasure decoding of a q-ary LDPC code with parity-check matrix $\boldsymbol{H}$ over the QEC yields the same block (symbol) error probability than iterative erasure decoding of a binary code with parity-check matrix $\boldsymbol{H}_b$ over the BEC if the positions of the nonzero entries of $\boldsymbol{H}$ and $\boldsymbol{H}_b$ coincide.

6.1.6 Iterative Decoding for Packet Erasures

Typical applications of binary erasure codes include transmitting data units (packets) of constant length L, which are either correctly received or erased by the channel. This situation may be modeled by the BLEC introduced in Section 3.3.2. The BLEC can be seen as a set of L parallel BECs, which are completely correlated. That is, if a bit of a packet is correctly received, then all bits of the packet are correctly received. If a bit of a packet is erased, then all packet bits are erased. In the simplest case, packets are erased with probability ϵ, and are correctly received with probability $1 - \epsilon$. We refer to this equivalent model as *packet erasure channel*. Consider in the following a binary (n, k) LDPC code with parity-check matrix $\boldsymbol{H}$, and assume that the code is used to *encode n packets out of k information packets*. The concept is illustrated in the following example.

Example 6.3 Consider a (6, 3) binary linear block code with parity-check matrix

$$\boldsymbol{H} = \begin{bmatrix} 1 & 1 & 0 & 1 & 0 & 0 \\ 1 & 0 & 1 & 0 & 1 & 0 \\ 0 & 1 & 1 & 0 & 0 & 1 \end{bmatrix}. \tag{6.21}$$

The code is used to encode three information packets

$$\boldsymbol{x}_1 = \begin{bmatrix} 1 \\ 0 \\ 1 \\ 1 \end{bmatrix} \quad \boldsymbol{x}_2 = \begin{bmatrix} 0 \\ 1 \\ 0 \\ 1 \end{bmatrix} \quad \boldsymbol{x}_3 = \begin{bmatrix} 1 \\ 1 \\ 0 \\ 1 \end{bmatrix} \tag{6.22}$$

into six packets. According to the first equation defined by $\boldsymbol{H}$, the first redundant packet is computed by adding bit-by-bit in GF(2) $\boldsymbol{x}_1$ and $\boldsymbol{x}_2$, that is,

$$\boldsymbol{x}_4 = \boldsymbol{x}_1 + \boldsymbol{x}_2 = \begin{bmatrix} 1 \\ 1 \\ 1 \\ 0 \end{bmatrix}. \tag{6.23}$$

Similarly, $\boldsymbol{x}_5$ and $\boldsymbol{x}_6$ are computed, according to $\boldsymbol{H}$, as

$$\boldsymbol{x}_5 = \boldsymbol{x}_1 + \boldsymbol{x}_3 = \begin{bmatrix} 0 \\ 1 \\ 1 \\ 0 \end{bmatrix} \quad \boldsymbol{x}_6 = \boldsymbol{x}_2 + \boldsymbol{x}_3 = \begin{bmatrix} 1 \\ 0 \\ 0 \\ 0 \end{bmatrix}. \tag{6.24}$$

After transmission over the packet erasure channel, erased packets may be recovered by applying BP or peeling decoding, processing *in parallel* all the bits in an erased packet. The following example clarifies the principle.

Example 6.4 Consider again the (6, 3) binary linear block code from Example 6.3. At the encoder side, the codeword

$$\boldsymbol{x} = \begin{bmatrix} \begin{bmatrix} 1 \\ 0 \\ 1 \\ 1 \end{bmatrix} & \begin{bmatrix} 0 \\ 1 \\ 0 \\ 1 \end{bmatrix} & \begin{bmatrix} 1 \\ 1 \\ 0 \\ 1 \end{bmatrix} & \begin{bmatrix} 1 \\ 1 \\ 1 \\ 0 \end{bmatrix} & \begin{bmatrix} 0 \\ 1 \\ 1 \\ 0 \end{bmatrix} & \begin{bmatrix} 1 \\ 0 \\ 0 \\ 0 \end{bmatrix} \end{bmatrix} \tag{6.25}$$

is generated and transmitted, column-wise, over a packet erasure channel. At the channel output, the decoder observes

$$\boldsymbol{y} = \begin{bmatrix} \begin{bmatrix} 1 \\ 0 \\ 1 \\ 1 \end{bmatrix} & \begin{bmatrix} \mathsf{E} \\ \mathsf{E} \\ \mathsf{E} \\ \mathsf{E} \end{bmatrix} & \begin{bmatrix} \mathsf{E} \\ \mathsf{E} \\ \mathsf{E} \\ \mathsf{E} \end{bmatrix} & \begin{bmatrix} 1 \\ 1 \\ 1 \\ 0 \end{bmatrix} & \begin{bmatrix} \mathsf{E} \\ \mathsf{E} \\ \mathsf{E} \\ \mathsf{E} \end{bmatrix} & \begin{bmatrix} 1 \\ 0 \\ 0 \\ 0 \end{bmatrix} \end{bmatrix}. \tag{6.26}$$

We compactly write the channel output as $\boldsymbol{y} = \begin{bmatrix} \boldsymbol{x}_1 & \mathsf{E} & \mathsf{E} & \boldsymbol{x}_4 & \mathsf{E} & \boldsymbol{x}_6 \end{bmatrix}$ where E denotes a packet erasure. The Tanner graph of the code is depicted in Figure 6.7. During the first peeling decoding iteration, the decoder uses the CN c_1 to recover $\boldsymbol{x}_2$ as the bit-wise sum

$$\boldsymbol{x}_2 = \boldsymbol{x}_1 + \boldsymbol{x}_4. \tag{6.27}$$

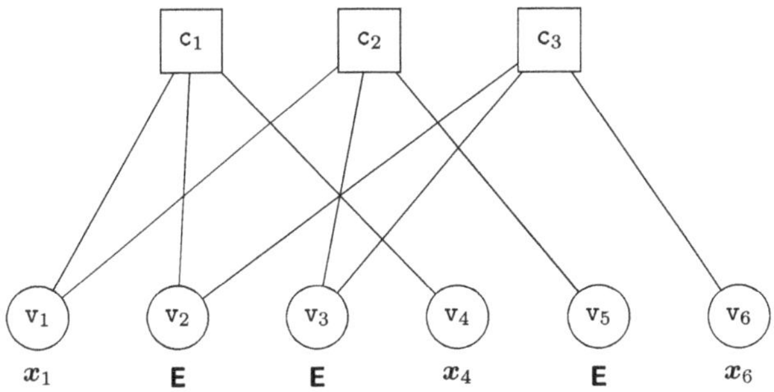

Figure 6.7 Tanner graph of the code from Example 6.4, before peeling decoding.

During the second iteration, $\boldsymbol{x}_3$ is computed through c_3 as

$$\boldsymbol{x}_3 = \boldsymbol{x}_2 + \boldsymbol{x}_6. \tag{6.28}$$

Finally, c_2 is used to obtain

$$\boldsymbol{x}_5 = \boldsymbol{x}_1 + \boldsymbol{x}_3. \tag{6.29}$$

6.2 Maximum-Likelihood Decoding

Block-wise ML erasure decoding of a linear block code over GF(q) (Section 2.4) requires solving the linear system

$$\boldsymbol{H}_{\mathcal{E}}\boldsymbol{x}_{\mathcal{E}}^{\mathsf{T}} = \boldsymbol{s}^{\mathsf{T}}, \tag{6.30}$$

where $\mathcal{E} = \{i_1, i_2, \ldots, i_e\}$ defines the indexes of the codeword symbols that are erased at the channel output, and where $\boldsymbol{x}_{\mathcal{E}} = \begin{bmatrix} x_{i_1} & x_{i_2} & \ldots & x_{i_e} \end{bmatrix}$ denotes the vector of erased symbols. Accordingly, $\boldsymbol{H}_{\mathcal{E}}$ is the submatrix of the code parity-check matrix with column indexes in $\mathcal{E}$. The right-hand side of (6.30) is

$$\boldsymbol{s}^{\mathsf{T}} = -\boldsymbol{H}_{\bar{\mathcal{E}}}\boldsymbol{x}_{\bar{\mathcal{E}}}^{\mathsf{T}} \tag{6.31}$$

where $\bar{\mathcal{E}}$ is the set of the indexes of nonerased codeword symbols, that is, the complement of $\mathcal{E}$. Extending Definition 2.20 to erasure channels, we refer to $\boldsymbol{s}$ as *syndrome*. The solution of (6.30) can be obtained by applying Gaussian elimination to the $(n-k) \times e$ matrix $\boldsymbol{H}_{\mathcal{E}}$. A unique solution can be found if and only if the rank of $\boldsymbol{H}_{\mathcal{E}}$ is equal to the number of erasures e. If the rank of $\boldsymbol{H}_{\mathcal{E}}$ is less than e, (6.30) admits multiple solutions. In this case, following the incomplete ML decoding principle discussed in Remark 2.2, the decoder outputs an error flag, declaring a decoding failure.

Example 6.5 (Gaussian elimination) Consider the system of linear equations in GF(2)

$$\begin{bmatrix} 1 & 0 & 1 \\ 1 & 1 & 0 \\ 1 & 1 & 1 \end{bmatrix} \begin{bmatrix} x_1 \\ x_2 \\ x_3 \end{bmatrix} = \begin{bmatrix} 0 \\ 1 \\ 0 \end{bmatrix}. \tag{6.32}$$

By adding the first row to the second and the third rows, we obtain

$$\begin{bmatrix} 1 & 0 & 1 \\ 0 & 1 & 1 \\ 0 & 1 & 0 \end{bmatrix} \begin{bmatrix} x_1 \\ x_2 \\ x_3 \end{bmatrix} = \begin{bmatrix} 0 \\ 1 \\ 0 \end{bmatrix}. \tag{6.33}$$

We now add the second row to the third row, resulting in

$$\begin{bmatrix} 1 & 0 & 1 \\ 0 & 1 & 1 \\ 0 & 0 & 1 \end{bmatrix} \begin{bmatrix} x_1 \\ x_2 \\ x_3 \end{bmatrix} = \begin{bmatrix} 0 \\ 1 \\ 1 \end{bmatrix}, \tag{6.34}$$

where the system matrix has been transformed into row-echelon form. Then, back-substitution yields $x_3 = 1$, $x_2 = 0$, and $x_1 = 1$.

As anticipated in Remark 6.7, ML decoding may allow resolving erasure patterns not decodable by iterative decoders, as shown in the following example.

Example 6.6 Consider again the binary linear block code of Example 6.2 and the channel observation of (6.18), which was shown to be unsolvable via iterative decoding. ML decoding requires the solution of

$$\begin{bmatrix} 1 & 0 & 0 & 1 \\ 0 & 0 & 1 & 1 \\ 1 & 1 & 1 & 1 \\ 1 & 0 & 1 & 0 \\ 0 & 1 & 0 & 0 \end{bmatrix} \begin{bmatrix} x_3 \\ x_5 \\ x_7 \\ x_8 \end{bmatrix} = \begin{bmatrix} 0 \\ 0 \\ 0 \\ 0 \\ 0 \end{bmatrix}. \tag{6.35}$$

Observe that the value of x_5 can be readily found through the last equation ($x_5 = 0$), reducing the problem to solving

$$\begin{bmatrix} 1 & 0 & 1 \\ 0 & 1 & 1 \\ 1 & 1 & 1 \\ 1 & 1 & 0 \end{bmatrix} \begin{bmatrix} x_3 \\ x_7 \\ x_8 \end{bmatrix} = \begin{bmatrix} 0 \\ 0 \\ 0 \\ 0 \end{bmatrix}. \tag{6.36}$$

The system matrix on the right-hand side has a rank equal to three, and by applying Gaussian elimination, we find $x_3 = x_7 = x_8 = 0$.

Remark 6.9 (Block-wise and bit-wise ML erasure decoding). Consider a block-wise ML decoder that attempts the solution of (6.30), for example, via Gaussian elimination. If the rank of $\boldsymbol{H}_{\mathcal{E}}$ is less than the number of erasures e, the incomplete block-wise ML decoder will declare a decoding failure. However, a subset of the erased symbols may still be decoded. This happens if all solutions of (6.30) agree on the value of some erased codeword symbols (see Example 6.7). By solving all erasures symbols that admit a unique solution, a decoder based on (6.30) would realize an *incomplete* symbol-wise ML decoder, that is, a decoder that implements the decision rule from Definition 2.13 where ties result in the declaration of a symbol erasure. Owing to the typical assumption of uniform a priori distribution of codeword symbols, the decoder may be equivalently referred to as *incomplete* symbol-wise MAP decoder (see also Definition 2.12).

Example 6.7 Consider a (6, 3) binary linear block code with parity-check matrix

$$\boldsymbol{H} = \begin{bmatrix} 1 & 1 & 0 & 1 & 0 & 0 \\ 1 & 0 & 1 & 0 & 1 & 0 \\ 0 & 1 & 1 & 0 & 0 & 1 \end{bmatrix} \tag{6.37}$$

and assume that the decoder observes $\boldsymbol{y} = \begin{bmatrix} 0 & \mathsf{E} & \mathsf{E} & \mathsf{E} & 1 & \mathsf{E} \end{bmatrix}$. Following (6.30), we attempt the solution of

$$\begin{bmatrix} 1 & 0 & 1 & 0 \\ 0 & 1 & 0 & 0 \\ 1 & 1 & 0 & 1 \end{bmatrix} \begin{bmatrix} x_2 \\ x_3 \\ x_4 \\ x_6 \end{bmatrix} = \begin{bmatrix} 0 & 1 & 0 \end{bmatrix}. \tag{6.38}$$

Note that the system is obviously rank deficient: only three equations are available to solve four unknowns. The block-wise ML decoder will hence declare a failure. Nevertheless, the value of x_3 can be resolved using the second equation, resulting in $x_3 = 1$. No further erasure can be resolved. In fact, (6.38) admits two solutions,

$$x_2 = 0 \quad x_3 = 1 \quad x_4 = 0 \quad x_6 = 1 \tag{6.39}$$

and

$$x_2 = 1 \quad x_3 = 1 \quad x_4 = 1 \quad x_6 = 0 \tag{6.40}$$

and the only bit value on which the two solutions agree is $x_3 = 1$.

The solution of (6.30) via Gaussian elimination has, in general, cubic complexity in the number of erasures. Assuming the number of erasures to scale linearly in the blocklength n implies a number of operations $\mathcal{O}\left(n^3\right)$. For LDPC codes, though, the complexity of ML erasure decoding can be drastically reduced by exploiting the sparsity of the parity-check matrix. The low density of nonzero entries in $\boldsymbol{H}$ can be used to simplify Gaussian elimination, leading to a class of algorithms that appeared in literature under various names (*smart* or *structured Gaussian elimination*, and *inactivation decoding*, to name a few).

6.2.1 Structured Gaussian Elimination

Structured Gaussian elimination exploits the sparsity of the parity-check matrix of LDPC codes to reduce the complexity of Gaussian elimination when applied to solve (6.30). In particular, the idea is to identify a (hopefully) small subset of erasures – referred to as *pivots* or *reference variables* – that, when revealed, would allow the use of low-complexity iterative decoding to recover the remaining erasures. The solution

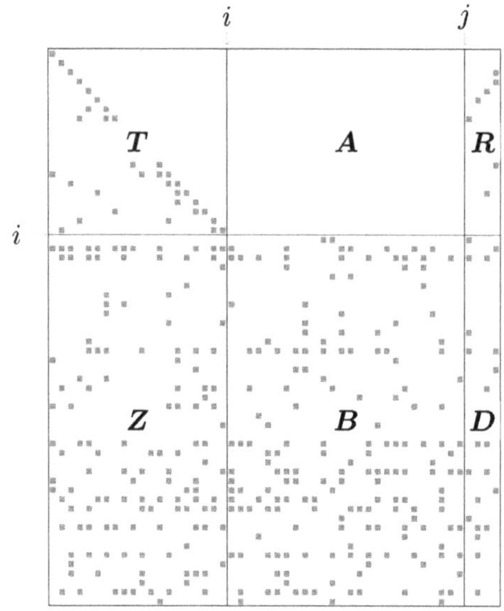

Figure 6.8 Matrix $\boldsymbol{H}_{\mathcal{E}}$ during triangulation step.

of the reference variables is performed with Gaussian elimination. A small number of reference variables allows us to keep the complexity of this step low. Structured Gaussian elimination is based on three steps, which are summarized below.

1. **Triangulation and Pivoting.**
   Put the $(n-k) \times e$ system matrix $\boldsymbol{H}_{\mathcal{E}}$ in approximate lower triangular form by applying row and column permutation only. The system matrix takes the form

$$\begin{bmatrix} \boldsymbol{T} & \boldsymbol{A} & \boldsymbol{R} \\ \boldsymbol{Z} & \boldsymbol{B} & \boldsymbol{D} \end{bmatrix} \tag{6.41}$$

   depicted in Figure 6.8, where $\boldsymbol{T}$ is lower-triangular and $\boldsymbol{A}$ is a zero matrix. The matrix form (6.41) is obtained iteratively; that is, at every iteration, the submatrix

$$\begin{bmatrix} \boldsymbol{T} \\ \boldsymbol{Z} \end{bmatrix} \tag{6.42}$$

   is augmented by appending one column. The triangulation process may halt if no column in the submatrix

$$\begin{bmatrix} \boldsymbol{A} \\ \boldsymbol{B} \end{bmatrix} \tag{6.43}$$

   can be found to augment the submatrix in (6.42), keeping $\boldsymbol{T}$ lower-triangular, that is, if the submatrix $\boldsymbol{B}$ does not contain any row with Hamming weight 1. When this happens, pivoting takes place by removing one or more columns of the submatrix in (6.43). The removed columns are appended to the submatrix

$$\begin{bmatrix} \boldsymbol{R} \\ \boldsymbol{D} \end{bmatrix}. \tag{6.44}$$

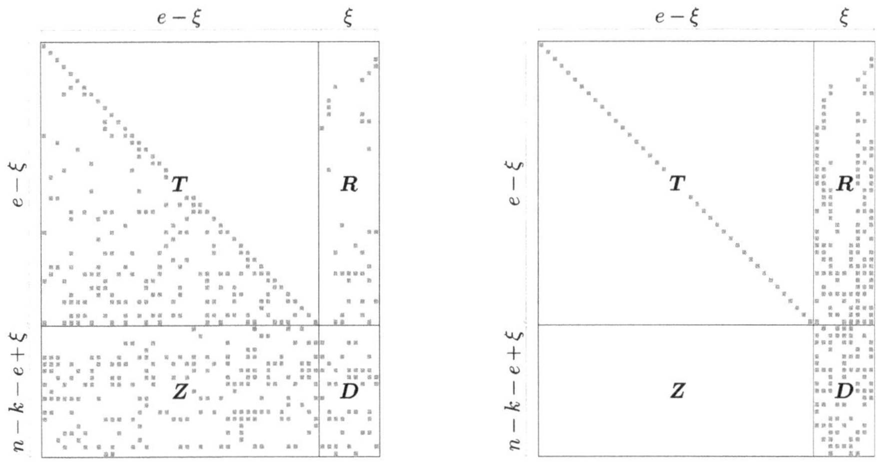

(a) $\boldsymbol{H}_{\mathcal{E}}$ after triangulation step. (b) $\boldsymbol{H}_{\mathcal{E}}$ after zeroing out step.

Figure 6.9 Illustration of $\boldsymbol{H}_{\mathcal{E}}$ during structured Gaussian elimination. Black squares illustrate nonzero entries in $\boldsymbol{H}_{\mathcal{E}}$.

Note that by removing columns from (6.43), the Hamming weight of some rows of $\boldsymbol{B}$ gets reduced. Hence, the triangulation process may possibly be resumed. The triangulation/pivoting step terminates when the submatrix (6.43) disappears, resulting in a system matrix in the form

$$\begin{bmatrix} \boldsymbol{T} & \boldsymbol{R} \\ \boldsymbol{Z} & \boldsymbol{D} \end{bmatrix}, \tag{6.45}$$

where $\boldsymbol{T}$ has dimension $(e-\xi)\times(e-\xi)$, $\boldsymbol{Z}$ has dimension $(n-k-e+\xi)\times(e-\xi)$, $\boldsymbol{D}$ is a $(n-k-e+\xi)\times\xi$ matrix, and $\boldsymbol{R}$ is a $(e-\xi)\times\xi$ matrix (see Figure 6.9(a)). Observe that the matrix (6.45) has the same density as the original system matrix $\boldsymbol{H}_{\mathcal{E}}$, since only row and column permutations are used during this step. The ξ codeword symbols associated with the columns forming the submatrix in (6.44) are the pivots.

2. **Nullification.**
   The lower triangular submatrix $\boldsymbol{T}$ is transformed into an identity matrix by row additions. Moreover, $\boldsymbol{Z}$ is made a zero matrix by row additions, leading to the matrix depicted in Figure 6.9(b). Due to the row additions, both $\boldsymbol{R}$ and $\boldsymbol{D}$ may not be sparse anymore.
3. **Pivots Resolution.**
   Gaussian elimination is applied to the $(n-k-e+\xi)\times\xi$ submatrix $\boldsymbol{D}$. If the rank of $\boldsymbol{D}$ equals ξ, then the value of the ξ pivots can be resolved, allowing the solution of the remaining unknowns either by back substitution via the submatrix $\begin{bmatrix} \boldsymbol{T} & \boldsymbol{R} \end{bmatrix}$, or by feeding the pivots values back to the iterative erasure decoder operating over the Tanner graph defined by the code parity-check matrix.

Note that all row operations (permutations and additions) applied to the system matrix must be applied, in parallel, to the syndrome $\boldsymbol{s}$.

Assuming the number of erasures e and the number of pivots ξ to be linear in the blocklength n, for large enough n, the last step dominates the complexity, as it requires a number of operations that grows as $\mathcal{O}(\xi^3)$, that is $\mathcal{O}(n^3)$. Structured Gaussian elimination, hence, still maintains the cubic decoding complexity of (standard) Gaussian elimination. Nevertheless, the number of variables that are resolved with cubic complexity is reduced from e to ξ. It is, therefore, crucial to keep the *average* number of pivots $\bar{\xi}$ as small as possible. This can be achieved by resorting to a smart pivoting strategy and designing codes well-suited for structured Gaussian elimination decoding. Section 8.3 will address the former topic. We focus next on the latter problem, that is, how to devise effective pivoting techniques.

6.2.1.1 Pivoting Algorithms

The selection of the pivots is typically performed by resorting to heuristic greedy criteria, aiming at minimizing the number of pivots at the end of the triangulation and pivoting step. The criteria can be based on the exploration of the structure of the submatrix $\boldsymbol{B}$, identifying columns whose removal from (6.43) allows to increase, as much as possible, the size of the lower-triangular matrix $\boldsymbol{T}$, before the triangulation process halts again. Some definitions that can be used to define the metrics adopted in the selection of the pivots are outlined in the following.

Definition 6.3 (Incident columns of a matrix row) The i column of a matrix $\boldsymbol{B}$ is said to be incident to the j row of $\boldsymbol{B}$ if $b_{j,i}$ is nonzero.

Definition 6.4 (Accumulated column weight of a row) The accumulated column weight of a row in a matrix is the sum of the Hamming weights columns that are incident to the row.

Definition 6.5 (Cluster of a bipartite graph) A cluster of a bipartite graph is connected subgraph induced by a subset of CNs, which includes at least one VN, called *key node*, with the following property: If the key node is removed from the subgraph, then the set of VNs remaining in the subgraph does not contain any stopping set. The number of VNs belonging to the cluster is referred to as cluster size.

Definition 6.6 (Component of a bipartite graph) A component[2] of a bipartite graph is a cluster, where the degree of each CN in the cluster is two. The number of VNs belonging to the component is referred to as component size.

Examples of pivoting algorithms are outlined below.

Maximum Column Weight Pivoting. The erased symbol associated with a column of (6.43) with largest Hamming weight is selected as pivot. The heuristic motivation

[2] The definition of a component of a bipartite graph follows directly from the definition, in graph theory, of a component of a (nonbipartite) graph. Given a bipartite graph $\mathcal{G}$, consider the subgraph formed by its degree-2 CNs and their neighboring VNs. The subgraph can be equivalently represented by a $\mathcal{G}'$ graph with a single set of nodes, where each node corresponds to a VN in the subgraph of $\mathcal{G}$, and where an edge connects two nodes in $\mathcal{G}'$ if and only if the corresponding VNs in the subgraph of $\mathcal{G}$ are connected to the same degree-2 CN in the subgraph. A component of $\mathcal{G}'$ is a *maximal connected subgraph* of $\mathcal{G}'$, that is, a connected subgraph of $\mathcal{G}'$ with no neighbors outside the subgraph.

of this approach follows by observing that removing a column of large Hamming weight from (6.43) minimizes the density of the residual submatrix $\boldsymbol{B}$, hence increasing the probability of having, after removing the column, weight-one rows in the matrix.

Maximum Accumulated Weight Pivoting. Among the minimum-weight rows of $\boldsymbol{B}$, select the one with the largest accumulated weight. Denoting by w the Hamming weight of the selected row, define as pivots the symbols associated with $w - 1$ columns among those that are incident to the selected row. The $w - 1$ columns are chosen uniformly at random among the w incident columns. Note that, differently from the maximum column weight pivoting algorithm, the maximum accumulated weight strategy may select more than one pivot at once, ensuring the presence, in the residual submatrix $\boldsymbol{B}$, of a weight-one row.

Maximum Component Size Pivoting. If the minimum Hamming weight of the rows of $\boldsymbol{B}$ is larger than two, then select the pivots according to the maximum accumulated weight strategy. If $\boldsymbol{B}$ possesses one or more weight-two rows, then identify the largest component in the bipartite graph representation of $\boldsymbol{B}$, and select as pivot a random symbol among the ones associated with the VNs participating in the largest component. Denote by w the size of the component. Since each VN in a component is a key node, selecting a random VN in the component allows resuming the triangulation step for at least $w - 1$ iterations.

Maximum Cluster Size Pivoting. If the minimum Hamming weight of the rows of $\boldsymbol{B}$ is larger than two, then select the pivots according to the maximum accumulated weight strategy. If $\boldsymbol{B}$ possesses one or more weight-two rows, then identify the largest cluster in the bipartite graph representation of $\boldsymbol{B}$, and select as pivot a random symbol among the ones associated with the key VNs of the cluster. Maximum cluster size pivoting generalizes the maximum component size strategy. In particular, for a given submatrix $\boldsymbol{B}$, maximum cluster pivoting ensures resuming as the triangulation step for a number of iterations that is at least as large as the one of maximum component size pivoting. This fact follows by observing that each component is a cluster, but not all clusters are also components. Hence, the size of the largest component is always equal to or lower than the size of the largest cluster.

The average number of pivots obtained by means to various pivoting algorithms is summarized in Table 6.1. The results are provided for two (2048, 1024) LDPC codes: a regular (3, 6) LDPC code, and an irregular one with edge-oriented degree distributions

$$\begin{aligned}\lambda(x) &= 0.2221x + 0.2162x^2 + 0.0289x^6 + 0.1927x^7 \\ &\quad + 0.0508x^{17} + 0.0784x^{18} + 0.2110x^{53} \\ \rho(x) &= 0.0434x^7 + 0.9083x^8 + 0.0445x^9 + 0.0012x^{10} \\ &\quad + 0.0026x^{11}.\end{aligned} \tag{6.46}$$

The average number of pivots has been obtained through Monte Carlo simulations, fixing the number of erasures at the input of the decoder. The more sophisticated pivoting strategies – that are maximum component/cluster size – yield remarkable

Table 6.1 Average number of pivots measured for various numbers of erasures at the erasure channel output. Different pivoting strategies. The codes under consideration are a $(3,6)$ regular LDPC code and an irregular LDPC code, each with $n = 2\,048$ and $k = 1\,024$

Code	Pivoting algorithm	Number of erasures 1016	1000	984	968	952
Regular	Maximum column weight	94.3	89.9	80.4	67.8	54.4
Regular	Maximum accumulated weight	81.1	77.5	68.4	57.2	45.4
Regular	Maximal component size	67.1	64.2	55.9	45.1	34.4
Regular	Maximum cluster size	66.5	63.8	55.4	44.6	33.9
Irregular	Maximum column weight	37.0	26.5	17.3	9.7	5.1
Irregular	Maximum accumulated weight	35.2	24.8	16.3	9.3	4.7
Irregular	Maximum component size	33.5	23.3	15.6	8.6	4.4
Irregular	Maximum cluster size	33.0	22.7	14.9	8.3	4.2

savings in the average number of pivots for both codes. For the regular LDPC code case, the average number of pivots under maximum cluster size pivoting is nearly half of the one obtained with random pivoting. The reduction in number of pivots is even more tangible in the irregular LDPC code case.

6.2.2 Guessing Decoder

An intuitive way to modify the peeling decoder to achieve the ML decoding performance is provided by introducing the possibility to *guess* an erased bit value whenever the peeling decoder gets stuck in a stopping set. This modification of the peeling decoder is referred to as *guessing* decoder. The decoding algorithm works as follows. Peeling decoding is applied to the subgraph induced by the erasure pattern. If, at a given iteration, the peeling decoder cannot find CNs with residual degree one – that is, the VNs in the induced subgraph at the beginning of the iteration form a stopping set – a random VN in the stopping set is chosen, and two branches of the peeling decoder are instantiated: In one copy, the bit associated with the chosen VN is set to 0, and in the other copy the value of the VN bit is set to 1. This action is referred to as *guessing*. The process is repeated every time an instance of the peeling decoder gets stuck in a stopping set, possibly resulting in multiple peeling decoding branches. Note that if the decoder performed γ guesses till iteration ℓ, at that iteration, the number of running peeling decoder instances is upper bounded by 2^{ℓ}. It may happen that during the decoding of one branch, a VN is connected to two or more residual degree-1 CNs, which carry contradicting decisions on the VN bit value. In this case, the branch can be pruned since it is associated with guessed bits that are incompatible with code parity-check constraints. These events are referred to as contradictions. The branches that survive until the last decoding iterations yield a list with all the codewords consistent with the channel output. If only one branch survives, then the solution is unique.

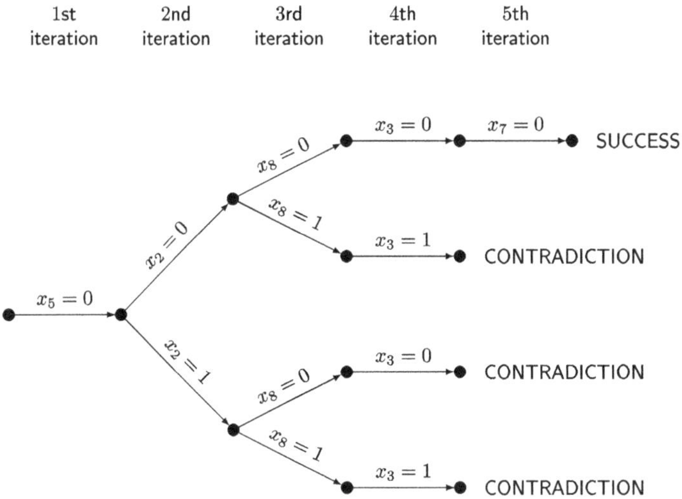

Figure 6.10 Guessing decoder from Example 6.8.

Example 6.8 (Guessing decoder) Consider again the binary linear block code of Example 6.2, with parity-check matrix

$$\boldsymbol{H} = \begin{bmatrix} 1 & 0 & 1 & 0 & 0 & 1 & 0 & 1 \\ 1 & 1 & 0 & 1 & 0 & 0 & 1 & 1 \\ 0 & 1 & 1 & 0 & 1 & 0 & 1 & 1 \\ 0 & 0 & 1 & 1 & 0 & 1 & 1 & 0 \\ 1 & 0 & 0 & 1 & 1 & 1 & 0 & 0 \end{bmatrix} \tag{6.47}$$

At the channel output, the peeling decoder observes

$$\boldsymbol{y} = \begin{bmatrix} 0 & \mathsf{E} & \mathsf{E} & 0 & \mathsf{E} & 0 & \mathsf{E} & \mathsf{E} \end{bmatrix}. \tag{6.48}$$

Peeling decoding succeeds in revealing $x_5 = 0$, but the decoding process halts after the first iteration since the remaining erased VNs form a stopping set. The guessing decoder proceeds by selecting a random VN in the stopping set. Consider for example, v_2: Two instances of peeling decoding are created, one for $x_2 = 0$ and one for $x_2 = 1$. The creation of the two branches is visualized in Figure 6.10. Unfortunately, the choice of v_2 is insufficient to unlock the peeling decoder since the residual set of erased VNs still forms a stopping set. Another guess is required: The decoder selects v_8, bifurcating each active branch according to $x_8 = 0$ and $x_8 = 1$. This second guess allows the four instances of the peeling decoder to reach the last iteration, where three branches result in a contradiction on VN v_7. Only one branch survives, leading to a unique ML decision.

The guessing principle can also be applied to BP decoding. Some critical observations follow.

Remark 6.10. The guessed VNs play for the guessing decoder the role that pivots play in the structured Gaussian elimination. In particular, if a set of guessed VNs allows the guessing decoder to proceed till the end, then the VNs represent a valid set of pivots for structured Gaussian elimination. Their selection as pivots is sufficient to obtain a system matrix in the form of Figure 6.9(b).

Remark 6.11. Although intuitive, the guessing decoder is not efficient since it may require running a number of parallel instances of the peeling decoder which grows exponentially in the number of guesses. In contrast, structured Gaussian elimination has a complexity that grows only cubically with the number of pivots.

6.3 Historical Notes

Message-passing decoding algorithms for LDPC codes – including the algorithm of Section 6.1.1 – were introduced by R. Gallager in his seminal work [1]. M. Tanner introduced bipartite graphs as representations of linear block codes [2] enabling a generalization of Gallager's message-passing decoder to a large class of codes. The definition of BP as an efficient algorithm to perform inference in probabilistic graphical models is due to J. Pearl [3,4] (see [5] for a complete reference on graphical models used in inference problems). Following the introduction of turbo codes [6], it was recognized that both iterative decoding of turbo codes and the message-passing decoding of LDPC codes are special instances of Pearl's BP algorithm [7–9]. Factor graphs were independently developed as general graphical models for probabilistic inference in various works [10–13] that largely extended the framework of Tanner graphs (see also [14]). When cast over a factor graph, BP is also known as sum–product algorithm (SPA). The implications on the code minimum distance of the cycle-free condition in a code Tanner graph is discussed in [15]. For a comprehensive treatment of iterative decoding, extended to general channels and to various classes of codes, we refer the reader to [16–18].

Peeling erasure decoding of LDPC codes was introduced in [19], and later used in [20, 21] to analyze the erasure correction capability of (irregular) LDPC codes. Peeling decoding also plays a fundamental role in the finite-length analysis of LDPC code ensembles [17, Sec. 3.19].

Stopping sets were introduced in [22, Appendix C]. Examples of effective algorithms that empirically allow limiting the number of small-size stopping sets when designing a Tanner graph are the PEG [23] and the ACE algorithms [24].

Structured Gaussian elimination algorithms for sparse linear systems were devised and studied in the context of cryptographic applications [25–27] as well as in coding theory. The fact that sparse linear systems can be solved via Gaussian elimination with reduced complexity was observed in [28, Sec. 2.6]. Efficient ML decoding algorithms based on structured Gaussian elimination were proposed for LDPC codes [22, 29–31], Raptor codes [32], and, more recently for polar-like codes [33]. A comparison of various pivoting strategies for structured Gaussian elimination decoding of LDPC codes is provided in [34], demonstrating the practicality of ML erasure decoding of LDPC codes at high data rates.

Guessing as a means to improve the performance of iterative decoding over the BEC was proposed in [29]. The principle of iterative erasure decoding with guessing was used in [35] to unveil fundamental properties that link the performance of LDPC codes under BP and MAP erasure decoding, in the limit of large blocklengths.

6.4 Exercises

Exercise 6.1 (Sufficient statistics) Consider a binary-input DMC with channel transition probability $P(y|x)$. Show that the LLR (6.4) is a *sufficient statistic* for estimating X given Y, that is, X is independent on Y given L. *Hint:* The proof follows by applying the Fisher-Neyman factorization theorem, by which it is sufficient to prove that $P(y|x)$ can be expressed as $a(x,L)b(y)$ for suitably defined functions $a(x,L)$ and $b(y)$.

Exercise 6.2 Following Exercise 6.1, prove that

$$L = \log \frac{P(L|X=0)}{P(L|X=1)}. \tag{6.49}$$

Exercise 6.3 (VN update rule) Derive the log-probability VN update rule (6.9). *Hint*: With uniform priors ($P(X_i = 0) = P(X_i = 1) = 1/2$) and by applying Bayes' rule, express (6.5) as

$$L_{i\to j} = \log \frac{P\big(L_i, \{L_{i\leftarrow z}\}_{z\neq j} | X_i = 0\big)}{P\big(L_i, \{L_{i\leftarrow z}\}_{z\neq j} | X_i = 1\big)}. \tag{6.50}$$

Assuming all incoming messages and the channel LLR to be independent, conditioned to X_i, proceed by writing

$$L_{i\to j} = \log \frac{P(L_i|X_i = 0)}{P(L_i|X_i = 1)} + \sum_{z:cnz\in\mathcal{N}(v_i)\setminus c_j} \log \frac{P(L_{i\leftarrow z}|X_i = 0)}{P(L_{i\leftarrow z}|X_i = 1)}. \tag{6.51}$$

What is left to prove is that first term of the summation on the right-hand side of (6.51) is L_i, whereas

$$\log \frac{P(L_{i\leftarrow z}|X_i = 0)}{P(L_{i\leftarrow z}|X_i = 1)} = L_{i\leftarrow z} \tag{6.52}$$

(see Exercise 6.2).

Exercise 6.4 (CN update rule) Consider $d - 1$ independent binary r.v.s X_1, $X_2, \ldots, X_{d-1}$ where

$$p_i = P(X_i = 1). \tag{6.53}$$

Denote their sum over GF(2) as $X_d = X_1 + X_2 + \ldots + X_{d-1}$. Show that

$$P(X_d = 0) = \frac{1}{2} + \frac{1}{2}\prod_{i=1}^{d-1}(1 - 2p_i). \tag{6.54}$$

Denote the log-probability ratio for X_i as

$$L_i = \log \frac{P(X_i = 0)}{P(X_i = 1)} \tag{6.55}$$

for $i = 1, \ldots, d$. Show that

$$L_d = 2 \tanh^{-1} \left[\prod_{i=1}^{d-1} \tanh\left(\frac{L_i}{2} \right) \right]. \tag{6.56}$$

Exercise 6.5 (BP over cycle-free graphs) Consider a (4,2) code with parity-check matrix

$$\boldsymbol{H} = \begin{pmatrix} 1 & 1 & 1 & 0 \\ 0 & 0 & 1 & 1 \end{pmatrix}. \tag{6.57}$$

The code is used to transmit over a BSC channel with error probability $p = 0.1$, resulting in a channel output $\boldsymbol{y} = \begin{bmatrix} 0 & 1 & 1 & 0 \end{bmatrix}$. Draw the code Tanner graph, and perform BP decoding. Show that the algorithm returns the exact log-APP ratios for each codeword bit, that is,

$$\begin{aligned} L_1^{\mathsf{APP}} &= 2.1972 & L_2^{\mathsf{APP}} &= -2.1972 \\ L_3^{\mathsf{APP}} &= -1.5163 & L_4^{\mathsf{APP}} &= -1.5163. \end{aligned} \tag{6.58}$$

Exercise 6.6 (Stopping sets) Consider the parity-check matrix

$$\boldsymbol{H} = \begin{bmatrix} 1 & 1 & 0 & 0 \\ 1 & 1 & 1 & 0 \\ 1 & 0 & 1 & 1 \end{bmatrix}. \tag{6.59}$$

Determine the information length k, blocklength n and list all codewords. List the smallest stopping set(s). Argue that an ML decoder would recover the codeword upon the channel observation $\boldsymbol{y} = \begin{bmatrix} \mathsf{E} & \mathsf{E} & \mathsf{E} & 1 \end{bmatrix}$, while an iterative decoder would not.

Exercise 6.7 Consider the (8,5) binary linear block code of Example 6.2. What is the code minimum distance? What is the minimum stopping set size of the Tanner graph associated with the parity-check matrix in (6.17)?

Exercise 6.8 Consider the (4,2) binary linear block code of Exercise 6.5. List all stopping sets associated with the parity-check matrix in the example, and all nonzero codewords. Verify that all stopping sets contain the support of a nonzero codeword. Use this observation to prove that iterative erasure decoding applied to the cycle-free Tanner graph of the code resolves all erasure patterns that can be recovered by the ML decoder.

Exercise 6.9 (Complexity of Gaussian elimination) Consider a dense nonsingular $n \times n$ binary system matrix. By dense we refer to the fact that roughly half of the entries is one and the other half is zero. Derive the number of additions (worst case) for the different steps of Gaussian elimination, that is, the forward elimination step and the back-substitution step.

Exercise 6.10 (Structured Gaussian elimination) Consider the system

$$\begin{bmatrix} 1 & 1 & 1 & 1 & 0 & 0 \\ 0 & 0 & 0 & 0 & 1 & 0 \\ 0 & 1 & 1 & 1 & 1 & 0 \\ 1 & 1 & 0 & 1 & 0 & 1 \\ 1 & 1 & 1 & 1 & 1 & 0 \\ 1 & 1 & 0 & 0 & 1 & 0 \end{bmatrix} \boldsymbol{x}_{\mathcal{E}}^{\mathsf{T}} = \begin{bmatrix} 1 \\ 0 \\ 0 \\ 0 \\ 0 \\ 1 \end{bmatrix}.$$

Apply structured Gaussian elimination with a pivoting strategy of your choice to recover the unknowns. Does the system have a unique solution? What is the number of pivots?

References

[1] R. Gallager, *Low-density parity-check codes.* Cambridge, MA, USA: MIT Press, 1963.

[2] M. Tanner, "A recursive approach to low complexity codes," *IEEE Trans. Inf. Theory*, vol. 27, no. 5, pp. 533–547, Sep. 1981.

[3] J. Pearl, "Reverend Bayes on inference engines: A distributed hierarchical approach," in *Nat. Conf. Artificial Intelligence*, Pittsburgh, PA,USA, Aug. 1982, pp. 133–136.

[4] J. Pearl, *Probabilistic reasoning in intelligent systems: Networks of plausible inference.* San Francisco, CA, USA: Morgan Kaufmann Publishers Inc., 1988.

[5] D. Koller and N. Friedman, *Probabilistic graphical models*, ser. Adaptive Computation and Machine Learning series. London, England: MIT Press, Jul. 2009.

[6] C. Berrou, A. Glavieux, and P. Thitimajshima, "Near Shannon limit error-correcting coding and decoding: Turbo-codes," in *Proc. IEEE Int. Conf. Commun.*, Geneva, Switzerland, May 1993.

[7] D. MacKay and R. Neal, "Good codes based on very sparse matrices," in *5th IMA Conference Cirencester*, ser. Lecture Notes in Computer Science, C. Boyd, Ed., vol. 1025. Springer Berlin Heidelberg, Dec. 1995, pp. 100–111.

[8] R. McEliece, D. MacKay, and J.-F. Cheng, "Turbo decoding as an instance of Pearl's 'belief propagation' algorithm," *IEEE J. Sel. Areas Commun.*, vol. 16, no. 2, pp. 140–152, Feb. 1998.

[9] D. MacKay, "Good error-correcting codes based on very sparse matrices," *IEEE Trans. Inf. Theory*, vol. 45, no. 2, pp. 399–431, Mar. 1999.

[10] N. Wiberg, H. Loeliger, and R. Kötter, "Codes and iterative decoding on general graphs," *European Trans. Telecommun.*, vol. 6, no. 5, p. 513–525, Sep. 1995.

[11] N. Wiberg, "Codes and decoding on general graphs," Ph.D. dissertation, Linköping University, Linköping, Sweden, Apr. 1996.

[12] G. Forney, "Codes on graphs: normal realizations," *IEEE Trans. Inf. Theory*, vol. 47, no. 2, pp. 520–548, Feb. 2001.

[13] F. Kschischang, B. Frey, and H.-A. Loeliger, "Factor graphs and the sum-product algorithm," *IEEE Trans. Inf. Theory*, vol. 47, no. 2, pp. 498–519, Feb. 2001.

[14] S. Aji and R. McEliece, "The generalized distributive law," *IEEE Trans. Inf. Theory*, vol. 46, no. 2, pp. 325–343, Mar. 2000.

[15] T. Etzion, A. Trachtenberg, and A. Vardy, "Which codes have cycle-free Tanner graphs?" *IEEE Trans. Inf. Theory*, vol. 45, no. 6, pp. 2173–2181, Sep. 1999.

[16] D. MacKay, *Information theory, inference & learning algorithms.* New York, NY, USA: Cambridge University Press, 2002.

[17] T. Richardson and R. Urbanke, *Modern coding theory.* New York, NY, USA: Cambridge University Press, 2008.

[18] W. Ryan, S. Lin, and S. Wilson, *Channel codes – Classical and modern.* New York, NY, USA: Cambridge University Press, 2024.

[19] V. V. Zyablov and M. S. Pinsker, "Decoding complexity of low-density codes for transmission in a channel with erasures," *Probl. Peredachi Inf.*, vol. 10, no. 1, pp. 15–28, 1974.

[20] M. Luby, M. Mitzenmacher, A. Shokrollahi, D. Spielman, and V. Stemann, "Practical loss-resilient codes," in *Proc. 29th Symp. Theory Computing*, El Paso, TX, USA, May 1997, pp. 150–159.

[21] M. Luby, M. Mitzenmacher, A. Shokrollahi, and D. Spielman, "Efficient erasure correcting codes," *IEEE Trans. Inf. Theory*, vol. 47, no. 2, pp. 569–584, Feb. 2001.

[22] T. Richardson and R. Urbanke, "Efficient encoding of low-density parity-check codes," *IEEE Trans. Inf. Theory*, vol. 47, no. 2, pp. 638–656, Feb. 2001.

[23] X.-Y. Hu, E. Eleftheriou, and D. Arnold, "Regular and irregular progressive edge-growth Tanner graphs," *IEEE Trans. Inf. Theory*, vol. 51, no. 1, pp. 386–398, Jan. 2005.

[24] T. Tian, C. Jones, J. D. Villasenor, and R. Wesel, "Selective avoidance of cycles in irregular LDPC code construction," *IEEE Trans. Commun.*, vol. 52, no. 8, pp. 1242–1247, Aug. 2004.

[25] A. M. Odlyzko, "Discrete logarithms in finite fields and their cryptographic significance," in *Proc. EUROCRYPT Workshop on Advances in Cryptology: Theory and Application of Cryptographic Techniques*, Paris, France, Apr. 1984, pp. 224–314.

[26] B. LaMacchia and A. Odlyzko, "Solving large sparse linear systems over finite fields," in *Proc. 10th Ann. Int. Crypt. Conf. Adv. in Crypt.*, Santa Barbara, CA, USA, Aug. 1990, pp. 109–133.

[27] C. Pomerance and J. W. Smith, "Reduction of huge, sparse matrices over finite fields via created catastrophes," *Experimental Mathematics*, vol. 1, no. 2, pp. 89–94, 1992.

[28] E. Berlekamp, *Algebraic coding theory.* New York, NY, USA: McGraw-Hill, 1968.

[29] H. Pishro-Nik and F. Fekri, "On decoding of low-density parity-check codes over the binary erasure channel," *IEEE Trans. Inf. Theory*, vol. 50, no. 3, pp. 439–454, Mar. 2004.

[30] D. Burshtein and G. Miller, "An efficient maximum likelihood decoding of LDPC codes over the binary erasure channel," *IEEE Trans. Inf. Theory*, vol. 50, no. 11, pp. 2837–2844, Nov. 2004.

[31] B. Vellambi and F. Fekri, "Results on the improved decoding algorithm for low-density parity-check codes over the binary erasure channel," *IEEE Trans. Inf. Theory*, vol. 53, no. 4, pp. 1510–1520, Apr. 2007.

[32] M. Shokrollahi, S. Lassen, and R. Karp, "Systems and processes for decoding chain reaction codes through inactivation," Feb. 2005, US Patent 6,856,263.

[33] M. C. Coşkun, J. Neu, and H. D. Pfister, "Successive cancellation inactivation decoding for modified Reed-Muller and eBCH codes," in *Proc. IEEE Int. Symp. Inf. Theory.* Los Angeles, CA, USA: IEEE, June 2020, pp. 437–442.

[34] E. Paolini, G. Liva, B. Matuz, and M. Chiani, "Maximum likelihood erasure decoding of LDPC codes: Pivoting algorithms and code design," *IEEE Trans. Commun.*, vol. 60, no. 11, pp. 3209–3220, Nov. 2012.

[35] C. Measson, A. Montanari, and R. Urbanke, "Maxwell construction: The hidden bridge between iterative and maximum a posteriori decoding," *IEEE Trans. Inf. Theory*, vol. 54, no. 12, pp. 5277–5307, Dec. 2008.

7 Iterative LDPC Decoder Analysis

This chapter delves into LDPC code ensembles under iterative erasure decoding. In the limit where the blocklength tends to infinity, the symbol erasure probability after iterative decoding usually exhibits an "on–off" threshold phenomenon, that is, there exists a channel erasure probability below which the symbol erasure probability vanishes and above which it is bounded away from zero. This *iterative decoding threshold* represents a key parameter for code design, even in the finite blocklength regime. In this regime, the slope of error probability curve may visibly flatten out for small erasure probabilities. This phenomenon, referred to as *error floor*, is due to stopping sets of small size. The chapter discusses how to obtain the iterative decoding threshold of unstructured, partly structured, and protograph LDPC code ensembles, with a focus on *binary* ensembles. The results directly hold for, or can be easily extended to, nonbinary ones. The chapter then introduces a tight union bound on the block error probability to predict the error floor. Finally, it provides some practical insights on the code design of LDPC codes for erasure channels with and without memory.

7.1 Density Evolution Analysis for Unstructured Ensembles

Density evolution is a tool for asymptotic analysis of LDPC code ensembles under BP decoding. It consists of letting the ensemble blocklength n become arbitrarily large, of modeling the messages exchanged among VNs and CNs as r.v.s, and of tracking the evolution of their p.m.f.s over the decoding iterations. For binary LDPC code ensembles over the BEC (and for nonbinary ones over the QEC), the message distribution can be parametrized by the message erasure probability. For ease of illustration, we first describe density evolution for unstructured regular LDPC code ensembles. We then extend the description irregular and (partly) structured ensembles.

7.1.1 Extrinsic Erasure Probability Evolution on Regular and Irregular Trees

7.1.1.1 Regular Tree

Consider the tree (i.e., connected and acyclic graph) depicted in Figure 7.1 and denoted as T_ℓ. It contains two different types of nodes, represented as squares and circles. Nodes in the tree are arranged into depth levels, such that there is exactly one layer of circle nodes and one layer of square nodes at each depth. The total number of depth levels is $\ell + 1$, indexed from 0 to ℓ, with the root (a circle) located at level ℓ. Level 0 only includes one layer of circle nodes, representing the leaves

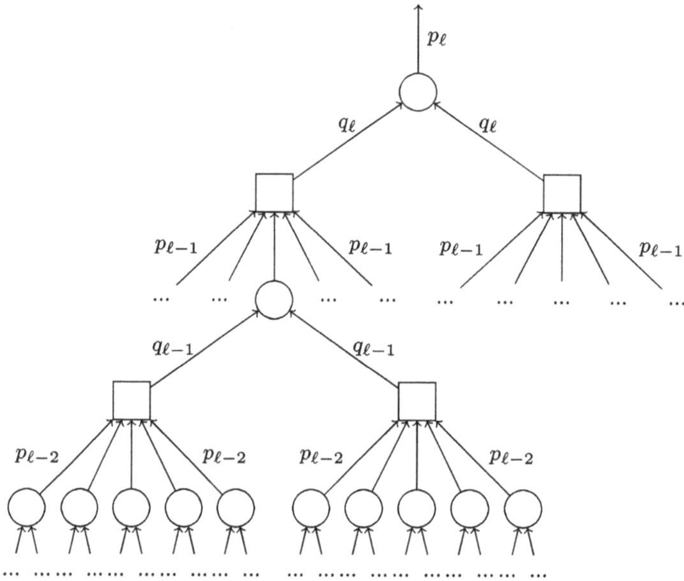

Figure 7.1 Tree graph T_ℓ with node degrees $\mathsf{d}_v = 3$ and $\mathsf{d}_c = 6$. The figure shows nodes at depth ℓ (which includes the root), depth $\ell - 1$, and the bit nodes at depth $\ell - 2$.

of the tree. We let each square node have a degree d_c and each circle node have a degree d_v, except for the leaf nodes, which by definition have a degree 1. In this sense, T_ℓ is said to be regular. Also, the degree of the root is d_v, although one of its edges is depicted as floating: We assume it is connected to some external entity with the role of a "sink." Next, we interpret every circle node as a bit and every square node as a parity-check constraint imposed on the d_c bits associated with the circle nodes connected to it. Therefore, every bit node (apart from the root and the leaves) is subject to d_v parity-check constraints.

Assume now that every bit corresponding to a circle node in the tree is initially known (initial value 0 or 1) with probability $1 - \epsilon$ and unknown (initial value E) with probability ϵ, independently from bit to bit. Next, consider the following iterative algorithm over T_ℓ, which propagates messages from the leaves to the root. At the beginning, every leaf bit node at depth 0 communicates its value (0, 1, or E) to the SPC node at depth 1 where it is connected to. If none of the $\mathsf{d}_c - 1$ messages received by an SPC node is equal to E, the SPC node determines the value of the bit node at level 1 where it is connected to and communicates this value to the bit node. Otherwise, it sends to the bit node a message equal to E. Next, if a bit node at depth 1 receives at least one message different from E out of its $\mathsf{d}_v - 1$ received messages, or if the corresponding bit was initially known, the bit node determines its own value and sends it to the SPC node at depth 2 where it is connected to. Otherwise, it sends a message equal to E. This process is repeated in the next layers, until the root receives its $\mathsf{d}_v - 1$ messages, possibly determines its value, and sends the corresponding message to the sink.

We are interested in the probability that the root node sends an erasure message to the sink. To this aim, we denote by q_t the probability that an SPC node at depth t sends an E message to the bit node at depth t where it is connected to. Similarly, we denote by p_t the probability that a bit node at depth t sends an E message to the SPC node at depth $t+1$ where it is connected to. Since, due to the tree structure of the graph, all messages sent during a parallel processing step are independent random variables, it is immediate to recognize that

$$q_t = 1 - (1 - p_{t-1})^{\mathsf{d}_c - 1} \tag{7.1}$$

and

$$p_t = \epsilon q_t^{\mathsf{d}_v - 1} \tag{7.2}$$

for all $t \in \{0, 1, \ldots, \ell\}$, with $p_0 = \epsilon$. The sought probability, p_ℓ, can therefore be computed by setting $p_0 = \epsilon$ and by recursively applying (7.1) and (7.2) ℓ times. Moreover, considering the same problem for the two tree graphs T_ℓ and $T_{\ell-1}$ and exploiting (7.1) and (7.2), we obtain the recursion

$$p_\ell = \epsilon\left(1 - (1 - p_{\ell-1})^{\mathsf{d}_c - 1}\right)^{\mathsf{d}_v - 1} \tag{7.3}$$

with starting point $p_0 = \epsilon$.

Remark 7.1. In the above-described computation algorithm over the regular tree, every message that a node sends along one edge of the tree is computed based on the messages the node receives along the other edges (plus, for bit nodes, the initial value). Stated equivalently, only *extrinsic information* floods through the tree, from the leaves to the root. The erasure probabilities expressed by (7.1) and (7.2) are referred to as *extrinsic erasure probabilities*. Similarly, p_ℓ is the extrinsic erasure probability from the root node to the sink.

7.1.1.2 Irregular Tree

Next, we consider the same computation algorithm on an irregular tree. The tree structure is the same as before, with $\ell + 1$ depth levels, each level including one layer of bit nodes and one layer of SPC nodes, and only degree-1 bit nodes (leaves) at depth 0. Unlike the previous case, however, the tree is built in a probabilistic way from the root to the leaves, as follows. The number of children of the root node is drawn randomly by sampling a discrete random variable $J \in \{1, \ldots, \mathsf{d}_v\}$ such that $P(J = j) = \lambda_j$. The number of children of each SPC node connected to the root is also drawn randomly by sampling a discrete random variable $I \in \{1, \ldots, \mathsf{d}_c\}$ with $P(I = i) = \rho_i$. This is done independently for each SPC node. This random tree generation procedure is repeated for all the subsequent layers down to the leaves by sampling the random variables J and I for the bit nodes and the SPC nodes, respectively, independently for each node. As for the regular case, due to the tree structure all messages sent during a parallel processing step are independent random variables.

Over the irregular tree, we still introduce two probabilities q_t and p_t, with a slightly different meaning: q_t is the average probability that an SPC node at depth t sends an E message to the bit node at depth t where it is connected to; p_t is the average probability that a bit node at depth t sends an E message to the SPC node at depth $t+1$ where it is connected to. The "average" is calculated with respect to node degrees, which are random variables according to the probabilistic tree construction. We extend (7.1) and (7.2) to the irregular case by replacing all message erasure probabilities at the input of the nodes belonging to the same layer with the corresponding average probability. Hence, all messages at the input of SPC nodes at depth t have a probability p_{t-1} of being equal to E, and all messages at the input of bit nodes at depth t have a probability q_t of being equal to E, independent of the message.

Next, let $\lambda(x) = \sum_{j=1}^{\mathsf{d}_v} \lambda_j x^{j-1}$ and $\rho(x) = \sum_{i=1}^{\mathsf{d}_c} \rho_i x^{i-1}$. Using these two polynomials, the equations generalizing (7.1) and (7.2) to the irregular case can be expressed as

$$q_t = 1 - \rho(1 - p_{t-1}) \tag{7.4}$$

and

$$p_t = \epsilon\lambda(q_t)\,. \tag{7.5}$$

The average extrinsic erasure probability from the root to the sink node, p_ℓ, can be obtained setting $p_0 = \epsilon$ and recursively applying (7.4) and (7.5) ℓ times. Furthermore, considering two irregular trees of depths ℓ and $\ell - 1$, we can generalize the recursion (7.3) as

$$p_\ell = \epsilon\lambda(1 - \rho(1 - p_{\ell-1})) \tag{7.6}$$

with starting point $p_0 = \epsilon$.

In Section 7.1.2, we discuss how the recursions (7.3) and (7.6) can be used to analyze the iterative erasure decoder over the BEC (as introduced in Section 6.1.2) for unstructured binary regular and irregular LDPC code ensembles, in the asymptotic regime where n tends to infinity. As mentioned at the beginning of the chapter, the reasoning remains valid with no modifications for ensembles constructed over GF(q) and used to transmit over the QEC.

7.1.2 Density Evolution for Unstructured LDPC Code Ensembles

Consider an LDPC code belonging to the regular ensemble $\mathscr{C}^n_{\mathsf{d}_v,\mathsf{d}_c}$ (as defined in Section 4.2.1) and a specific edge of its Tanner graph, for example, the edge connecting VN v_j and CN c_i. Then, consider the following graph, that we call the neighborhood of the edge $(\mathsf{v}_j, \mathsf{c}_i)$ up to depth ℓ, denoting it as $\mathcal{N}_\ell(\mathsf{v}_j, \mathsf{c}_i)$. It is obtained by "unfolding" the bipartite graph starting from v_j, as follows. Depth level ℓ includes v_j, the $\mathsf{d}_v - 1$ CNs connected to v_j, excluding c_i, and the $\mathsf{d}_v - 1$ edges connecting v_j to these CNs. Depth level $\ell - 1$ includes the VNs connected to the CNs at level ℓ, excluding the VNs at level ℓ (that is v_j), the CNs connected to these VNs, excluding the CNs at level ℓ, and the edges connecting such VNs and CNs. The subsequent layers, up to layer 1, are built with the same procedure. Depth level 0 includes only a layer of VNs, playing the role of leaves.

Let the LDPC code be transmitted over a BEC with erasure probability ϵ and be decoded iteratively according to Algorithm 1 introduced in Chapter 6. Consider the message from v_j to c_i at decoding iteration ℓ: It is immediate to recognize that this message coincides with the one we obtain by running on $\mathcal{N}_\ell(\mathsf{v}_j, \mathsf{c}_i)$ the computation algorithm over a regular tree described in Section 7.1, with c_i playing the role of the "sink." Note that, despite this connection, the recursion (7.3) cannot be extended trivially to $\mathcal{N}_\ell(\mathsf{v}_j, \mathsf{c}_i)$: This is because the LDPC code Tanner graph contains cycles, and therefore the same VN or CN may appear multiple times when unfolding the Tanner graph to build up $\mathcal{N}_\ell(\mathsf{v}_j, \mathsf{c}_i)$.

Some important results, however, allow us to apply (7.3) to unstructured regular LDPC code ensembles in the limit where the blocklength n tends to infinity, which is sometimes called the "asymptotic regime." The main such result is stated by Theorem 7.1, where we recall that the procedure to draw a specific Tanner graph randomly from the ensemble $\mathscr{C}^n_{\mathsf{d}_v,\mathsf{d}_c}$ is the one detailed in Section 4.2.1.

Theorem 7.1 *Consider the probability space of all Tanner graphs from* $\mathscr{C}^n_{\mathsf{d}_v,\mathsf{d}_c}$ *and all BEC realizations. Let* Z_ℓ *be the fraction of erasure VN-to-CN messages after* ℓ *rounds of iterative decoding and* p_ℓ *be the erasure probability from* (7.3). *Then, for any* $\delta > 0$ *there exists* $\bar{n} = \bar{n}(\ell, \mathsf{d}_v, \mathsf{d}_c, \delta)$ *such that for all* $n > \bar{n}$

$$P(|Z_\ell - p_\ell| > \delta) < 2\exp\left(-\beta\delta^2 n\right), \tag{7.7}$$

where $\beta = \beta(\ell, \mathsf{d}_v, \mathsf{d}_c) > 0$.

Broadly speaking, Theorem 7.1 states that, for n large enough, the probability that an edge carries an erasure message from a VN to a CN after ℓ rounds of iterative decoding for a code randomly picked from the ensemble $\mathscr{C}^n_{\mathsf{d}_v,\mathsf{d}_c}$ can be predicted by (7.3) almost surely. Theorem 7.1 follows from two intermediate results. The first is a concentration result which says that (by a martingale argument) Z_ℓ concentrates around $\mathsf{E}[Z_\ell]$; formally, $P(|Z_\ell - \mathsf{E}[Z_\ell]| > \delta/2) \leq 2\exp(-\beta\delta^2 n)$ for all $n > \bar{n}$. This means that, for n large enough, the performance of a code randomly picked from the ensemble $\mathscr{C}^n_{d_v,d_c}$ is close to the ensemble average with high probability. The second result says that $\mathsf{E}[Z_\ell]$ converges to p_ℓ; formally, for all $n > 2\gamma/\delta$ we have $|\mathsf{E}[Z_\ell] - p_\ell| < \delta/2$, where $\gamma = \gamma(\ell, \mathsf{d}_v, \mathsf{d}_c) > 0$. This latter convergence result relies on the fact that, for fixed $\ell > 0$, for a randomly drawn Tanner graph in $\mathscr{C}^n_{\mathsf{d}_v,\mathsf{d}_c}$, the probability that $\mathcal{N}_\ell(\mathsf{v}_j, \mathsf{c}_i)$ is not tree-like is upper bounded by γ/n.

7.1.3 Density Evolution for Unstructured Irregular LDPC Code Ensembles

So far we have discussed how the recursion (7.3) is useful to analyze the performance of unstructured regular LDPC code ensembles under iterative erasure decoding, in the asymptotic regime where the blocklength tends to infinity. In a similar way, we can employ the recursion (7.6) to analyze unstructured irregular LDPC ensembles under iterative erasure decoding, again in the asymptotic regime. (We recall that the codes in the ensemble obey the constraints imposed by the edge-oriented degree distribution polynomials $\lambda(x)$ and $\rho(x)$ defined in (4.3) and (4.4); an irregular Tanner graph is exemplified in Figure 7.2, where $\mathbf{\Pi}$ symbolizes an edge interleaver.)

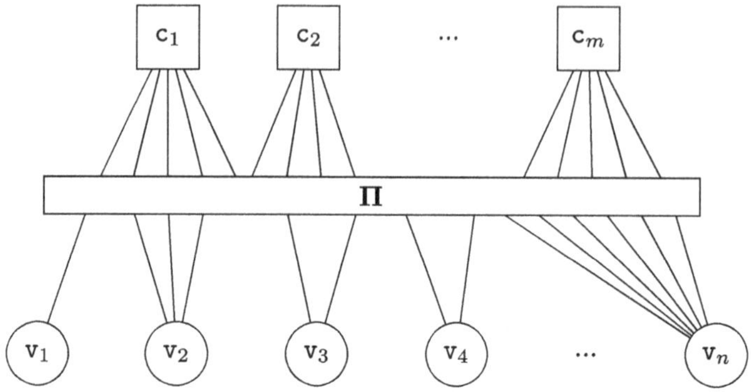

Figure 7.2 Tanner graph of an irregular unstructured LDPC code.

Considering an unstructured irregular ensemble $\mathscr{C}^n_{\lambda,\rho}$, letting p_ℓ be the erasure probability from (7.6) and the two polynomials $\lambda(x)$ and $\rho(x)$ in (7.6) coincide with the edge-oriented VN and CN degree distribution polynomials defining the irregular ensemble, the result expressed by Theorem 7.1 remains essentially valid, stating concentration around p_ℓ of the fraction of erasure VN-to-CN messages for a random code drawn from $\mathscr{C}^n_{\lambda,\rho}$, for n large enough. As such, in the asymptotic regime the probability that an edge carries an erasure message from a VN to a CN after ℓ rounds of iterative decoding for a code randomly picked in $\mathscr{C}^n_{\lambda,\rho}$ can be predicted by (7.6) almost surely.

Remark 7.2. The recursions (7.3) and (7.6) are usually referred to as *density evolution recursions* for unstructured regular and irregular LDPC code ensembles, respectively. Note that (7.6) boils down to (7.3) in the regular case. Therefore, hereafter we will only refer to (7.6), which includes (7.3) as a special case.

7.1.4 Iterative Decoding Threshold for Unstructured LDPC Code Ensembles

Consider an LDPC code ensemble $\mathscr{C}^n_{\lambda,\rho}$ over the BEC and let $n \to \infty$. Consider the density evolution recursion (7.6), which describes almost surely the evolution of the fraction of erasure messages from VNs to CNs over decoding iterations, and observe that p_ℓ is a function of ϵ. Next, let the decoding iteration index $\ell \to \infty$. The largest value of the channel erasure probability ϵ such that $p_\ell(\epsilon)$ converges to zero as ℓ goes to infinity is termed *iterative decoding threshold* and is indicated as $\epsilon^\star_{\mathsf{BP}}$. We formalize this in Definition 7.1.

Definition 7.1 (Iterative decoding threshold) Consider an unstructured LDPC code ensemble defined by its edge-oriented degree distribution polynomials $\lambda(x)$ and $\rho(x)$. Let $n \to \infty$ and $p_\ell(\epsilon)$ be given in (7.6). Then, the ensemble's iterative decoding threshold $\epsilon^\star_{\mathsf{BP}}$ is

$$\epsilon_{\mathsf{BP}}^{\star} = \sup\left\{\epsilon \in [0,1] : p_\ell(\epsilon) \xrightarrow{\ell \to \infty} 0\right\}. \tag{7.8}$$

Remark 7.3. The density evolution recursion (7.6) is independent of the field order. Thus, for a q-ary LDPC code ensemble over the QEC, the iterative decoding threshold does not depend upon the choice of q.

Remark 7.4. The definition of $\epsilon_{\mathsf{BP}}^{\star}$ relies on two limits, one as $n \to \infty$ and the other as $\ell \to \infty$. The order of application of the two limits is very important, that is, we must let the blocklength tend to infinity at first (asymptotic regime) and subsequently the number of decoding iterations tend to infinity.

An alternative definition of iterative decoding threshold for unstructured code ensembles is based on the observation that the function

$$f(x,\epsilon) = \epsilon\lambda(1 - \rho(1 - x)), \tag{7.9}$$

is monotonically increasing in both its arguments ϵ and x. Then,

$$\epsilon_{\mathsf{BP}}^{\star} = \sup\{\epsilon \in [0,1] : x > f(x,\epsilon)\ \forall x \in (0,1]\}. \tag{7.10}$$

Example 7.1 (Threshold computation) Consider the regular (3,6) code ensemble with $\lambda(x) = x^2$ and $\rho(x) = x^5$. Evaluation of the density evolution recursion yields an iterative decoding threshold of $\epsilon_{\mathsf{BP}}^{\star} \approx 0.42944$.

7.1.5 Stability Condition for Unstructured Ensembles

The stability condition indicates whether the fixed point zero of the density evolution recursion is stable or not. Importantly, the stability conditions yields an upper bound on the code ensemble's iterative decoding threshold.

Let us write the right-hand side of (7.6) as a Taylor series around the point zero. For p_ℓ close to zero, we truncate the series at the first-order derivative and obtain

$$p_\ell \approx \epsilon\lambda_1 + \epsilon\lambda'(0)\rho'(1)p_{\ell-1}\,. \tag{7.11}$$

We make the following observations:

a) When $\lambda_1 > 0$, p_ℓ will be zero only if ϵ is zero. Hence, next we only consider LDPC code ensembles with $\lambda_1 = 0$.
b) When $\lambda_1 = 0$, $p_{\ell-1} = 0$ yields $p_\ell = 0$. We say that zero is a fixed point of the density evolution recursion in (7.6).
c) When $\lambda_1 = 0$, convergence to zero (for small p_ℓ) will depend on the value of $\epsilon\lambda'(0)\rho'(1)$.

From the previous observation, we shall derive an upper bound on the iterative decoding threshold. We note that for small p_ℓ, convergence is not possible if

$$\frac{p_\ell}{p_{\ell-1}} = \epsilon\lambda'(0)\rho'(1) > 1 \tag{7.12}$$

or equivalently

$$\epsilon > \frac{1}{\lambda'(0)\rho'(1)} = \epsilon^{\mathsf{stab}}. \tag{7.13}$$

As a consequence, the code ensemble's iterative decoding threshold $\epsilon^{\star}_{\mathsf{BP}}$ must not be larger than ϵ^{stab}, which yields the upper bound $\epsilon^{\star}_{\mathsf{BP}} \leq \epsilon^{\mathsf{stab}}$. This upper bound turns out to be tight for the capacity achieving sequences that will be presented in Section 7.5.1. When (7.13) holds, we say that the fixed point zero of the density evolution recursion is unstable. When ϵ is smaller than the right-hand side of (7.13), the fixed point is called stable.

Example 7.2 Consider the degree distribution pair $\lambda(x) = \frac{1}{3}x + \frac{2}{3}x^3$ and $\rho(x) = x^5$. From (7.12) we obtain $\epsilon^{\mathsf{stab}} = \frac{3}{5}$. Thus, the code ensemble's iterative decoding threshold is at most $\frac{3}{5}$.

7.2 Extrinsic Information Transfer Analysis

7.2.1 EXIT Functions

An alternative technique to compute a code ensemble's iterative decoding threshold is extrinsic information transfer (EXIT) analysis. While for most communication channels, EXIT analysis yields an approximation of density evolution analysis, over the erasure channel the two techniques are equivalent. The EXIT analysis also provides a useful graphical interpretation of the iterative decoder convergence properties. The building blocks of EXIT analysis are the so-called EXIT functions which are defined next. Transmission of a binary code $\mathcal{C}$ over a BEC is assumed.

Definition 7.2 (EXIT function) Let $\boldsymbol{X}$ be a codeword of $\mathcal{C}$ and $\boldsymbol{Y}$ denote the observed sequence after transmission over a BEC with erasure probability ϵ. The EXIT function associated with the ith bit of $\mathcal{C}$ is

$$h_i(\epsilon) = \mathsf{H}(X_i|\boldsymbol{Y}_{\sim i}(\epsilon)), \tag{7.14}$$

where we used the notation $\boldsymbol{Y}_{\sim i}$ to denote the vector $\boldsymbol{Y}$ without its ith element.

Definition 7.3 (Average EXIT function) The average EXIT function is

$$h(\epsilon) = \frac{1}{n}\sum_{i=1}^{n} h_i(\epsilon). \tag{7.15}$$

EXIT functions can be alternatively stated in terms of mutual information (MI) between a code symbol i and the observed sequence $\boldsymbol{Y}$ (excluding the ith observation) as

$$\mathsf{I}(X_i; \boldsymbol{Y}_{\sim i}(\epsilon)) = 1 - \mathsf{H}(X_i | \boldsymbol{Y}_{\sim i}(\epsilon)). \tag{7.16}$$

In terms of MI, the average EXIT function over all n code symbols is defined as

$$\mathsf{I}(\epsilon) = \frac{1}{n} \sum_{i=1}^{n} \mathsf{I}(X_i; \boldsymbol{Y}_{\sim i}(\epsilon)). \tag{7.17}$$

In the following equations, we provide the EXIT functions for VNs and CNs, the two building blocks of LDPC codes. We denote the average MI between an incoming VN message and the associated symbol by I_{AV}; between an outgoing VN message and the associated symbol by I_{EV}; between an incoming CN message and the associated symbol by I_{AC}; and between an outgoing CN message and the associated symbol by I_{EC}.

Over a BEC, the mutual information between an input bit and the corresponding output bit is equal to $1-\epsilon$. With this in mind, if we assume that VNs and CNs exchange messages over BECs with erasure probabilities q and p, respectively, (7.4) can be restated in terms of MI. This yields the CN EXIT function

$$\mathsf{I}_{\mathrm{EC}} = \rho(\mathsf{I}_{\mathrm{AC}}) \tag{7.18}$$

$$= \mathsf{T}_{\mathrm{C}}(\mathsf{I}_{\mathrm{AC}})\,. \tag{7.19}$$

Similarly, from (7.5), we obtain the VN EXIT function

$$\mathsf{I}_{\mathrm{EV}} = 1 - \epsilon\lambda(1 - \mathsf{I}_{\mathrm{AV}}) \tag{7.20}$$

$$= \mathsf{T}_{\mathrm{V}}(\mathsf{I}_{\mathrm{AV}}, \epsilon)\,. \tag{7.21}$$

Using the identities $\mathsf{I}_{\mathrm{AV}} = \mathsf{I}_{\mathrm{EC}}$ and $\mathsf{I}_{\mathrm{AC}} = \mathsf{I}_{\mathrm{EV}}$, we may now rewrite the density evolution recursion in (7.6) as

$$\mathsf{I}_{\mathrm{EV},\ell} = 1 - \epsilon\lambda(1 - \rho(\mathsf{I}_{\mathrm{EV},\ell-1}))\,. \tag{7.22}$$

7.2.2 EXIT Charts

Further insights on the decoder convergence can be gained by visualizing the VN and CN transfer functions in one plot, resulting in a so-called EXIT chart. More precisely, we plot $\mathsf{T}_{\mathrm{V}}(\mathsf{I}_{\mathrm{AV}}, \epsilon)$, where ϵ is a fixed parameter, and $\mathsf{I}_{\mathrm{EC}} = \mathsf{T}_{\mathrm{C}}(\mathsf{I}_{\mathrm{AC}})$. Recall that the output of a VN serves as input to a CN and vice versa. For this, practically one of the curves needs to be inverted.

Example 7.3 (EXIT charts) Figure 7.3 shows the EXIT chart of a (3,6) regular LDPC code ensemble. The extrinsic MI at the VN output is obtained from (7.20) as

$$\mathsf{I}_{\mathrm{EV}} = \mathsf{T}_{\mathrm{V}}(\mathsf{I}_{\mathrm{AV}}, \epsilon) = 1 - \epsilon(1 - \mathsf{I}_{\mathrm{AV}})^2\,. \tag{7.23}$$

Similarly, the extrinsic MI at the CN output is given by (7.18)

$$\mathsf{I}_{EC} = \mathsf{T}_C(\mathsf{I}_{AC}) = (\mathsf{I}_{AC})^5 . \tag{7.24}$$

The three nonsolid lines show $\mathsf{I}_{EV} = \mathsf{T}_V(\mathsf{I}_{AV}, \epsilon)$ for channel erasure probabilities $\epsilon \in \{0.399, 0.429, 0.459\}$. Note that the abscissa represents $\mathsf{I}_{AV} = \mathsf{I}_{EC}$, while the ordinate $\mathsf{I}_{EV} = \mathsf{I}_{AC}$. In the same figure, we plotted the inverted transfer function for the CN operators, $\mathsf{I}_{AC} = \mathsf{T}_C^{-1}(\mathsf{I}_{EC})$ (solid line). The inlay figure provides a zoom around the coordinate (0.23, 0.75) where the two EXIT charts come close to each other.

Next, we discuss some observations regarding EXIT charts.

7.2.2.1 Convergence

We can provide a necessary condition for decoding convergence. Observe that, by definition, the edge-oriented degree distribution polynomials $\lambda(x)$ and $\rho(x)$ are strictly monotonically increasing functions for $x \in [0, 1]$. Following (7.22), dropping the iteration index we introduce the function

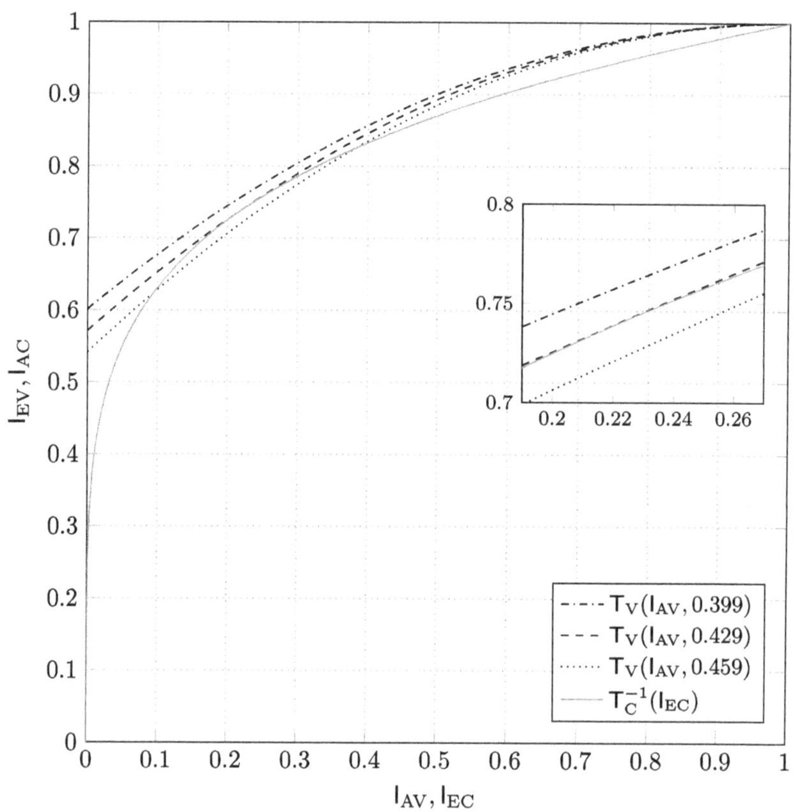

Figure 7.3 EXIT charts for (3, 6) regular LDPC code ensemble for different channel erasure probabilities.

$$f(\mathsf{I}_{\mathrm{EV}}, \epsilon) = 1 - \epsilon\lambda(1 - \rho(\mathsf{I}_{\mathrm{EV}}))\,. \tag{7.25}$$

It can be shown that $f(\mathsf{I}_{\mathrm{EV}}, \epsilon)$ is monotonically increasing in both its arguments ϵ and I_{EV}. For ensembles without degree-1 VNs, the point $(1, 1)$ can be reached by requiring an increase of MI at every iteration,

$$\mathsf{I}_{\mathrm{EV}} < f(\mathsf{I}_{\mathrm{EV}}, \epsilon)\ \forall\ \mathsf{I}_{\mathrm{EV}} \in [0, 1). \tag{7.26}$$

Exploiting

$$f(\mathsf{I}_{\mathrm{EV}}, \epsilon) = \mathsf{T}_{\mathrm{V}}(\mathsf{T}_{\mathrm{C}}(\mathsf{I}_{\mathrm{EV}}), \epsilon) \tag{7.27}$$

we may restate (7.26) as

$$\mathsf{T}_{\mathrm{V}}^{-1}(\mathsf{I}_{\mathrm{EV}}, \epsilon) < \mathsf{T}_{\mathrm{C}}(\mathsf{I}_{\mathrm{EV}})\,. \tag{7.28}$$

Thus, for convergence the inverted VN EXIT function must be below the CN EXIT function. Equivalently, the VN EXIT function must be above the inverted CN function (see Figure 7.3).

Example 7.4 (Convergence) Consider a $(3, 6)$ code ensemble with edge-oriented degree distribution pair $\lambda(x) = x^2$ and $\rho(x) = x^5$. The recursion in (7.22) is $\mathsf{I}_{\mathrm{EV},\ell} = 1 - \epsilon(1 - (\mathsf{I}_{\mathrm{EV},\ell-1})^5)^2$. We initialize $\mathsf{I}_{\mathrm{EV},0} = 0$. By numerical evaluation we observe that after a large number of iterations (1 000 in our case) $\mathsf{I}_{\mathrm{EV},\ell}$ closely approaches one for erasure probabilities not larger than 0.429. This is consistent with Figure 7.3.

7.2.2.2 Stability Condition

We can also provide a graphical interpretation of stability. The derivative of the CN EXIT function, evaluated in $\mathsf{I}_{\mathrm{AC}} = 1$, is $\rho'(1)$. Thus, the derivative of the inverted CN EXIT function in $\mathsf{I}_{\mathrm{EC}} = 1$ is $\frac{1}{\rho'(1)}$. The derivative of the VN EXIT function evaluated in $\mathsf{I}_{\mathrm{AV}} = 1$ is $\epsilon\lambda'(0)$. A necessary condition for convergence is that two EXIT functions do not cross nearby $(1, 1)$. Hence, the slope of the VN EXIT function has to be smaller than the slope of the inverted CN function in the vicinity of $(1, 1)$, possibly with equality in $(1, 1)$.[1] Formally,

$$\epsilon\lambda'(0) \leq \frac{1}{\rho'(1)}\,. \tag{7.29}$$

This yields the same upper bound on the iterative decoding threshold as in Section 7.1.5.

[1] Observe that the VN EXIT function has to be always greater than the inverted CN EXIT function, except in $(1, 1)$.

Example 7.5 (Stability condition) Consider again the (3,6) regular LDPC code ensemble with degree distribution pair $\lambda(x) = x^2$ and $\rho(x) = x^5$, VN and CN EXIT functions given in (7.23) and (7.24). The derivative of the VN EXIT function with respect to I_{AV} is $\epsilon 2(1 - \mathsf{I}_{\mathrm{AV}})$. At $\mathsf{I}_{\mathrm{AV}} = 1$ this yields zero. Hence, condition in (7.29) is fulfilled for all ϵ. In this case, the upper bound on the iterative decoding threshold is one.

7.2.2.3 Area Difference

We discuss the area difference below the VN and inverted CN function. For ease of presentation, we consider only the case of regular code ensembles. The area under the VN EXIT curve is

$$A_{\mathrm{V}} = \int_0^1 \left(1 - \epsilon(1-x)^{\mathsf{d}_v - 1}\right)\mathrm{d}x \tag{7.30}$$

$$= 1 - \frac{\epsilon}{\mathsf{d}_v} \tag{7.31}$$

$$= 1 - \frac{1 - \mathsf{C}(\epsilon)}{\mathsf{d}_v}, \tag{7.32}$$

where $\mathsf{C}(\epsilon) = 1 - \epsilon$ is the capacity of the BEC with erasure probability ϵ. The area under the inverted CN EXIT curve can be computed as one minus the area below the CN curve, that is,

$$A_{\mathrm{C}^{-1}} = 1 - \int_0^1 x^{\mathsf{d}_c - 1}\mathrm{d}x \tag{7.33}$$

$$= 1 - \frac{1}{\mathsf{d}_c}. \tag{7.34}$$

We can state the following:

(a) The area above the VN curve plus that below the inverted CN curve must be less or equal to one. Thus,

$$1 - A_{\mathrm{V}} + A_{\mathrm{C}^{-1}} \leq 1 \tag{7.35}$$

$$1 - \frac{1 - \mathsf{C}(\epsilon)}{\mathsf{d}_v} + 1 - \frac{1}{\mathsf{d}_c} \leq 1 \tag{7.36}$$

$$1 - \frac{\mathsf{d}_v}{\mathsf{d}_c} \leq \mathsf{C}(\epsilon) \tag{7.37}$$

$$R \leq \mathsf{C}(\epsilon), \tag{7.38}$$

which is consistent with Shannon's coding theorem, that is, the design rate of any regular LDPC code cannot surpass capacity.

(b) Any area difference below the VN and the inverted CN EXIT curves yields a loss in rate with respect to channel capacity. To see this, let the area difference $A_\delta = A_{\mathrm{V}} - A_{\mathrm{C}^{-1}}$. We have

$$A_\delta = \frac{1}{\mathsf{d}_c} - \frac{1 - \mathsf{C}(\epsilon)}{\mathsf{d}_v} \tag{7.39}$$

and

$$\mathsf{d}_v A_\delta = \mathsf{C}(\epsilon) - R \geq 0. \tag{7.40}$$

Observe that equality is only obtained if the area difference is zero (or if the VN degree is 0).

(c) The number of required decoding iterations increases with decreasing gap between the two curves (i.e., with narrowing of the tunnel). This is because the increase of MI at the output (or input) of a node from one iteration to another depends on the gap between the two EXIT functions. A big gap yields a large increase in MI at each iteration.

Remark 7.5. To approach capacity under iterative decoding, one would need to "match" the VN and CN EXIT functions at the threshold, that is, at $\epsilon = \epsilon^{\star}_{\mathsf{BP}}$. Since by doing so the gap area A_δ will vanish, we can conclude that an ensemble approaching capacity would require a very large number of decoding iterations when used to transmit over a BEC with erasure probability close to the ensemble iterative decoding threshold.

Example 7.6 (Evolution of MI) The evolution of MI for iterative decoding is visualized in Figure 7.4 for the regular (3,6) LDPC code ensemble. The arrows visualize the evolution of the MI for recursive application of the VN and CN transfer functions, sometimes referred to as decoding trajectory. We assume $\epsilon = 0.399$.

7.2.2.4 Duality

Let $\mathcal{C}$ be a linear code and let us denote its EXIT function by $\mathsf{T}_{\mathcal{C}}(x)$. Similarly, let $\mathcal{C}^{\perp}$ be its dual code with EXIT function $\mathsf{T}_{\mathcal{C}^{\perp}}(x)$. Then, for the BEC we have $\mathsf{T}_{\mathcal{C}}(x) + \mathsf{T}_{\mathcal{C}^{\perp}}(1-x) = 1$, which implies that knowing one EXIT function, the other can be easily derived.

First, observe that

$$\left.\frac{\partial \mathsf{H}(\boldsymbol{X}|\boldsymbol{Y}(\epsilon_1,\epsilon_2,\ldots,\epsilon_n))}{\partial \epsilon_i}\right|_{\epsilon_i=\epsilon} = \mathsf{H}(X_i|\boldsymbol{Y}_{\sim i}(\epsilon)) \tag{7.41}$$

$$= 1 - \mathsf{I}(X_i,\boldsymbol{Y}_{\sim i}(\epsilon))\,. \tag{7.42}$$

Equation (7.41) is proven in Section 8.1, while (7.42) results from uniformly distributed X_i.

Recall from Section 6.2 that the set of codeword indices where an erasure occurs is denoted by $\mathcal{E}$, with $e = |\mathcal{E}|$, and the complementary set by $\bar{\mathcal{E}}$. Following (6.30), with slight change of notation, an ML decoder aims to solve the system of equations

$$\boldsymbol{H}_{\mathcal{E}}\boldsymbol{x}_{\mathcal{E}}^{\mathsf{T}} = \boldsymbol{H}_{\bar{\mathcal{E}}}\boldsymbol{x}_{\bar{\mathcal{E}}}^{\mathsf{T}}. \tag{7.43}$$

Remind that $\boldsymbol{x}_{\bar{\mathcal{E}}}$ are the known codeword symbols at the decoder input, $\boldsymbol{x}_{\mathcal{E}}$ the codeword symbols that were erased by the channel, and $\boldsymbol{H}_{\bar{\mathcal{E}}}$ and $\boldsymbol{H}_{\mathcal{E}}$ the corresponding submatrices of the code's parity-check matrix $\boldsymbol{H}$. The dimension of the solution space is $e - \text{rank}(\boldsymbol{H}_{\mathcal{E}})$. Starting from the code's generator matrix $\boldsymbol{G}$ and the information

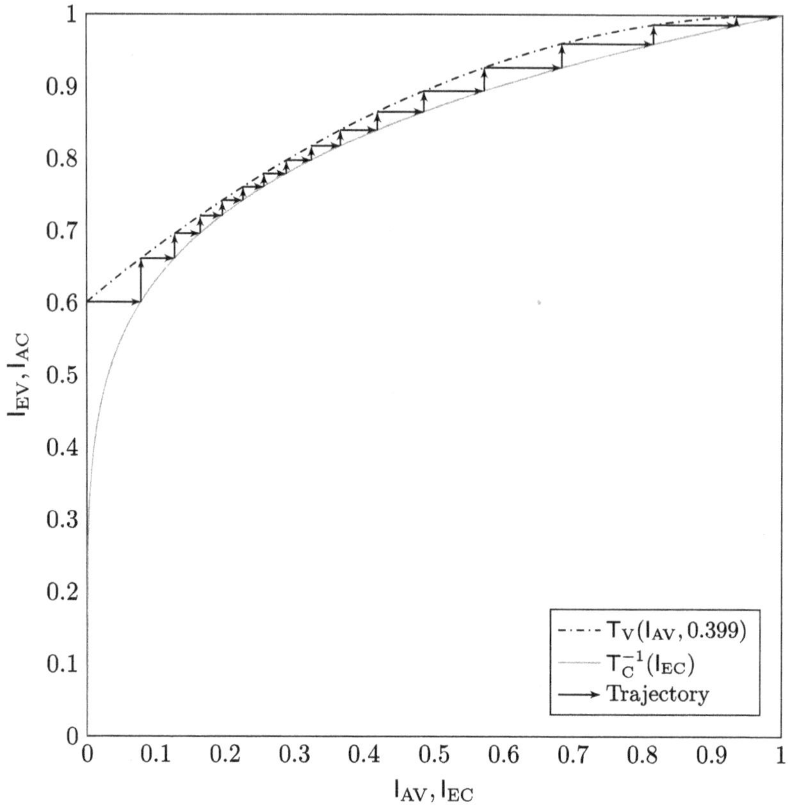

Figure 7.4 EXIT charts for (3,6) regular LDPC code ensemble for $\epsilon = 0.399$ and decoding trajectory.

word $\boldsymbol{u}$, we can also write

$$\boldsymbol{u}\boldsymbol{G}_{\bar{\mathcal{E}}} = \boldsymbol{x}_{\bar{\mathcal{E}}}. \tag{7.44}$$

The dimension of the solution space is $k - \text{rank}(\boldsymbol{G}_{\bar{\mathcal{E}}})$, which must also match $e - \text{rank}(\boldsymbol{H}_{\mathcal{E}})$. Under the assumption of uniformly distributed codewords, the solution spaces have full entropy, with

$$\begin{aligned} \mathsf{H}(\boldsymbol{X}|\boldsymbol{Y} = \begin{bmatrix}\boldsymbol{x}_{\mathcal{E}} & \boldsymbol{x}_{\bar{\mathcal{E}}}\end{bmatrix}) &= k - \text{rank}(\boldsymbol{G}_{\bar{\mathcal{E}}}) & (7.45)\\ &= e - \text{rank}(\boldsymbol{H}_{\mathcal{E}}). & (7.46)\end{aligned}$$

Averaging over all possible erasure patterns, we obtain

$$\begin{aligned} \mathsf{H}(\boldsymbol{X}|\boldsymbol{Y}(\epsilon_1,\epsilon_2,\ldots,\epsilon_n)) &= k - \sum_{\mathcal{E}}\left(\prod_{i\in\mathcal{E}}\epsilon_i\right)\left(\prod_{i\in\bar{\mathcal{E}}}(1-\epsilon_i)\right)\text{rank}(\boldsymbol{G}_{\bar{\mathcal{E}}}) & (7.47)\\ &= \sum_{i=1}^{n}\epsilon_i - \sum_{\mathcal{E}}\left(\prod_{i\in\mathcal{E}}\epsilon_i\right)\left(\prod_{i\in\bar{\mathcal{E}}}(1-\epsilon_i)\right)\text{rank}(\boldsymbol{H}_{\mathcal{E}}). & (7.48)\end{aligned}$$

Let $\mathsf{H}^{\perp}(\boldsymbol{X}|\boldsymbol{Y})$ be the conditional entropy when $\boldsymbol{X}$ is chosen from the dual code assuming uniformly distributed codewords. Using (7.47) and (7.48), it can be shown that

$$\mathsf{H}^{\perp}(\boldsymbol{X}|\boldsymbol{Y}(\epsilon_1,\epsilon_2,\ldots,\epsilon_n) = \mathsf{H}(\boldsymbol{X}|\boldsymbol{Y}(1-\epsilon_1,1-\epsilon_2,\ldots,1-\epsilon_n) - k + \sum_{i=1}^{n} \epsilon_i. \quad (7.49)$$

The proof of (7.49) is left to Exercise 7.4. Taking the derivative with respect to ϵ_i and making use of (7.41) and (7.42), we obtain

$$\mathsf{I}^{\perp}(X_i,\boldsymbol{Y}_{\sim i}(\epsilon)) = 1 - \mathsf{I}(X_i,\boldsymbol{Y}_{\sim i}(1-\epsilon)), \quad (7.50)$$

where $\mathsf{I}^{\perp}(X_i,\boldsymbol{Y}_{\sim i}(\epsilon))$ is the MI between code symbol i of the dual code and $\boldsymbol{Y}_{\sim i}(\epsilon)$. The identity in (7.50) is referred to as the duality property of EXIT functions.

Example 7.7 (Duality of SPC and repetition codes) Consider a $(4,3)$ SPC code and its dual $(4,1)$ repetition code. The EXIT function for the SPC code assuming a BEC with erasure probability p (or mutual information equal to $1-p$) is

$$\mathsf{I}(X_i,\boldsymbol{Y}_{\sim i}(p)) = (1-p)^3\,. \quad (7.51)$$

Then, the EXIT function for the dual repetition code assuming a BEC with erasure probability p (or mutual information equal to $1-p$) is

$$\mathsf{I}^{\perp}(X_i,\boldsymbol{Y}_{\sim i}(p)) = 1-(1-(1-p))^3 = 1-p^3\,. \quad (7.52)$$

7.3 Density Evolution Analysis for Structured Ensembles

Chapter 4 discussed various ways to impose structures on LDPC code ensembles. In Section 7.3.1, we extend the density evolution analysis to code ensembles with structure, focusing on IRA code ensembles and protograph LDPC code ensembles. All results can be restated equivalently in terms of evolution of the MI. Likewise, we present results on the stability condition for structured ensembles.

7.3.1 Analysis of IRA Ensembles

For IRA codes, the submatrix of the parity-check matrix corresponding to the redundant symbols is composed of columns of weight 2 which form a double diagonal structure. This is discussed in detail in Section 4.2.2 and depicted in Figure 4.5. Density evolution analysis of IRA code ensembles treats messages from VNs associated with information symbols and messages from VN associated with redundant symbols separately.

Next, we adopt the notation of Section 4.2.2, which we briefly recall hereafter. The k VNs corresponding to the information symbols are said to be of type $\mathtt{V}_1$ and are connected to CNs by edges of type $\mathtt{E}_1$. Similarly, the $n-k$ VNs corresponding to the redundant symbols are said to be of type $\mathtt{V}_2$ and are connected to CNs by edges of type $\mathtt{E}_2$. It is appropriate to define the degree distributions considering the VNs of type $\mathtt{V}_1$

only, since the parity part is fixed (double diagonal). Recall from Section 4.2.2 that ψ_j denotes the fraction of edges of type $\mathtt{E}_1$ connected to VNs of type $\mathtt{V}_1$ of degree j, yielding the edge-oriented degree distribution polynomial $\psi(x) = \sum_{j=1}^{\mathsf{d}_v} \psi_j x^{j-1}$. As pointed out in Remark 4.6, we can allow CNs of different degrees and denote by ϕ_i the fraction of edges of type $\mathtt{E}_1$ connected to CNs of degree $i+2$. We obtain the edge-oriented degree distribution polynomial $\phi(x) = \sum_{i=1}^{\mathsf{d}_c-2} \phi_i x^{i-1}$ and the corresponding node-oriented degree distribution polynomial $\Phi(x) = \sum_{i=1}^{\mathsf{d}_c-2} \Phi_i x^i$, with $\Phi_i = (\phi_i/i)/(\sum_d \phi_d/d)$. The rate of the IRA code ensemble $\mathscr{I}^n_{\psi,\phi}$ is given by (4.21).

Next, we assume that all VNs of type $\mathtt{V}_2$ have degree 2. Note that, actually, there exists a single VN with degree 1 (see also Figure 4.5); its contribution, however, vanishes in the setting where n goes to infinity. The expressions of the average extrinsic erasure probabilities at the VN output on edges $\mathtt{E}_1$ and $\mathtt{E}_2$ are

$$p^{(\mathtt{E}_1)} = \epsilon\psi\left(q^{(\mathtt{E}_1)}\right) \tag{7.53}$$

$$p^{(\mathtt{E}_2)} = \epsilon q^{(\mathtt{E}_2)} . \tag{7.54}$$

Likewise, the average extrinsic erasure probabilities at the CN output on edges $\mathtt{E}_1$ and $\mathtt{E}_2$ are

$$q^{(\mathtt{E}_1)} = 1 - \phi\left(1 - p^{(\mathtt{E}_1)}\right)\left(1 - p^{(\mathtt{E}_2)}\right)^2 \tag{7.55}$$

$$q^{(\mathtt{E}_2)} = 1 - \Phi\left(1 - p^{(\mathtt{E}_1)}\right)\left(1 - p^{(\mathtt{E}_2)}\right). \tag{7.56}$$

Incorporating (7.55) into (7.53) and (7.56) into (7.54), we obtain the two simultaneous density evolution recursions

$$p_\ell^{(\mathtt{E}_1)} = \epsilon\psi\left(1 - \phi\left(1 - p_{\ell-1}^{(\mathtt{E}_1)}\right)\left(1 - p_{\ell-1}^{(\mathtt{E}_2)}\right)^2\right) \tag{7.57}$$

$$p_\ell^{(\mathtt{E}_2)} = \epsilon\left(1 - \Phi\left(1 - p_{\ell-1}^{(\mathtt{E}_1)}\right)\left(1 - p_{\ell-1}^{(\mathtt{E}_2)}\right)\right). \tag{7.58}$$

Example 7.8 (Density evolution recursions for IRA code ensemble) Consider an IRA ensemble with $\psi(x) = x^3$ and $\phi(x) = x^3$. It is easy to verify that $\Phi(x) = x^4$ and that $R = 1/2$. From (7.57) and (7.58), we obtain

$$p_\ell^{(\mathtt{E}_1)} = \epsilon\left(1 - \left(1 - p_{\ell-1}^{(\mathtt{E}_1)}\right)^3\left(1 - p_{\ell-1}^{(\mathtt{E}_2)}\right)^2\right)^3 \tag{7.59}$$

$$p_\ell^{(\mathtt{E}_2)} = \epsilon\left(1 - \left(1 - p_{\ell-1}^{(\mathtt{E}_1)}\right)^4\left(1 - p_{\ell-1}^{(\mathtt{E}_2)}\right)\right). \tag{7.60}$$

One may obtain an upper bound ϵ^{stab} on the IRA iterative decoding threshold by computing the eigenvalues of the two-dimensional discrete dynamical system described by (7.57) and (7.58). A more detailed discussion follows in Section 7.3.2.

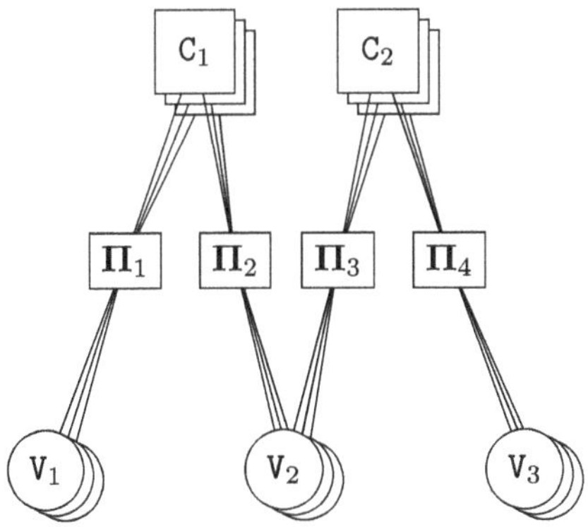

Figure 7.5 Tanner graph of a protograph LDPC ensemble. The three different layers in the figure represent three copies of the original protograph.

7.3.2 Protograph Ensembles

As described in Section 4.2.3, a protograph is a small bipartite graph that is used to represent an ensemble of LDPC codes with a partial structure. To draw a specific code from the protograph ensemble, the protograph is copied several times and the copies are connected by permuting their edges among each other as described in Section 4.2.3. These "copy-and-permute" operations are executed preserving the neighborhoods of the VNs and CNs. In other words, if in the protograph a node of type i was connected through an edge to a node of type j, any of the l copies of the node of type i must be connected to one of the l copies of the node of type j after the permutation. This constraint is captured by the protograph density evolution analysis presented next.

Example 7.9 (Lifting of protograph) Consider the protograph with base matrix

$$\boldsymbol{B} = \begin{bmatrix} 1 & 1 & 0 \\ 0 & 1 & 1 \end{bmatrix}.$$

For illustration purposes, we expand the base matrix by a lifting factor $Q = 3$. A Tanner graph representation of the resulting $(9, 3)$ code is depicted in Figure 7.5, where the interleavers $\boldsymbol{\Pi}_1$, $\boldsymbol{\Pi}_2$, and $\boldsymbol{\Pi}_3$ indicate some fixed permutations of the edges. The corresponding protograph code ensemble is given by all possible interleaver choices, each with a uniform probability measure.

For the analysis, we enumerate all VNs and CNs of the protograph corresponding to the $n_c \times n_v$ base matrix $\boldsymbol{B} = \{B_{i,j}\}$. Each VN is connected to a neighboring CN through a specific edge of type $\mathrm{E}_{i,j}$. For each edge type, we need to compute extrinsic erasure probabilities. Letting ℓ be the iteration index, we formulate the analysis in terms of erasure probabilities and denote by

- $p_\ell^{(i,j)}$ the extrinsic erasure probability of a message sent by a VN of type $\mathtt{V}_j$ to a CN of type C_i on the edge of type $\mathtt{E}_{i,j}$;
- $q_\ell^{(i,j)}$ the extrinsic erasure probability of a message sent by a type-i CN C_i to a type-j $\mathtt{V}_j$ on the edge of type $\mathtt{E}_{i,j}$;
- $p_\ell^{(j)}$ the a posteriori erasure probability for VN of type j.

Assuming that no protograph VN is punctured, we can express the erasure probability of a message sent by a type-j VN, $j \in \{1,\ldots,n_v\}$, to a type-i CN, $i \in \{1,\ldots,n_c\}$, as

$$p_\ell^{(i,j)} = \epsilon \prod_{s, s\neq i} \left(q_{\ell-1}^{(s,j)}\right)^{b_{s,j}} \left(q_{\ell-1}^{(i,j)}\right)^{b_{i,j}-1}. \tag{7.61}$$

Similarly, we can express the erasure probability of a message sent by a type-i CN to a type-j VN as

$$q_\ell^{(i,j)} = 1 - \prod_{s, s\neq j} \left(1 - p_\ell^{(i,s)}\right)^{b_{i,s}} \left(1 - p_\ell^{(i,j)}\right)^{b_{i,j}-1}. \tag{7.62}$$

The initial condition is $q_0^{(i,j)} = 1$ for all pairs (i,j). At the end of each iteration, which consists of executing (7.61) for all pairs (i,j) and then of executing (7.62) for all pairs (i,j), we can compute the a posteriori erasure probability for a type-j VN, $j \in \{1,2,\ldots,n_v\}$ as

$$p_\ell^{(j)} = \epsilon \prod_{s} \left(q_\ell^{(s,j)}\right)^{b_{s,j}}. \tag{7.63}$$

Definition 7.4 (Iterative decoding threshold of protograph ensemble) The iterative decoding threshold of a protograph LDPC code ensemble is defined as

$$\epsilon_{\mathsf{BP}}^{\star} = \sup\left\{\epsilon \in [0,1] : p_\ell^{(j)}(\epsilon) \xrightarrow{\ell\to\infty} 0 \; \forall j \in \{1,2,\ldots,n_v\}\right\}. \tag{7.64}$$

Remark 7.6. In Section 4.2.3, it was pointed out how protographs may include punctured VNs. Puncturing may be easily taken into account in density evolution. Specifically, if type-j VNs are punctured ones, (7.61) and (7.63) should be updated by simply replacing ϵ with 1.

Example 7.10 (Threshold computation for protograph ensemble) Consider the protograph with base matrix

$$\boldsymbol{B} = \begin{bmatrix} 1 & 1 & 1 \\ 2 & 1 & 1 \end{bmatrix}.$$

We have two CN types and three VN types. Assume that type-1 VNs, corresponding to first column of $\boldsymbol{B}$, are punctured, yielding a design rate $\mathrm{R} = 1/2$. Protograph density evolution yields an iterative decoding threshold $\epsilon_{\mathsf{BP}}^{\star} = 0.45$. Note that in principle, for this ensemble density evolution entails computation of 6 different messages, one per edge type, per iteration. However, both type-2 and type-3 VNs are connected once to CNs of types 1 and 2. This symmetry yields the same message distributions on the edge types $\mathrm{E}_{1,2}$ and $\mathrm{E}_{1,3}$, and $\mathrm{E}_{2,2}$ and $\mathrm{E}_{2,3}$, respectively. This reduces the number of different messages to be computed per iteration from 6 to 4. Still, density evolution of protograph ensembles remains more computationally expensive than for the unstructured case, especially when the number of edge types becomes large.

7.3.2.1 Stability

In density evolution for protograph ensembles, we track the erasure probabilities $p_\ell^{(i,j)}$ for all $i \in \{1,\dots,n_c\}$ and $j \in \{1,\dots,n_v\}$ over the iteration index ℓ. Similar to Section 7.2.2.2, we seek for a linear approximation of the system around the fixed point $\boldsymbol{p} = \begin{bmatrix} p^{(1,1)} & \dots & p^{(n_c,n_v)} \end{bmatrix} = \mathbf{0}$ of the density evolution recursion. This is obtained by developing a recursion in the form $p_\ell^{(i,j)} = f_{i,j}(\boldsymbol{p}_{\ell-1})$ for all pairs (i,j), and by computing the Jacobian matrix of the vector function $\begin{bmatrix} f_{1,1}(\boldsymbol{p}) & f_{n_c,n_v}(\boldsymbol{p}) \end{bmatrix}$, denoted as $\boldsymbol{J}(\boldsymbol{p})$. Then, local stability of the fixed point $\boldsymbol{p} = \mathbf{0}$ depends on the eigenvalues t_i of the matrix $\boldsymbol{J}(\mathbf{0})$. The fixed point is unstable if the spectral radius of $\boldsymbol{J}(\mathbf{0})$ is greater than 1, that is, if $\max_i |t_i| > 1$ and is stable if $\max_i |t_i| < 1$. If $\max_i |t_i| = 1$, the fixed point might be stable. We illustrate the computation of stability by means of an example.

Example 7.11 (Stability condition for protograph ensemble) Consider a protograph with base matrix

$$\boldsymbol{B} = \begin{bmatrix} 4 & 2 \end{bmatrix},$$

and let both protograph VNs be unpunctured. We have two different VN types, $\mathtt{V}_1$ and $\mathtt{V}_2$, and one CN type $\mathtt{C}_1$. Further, we have two different edge types, $\mathtt{E}_{1,1}$ and $\mathtt{E}_{1,2}$. We aim at computing ϵ^{stab}. The density evolution equations are

$$p_\ell^{(1,1)} = \epsilon\left(q_{\ell-1}^{(1,1)}\right)^3 \tag{7.65}$$

$$p_\ell^{(1,2)} = \epsilon\, q_{\ell-1}^{(1,2)} \tag{7.66}$$

$$q_\ell^{(1,1)} = 1 - \left(1 - p_\ell^{(1,1)}\right)^3\left(1 - p_\ell^{(1,2)}\right)^2 \tag{7.67}$$

$$q_\ell^{(1,2)} = 1 - \left(1 - p_\ell^{(1,1)}\right)^4\left(1 - p_\ell^{(1,2)}\right). \tag{7.68}$$

By substitution, we obtain the discrete dynamical system

$$p_\ell^{(1,1)} = \epsilon\left(1 - \left(1 - p_{\ell-1}^{(1,1)}\right)^3\left(1 - p_{\ell-1}^{(1,2)}\right)^2\right)^3 = f_{1,1}(p_{\ell-1}^{(1,1)}, p_{\ell-1}^{(1,2)}) \tag{7.69}$$

$$p_\ell^{(1,2)} = \epsilon\left(1 - \left(1 - p_{\ell-1}^{(1,1)}\right)^4\left(1 - p_{\ell-1}^{(1,2)}\right)\right) = f_{1,2}(p_{\ell-1}^{(1,1)}, p_{\ell-1}^{(1,2)}). \tag{7.70}$$

Next, we consider the vector function $\begin{bmatrix} f_{1,1}(p^{(1,1)}, p^{(1,2)}) & f_{1,2}(p^{(1,1)}, p^{(1,2)}) \end{bmatrix}$, we compute its Jacobian matrix $\boldsymbol{J}(p^{(1,1)}, p^{(1,2)})$, we evaluate it at the fixed point $(0,0)$, and we find the eigenvalues of $\boldsymbol{J}(0,0)$ as the solutions of $\det(\boldsymbol{J}(0,0) - t\boldsymbol{I}) = 0$, where $\boldsymbol{I}$ is the 2×2 identity matrix. We obtain

$$\det\begin{pmatrix} -t & 0 \\ 4\epsilon & \epsilon - t \end{pmatrix} = t(t - \epsilon) = 0, \tag{7.71}$$

so $t_0 = 0$ and $t_1 = \epsilon$, which is the eigenvalue with the largest magnitude. Recall that ϵ^{stab} is the largest value of ϵ such that the eigenvalue with the maximum magnitude is not greater than 1, that is, $|t_1| \leq 1$. Thus, $\epsilon^{\mathsf{stab}} = 1$.

7.4 Finite-Length Analysis

The iterative decoding threshold predicts the code performance in the asymptotic regime where we let the blocklength tend to infinity and then the number of decoding iterations tends to infinity. For finite blocklengths, cycles need to be accounted for, in particular when an iterative decoder is used. The effect is visible as a flattening of the error rate curve that takes place at low erasure probabilities. This phenomenon is also referred to as error floor. Error floors under iterative decoding are attributed to cycles in the graph which form stopping sets (see Section 6.1.4). One approach to predict the block error probability at moderate to low erasure probabilities is based on the concept of *stopping set size enumerators.*

7.4.1 Stopping Set Size Enumerating Function

The stopping set size enumerating function of an LDPC code represents the extension to stopping sets of the concept of weight enumerating function introduced in Definition 2.17. This is defined as $S(X) = \sum_{t=1}^{n} S_t X^t$, where S_t is the number of stopping sets of size t.

While the stopping set size enumerating function refers to a single LDPC code, for LDPC code ensembles the average stopping set size enumerating function over the ensemble is of interest. This is defined as $\bar{S}(X) = \sum_{t=1}^{n} \bar{S}_t X^t$, where $\bar{S}_t$ is the expectation of the random variable representing the number of stopping sets of size t of a random code whose realizations belong to the ensemble. The expected stopping set size enumerator $\bar{S}_t$ can be computed using approaches analogous to the ones employed to calculate the average weight enumerators. The differences are represented by updated definitions of (valid) CN socket assignment and VN assignment, previously introduced in Definitions 5.1 and 5.2. When dealing with stopping sets, a valid CN socket assignment is defined as a subset of the CN sockets such that each CN has a null or at least two sockets in the subset; a VN assignment of size t is defined as a subset of t VNs. Note that these definitions are independent of the field size q. As a baseline, we assume $q = 2$ in the sequel.

For a (binary) irregular LDPC code ensemble $\mathscr{C}^n_{\lambda,\rho}$, we obtain an expression of $\bar{S}_t$ analogous to (5.30) with $q = 2$, namely,

$$\bar{S}_t = \sum_{t \in \mathcal{T}_t} \frac{\mathrm{coeff}\left(\prod_{i=2}^{\mathsf{d}_c} (\Theta_i(Z))^{P_i m}, Z^{\sum_{j=2}^{\mathsf{d}_v} t_j j}\right)}{\binom{\sum_{j=2}^{\mathsf{d}_v} \Lambda_j j n}{\sum_{j=2}^{\mathsf{d}_v} t_j j}} \prod_{j=2}^{\mathsf{d}_v} \binom{\Lambda_j n}{t_j}, \tag{7.72}$$

where $\mathcal{T}_t$ is the set of all $\boldsymbol{t} = (t_2, t_3 \ldots, t_{\mathsf{d}_v})$ such that each t_j is a nonnegative integer, $0 \le t_j \le \Lambda_j n$, for all $j \in \{2,3,\ldots,\mathsf{d}_v\}$, and $\sum_{j=2}^{\mathsf{d}_v} t_j = t$. The polynomial $\Theta_i(Z)$ is defined as $\Theta_i(Z) = 1 + \sum_{v=2}^{i} \binom{i}{v} Z^v$ and represents the stopping set enumerating function of a CN of degree i. Note that the two expressions (5.30) with $q = 2$ and (7.72) are formally identical and only differ in the fact that the polynomial $P_{2,i}(Z)$ is now replaced by $\Theta_i(Z)$.

In the context of stopping sets, the analogous of the weight spectral shape introduced in (5.36) is represented by the stopping set size spectral shape,

$$\Gamma(\tau) = \lim_{n \to \infty} \frac{1}{n} \log_2 \bar{S}_{\tau n} \, .$$

Applying to (7.72), the same approach described in Section 5.6, we obtain an expression of $\Gamma(\tau)$ similar to (5.84). In particular, the stopping set size spectral shape of an unstructured LDPC code ensemble is given by

$$\Gamma(\tau) = \frac{1}{\ln 2} \left[(1 - \mathrm{R}) \sum_{i=2}^{\mathsf{d}_c} P_i \ln \Theta_i(z_0) - \omega \ln x_0 \right.$$
$$\left. + \sum_{j=2}^{\mathsf{d}_v} \Lambda_j \ln \Phi_j(x_0, y_0) + \frac{1}{\int \lambda} \ln(1 - \beta \textstyle\int \lambda) \right],$$

where $\Phi_j(X,Y) = 1 + XY^j$ and x_0, y_0, z_0, and β are the unique positive real roots of the four simultaneous equations

$$(1 - \mathrm{R})\, z_0 \sum_{i=2}^{\mathsf{d}_c} P_i \frac{\frac{\partial \Theta_i(Z)}{\partial Z}(z_0)}{\Theta_i(z_0)} = \beta \, . \tag{7.73}$$

$$y_0 = \frac{1}{z_0} \frac{\beta \int \lambda}{(1 - \beta \int \lambda)} \, . \tag{7.74}$$

$$x_0 \sum_{j=2}^{\mathsf{d}_v} \Lambda_j \frac{\frac{\partial \Phi_j}{\partial X}(x_0, y_0)}{\Phi_j(x_0, y_0)} = \tau \, . \tag{7.75}$$

$$y_0 \sum_{j=2}^{\mathsf{d}_v} \Lambda_j \frac{\frac{\partial \Phi_j}{\partial Y}(x_0, y_0)}{\Phi_j(x_0, y_0)} = \beta \, . \tag{7.76}$$

In the unstructured regular case, the expression of $\Gamma(\tau)$ simplifies to

$$\Gamma(\tau) = \frac{1}{\ln 2} \left[(1 - \mathrm{R}) \ln \Theta_{\mathsf{d}_c}(z_0) - \tau\, \mathsf{d}_v \ln z_0 - (\mathsf{d}_v - 1) \mathsf{H}_{\mathsf{e}}(\omega) \right],$$

where $z_0 = z_0(\tau)$ is the unique positive real root of

$$\frac{z_0 \frac{\mathrm{d} \ln \Theta_{\mathsf{d}_c}}{\mathrm{d}Z}(z_0)}{\Theta_{\mathsf{d}_c}(z_0)} = \mathsf{d}_c\, \tau \, . \tag{7.77}$$

For an ensemble characterized by stopping set size spectral shape $\Gamma(\tau)$, we define the critical fractional stopping set size as $\tau^* = \inf\{\tau > 0 : \Gamma(\tau) > 0\}$. In case of an irregular LDPC code ensemble, this parameter may be efficiently computed by solving the 5×5 system of equations given by (7.73), (7.74), (7.75) (with $\tau = \tau^*$), and (7.76), together with $\Gamma(\tau^*) = 0$. For a regular unstructured ensemble, it may be efficiently computed by solving the 2×2 system of equations given by (7.77) (with $\tau = \tau^*$) and $\Gamma(\tau^*) = 0$.

Remark 7.7. Since for any LDPC code, the minimum stopping set size is equal or less than its minimum distance, for any ensemble (unstructured or structured), we have $\tau^* \leq \omega^*$ where ω^* is the critical fractional codeword weight (see Remark 5.5).

Remark 7.8. It is possible to extend the concepts of stopping set enumerating function and stopping set size spectral shape also to structured ensembles.

7.4.2 Union Bound on the Block Error Probability

We provide a union bound on the block error probability of an LDPC code under iterative decoding. The bound requires knowledge of the stopping set size enumerating function, which depends on the parity-check matrix $\boldsymbol{H}$ used to represent the code constraints. In general, there are many possible choices of $\boldsymbol{H}$ (see also Exercise 7.7) so that, strictly speaking, we provide an upper bound of the block error probability of a specific representation of an LDPC code. The bound can be also cast for an ensemble of codes.

Theorem 7.2 (Upper bound on the block error probability) *Consider an (n, k) LDPC code $\mathcal{C}$ defined over* GF(2), *and let S_t, $t \in \{1, \dots, n\}$ be the stopping set size enumerators corresponding to its parity-check matrix representation. Assume transmission over a BEC with erasure probability ϵ. Then, an upper bound on the block error probability of $\mathcal{C}$ under iterative decoding is*

$$P_B(\mathcal{C}) \leq \sum_{e=n-k+1}^{n} \binom{n}{e} \epsilon^e (1-\epsilon)^{n-e} + \sum_{e=1}^{n-k} \binom{n}{e} \epsilon^e (1-\epsilon)^{n-e} \min\left\{1, \sum_{t=t_{\min}}^{e} \binom{e}{t} \frac{S_t}{\binom{n}{t}}\right\}, \qquad (7.78)$$

where $t_{\min}$ is the minimum stopping set size.

Proof The upper bound consists of two addends. The first one is the Singleton bound $P_B^{(\mathrm{S})}$ (see Chapter 3), which states that decoding fails whenever there are more than $n-k$ erasures, irrespective of the decoding algorithm. The second addend represents the probability of decoding failure in case there are $n-k$ or less erasures.

Let $P(\text{error})$ be the probability that an erasure pattern $\mathcal{E}$ includes a stopping set $\mathcal{S}$, that is, $P(\text{error}) = P(\exists\, \mathcal{S} | \mathcal{S} \subseteq \mathcal{E})$. We may write

$$P_B(\mathcal{C}) = P_B^{(\mathrm{S})} + \sum_{e=1}^{n-k} P(|\mathcal{E}| = e) P(\text{error} \| \mathcal{E}| = e) \qquad (7.79)$$

$$= P_B^{(\mathrm{S})} + \sum_{e=1}^{n-k} \binom{n}{e} \epsilon^e (1-\epsilon)^{n-e} P\left(\bigcup_{\mathcal{S}} \{\mathcal{S} \subseteq \mathcal{E}\} \| \mathcal{E}| = e\right) \qquad (7.80)$$

$$\leq P_B^{(\mathsf{S})} + \sum_{e=1}^{n-k} \binom{n}{e} \epsilon^e (1-\epsilon)^{n-e} \min\left\{1, \sum_{\mathcal{S}} P(\mathcal{S} \subseteq \mathcal{E} \| \mathcal{E}| = e)\right\} \tag{7.81}$$

$$= P_B^{(\mathsf{S})} + \sum_{e=1}^{n-k} \binom{n}{e} \epsilon^e (1-\epsilon)^{n-e} \min\left\{1, \sum_{t=t_{\min}}^{e} \binom{e}{t} \frac{S_t}{\binom{n}{t}}\right\}. \tag{7.82}$$

In the above development, (7.81) is due to the union bound. Moreover, (7.82) follows since, for any erasure pattern of size e, there are $\binom{e}{t}$ ways to erase a set of t VNs and since the probability that those t VNs form a stopping set is $\frac{S_t}{\binom{n}{t}}$. □

Remark 7.9. The parameter $t_{\min}$ is sometimes referred to as the stopping set distance, or simply as the stopping distance.

Theorem 7.2 can be easily extended to linear LDPC code ensembles, as follows.

Theorem 7.3 *Consider an ensemble $\mathscr{C}$ of (n,k) LDPC codes over $GF(2)$. Let $\bar{S}_t$ be the expected multiplicity of stopping sets of size t in a random code $C \in \mathscr{C}$. Then, the expected block error probability when C is used to transmit over a BEC with erasure probability ϵ can be upper bounded as*

$$P_B(\mathscr{C}) = \mathsf{E}[P_B(C)] \tag{7.83}$$

$$\leq \sum_{e=n-k+1}^{n} \binom{n}{e} \epsilon^e (1-\epsilon)^{n-e}$$

$$+ \sum_{e=1}^{n-k} \binom{n}{e} \epsilon^e (1-\epsilon)^{n-e} \min\left\{1, \sum_{t=t_{min}}^{e} \binom{e}{t} \frac{\bar{S}_t}{\binom{n}{t}}\right\}. \tag{7.84}$$

Remark 7.10. The proof of Theorem 7.3, here omitted, follows easily from linearity of the expectation. Note also that both Theorems 7.2 and 7.3 hold for LDPC codes and LDPC code ensembles defined over GF(q) and transmitted over a QEC with erasure probability ϵ.

The block error probability in (7.84) depends on $t_{\min}$. Following the ideas in Section 5.7, we may consider expurgated code ensembles which only contain LDPC codes whose minimum stopping set size $t_{\min}$ is greater than or equal to a target value. Then, Theorem 5.10 can be modified to obtain an upper bound on the expected stopping set size of the expurgated ensemble.

Theorem 7.4 *Let $\mathscr{C}$ be an ensemble of linear block codes, $\bar{S}_t$ be the expected multiplicity of stopping sets of size t for a random code drawn from $\mathscr{C}$, and $F(d) = \min\{1, \sum_{t=2}^{d} \bar{S}_t\}$. For any $0 < p^* < 1$, let $d^* = \max\{d > 0 : F(d) \leq p^*\}$ and $\mathscr{C}^{\mathrm{ex}}$ be the expurgated ensemble with respect to d^*. Then, the expected multiplicity of stopping sets of weight t of a random code drawn from $\mathscr{C}^{\mathrm{ex}}$, $\bar{S}_t^{\mathrm{ex}}$, fulfills*

$$\bar{S}_t^{\mathrm{ex}} = 0 \quad \textit{if } 1 \leq t \leq d^*$$

and

$$\bar{S}_t^{\mathrm{ex}} \leq \frac{1}{1-p^*} \bar{S}_t \quad \textit{if } d^* < t \leq n\,.$$

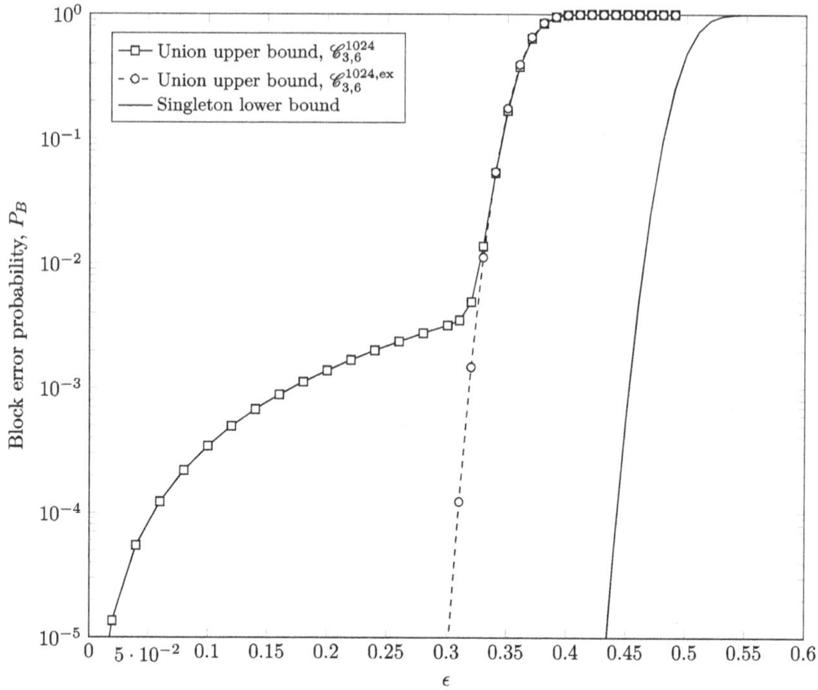

Figure 7.6 Union upper bound on the block error probability versus erasure probability ϵ of a binary regular (3,6) LDPC code ensemble $\mathscr{C}_{3,6}^{1024}$ and of an expurgated ensemble $\mathscr{C}_{3,6}^{1024,\text{ex}}$ with minimum stopping set size of 21.

The proof of Lemma 7.4 goes in line with that of Theorem 5.10. Next we exemplify the upper bound on the error probability for the expurgated code ensemble.

Example 7.12 (Block error probability of expurgate LDPC ensemble) Consider the ensemble $\mathscr{C}_{3,6}^{1024}$ of binary linear LDPC codes with constant VN degree $\mathsf{d}_v = 3$ and constant CN degree $\mathsf{d}_c = 6$. In $\mathscr{C}_{3,6}^{1024}$, the expected number of stopping sets of size smaller than 21 can be determined to be 0.45. Thus, there must be codes in the ensemble $\mathscr{C}_{3,6}^{1024}$ which have stopping set size of 21 or larger. Considering only those codes, we obtain an expurgated code ensemble $\mathscr{C}_{3,6}^{1024,\text{ex}}$. Theorem 7.4 yields an upper bound on the average stopping set size enumerators for the expurgated ensemble. We plot the union upper bound on the block error probability for both ensembles in Figure 7.6. We can observe the beneficial effect of expurgation for small erasure probabilities.

7.5 Code Design for the Belief Propagation Decoder

In the sequel, we discuss how to obtain sequences of degree distribution pairs (λ, ρ) which are capacity achieving on erasure channels. Despite their theoretical importance, these sequences are less relevant for practical applications since for many blocklengths of interest, they show disadvantages in decoding complexity and/or performance with respect to designs obtained by numerical optimization. Hence, we will then address how to obtain "good" LDPC codes by means of numerical optimization.

7.5.1 Capacity Achieving Sequences

When dealing with capacity achieving sequences, the design goal is to find LDPC code ensembles whose asymptotic iterative decoding threshold is arbitrarily close to the BEC (or QEC) capacity. The concept of capacity achieving sequence of some fixed rate R is rigorously reviewed hereafter.

Definition 7.5 (Capacity achieving sequence of rate R) Consider a degree distribution pair $(\lambda^{(d)}(x), \rho^{(d)}(x))$ that depends on a natural parameter d. Regarding d as an index, consider the corresponding sequence of degree distribution pairs $\{(\lambda^{(d)}(x), \rho^{(d)}(x))\}$. Let the design rate

$$\mathrm{R} = 1 - \frac{\int_0^1 \rho^{(d)}(x)\mathrm{d}x}{\int_0^1 \lambda^{(d)}(x)\mathrm{d}x} \tag{7.85}$$

be independent of d. The sequence $\{(\lambda^{(d)}(x), \rho^{(d)}(x))\}$ is said to be a capacity achieving sequence of rate R if for all $\epsilon < 1 - \mathrm{R}$ there exists $d_0 \in \mathbb{N}$ such that for all $d \geq d_0$ we have

$$\epsilon\lambda^{(d)}\left(1 - \rho^{(d)}(1-x)\right) < x \ \forall x \in (0,1]\,. \tag{7.86}$$

Remark 7.11. Recalling (7.9) and (7.10), we see that the sequence of iterative decoding thresholds corresponding to a sequence of degree distribution pairs, fulfilling Definition 7.5, converges to $1 - \mathrm{R}$. In this sense, the sequence achieves capacity under iterative decoding as d tends to infinity.

Several families of capacity-achieving sequences have been constructed to date. A prominent example, called the heavy-tail/Poisson sequence, is provided in Example 7.13. The construction exploits the following fact. Let $\lambda(x) = \sum_{i=2}^{\infty} \lambda_i x^{i-1}$ be the power series expansion of some analytic function. Let $\lambda_i \geq 0$ for all $i \geq 2$ and $\lambda(1) = 1$. Moreover, the truncated power series is defined as

$$\hat{\lambda}^{(d)}(x) = \sum_{i=2}^{d+1} \lambda_i x^{i-1}, \tag{7.87}$$

and its normalized version as

$$\lambda^{(d)}(x) = \frac{\hat{\lambda}^{(d)}(x)}{\hat{\lambda}^{(d)}(1)}, \tag{7.88}$$

where we assume sufficiently large d such that $\hat{\lambda}^{(d)}(1)$ is nonzero. Then, we have

$$\lambda(x) \geq \lambda^{(d)}(x)\hat{\lambda}^{(d)}(1) \ \forall x \in [0,1]\,. \tag{7.89}$$

The result in (7.89) stems from the fact that dropping some nonnegative coefficients of the Taylor series representation of $\lambda(x)$ yields a lower bound on $\lambda(x)$, and from the equality in (7.88).

Example 7.13 (Heavy-tail/Poisson sequence) Consider the sequence of degree distribution pairs $\{(\lambda^{(d)}(x), \rho^{(d)}(x))\}$ defined by

$$\lambda^{(d)}(x) = a_d \sum_{i=2}^{d+1} \frac{1}{i-1} x^{i-1} \tag{7.90}$$

$$\rho^{(d)}(x) = \sum_{i=1}^{\infty} \frac{\exp(-b_d) b_d^{i-1}}{(i-1)!} x^{i-1} \tag{7.91}$$

and where

$$a_d = \frac{1}{\sum_{i=2}^{d+1} 1/(i-1)} = \frac{1}{\hat{\lambda}^{(d)}(1)} . \tag{7.92}$$

Note that $\rho^{(d)}(x) = \exp(b_d(x-1))$ and that $\lambda^{(d)}(x)$ is the truncated Taylor expansion of $\lambda(x) = -\ln(1-x)$. Imposing that all degree distribution pairs in the sequence have the same design rate R, b_d is determined by the rate constraint in (7.85), as the unique positive solution of

$$\frac{(1-\exp(-b_d))/b_d}{a_d d/(d+1)} = 1 - \mathrm{R} . \tag{7.93}$$

We now sketch that $\{(\lambda^{(d)}(x), \rho^{(d)}(x))\}$ is a capacity achieving sequence of rate R. We have

$$\begin{aligned}
\epsilon\lambda^{(d)}\left(1-\rho^{(d)}(1-x)\right) &\le \epsilon\lambda\left(1-\rho^{(d)}(1-x)\right)a_d && (7.94)\\
&= \epsilon\lambda(1-\exp(-b_d x))a_d && (7.95)\\
&= -\epsilon\ln(\exp(-b_d x))a_d && (7.96)\\
&= \epsilon b_d x a_d && (7.97)\\
&= \epsilon x \frac{(d+1)(1-\exp(-b_d))}{(1-\mathrm{R})d} && (7.98)\\
&\le \epsilon x \frac{d+1}{(1-\mathrm{R})d} . && (7.99)
\end{aligned}$$

Inequality (7.94) comes from (7.89). Equality (7.98) follows by replacing $b_d a_d$ using (7.93). Inequality (7.99) is obtained by observing that $b_d \ge 1$. A further discussion follows in Exercise 7.10. Combining (7.86) with the upper bound (7.99), a sufficient condition for successful decoding is

$$\epsilon x \frac{d+1}{d(1-\mathrm{R})} < x \tag{7.100}$$

or, equivalently,

$$\epsilon < \frac{d}{d+1}(1-\mathrm{R}) . \tag{7.101}$$

This leads us to $(1-\mathsf{R})d/(d+1) \leq \epsilon_{\mathsf{BP}}^{\star} \leq 1-\mathsf{R}$, which implies $\epsilon_{\mathsf{BP}}^{\star} \to 1-\mathsf{R}$ as $d \to \infty$.

The sequence in the example above has the practical drawback of unbounded average VN and CN degrees, leading to high decoding complexity. In fact, we can develop the average VN degree as

$$\frac{1}{\int_0^1 \lambda^{(d)}(x)\mathrm{d}x} = \frac{d+1}{d}\sum_{i=1}^{d}\frac{1}{i} = \frac{d+1}{d}\Big(\gamma + \ln(d) + \mathcal{O}(d^{-1})\Big), \tag{7.102}$$

where $\gamma \approx 0.577$ is the Euler–Mascheroni constant. The average VN degree (and, through the rate constraint, also the average CN degree) increases with d. This implies that, to operate at an erasure probability arbitrarily close to $1-\mathsf{R}$, the decoding complexity grows unbounded. There also exist capacity achieving sequences of (punctured) IRA code ensembles with bounded decoding complexity. However, for many practical blocklengths of interest, those sequences turn out to be less competitive in performance than code ensembles found by numerical optimization techniques.

7.5.2 LDPC Code Design via Numerical Optimization

We discuss the design of unstructured LDPC codes as a two-step approach. First, we aim at designing an LDPC code ensemble that is characterized by an iterative decoding threshold as good as possible. In this first phase, we might also target code designs for (efficient) ML decoding, which will be discussed in Chapter 8. Additionally, we may impose additional constraints on the ensemble, for example, on the weight or stopping set size spectral shape. Second, once a suitable degree distribution is found, some girth optimization algorithm is used to construct a finite blocklength code from the ensemble.

7.5.2.1 Search for Good Code Ensembles

We perform a search for code ensembles with excellent iterative decoding thresholds. First, we fix some constraints. This may concern the design rate R or the VN degrees. In particular, for complexity reasons, it is sensible to limit the maximum VN degree d_v. To simplify the task of the optimization algorithm, it is common practice to allow only a limited number of VN degrees. Further, we consider code ensembles with concentrated CN degrees for which the edge-oriented degree distribution polynomial is given by

$$\rho(x) = \rho_{\mathsf{d}_c-1}x^{\mathsf{d}_c-2} + \rho_{\mathsf{d}_c}x^{\mathsf{d}_c-1}. \tag{7.103}$$

In practice, such a limitation of the CN degrees does not cause a significant performance loss. We may add further constraints on the ensemble, including also specific structures as for IRA codes, which will be omitted for ease of presentation.

Once all constraints are fixed, we run an optimization algorithm that tests different edge-oriented degree distribution pairs and selects those with the best iterative decoding threshold. For a given $\lambda(x)$, $\rho(x)$ is directly obtained from the rate constraint in (4.9). We have

$$\mathsf{d}_c = \left\lceil \frac{1}{(1-\mathsf{R})\int_0^1 \lambda(x)\mathrm{d}x} \right\rceil, \tag{7.104}$$

where $\lceil\cdot\rceil$ is the ceil operator. Exploiting the identity $\rho_{\mathsf{d}_c} + \rho_{\mathsf{d}_c-1} = 1$, we obtain

$$\rho_{\mathsf{d}_c} = \mathsf{d}_c - \mathsf{d}_c(\mathsf{d}_c - 1)(1-\mathsf{R}) \int_0^1 \lambda(x)\mathrm{d}x \,. \tag{7.105}$$

A popular choice of the optimization algorithm is the evolutionary, parallel optimization algorithm known as differential evolution. It randomly generates a population P of edge-oriented VN degree distribution polynomials $\lambda(x)$ obeying some predefined constraints. The corresponding $\rho(x)$ is obtained from (7.104) and (7.105). For each pair $(\lambda(x), \rho(x))$ a gain function, that is, the iterative decoding threshold, is evaluated. Then, from each member of the population a new candidate solution is created by introducing modifications (i.e., mutation and recombination) in $\lambda(x)$. The candidate solution replaces the original polynomial if it shows an improved iterative decoding threshold. This step is done for all members of the population and is repeated several times.

Example 7.14 We perform a first round of optimization with the following constraints: We choose $\mathsf{R} = 0.5$, limit the maximum VN degree to 20, and require concentrated CN degrees. We find that most of the edges are connected to VNs of degrees $\{2, 3, 6, 20\}$. Hence, we perform a second round of optimization allowing only those VN degrees. As a result, we obtain the degree distribution pair

$$\lambda(x) = 0.297x + 0.175x^2 + 0.244x^5 + 0.284x^{19} \tag{7.106}$$

$$\rho(x) = 0.332x^6 + 0.668x^7 \tag{7.107}$$

with $\epsilon^\star_{\mathsf{BP}} = 0.494$.

7.5.2.2 Code Construction

We aim at constructing an LDPC code for a given target blocklength n (or target information length k), a design rate R, and a degree distribution pair $\lambda(x)$ and $\rho(x)$. For illustration purposes, we consider the code's Tanner graph consisting of n VNs and $m = (1-\mathsf{R})n$ CNs, where m is assumed to be an integer. With the help of the degree distribution pair, we can determine the number of VNs and CNs of certain degree. For this, we derive the fraction of VNs of degree d, Λ_d, and the fraction of CNs of degree d, P_d, using (4.7) and (4.8). Then, the number of VNs of degree d is $n_d = \mathrm{round}(\Lambda_d\, n)$ and the number CNs of degree d is $m_d = \mathrm{round}(P_d\, m)$. Note that further adjustments of n_d and/or m_d might be necessary such that the number of edges emanating edges from the VNs are equal to those emanating from the CNs. The following condition must be fulfilled

$$\sum_{d=1}^{n} n_d d = \sum_{d=1}^{m} m_d d. \tag{7.108}$$

Once n_d and m_d are fixed, we need to place the $E = \sum_{d=1}^{n} n_d d$ edges to connect VNs of different degrees and to the CNs of different degrees. The placement of the edges shall be done to avoid harmful loops in the graph since they are building blocks of stopping sets. Examples of algorithms that address the construction of Tanner graphs aiming at large girths are the PEG and ACE algorithms, which have been mentioned in Section 6.1.4.

7.6 Code Design for Erasure Channels with Memory

Many LDPC codes have excellent erasure recovering capabilities under iterative decoding on memoryless channels, such as the BEC. However, they might perform poorly on *correlated* erasure channels, that is, on channels where erasures tend to appear consecutively which is typical for practical communication systems. The penalty in performance is owing to the interplay between certain stopping sets and erasure bursts. To give an example, consider a rate-1/2 IRA code whose parity-check matrix can be divided into two parts: the rightmost $n/2$ columns corresponding to parity symbols forming the double diagonal and the leftmost $n/2$ columns corresponding to information symbols. If an erasure burst hits only the parity symbols, then decoding will be always successful (since for IRA codes the parity-check matrix is used for encoding). By contrary, if a fraction of the, information symbols are erased, decoding may fail.

To increase reliability with respect to consecutive erasures under iterative decoding, one may make use of *interleavers* which are either embedded in the code's parity-check matrix (permutation of columns) or placed as a separate block after the encoder. The role of the interleaver is to spread the erasures uniformly within a codeword, that is, to mimic transmission over a channel which introduces independent erasures, symbol by symbol.

The analysis of channels with memory is an involved task. However, often random interleavers or structured interleavers (such as row–column interleavers, or convolutional interleavers) turn out to be good choices. In Sections 7.6.1–7.6.3, we will discuss interleaver design for specific erasure channels with memory described in Chapter 3 assuming iterative decoding. Interleaver design for the ML decoder is discussed in Chapter 8.

7.6.1 Single Burst Erasure Channel

Here, we consider an acyclic SBUEC that introduces a single erasure burst per codeword, that is, it erases a set of consecutive codeword symbols as discussed in Section 3.3. It can be seen as a toy model for correlated communication channels with the advantage that optimization of the channel code for the SBUEC is fairly simple. We recall an important parameter which is the MTBL $L_{\max}$, the maximum length of a single erasure burst that is correctable by a decoder regardless of its position in the codeword. We also recall that for any (n, k) linear block code, $L_{\max}$ cannot exceed $n - k$.

7.6.2 Stopping Set Pivots and Their Span

We start by recapping some important concepts from Section 6.1:

- Stopping sets (Definition 6.2): A stopping set is any set of VNs such that any CN connected to this set is connected to it at least twice. If those VNs are erased, iterative decoding will fail. The union of stopping sets is also a stopping set. Hence, one can define a maximal stopping set included in a subset of the VNs as the union of all the stopping sets included in the subset.
- Peeling decoding (Section 6.1.3): A peeling decoder can be considered as a particular iterative decoder which resolves erased VNs one-by-one by exploiting a parity-check equation with one unknown only. All the following results assume a peeling decoder.
- Induced subgraph (Definition 6.1): A subgraph $\mathcal{G}$ induced by the stopping set $\mathcal{S}$ is a bipartite graph composed of the VNs in $\mathcal{S}$, the CNs neighboring to the VNs in $\mathcal{S}$, and the respective edges connecting these CNs to the VNs in $\mathcal{S}$.

Definition 7.6 (Pivot of a stopping set) A VN $\mathrm{v} \in \mathcal{S}$ is called a pivot of the stopping $\mathcal{S}$ if $\mathcal{S} \setminus \{\mathrm{v}\}$ does contain any stopping sets.

We make the following observations regarding pivots of stopping sets:

(a) If the subgraph induced by a stopping set is composed of disjoint subgraphs, there exists no pivot for the stopping set. This is because all disjoint graphs must be stopping sets on their own.

(b) A necessary condition for a VN to be a pivot is to be connected to at least one CN of degree 2 in the subgraph $\mathcal{G}$. This is because to start peeling decoding, we require at least one CN connected to one unknown VN only.

(c) Assume that a VN v_i is a pivot. As discussed, it must be connected to a degree-2 CN in the induced subgraph. Then, the other VN v_j connected to that CN is also a pivot.

(d) A stopping set has no pivots or at least two pivots. An example for the former case is when all CNs in $\mathcal{G}$ are of degree 3 or more. The latter case results from the previous observation.

Example 7.15 (Pivots of stopping sets) Figure 7.7 illustrates a stopping set $\mathcal{S} = \{\mathrm{v}_1, \mathrm{v}_2, \mathrm{v}_3, \mathrm{v}_4\}$ along with its induced subgraph $\mathcal{G}$. It is easy to check that VNs v_1 and v_2 are pivots. For instance, if the value of v_1 is known and the values of all other VNs in $\mathcal{S}$ are unknown, a peeling decoder will succeed. It starts by recovering the value of v_2, then the value of v_3 and finally the value of v_4.

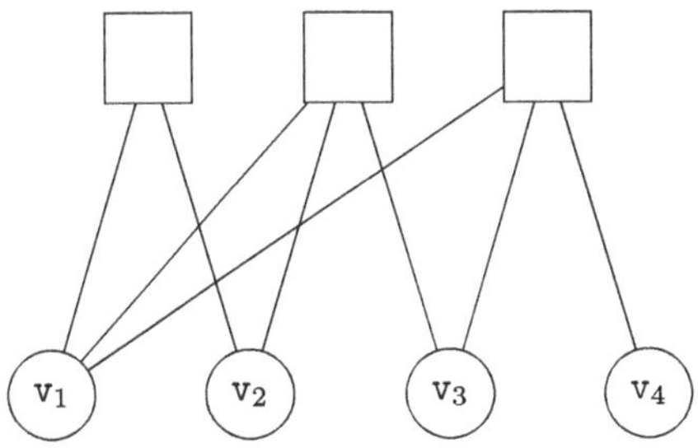

Figure 7.7 Example of a subgraph induced by a size-4 stopping set.

A further important figure on the SBUEC is the span of pivots.

Definition 7.7 (Span of pivots) Let $\mathcal{S}$ be a stopping set whose pivots form the nonempty set $\mathcal{P}$. Let v_l and v_h be the pivots with the lowest and highest index, respectively. Then, the span of pivots $\mu(\mathcal{P})$ is $\mu(\mathcal{P}) = h - l + 1$.

To make the code resistant against a single erasure burst of length $L_{\max}$ (starting at any position of the codeword), the set of pivots $\mathcal{P}$ of any stopping set must fulfill $\mu(\mathcal{P}) \geq L_{\max} + 1$. Otherwise, a peeling decoder would fail.

Example 7.16 (Span of pivots) Consider again the stopping set $\mathcal{S}$ from Example 7.15 and its induced subgraph $\mathcal{G}$ in Figure 7.7. We have $\mathcal{P} = \{\mathsf{v}_1, \mathsf{v}_2\}$ and $\mu(\mathcal{P}) = 2$.

7.6.3 Pivot Search and Swap Algorithm

The pivot search and swap (PSS) algorithm is a greedy optimization algorithm that permutes the columns of a parity-check matrix $\boldsymbol{H}$ in order to make the code resistant against single erasure bursts hitting a codeword. As illustrated later in this chapter, the algorithm turns out to be effective also on other types of erasure channels with memory. The goal of the PSS algorithm is to increase the span of the pivots, which requires a pivot search step. We can make the following statement.

Theorem 7.5 (Stopping set pivot theorem) *Let L_{max} be the MTBL of an LDPC code under peeling decoding, and let the erasure burst of length $L_{max}+1$ starting on the VN v_j and ending on the VN $\mathsf{v}_{j+L_{max}}$ be nonresolvable. Then, v_j and $\mathsf{v}_{j+L_{max}}$ are pivots of the maximal stopping set included in the burst.*

Proof Let $\mathcal{S}$ be the maximal stopping set included in the noncorrectable erasure burst of length $L_{\max} + 1$ and $\mathcal{G}$ be its induced subgraph. By hypothesis, an erasure burst of length $L_{\max}$, starting on v_j and ending on $\mathsf{v}_{j+L_{\max}-1}$ can be corrected by the peeling decoder. This implies that if the value of $\mathsf{v}_{j+L_{\max}}$ is known, then the decoder applied to $\mathcal{G}$ is able to successfully correct all VNs in the stopping set. Hence, by definition

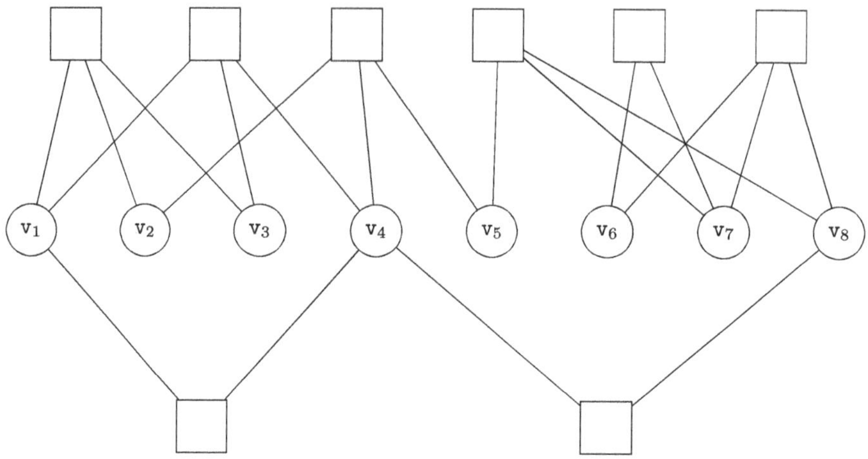

Figure 7.8 Example of a subgraph induced by a size-8 stopping set.

$\mathsf{v}_{j+L_{\max}}$ is a pivot of $\mathcal{S}$. Analogously, an erasure burst of length $L_{\max}$, starting on v_{j+1} and ending on $\mathsf{v}_{j+L_{\max}}$, can be corrected by an iterative decoder. By reasoning in the same way as the previous one, we can prove the argument that v_j is a pivot of $\mathcal{S}$. □

Further pivots might be identified as follows. Starting from the set of pivots $\mathcal{P}_0 = \{\mathsf{v}_j, \mathsf{v}_{j+L_{\max}}\}$, we search for further pivots by identifying the degree-2 CNs connected to the elements of $\mathcal{P}_0$. Following observation (b) in the previous section, a VN connected to a pivot by a degree-2 CN is also a pivot. Let the newly found pivots form the set $\hat{\mathcal{P}}_0$. We include them in the set of pivots and obtain $\mathcal{P}_1 = \mathcal{P}_0 \cup \hat{\mathcal{P}}_0$. We apply this recursion till the set of pivots increases. As the following example shows the pivot searching algorithm sketched earlier is suboptimal in the sense that it is in general not able to identify all pivots of a $\mathcal{S}$. It provides however a practical and efficient way of identifying pivots.

Example 7.17 (Pivot search) Consider a size-8 stopping set $\mathcal{S}$ and its induced Tanner graph $\mathcal{G}$ as depicted in Figure 7.8. Assume that, by Theorem 7.5, the pivots $\mathcal{P}_0 = \{\mathsf{v}_1, \mathsf{v}_8\}$ are known. Starting from v_1 we identify v_4 as a new pivot, since it is connected to v_1 by a degree-2 CN. If we repeat the procedure for v_8, we obtain also v_4 as new pivot. Both degree-2 CNs connecting v_8 and v_1 to v_4, respectively, are plotted in the lower part of Figure 7.8 for convenience. Thus, $\mathcal{P}_1 = \{\mathsf{v}_1, \mathsf{v}_4, \mathsf{v}_8\}$. Here, the procedure stops. Note that v_6 and v_7 are also pivots which are not identified by the procedure.

Now we sketch the PSS algorithm. Let's assume an MTBL of $L_{\max}$. Our goal is to increase the MTBL to $L_{\max} + 1$. First, the algorithm identifies all nonresolvable single bursts of length $L_{\max} + 1$ whose number is N_B. Let $\mathcal{S}^{(i)}$ be the maximal stopping set included in the ith nonresolvable burst, $\mathcal{B}^{(i)}$ the set of all VNs in the burst, and let $\mathrm{v}_f^{(i)}$, as well as $\mathrm{v}_l^{(i)}$ be the first and the last VN in the ith burst, respectively. Both VNs are pivots by Theorem 7.5. Then, the algorithm identifies further pivots, for example, as discussed previously, yielding a (possibly not exhaustive) set of pivots $\mathcal{P}^{(i)}$. Next, the algorithm performs pivot swapping, that is, one of the pivots in $\mathcal{P}^{(i)}$ is swapped with a VN $\bar{\mathrm{v}}^{(i)}$ not included in $\mathcal{B}^{(i)}$. Hence, the burst will become resolvable by the peeling decoder. Moreover, if we require that the span of pivots $\mu(\mathcal{P}^{(i)})$ after the swapping operation is larger than $L_{\max} + 1$, then for *any* erasure burst of length $L_{\max} + 1$ a pivot of $\mathcal{S}^{(i)}$ will be known.

The PSS procedure is applied to every maximal stopping set $\mathcal{S}^{(i)}$, where $i \in \{1, \ldots, N_B\}$. This does not necessarily mean that any erasure burst of length $L_{\max} + 1$ will be resolvable, since any swap could reduce the span of pivots of some other stopping set. However, this approach has been shown to be effective in practice. We will underpin this statement by numerical results later on. Once the procedure succeeds for $L_{\max} + 1$, it is executed for $L_{\max} + 2$ and so on. Once a failure is observed, all permutations are refused and a new sequence of N_B permutations is performed, at most for F times, where F is a user-defined parameter. Starting from a parity-check matrix $\boldsymbol{H}$, the algorithm provides a permuted version of the parity-check matrix $\boldsymbol{H}_\pi$ so that the permuted code is more resistant to single erasure bursts than the original channel code. The PSS algorithm is formalized in Algorithm 2.

Some remarks on the PSS algorithm follow. Once a good LDPC code for the BEC under iterative decoding is found, the PSS can be used to make it robust against a single burst of erasures. One may use the PSS algorithm either to generate an interleaver that permutes the code symbols or to directly permute the columns of the parity-check matrix yielding an equivalent code (see Section 2.3). Complexity of the algorithm is dominated by the search step for unresolvable erasure patterns. In order to increase the MTBL from $L_{\max}$ to $L_{\max} + 1$, we require $n - L_{\max}$ calls of the peeling decoder, each of them needs to recover $L_{\max} + 1$ unknowns. Being $L_{\max} \le n(1_R)$ we conclude that the complexity of the search step is quadratic in n.

Table 7.1 provides $L_{\max}$ values obtained before and after applying the PSS algorithm for different codes. As a benchmark for the maximum value of $L_{\max}$ of a length-n LDPC code, we will use the estimate $n\epsilon^\star_{\mathsf{BP}}$. It comes from the observation that decoding of a very long LDPC code will succeed with high probability if the erasure probability is not larger than $\epsilon^\star_{\mathsf{BP}}$ or equivalently if the number of erasures in a codeword is less than $n\epsilon^\star_{\mathsf{BP}}$. The codes under consideration are a (2640, 1320) Margulis code which is a quasi-cyclic (3, 6) regular LDPC code, a (4, 32) regular LDPC code with code parameters (4608, 4033), and a (2000, 1000) IRA LDPC code with degree distribution pair $\psi(x) = x^4$ and $\phi(x) = x^4$. The last two codes were generated using the PEG algorithm. Observe from the table that the PSS algorithm increases the $L_{\max}$ considerably. In all cases, $L_{\max}$ is close to $n\epsilon^\star_{\mathsf{BP}}$.

Algorithm 2: Pivot search and swapping algorithm

//Initialization:
set burst length $L = L_{\max} + 1$.
//Pivot search step:
find all noncorrectable erasure bursts of length L and let their number be N_B.
for $i = 1 : N_B$ **do**
  Identify a set of pivots $\mathcal{P}^{(i)}$ of $\mathcal{S}^{(i)}$.
end
//Pivot swap step:
for $f = 1 : F$ **do**
  for $i = 1 : N_B$ **do**
    Randomly choose a pivot $\mathrm{v}_p^{(i)} \in \mathcal{P}^{(i)}$.
    Choose a VN $\bar{\mathrm{v}}^{(i)}$ from the set

$$\mathcal{V} \setminus \left\{ \mathcal{B}^{(i)} \cup \left(\bigcup_{j \neq i} \mathcal{P}^{(j)} \right) \cup \left\{ \bar{\mathrm{v}}^{(0)}, \ldots, \bar{\mathrm{v}}^{(i-1)} \right\} \right\} \tag{7.109}$$

    and swap $\bar{\mathrm{v}}^{(i)}$ and $\mathrm{v}_p^{(i)}$ such that the span of pivots after the swapping operation is at least $L_{\max} + 2$.
  end
  //Check:
  Run iterative decoder and check if erasure burst of length L is resolvable.
  if *erasure burst of length L is resolvable* **then**
    increase L by one and move to pivot search step.
  else
    undo all VNs swaps for the current L.
  end
end

Table 7.1 $L_{\max}$ values for various LDPC codes before and after applying the PSS algorithm

	Margulis code	IRA code (PEG)	(4,32) LDPC code (PEG)
R	0.5	0.5	0.875
$L_{\max}$ before PSS	1033	403	287
$L_{\max}$ after PSS	1135	914	425
$\lfloor n\epsilon^{\star}_{\mathrm{BP}} \rfloor$	1133	946	445

7.6.4 Results for the Block and Gilbert Erasure Channels

While the MTBL plays an important rule for the SBUEC, there is no guarantee that an LDPC code with a large MTBL will perform well on other correlated erasure channels. However, there is empirical evidence that the MTBL is often a good proxy,

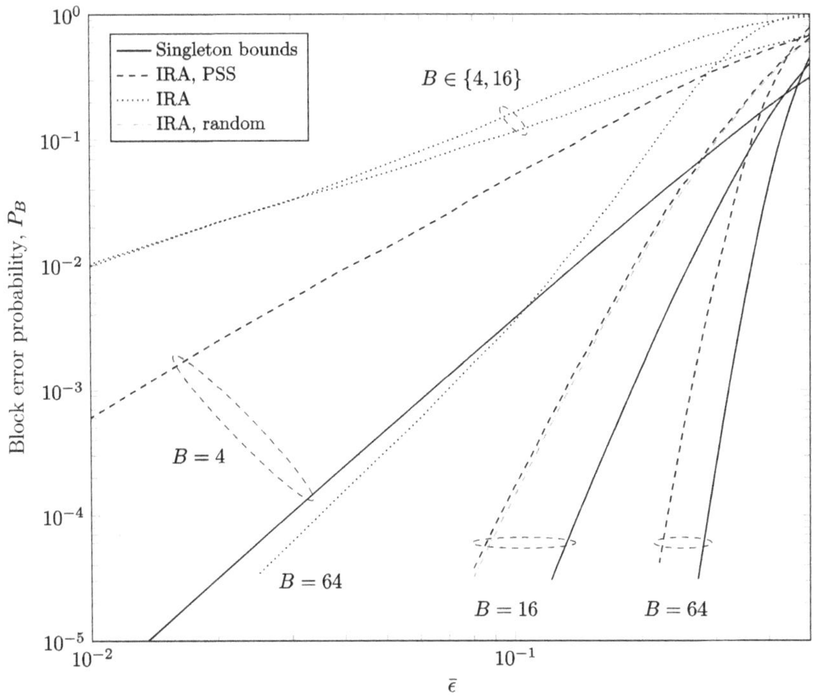

Figure 7.9 Block error rate versus ϵ on BLEC for IRA code, permuted IRA code obtained via the PSS algorithm, and randomly permuted IRA code (only shown for $B = 16$). As benchmarks, Singleton lower bounds on the block error probability are plotted.

especially for channels introducing long erasure bursts. Next, we will illustrate that the PSS technique described in Section 7.6.3 can provide good results on other erasure channels with memory, namely the BLEC (see Section 3.3.2) and the GEC (see Section 3.3.3).

The BLEC divides the codeword into B blocks of length $L = n/B$ and assumes that each of the blocks is erased independently with probability ϵ. In the last part of Section 7.6.4, we will show simulation results of a code on the BLEC which was optimized for the simpler (acyclic) SBUEC. To this end, we consider an IRA code with blocklength $n = 1024$, information length $k = 512$, and rate $R = 0.5$. The edge-oriented degree distribution polynomials are (see Section 4.2.2)

$$\psi(x) = 0.2779x^2 + 0.0371x^6 + 0.2478x^7 + 0.0653x^{17} + 0.1007x^{19} + 0.2712x^{53} \tag{7.110}$$

$$\phi(x) = 0.0151x^5 + 0.9648x^6 + 0.0201x^7 . \tag{7.111}$$

We apply the PSS algorithm and obtain an equivalent code. For both codes, block error rates on the BLEC are obtained via Monte Carlo simulations and are depicted in Figure 7.9 for block sizes $B \in \{4, 16, 64\}$. For $B = 16$ the performance of an equivalent IRA code obtained by a random permutation is also depicted. Observe from the figure that the standard IRA code performs poorly on the BLEC for all values of B. By contrast, both equivalent codes (obtained via the PSS algorithm and via a

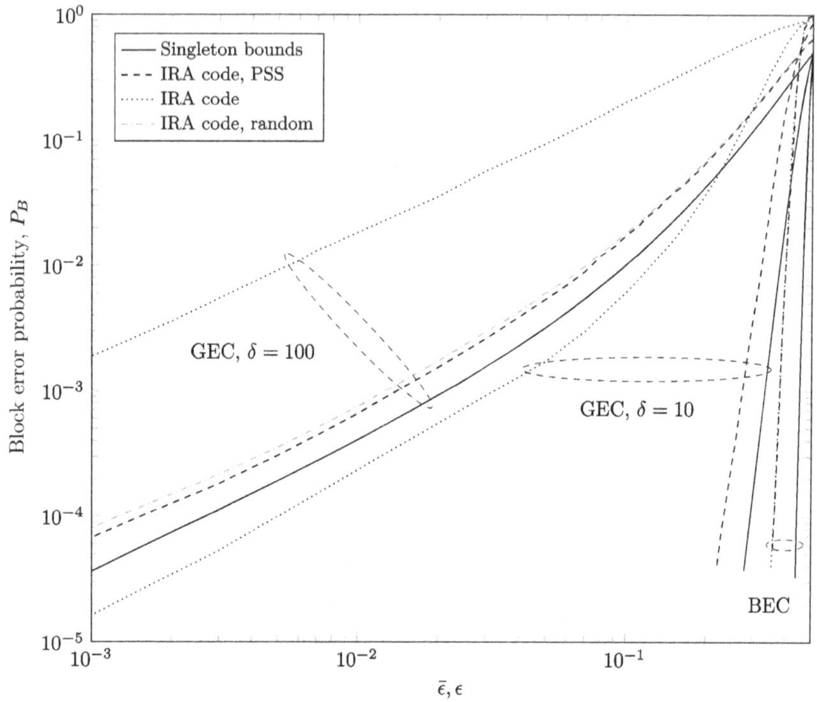

Figure 7.10 Block error rate versus $\bar{\epsilon}$ or ϵ on Gilbert erasure channel/BEC for IRA code, permuted IRA code obtained via the PSS algorithm, and randomly permuted IRA code (only shown for $\delta = 100$). As benchmarks, Singleton lower bounds on the block error probability are plotted. From left to right: block error rate/probability for the GEC with $\delta = 100$, for the GEC with $\delta = 10$, and for the BEC.

random permutation) perform reasonably close to the respective Singleton bounds for $B \in \{16, 64\}$.

Figure 7.9 shows simulation results of the above IRA codes on the GEC (see Section 3.3.3). The GEC is fully characterized by the average burst duration δ and the average erasure probability $\bar{\epsilon}$. For the results in the figure, we consider a BEC with erasure probability ϵ, a GEC with average erasure probability $\bar{\epsilon}$ and average erasure burst length $\delta = 10$, and a GEC with average erasure probability $\bar{\epsilon}$ and average erasure burst length $\delta = 100$. As benchmarks, the corresponding Singleton lower bounds are also shown (see Section 3.3.3). As expected, on the BEC both codes perform the same. Observe that the permuted IRA code designed for the SBUEC outperforms the IRA code on the GEC, with performance close to the respective Singleton bound. This is an example where a code designed for the SBUEC also performs well on another correlated channel while very poor performance is obtained when no permutation is applied at all.

Another option to make the original IRA code robust against correlated erasures of the GEC is a random permutation of the code symbols. We only present simulation results for $\delta = 100$. For smaller δ, they virtually coincide with the performance of the permuted IRA code with the PSS algorithm. The result in Figure 7.10 illustrates that also a random permutation is a good choice for the GEC.

7.7 Historical Notes

Already in his doctoral thesis, Gallager gave expressions for the evolution of the message error probability versus the iteration index on a binary symmetric channel [1]. He considered regular codes and assumed that the corresponding code's graph was locally a tree. Luby et al. derived an analysis for peeling decoding of irregular code ensembles by means of differential equations in [2,3] and a modified analysis yielding the density evolution equations for the erasure channel [4]. Richardson et al. extended the results to a large class of channels [5] and demonstrated capacity approaching code constructions on Gaussian channels in [6]. Proofs on the concentration and local tree behavior can be found in [3–5]. Stability for unstructured code ensembles is discussed in [6,7]. For structured ensembles, the density evolution recursions describe a discrete dynamical system which can be linearized and its stability analyzed by standard tools from the literature (see, e.g., [8]).

EXIT charts were introduced by ten Brink in [9] to gain more insights into the convergence behavior of iterative decoding algorithms. Basic properties such as duality or the area property were studied in [10] for erasure channels. Stopping sets were originally introduced in [11]. Based on [11], exact expressions for the error probability of finite-length regular LDPC ensembles were derived in [12]. Extensions followed in [13]. A different kind of finite-length analysis, which is based on scaling laws [14], was proposed in [15]. Here, the focus is on how the waterfall region performance of iterative decoders scales with the blocklength.

Accumulator-based LDPC codes were first introduced in [16]. As member of this class, IRA codes [17] (see Chapter 4) possess highly efficient encoders combined with remarkable code performance. Iterative threshold computation for protograph ensembles [18] (see Chapter 4) can be found in [19]. Tools for the design of structured/unstructured LDPC codes are manifold. The differential evolution algorithm for unstructured code ensembles was presented in [20]. The PEG algorithm was introduced in [21]. An important extension to which classifies cycles according to their connectivity was presented in [22]. Techniques for burst error correction were discussed in [23,24]. Although code design is mostly done by numerical optimization, the existence of capacity achieving sequences of code ensembles is a valuable result, since it answers the fundamental question of channel coding. The first capacity achieving sequence was introduced in [3], a systematic study was provided in [25] and extended in [26], while sequences with bounded decoding complexity were first presented in [27].

As a final remark, Gallager's work also inspired Zyablov and Pinsker. In the 1970s, they investigated erasure decoding of LDPC codes, thereby describing an algorithm that seeks for equations with one unknown that is recovered as the sum of all not erased symbols [28, Chapter 3]. This resembles of peeling decoding. Unfortunately, their work was published in Russian and hence mostly ignored by the mainstream channel coding community.

7.8 Exercises

Exercise 7.1 (Iterative decoding threshold) Consider the regular (3,4) LDPC code ensemble. Derive the expression for the density evolution recursion $p^{\ell} = f(p^{\ell-1}, \epsilon)$ which gives the evolution of the average erasure probability at a VN output. Set $p^{\ell} = p^{\ell-1}$ and determine the value of ϵ. This gives the ensemble's iterative decoding threshold $\epsilon^{\star}_{\mathsf{BP}}$.

Exercise 7.2 (Fixed point analysis) Consider the density evolution recursion $f(x, \epsilon) = \epsilon\lambda(1 - \rho(1 - x))$ for unstructured irregular LDPC code ensembles. Argue why stability of the fixed point $x = 0$ is a necessary but not sufficient condition for convergence of density evolution to zero.

Exercise 7.3 (EXIT recursion) Show that the function $f(\mathsf{I}_{\mathrm{EV}}, \epsilon)$ in Equation (7.25) is strictly monotonically increasing in I_{EV} for $\mathsf{I}_{\mathrm{EV}} \in [0, 1]$ and $\epsilon \in (0, 1]$.

Exercise 7.4 (Conditional entropy of dual code) Consider (7.49). Show that the conditional entropy $\mathsf{H}^{\perp}(\boldsymbol{X}|\boldsymbol{Y}(\epsilon_1, \epsilon_2, \ldots, \epsilon_n))$ for the dual code is

$$\mathsf{H}^{\perp}(\boldsymbol{X}|\boldsymbol{Y}(\epsilon_1, \epsilon_2, \ldots, \epsilon_n) = \mathsf{H}(\boldsymbol{X}|\boldsymbol{Y}(1 - \epsilon_1, 1 - \epsilon_2, \ldots, 1 - \epsilon_n) - k + \sum_{i=1}^{n} \epsilon_i \,. \quad (7.112)$$

Hint: Start with the definition of conditional entropy in (7.48), use the relation $\boldsymbol{H}^{\perp} = \boldsymbol{G}$, and observe that the second addend in (7.48) is invariant when $\mathcal{E}$ is replaced by $\bar{\mathcal{E}}$ and vice versa.

Exercise 7.5 (Stability condition 1) Consider a protograph with base matrix

$$\boldsymbol{B} = \begin{bmatrix} 3 & 3 \end{bmatrix}.$$

Derive the expression for the stability condition.

Exercise 7.6 (Stability condition 2) Derive the stability condition for the IRA code ensemble in Example 7.11.

Exercise 7.7 (Stopping set size enumerators) Consider two parity-check matrices

$$\boldsymbol{H}_1 = \begin{bmatrix} 1 & 1 & 0 & 0 \\ 1 & 0 & 1 & 0 \\ 1 & 0 & 0 & 1 \end{bmatrix} \text{ and } \boldsymbol{H}_2 = \begin{bmatrix} 1 & 1 & 0 & 0 \\ 1 & 0 & 1 & 0 \\ 1 & 1 & 1 & 1 \end{bmatrix}.$$

Argue that both matrices describe the same $(4, 1)$ repetition code. Derive the stopping set size enumerators for the two representations of the $(4, 1)$ repetition code. Which representation would you prefer when using a peeling decoder?

Exercise 7.8 (Stopping set pivot) Consider the matrix

$$A = \begin{bmatrix} 1 & 0 & 1 & 1 \\ 1 & 1 & 1 & 1 \\ 0 & 0 & 1 & 1 \end{bmatrix}.$$

Draw the corresponding bipartite graph and verify that the four VNs form a stopping set $\mathcal{S}$. Determine the pivots of the stopping set $\mathcal{S}$.

Exercise 7.9 (Pivot swapping) Consider the matrix

$$A = \begin{bmatrix} 1 & 0 & 0 & 0 & 0 & 1 & 1 & 0 \\ 1 & 1 & 0 & 0 & 0 & 0 & 1 & 0 \\ 0 & 1 & 0 & 0 & 0 & 1 & 0 & 1 \\ 0 & 0 & 0 & 1 & 1 & 0 & 0 & 1 \\ 0 & 0 & 1 & 1 & 0 & 0 & 0 & 0 \\ 0 & 0 & 1 & 1 & 1 & 0 & 0 & 0 \\ 1 & 1 & 0 & 0 & 0 & 0 & 0 & 0 \\ 0 & 1 & 0 & 0 & 1 & 0 & 0 & 0 \end{bmatrix}.$$

The VNs of the corresponding bipartite graph form a stopping set. Assume we identified some pivots $\mathcal{P} = \{\mathsf{v}_1, \mathsf{v}_2, \mathsf{v}_5\}$. Consider a single erasure burst of length 7. Which of the pivots in $\mathcal{P}$ must be swapped such that any erasure burst of length 7 becomes resolvable by a peeling decoder? What is the span of pivots after the swapping operation (for each of the possibilities)?

Exercise 7.10 (Heavy-tail/Poisson sequences) Inequality in (7.99) implies $b_d \geq 1$. Show that for unstructured code ensembles with minimum VN degree of 2, in fact $b_d \geq (1 - \exp(-0.5))/0.5 \approx 1.59$. Make use of (7.93).

References

[1] R. Gallager, *Low-density parity-check codes*. Cambridge, MA: MIT Press, 1963.

[2] M. Luby, M. Mitzenmacher, A. Shokrollahi, D. Spielman, and V. Stemann, "Practical loss-resilient codes," in *Proc. 29th Symp. Theory Computing*, El Paso, TX, USA, May 1997, pp. 150–159.

[3] M. Luby, M. Mitzenmacher, A. Shokrollahi, and D. Spielman, "Efficient erasure correcting codes," *IEEE Trans. Inf. Theory*, vol. 47, no. 2, pp. 569–584, Feb. 2001.

[4] M. Luby, M. Mitzenmacher, and A. Shokrollahi, "Analysis of random processes via and-or tree evaluation," in *Proc. 9th Ann. ACM-SIAM Symp. on Discrete Algorithms*, San Francisco, CA, USA, Jan. 1998, pp. 364–373.

[5] T. Richardson and R. Urbanke, "The capacity of low-density parity-check codes under message-passing decoding," *IEEE Trans. Inf. Theory*, vol. 47, no. 2, pp. 599–618, Feb. 2001.

[6] T. Richardson, A. Shokrollahi, and R. Urbanke, "Design of capacity-approaching irregular low-density parity-check codes," *IEEE Trans. Inf. Theory*, vol. 47, no. 2, pp. 619–637, Feb. 2001.

[7] M. A. Shokrollahi, "New sequences of linear time erasure codes approaching the channel capacity," in *Proc. Int. Symp. Applied Algebra, Algebraic Algorithms and Error-Correcting Codes*, Honolulu, HI, USA, Nov. 1999, pp. 65–76.

[8] S. Elyadi, *An introduction to difference equations, 3rd ed.* New York, NY: Springer, 2005.

[9] S. ten Brink, "Convergence behavior of iteratively decoded parallel concatenated codes," *IEEE Trans. Commun.*, vol. 49, no. 10, pp. 1727–1737, Oct. 2001.

[10] A. Ashikhmin, G. Kramer, and S. ten Brink, "Extrinsic information transfer functions: Model and erasure channel properties," *IEEE Trans. Inf. Theory*, vol. 50, no. 11, pp. 2657–2673, Nov. 2004.

[11] T. Richardson and R. Urbanke, "Efficient encoding of low-density parity-check codes," *IEEE Trans. Inf. Theory*, vol. 47, no. 2, pp. 638–656, Feb. 2001.

[12] C. Di, D. Proietti, I. Telatar, T. Richardson, and R. Urbanke, "Finite-length analysis of low-density parity-check codes on the binary erasure channel," *IEEE Trans. Inf. Theory*, vol. 48, no. 6, pp. 1570–1579, June 2002.

[13] J. Zhang and A. Orlitsky, "Finite-length analysis of LDPC codes with large left degrees," in *Proc. IEEE Int. Symp. Inf. Theory*, Lausanne, Switzerland, June 2002, p. 3.

[14] A. Montanari, "The glassy phase of Gallager codes," *Eur. Phys. J. B*, vol. 23, no. 1, pp. 121–136, Sep. 2001.

[15] A. Amraoui, A. Montanari, T. Richardson, and R. Urbanke, "Finite-length scaling for iteratively decoded LDPC ensembles," *IEEE Trans. Inf. Theory*, vol. 55, no. 2, pp. 473–498, Feb. 2009.

[16] D. Divsalar, H. Jin, and R. J. McEliece, "Coding theorems for turbo-like codes," in *Proc. Ann. Allerton Conf. on Commun., Control, and Comput.*, Monticello, IL, USA, Sep. 1998, pp. 201–210.

[17] H. Jin, A. Khandekar, and R. McEliece, "Irregular repeat-accumulate codes," in *Proc. IEEE Int. Symp. Turbo Codes and Rel. Topics*, Brest, France, Sep. 2000, pp. 1–8.

[18] J. Thorpe, "Low-density parity-check (LDPC) codes constructed from protographs," NASA JPL, Pasadena, CA, USA, IPN Progress Report 42-154, Aug. 2003.

[19] G. Liva and M. Chiani, "Protograph LDPC codes design based on EXIT analysis," in *Proc. IEEE Global Telecommun. Conf.*, Washington, DC, USA, Nov. 2007, pp. 3250–3254.

[20] A. Shokrollahi and R. Storn, "Design of efficient erasure codes with differential evolution," in *Proc. IEEE Int. Symp. Inf. Theory*, Sorrento, Italy, June 2000, p. 5.

[21] X.-Y. Hu, E. Eleftheriou, and D. Arnold, "Regular and irregular progressive edge-growth Tanner graphs," *IEEE Trans. Inf. Theory*, vol. 51, no. 1, pp. 386–398, Jan. 2005.

[22] T. Tian, C. Jones, J. D. Villasenor, and R. Wesel, "Selective avoidance of cycles in irregular LDPC code construction," *IEEE Trans. Commun.*, vol. 52, no. 8, pp. 1242–1247, Aug. 2004.

[23] M. Fossorier, "Universal burst error correction," in *Proc. IEEE Int. Symp. Inf. Theory*, 2006, pp. 1969–1973.

[24] E. Paolini and M. Chiani, "Construction of near-optimum burst erasure correcting low-density parity-check codes," *IEEE Trans. Commun.*, vol. 57, no. 5, pp. 1320–1328, May 2009.

[25] P. Oswald and A. Shokrollahi, "Capacity-achieving sequences for the erasure channel," *IEEE Trans. Inf. Theory*, vol. 48, no. 12, pp. 364–373, Dec. 2002.

[26] A. Orlitsky, K. Viswanathan, and J. Zhang, "On capacity-achieving sequences of degree distributions," in *Proc. IEEE Int. Symp. Inf. Theory*, Yokohama, Japan, June 2003, pp. 269–269.

[27] H. D. Pfister, I. Sason, and R. Urbanke, "Capacity-achieving ensembles for the binary erasure channel with bounded complexity," *IEEE Trans. Inf. Theory*, vol. 51, no. 7, pp. 2352–2379, July 2005.

[28] V. V. Zyablov and M. S. Pinsker, "Decoding complexity of low-density codes for transmission in a channel with erasures," *Probl. Peredachi Inf.*, vol. 10, no. 1, pp. 15–28, 1974.

8 Maximum-Likelihood LDPC Decoder Analysis

In this chapter, the performance of LDPC codes under ML decoding is analyzed. ML decoding is intended here either as the block-wise or the symbol-wise decoding criterion (see Section 2.2). More specifically, the asymptotic analysis of the ML decoding threshold addresses the performance in terms of symbol-wise ML decoding, whereas finite-length bounds are provided for the block error probability under block-wise ML decoding. While the focus is on unstructured LDPC code ensembles, the results in this chapter can be considered to a large extent valid for other LDPC code ensembles.

8.1 Asymptotic Performance Analysis

We have seen in Chapter 7 that the performance of an LDPC code ensemble under BP decoding can be conveniently analyzed, in the limit of large block-lengths, via EXIT (density evolution) analysis. This allows characterizing the erasure correction capability of an ensemble through a single parameter, called BP decoding threshold. In this section, we are interested in a similar characterization of an LDPC ensemble performance under ML decoding. Remarkably, we will see that also the derivation of decoding threshold under ML decoding can be based on EXIT analysis. The main building block for the computation of an ensemble ML decoding threshold is provided by Theorem 8.1 (Area Theorem).

Theorem 8.1 (Area Theorem) *Consider a rate-R, length-n binary linear block code $\mathcal{C}$, the transmitted codeword $\boldsymbol{X}$ and the observed n-tuple $\boldsymbol{Y}$ at the output of a BEC with erasure probability ϵ. Denote by $h(\epsilon)$ the EXIT function of $\mathcal{C}$ as defined in* (7.15). *We have that the area under $h(\epsilon)$ equals the code rate, that is*

$$A = R \tag{8.1}$$

with

$$A = \int_0^1 h(\epsilon)d\epsilon. \tag{8.2}$$

Proof Assume the transmission of the n codeword bits over n parallel BECs with erasure probabilities $\epsilon_1, \epsilon_2, \ldots, \epsilon_n$, where $\epsilon_i = \epsilon$ for all $i = 1, \ldots, n$. We can write

$$\frac{\mathrm{dH}(X_1, X_2, \ldots, X_n | Y_1(\epsilon_1), Y_2(\epsilon_2), \ldots, Y_n(\epsilon_n))}{\mathrm{d}\epsilon} = \tag{8.3}$$

$$= \sum_{i=1}^{n} \left(\frac{\partial \mathrm{H}(X_1, X_2, \ldots, X_n | Y_1(\epsilon_1), Y_2(\epsilon_2), \ldots, Y_n(\epsilon_n))}{\partial \epsilon_i} \right)_{\epsilon_1 = \epsilon_2 = \ldots = \epsilon_n = \epsilon} \tag{8.4}$$

$$= \sum_{i=1}^{n} \left(\frac{\partial \mathrm{H}(X_i | Y_1(\epsilon_1), Y_2(\epsilon_2), \ldots, Y_n(\epsilon_n))}{\partial \epsilon_i} + \right.$$
$$\left. + \frac{\partial \mathrm{H}(X_{\sim i} | X_i, Y_1(\epsilon_1), Y_2(\epsilon_2), \ldots, Y_n(\epsilon_n))}{\partial \epsilon_i} \right)_{\epsilon_1 = \epsilon_2 = \ldots = \epsilon_n = \epsilon} \tag{8.5}$$

$$= \sum_{i=1}^{n} \left(\frac{\partial \mathrm{H}(X_i | Y_1(\epsilon_1), Y_2(\epsilon_2), \ldots, Y_n(\epsilon_n))}{\partial \epsilon_i} \right)_{\epsilon_1 = \epsilon_2 = \ldots = \epsilon_n = \epsilon} \tag{8.6}$$

$$= \sum_{i=1}^{n} \frac{\partial (\epsilon_i \mathrm{H}(X_i | \boldsymbol{Y}_{\sim i}(\epsilon)))}{\partial \epsilon_i} \tag{8.7}$$

$$= \sum_{i=1}^{n} \mathrm{H}(X_i | \boldsymbol{Y}_{\sim i}(\epsilon)) \tag{8.8}$$

$$= n h(\epsilon), \tag{8.9}$$

where (8.5) is a consequence of the chain rule of conditional entropy, (8.6) is obtained by observing that the second addend in (8.5) does not depend on ϵ_i as the X_i is given, and (8.7) follows from writing

$$\mathrm{H}(X_i | Y_1, Y_2, \ldots, Y_n) = \epsilon_i \mathrm{H}(X_i | Y_1, Y_2, \ldots, Y_i = E, \ldots, Y_n) +$$
$$+ (1 - \epsilon_i) \mathrm{H}(X_i | Y_1, Y_2, \ldots, Y_i = X_i, \ldots, Y_n) \tag{8.10}$$

and noting that the second addend is zero since there is no uncertainty on X_i upon observing it. Finally, observe that $\mathrm{H}(X_1, X_2, \ldots, X_n | Y_1(\epsilon), Y_2(\epsilon), \ldots, Y_n(\epsilon))$ equals 0 for $\epsilon = 0$ while it equals k for $\epsilon = 1$. Hence, by integrating (8.9) between 0 and 1, we get

$$n \int_0^1 h(\epsilon) \mathrm{d}\epsilon = \mathrm{H}(\boldsymbol{X} | \boldsymbol{Y}(1)) - \mathrm{H}(\boldsymbol{X} | \boldsymbol{Y}(0)) \tag{8.11}$$

$$= k. \tag{8.12}$$

The result (8.1) follows immediately. □

Example 8.1 Consider the (4, 2) binary linear block code defined by the generator matrix

$$\boldsymbol{G} = \begin{bmatrix} 1 & 0 & 1 & 1 \\ 0 & 1 & 1 & 0 \end{bmatrix}. \tag{8.13}$$

To obtain its EXIT function, we apply the definition (7.15). We are hence concerned with finding the individual terms $\mathrm{H}(X_i | \boldsymbol{Y}_{\sim i}(\epsilon))$, for $i = 1, \ldots, 4$. A first approach consists of listing all possible channel outputs and summing the probabilities associated with the outputs that do not allow recovering the value of the specific codeword bit under analysis. Next, we do that for the first codeword bit (the process can be easily replicated for the remaining ones). The analysis is summarized in the following table, where we assume (without loss of generality) that the transmitted codeword is the all-zero one.

$\boldsymbol{y}_{\sim 1}$	$P(\boldsymbol{y}_{\sim 1})$	$\mathsf{H}(X_1 \mid \boldsymbol{Y}_{\sim 1} = \boldsymbol{y}_{\sim 1})$
$\begin{bmatrix} 0 & 0 & 0 \end{bmatrix}$	$(1-\epsilon)^3$	0
$\begin{bmatrix} \mathsf{E} & 0 & 0 \end{bmatrix}$	$\epsilon(1-\epsilon)^2$	0
$\begin{bmatrix} 0 & \mathsf{E} & 0 \end{bmatrix}$	$\epsilon(1-\epsilon)^2$	0
$\begin{bmatrix} 0 & 0 & \mathsf{E} \end{bmatrix}$	$\epsilon(1-\epsilon)^2$	0
$\begin{bmatrix} \mathsf{E} & \mathsf{E} & 0 \end{bmatrix}$	$\epsilon^2(1-\epsilon)$	0
$\begin{bmatrix} 0 & \mathsf{E} & \mathsf{E} \end{bmatrix}$	$\epsilon^2(1-\epsilon)$	1
$\begin{bmatrix} \mathsf{E} & 0 & \mathsf{E} \end{bmatrix}$	$\epsilon^2(1-\epsilon)$	1
$\begin{bmatrix} \mathsf{E} & \mathsf{E} & \mathsf{E} \end{bmatrix}$	ϵ^3	1

It follows that $\mathsf{H}(X_1 \mid \boldsymbol{Y}_{\sim 1}(\epsilon))$ is the sum of the probabilities of the last three events in the table, yielding $\mathsf{H}(X_1 \mid \boldsymbol{Y}_{\sim 1}(\epsilon)) = 2\epsilon^2(1-\epsilon) + \epsilon^3 = 2\epsilon^2 - \epsilon^3$. This approach does not scale well in n, since one needs to list, for every codeword bits, the 2^{n-1} possible outcomes for $\boldsymbol{Y}_{\sim i}$ (assuming the transmission of the all-zero codeword). A slightly more efficient approach relies on the inclusion–exclusion principle. For the case of two events S_1 and S_2, the principle states that $P(S_1 \cup S_2) = P(S_1) + P(S_2) - P(S_1 \cap S_2)$. We proceed as follows. We label the codewords as

$$\boldsymbol{x}_0 = \begin{bmatrix} 0 & 0 & 0 & 0 \end{bmatrix} \qquad \boldsymbol{x}_1 = \begin{bmatrix} 1 & 0 & 1 & 1 \end{bmatrix}$$
$$\boldsymbol{x}_2 = \begin{bmatrix} 0 & 1 & 1 & 0 \end{bmatrix} \qquad \boldsymbol{x}_3 = \begin{bmatrix} 1 & 1 & 0 & 1 \end{bmatrix}.$$

Assuming again the transmission of the all-zero codeword, we denote by $S_j^{(i)}$ the event where $\boldsymbol{x}_j$ cannot be distinguished from $\boldsymbol{x}_0$ after erasing the ith codeword bit and passing the remaining codeword bits through a BEC with erasure probability ϵ. For $i = 1$, we have

$$\mathsf{H}(X_1 \mid \boldsymbol{Y}_{\sim 1}(\epsilon)) = P\left(S_1^{(1)} \cup S_3^{(1)}\right) \tag{8.14}$$
$$= P\left(S_1^{(1)}\right) + P\left(S_3^{(1)}\right) - P\left(S_1^{(1)} \cap S_3^{(1)}\right) \tag{8.15}$$
$$= 2\epsilon^2 - \epsilon^3, \tag{8.16}$$

where (8.14) follows from the fact that we have to consider only the contribution of codewords that differ from the all-zero one in the ith coordinate. For the case above, since we are dealing with the first coordinate, the only two contributions come from $\boldsymbol{x}_1$ and $\boldsymbol{x}_3$, and where (8.15) is a direct application of the inclusion–exclusion principle. By applying the same technique to all the coordinates, we obtain

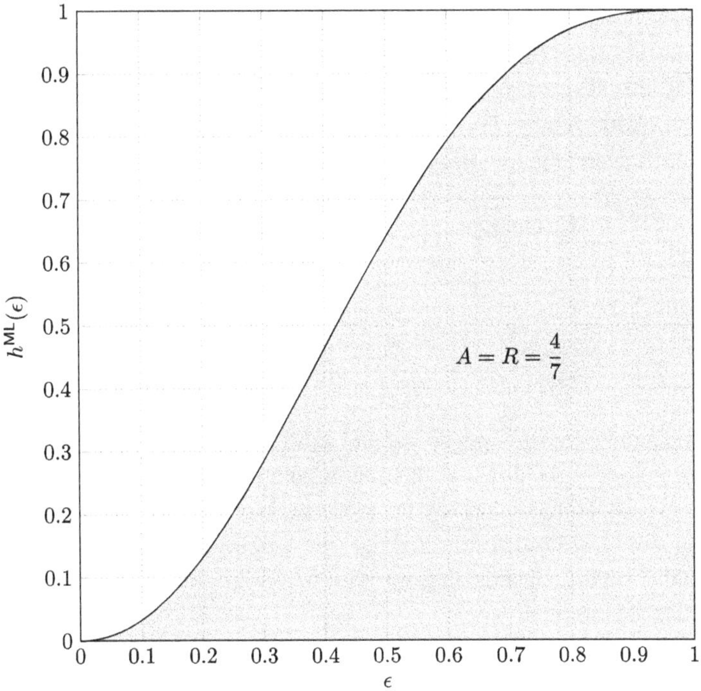

Figure 8.1 EXIT function of the (7,4) Hamming code.

$$h(\epsilon) = \frac{1}{2}\epsilon + \frac{3}{2}\epsilon^2 - \epsilon^3. \tag{8.17}$$

It is easy to verify that the integration of the EXIT function between 0 and 1 provides as result 1/2, that is, the rate of the code.

Example 8.2 (Hamming code) The EXIT function of the rate-4/7 (7,4) Hamming code is

$$h(\epsilon) = 3\epsilon^2 + 4\epsilon^3 - 15\epsilon^4 + 12\epsilon^5 - 3\epsilon^6, \tag{8.18}$$

and it is depicted in Figure 8.1. We have that

$$\int_0^1 h(\epsilon)\mathrm{d}\epsilon = 1 + 1 - 3 + 2 - \frac{3}{7} = \frac{4}{7}. \tag{8.19}$$

The validity of the theorem can be verified for SPC and repetition codes (Exercises 8.1 and 8.2), and it allows proving the converse of the channel coding theorem for LDPC codes (Exercise 8.6, see also Section 7.2.2). While the theorem is stated for binary linear block codes, its validity can be extended to linear block codes over GF(q) operated over the QEC by normalizing the EXIT function as

$$h(\epsilon) = \frac{1}{n \log_2 q} \sum_{i=1}^{n} \mathsf{H}(X_i | \boldsymbol{Y}_{\sim i}(\epsilon)). \tag{8.20}$$

Theorem 8.1 provides a simple method for deriving an upper bound to the threshold under ML decoding for an LDPC ensemble. Hereafter, we denote by $\mathscr{C}_{\lambda,\rho}$ the $\mathscr{C}^n_{\lambda,\rho}$ ensemble, in the $n \to \infty$ limit. For the regular case, we use $\mathscr{C}_{\mathsf{d_v},\mathsf{d_c}}$ to refer to the same asymptotic setting. For binary unstructured LDPC code ensembles $\mathscr{C}_{\lambda,\rho}$ under bit-wise ML decoding, the ML decoding threshold is defined as

$$\epsilon^\star_{\mathsf{ML}} = \sup\left\{\epsilon \in [0,1] : \lim_{n\to\infty} \mathsf{E}_{\mathscr{C}^n_{\lambda,\rho}}\left[P_b^{\mathsf{ML}}(\mathcal{C},\epsilon)\right] \to 0\right\} \tag{8.21}$$

or, equivalently,[1] as

$$\epsilon^\star_{\mathsf{ML}} = \sup\left\{\epsilon \in [0,1] : \lim_{n\to\infty} \mathsf{E}_{\mathscr{C}^n_{\lambda,\rho}}[h(\mathcal{C},\epsilon)] \to 0\right\}, \tag{8.22}$$

where the latter definition is general and valid for nonbinary LDPC code ensembles, too. Unfortunately, the direct evaluation of both $\mathsf{E}_{\mathscr{C}^n_{\lambda,\rho}}\left[P_b^{\mathsf{ML}}(\mathcal{C},\epsilon)\right]$ and $\mathsf{E}_{\mathscr{C}^n_{\lambda,\rho}}[h(\mathcal{C},\epsilon)]$ is tremendously complex already for moderate blocklengths, rendering the use of the definition above of limited practical use. We will hence resort to the derivation of an upper bound on $\epsilon^\star_{\mathsf{ML}}$. To this aim, we need first to introduce the BP EXIT function of an LDPC code ensemble $\mathscr{C}_{\lambda,\rho}$.

Definition 8.1 (BP EXIT function) Consider an LDPC ensemble $\mathscr{C}_{\lambda,\rho}$ and denote by

$$x_0(\epsilon) = \sup\{x \in (0,1) : x = \epsilon\lambda(1-\rho(1-x))\} \tag{8.23}$$

the largest fixed point of the EXIT recursion. The BP EXIT function of the ensemble is defined as

$$h^{\mathsf{BP}}(\epsilon) = \Lambda(1-\rho(1-x_0(\epsilon))). \tag{8.24}$$

The BP EXIT function definition given by (8.24) can be motivated by considering the root VN in the tree introduced in Section 7.1.1 to derive the density evolution recursions of BP decoding. For a given channel erasure probability ϵ, the VNs outbound message average erasure probability converges to $x_0(\epsilon)$, while the inbound message average erasure probability converges to $1-\rho(1-x_0(\epsilon))$. The root VN extrinsic estimate of the corresponding codeword bit will hence be an erasure with probability given by the right-hand side of (8.24). It is obviously true that $h^{\mathsf{BP}}(\epsilon) = 0$ when $\epsilon \le \epsilon^\star_{\mathsf{BP}}$. Moreover, due to the suboptimality of the BP decoder, we have that

$$h^{\mathsf{BP}}(\epsilon) \ge h(\epsilon) \tag{8.25}$$

[1] For binary linear block codes with transmission over the BEC, the equivalence can be established by observing that the error probability for the ith codeword bit under ML decoding can be obtained as $\epsilon\mathsf{H}(X_i|\boldsymbol{Y}_{\sim i})/2$. In fact, $\mathsf{H}(X_i|\boldsymbol{Y}_{\sim i})$ is the probability that X_i is still unknown (erased) after ML decoding, when at the decoder input $Y_i = \mathsf{E}$. The probability that X_i is still erased after ML decoding is thus $\epsilon\mathsf{H}(X_i|\boldsymbol{Y}_{\sim i})$. If, after decoding, the value of X_i is unknown, a decision $\hat{X}_i$ can be drawn by choosing a random bit value, resulting in a bit error probability of $1/2$. It follows that, for any positive ϵ, a vanishing small bit error probability can be achieved if and only if $h(\epsilon)$ vanishes.

for all $\epsilon \in (0,1)$. From (8.25), we have that

$$\int_0^1 h^{\mathsf{BP}}(\epsilon)\mathrm{d}\epsilon \geq \int_0^1 h(\epsilon)\mathrm{d}\epsilon \tag{8.26}$$

$$= \mathrm{R}. \tag{8.27}$$

Therefore, the area under the BP EXIT function is larger than the ensemble design rate R. Here, we implicitly assume that the rate of a random code in the ensemble converges in probability to the nominal rate (we refer to the discussion in Section 4.2.1). A simple upper bound on the ML decoding threshold $\epsilon^{\star}_{\mathsf{ML}}$ follows by introducing the integral of h^{BP} over the interval $(x,1)$

$$B(x) = \int_x^1 h^{\mathsf{BP}}(z)\mathrm{d}z \tag{8.28}$$

and by observing that

$$\epsilon^{\star}_{\mathsf{ML}} \geq B^{-1}(\mathrm{R}). \tag{8.29}$$

Thus, the ML decoding threshold is upper bounded by the erasure probability $\bar{\epsilon}_{\mathsf{ML}}$ satisfying

$$\int_{\bar{\epsilon}_{\mathsf{ML}}}^1 h^{\mathsf{BP}}(\epsilon)\mathrm{d}\epsilon = \mathrm{R}. \tag{8.30}$$

For most ensembles of interest, the threshold upper bound $\bar{\epsilon}_{\mathsf{ML}}$ is tight, that is, it provides the actual threshold value. This is the case, for instance, of regular LDPC code ensembles. Two visual examples are provided in Figure 8.2. The examples refer to the $\mathscr{C}_{3,6}$ and the $\mathscr{C}_{4,8}$ regular ensembles. The ML thresholds (upper bounds) are obtained by vertically cutting the BP EXIT functions at the value of ϵ for which the shaded areas equal the ensemble code rate $1/2$. ML thresholds for several regular LDPC code ensembles are provided in Table 8.1. As intuition suggests, increasingly dense regular ensembles tightly approach the Shannon limit under ML decoding, while suffering for large losses under BP decoding. This observation is confirmed when considering the performance of actual codes drawn from the respective ensembles. Figure 8.3 shows the block error probability of two $(2048, 1024)$ regular LDPC codes with variable node degrees 3 and 4, respectively. The block error probability is estimated via Monte Carlo simulations, under both BP and ML decoding. The results are consistent with the threshold predictions: While under BP decoding the $(3,6)$ LDPC code largely outperforms the $(4,8)$ LDPC code, the latter provides a slightly superior performance than the former one under ML decoding.

Remark 8.1. An underlying assumption in the application of (8.27) is that the design rate of an LDPC ensemble corresponds to the expected asymptotic rate. For regular ensembles, the assumption is easily verified (see Section 4.2.1). Fortunately, for most ensembles of interest the assumption holds true.

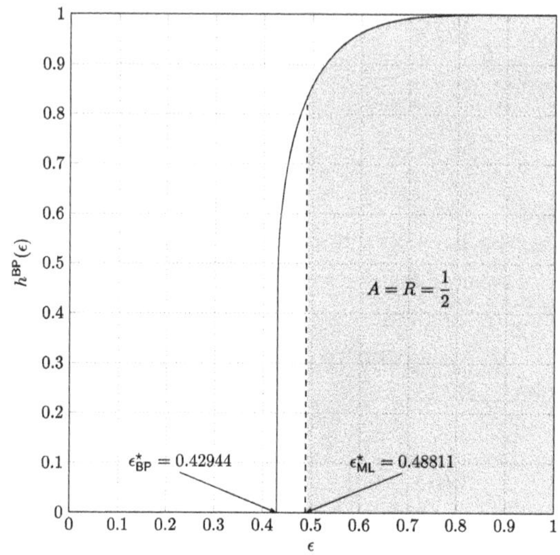

(a) $\mathscr{C}_{3,6}$ ensemble

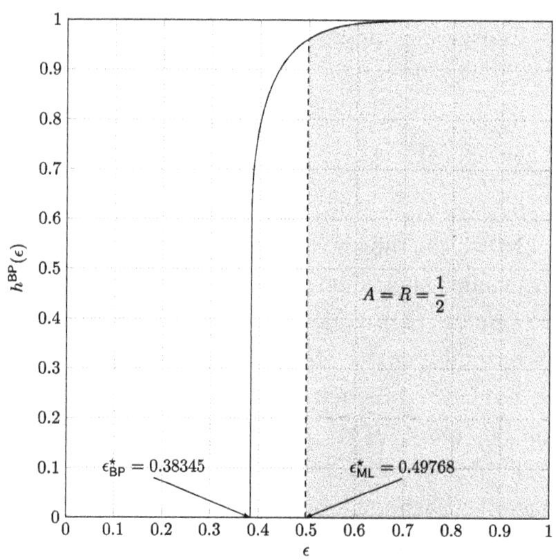

(b) $\mathscr{C}_{4,8}$ ensemble

Figure 8.2 BP EXIT functions with for two regular LDPC ensembles, with the corresponding BP and ML (upper bound) thresholds.

Remark 8.2. The principle used to obtain the ML threshold upper bound of (8.29) is general, and it can be applied to various LDPC code ensembles (including the generalized LDPC code ensembles that will be addressed in Chapter 9) by integrating the corresponding ensemble BP EXIT function.

Table 8.1 Thresholds for various regular LDPC ensembles with rate greater than or equal to 1/2

Ensemble	BP threshold $\epsilon^{\star}_{\mathrm{BP}}$	ML threshold $\epsilon^{\star}_{\mathrm{ML}}$	Shannon limit $1-R$
$\mathscr{C}_{3,6}$	0.42944	0.48811	1/2
$\mathscr{C}_{4,8}$	0.38344	0.49767	1/2
$\mathscr{C}_{5,10}$	0.34155	0.49943	1/2
$\mathscr{C}_{3,9}$	0.28283	0.31961	1/3
$\mathscr{C}_{4,12}$	0.25708	0.33019	1/3
$\mathscr{C}_{5,15}$	0.23032	0.33243	1/3
$\mathscr{C}_{3,12}$	0.21047	0.23732	1/4
$\mathscr{C}_{4,16}$	0.19309	0.24693	1/4
$\mathscr{C}_{5,20}$	0.17356	0.24913	1/4

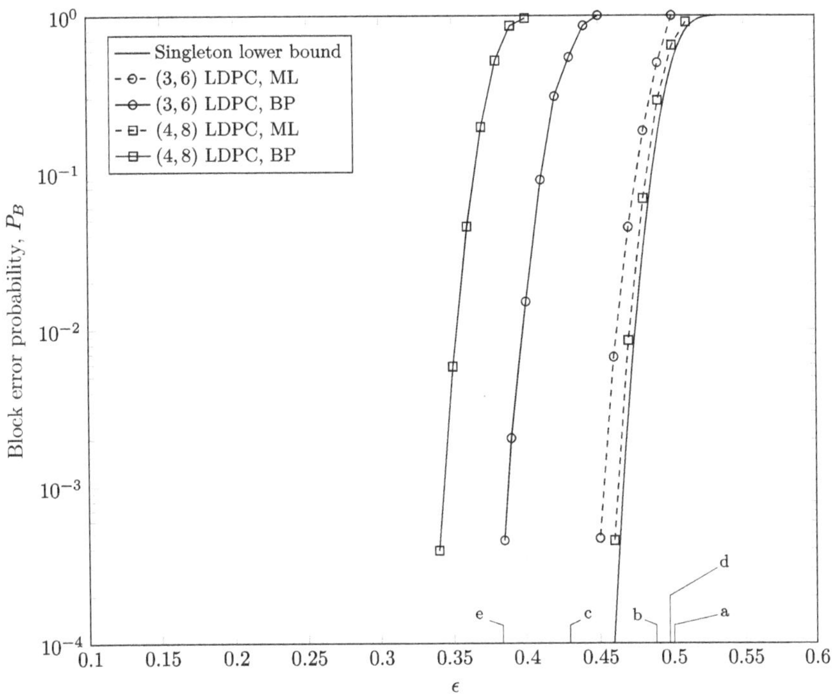

Figure 8.3 Comparison between two (2048, 1024) regular LDPC codes with variable and check node degree pairs (3,6) and (4,8). At the bottom of the chart, the Shannon limit (a), the ML decoding threshold for the (3,6) regular ensemble (b), the BP threshold for the (3,6) regular ensemble (c), the ML decoding threshold for the (4,8) regular ensemble (d), the BP threshold for the (4,8) regular ensemble (e).

8.2 Finite-Length Analysis

The performance of a linear block code under ML decoding, hence in particular of an LDPC code, can be well characterized by applying union bounding techniques, provided the code distance spectrum is known. A tight upper bound on the block error probability is stated by Theorem 8.2 for the general (nonbinary) case.

Theorem 8.2 *Consider an (n, k) linear block code $\mathcal{C}$ defined over* GF(q) *with weight enumerator A_w. The block error probability of $\mathcal{C}$ when used to transmit over a QEC with erasure probability ϵ can be upper bounded as*

$$P_B(\mathcal{C}) \leq \sum_{e=n-k+1}^{n} \binom{n}{e} \epsilon^e (1-\epsilon)^{n-e}$$
$$+ \sum_{e=1}^{n-k} \binom{n}{e} \epsilon^e (1-\epsilon)^{n-e} \min\left\{1, \frac{1}{q-1} \sum_{w=d_{\min}}^{e} \binom{e}{w} \frac{A_w}{\binom{n}{w}}\right\}. \quad (8.31)$$

Proof Let us assume the transmission of the all-zero codeword, denoted by $\boldsymbol{x}_1$. Indicating by $\mathcal{I}_E$ the set of the indexes of the erased codeword symbols at the channel output, we introduce the shorthand $P(\text{error})$ for the probability that an erasure pattern $\mathcal{I}_E$ erases the support of a nonzero codeword of $\mathcal{C}$. Formally,

$$P(\text{error}) = P(\text{supp}(\boldsymbol{x}) \subseteq \mathcal{I}_E \text{ for some } \boldsymbol{x} \in \mathcal{C} \setminus \boldsymbol{x}_1). \quad (8.32)$$

We have that

$$P_B(\mathcal{C}) = \sum_{e=n-k+1}^{n} \binom{n}{e} \epsilon^e (1-\epsilon)^{n-e} + \sum_{e=1}^{n-k} P(|\mathcal{I}_E| = e) P(\text{error}|\, |\mathcal{I}_E| = e) \quad (8.33)$$
$$= \sum_{e=n-k+1}^{n} \binom{n}{e} \epsilon^e (1-\epsilon)^{n-e} + \sum_{e=1}^{n-k} \binom{n}{e} \epsilon^e (1-\epsilon)^{n-e} P(\text{error}|\, |\mathcal{I}_E| = e) \quad (8.34)$$
$$= \sum_{e=n-k+1}^{n} \binom{n}{e} \epsilon^e (1-\epsilon)^{n-e} + \sum_{e=d_{\min}}^{n-k} \binom{n}{e} \epsilon^e (1-\epsilon)^{n-e}$$
$$\times P\left(\bigcup_{\boldsymbol{x} \in \mathcal{C} \setminus \boldsymbol{x}_1} \{\text{supp}(\boldsymbol{x}) \subseteq \mathcal{I}_E\} \Big|\, |\mathcal{I}_E| = e\right). \quad (8.35)$$

Here, (8.33) follows from the observation that the error probability can be decomposed into two contributions: The first contribution accounts for the probability that the number of erasures exceeds the number of linear constraints, whereas the second contribution is code-dependent, as it describes the probability of decoding error when the number of erasures is less than $(n-k)$. The last term in (8.35) can be bounded as

$$P\left(\bigcup_{\boldsymbol{x} \in \mathcal{C} \setminus \boldsymbol{x}_1} \{\text{supp}(\boldsymbol{x}) \subseteq \mathcal{I}_E\} \Big|\, |\mathcal{I}_E| = e\right)$$
$$\leq \min\left\{1, \sum_{\boldsymbol{x} \in \mathcal{C} \setminus \boldsymbol{x}_1} \frac{P(\text{supp}(\boldsymbol{x}) \subseteq \mathcal{I}_E \,|\, |\mathcal{I}_E| = e)}{|\mathcal{S}_{\boldsymbol{x}}|}\right\} \quad (8.36)$$

$$\leq \min\left\{1, \sum_{\boldsymbol{x}\in\mathcal{C}\backslash\boldsymbol{x}_1} \frac{P(\mathrm{supp}(\boldsymbol{x}) \subseteq \mathscr{I}_E \,|\, |\mathscr{I}_E| = e)}{q-1}\right\} \tag{8.37}$$

$$= \min\left\{1, \frac{1}{q-1} \sum_{w=d_{\min}}^{e} \binom{e}{w} \frac{A_w}{\binom{n}{w}}\right\}, \tag{8.38}$$

where in (8.36) $\mathcal{S}_{\boldsymbol{x}}$ is the *support set* of $\boldsymbol{x}$, defined as

$$\mathcal{S}_{\boldsymbol{x}} = \{\boldsymbol{x}' \in \mathcal{C} \,:\, \mathrm{supp}(\boldsymbol{x}') = \mathrm{supp}(\boldsymbol{x})\}, \tag{8.39}$$

and it is the set of codewords in $\mathcal{C}$ having the same support as $\boldsymbol{x}$ (including $\boldsymbol{x}$). Note that (8.36) is a direct application of the union bound, tightened by the observation that, for a linear block code over GF(q), several codewords share the same support. In particular, if $\boldsymbol{x}$ and $\boldsymbol{x}'$ share the same support, then

$$\{\mathrm{supp}(\boldsymbol{x}) \subseteq \mathscr{I}_E\} \cup \{\mathrm{supp}(\boldsymbol{x}') \subseteq \mathscr{I}_E\} = \{\mathrm{supp}(\boldsymbol{x}) \subseteq \mathscr{I}_E\}\,. \tag{8.40}$$

Furthermore, (8.37) stems from the following observation: For any nonzero codeword $\boldsymbol{x}$, we have $|\mathcal{S}_{\boldsymbol{x}}| \geq q-1$, where $|\mathcal{S}_{\boldsymbol{x}}|$ is the cardinality of $\mathcal{S}_{\boldsymbol{x}}$. Finally, (8.38) follows from

$$P(\mathrm{supp}(\boldsymbol{x}) \subseteq \mathscr{I}_E \,|\, |\mathscr{I}_E| = e) = P(\mathrm{supp}(\boldsymbol{x}') \subseteq \mathscr{I}_E \,|\, |\mathscr{I}_E| = e) \tag{8.41}$$

for any two codewords $\boldsymbol{x}$ and $\boldsymbol{x}'$ with the same Hamming weight, and by the fact that (by fixing the Hamming weight of $\boldsymbol{x}$ to w), the probability that the support of a length-n weight-w vector is punctured by an erasure pattern with e erasures is $\binom{e}{w}/\binom{n}{w}$. □

Theorem 8.2 can be easily extended to code ensembles, yielding the following results for generator and parity-check–based ensembles.

Theorem 8.3 *Consider an ensemble $\mathscr{C}$ of (n,k) linear block codes over* GF(q) *and a random code $\mathsf{C} \in \mathscr{C}$ with uniform distribution in $\mathscr{C}$. Then, the expected block error probability when C is used to transmit over a QEC with erasure probability ϵ can be upper bounded as*

$$P_B(\mathscr{C}) = \mathsf{E}[P_B(\mathsf{C})] \leq \sum_{e=n-k+1}^{n} \binom{n}{e} \epsilon^e (1-\epsilon)^{n-e}$$
$$+ \sum_{e=1}^{n-k} \binom{n}{e} \epsilon^e (1-\epsilon)^{n-e} \min\left\{1, \frac{1}{q-1} \sum_{w=1}^{e} \binom{e}{w} \frac{\bar{A}_w}{\binom{n}{w}}\right\}, \tag{8.42}$$

where $\bar{A}_w$ is the average weight enumerator of C.

Theorem 8.4 *Let $\mathscr{C}$ be an ensemble of linear block codes over* GF(q) *defined by $n \times m$ parity-check matrices. Consider a random code $\mathsf{C} \in \mathscr{C}$ with uniform distribution in $\mathscr{C}$. Then, the expected block error probability when C is used to transmit over a QEC with erasure probability ϵ can be upper bounded as in* (8.42) *by replacing $n-k$ with m.*

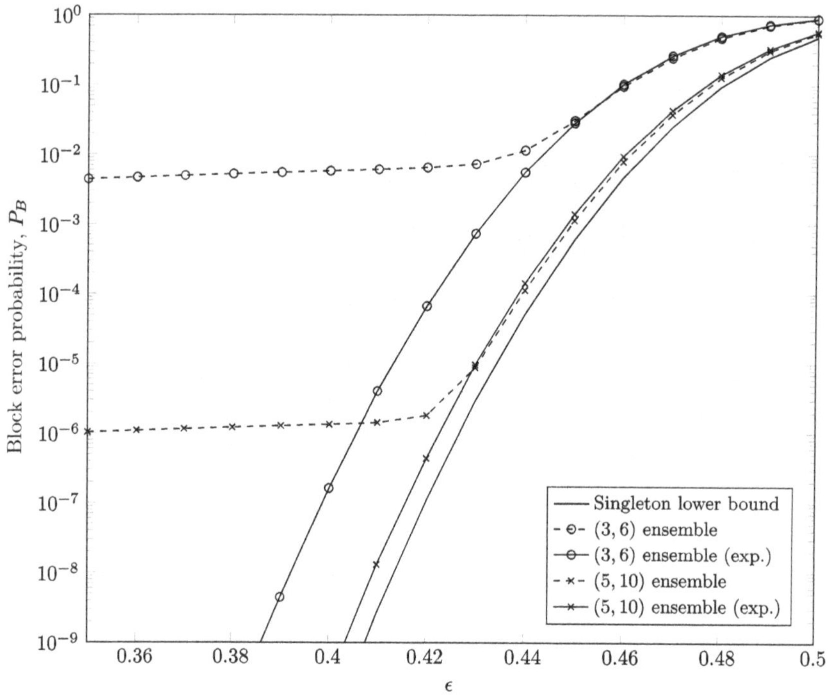

Figure 8.4 Union bounds according to Theorem 8.4 for two regular LDPC code ensembles. The result for the expurgated ensembles is provided too. Block length $n = 1024$.

The proof of Theorems 8.3, 8.4 follows by the linearity of the expectation (see Exercises 8.8 and 8.9). Remarkably, by applying Theorem 8.4 to the linear random ensembles $\mathscr{H}^q_{n,m}$ (from Definition 3.9), we obtain Berlekamp's random coding bound in its tighter version (3.42).

An example of the application of the upper bound in (8.42) is provided in Figure 8.4 for two binary regular LDPC ensembles with blocklength $n = 1024$. In particular, the analysis is applied to the (3,6) and the (5,10) ensembles, as well as to their expurgated versions (see Definition 5.9). The nonexpurgated ensembles performance is dominated, at low erasure probabilities, by a few codes with bad minimum distance properties. By expurgating the ensembles, the upper bounds show a dramatic improvement of the block error probability at low erasure probabilities.

8.3 Code Design for Maximum-Likelihood Decoders

The LDPC code design problem under BP decoding usually starts by choosing an ensemble with BP decoding threshold as close as possible to the Shannon limit. In the ML decoding case, one may obviously proceed in a similar way by replacing the BP decoding threshold with the ML one. It turns out that, when adopting this simple

Table 8.2 Thresholds and average number of pivots (random pivoting) for regular LDPC code ensembles

Ensemble	BP threshold	ML threshold	$\epsilon = 0.48$	$\epsilon = 0.45$	$\epsilon = 0.40$
$\mathscr{C}_{3,6}$	0.4294	0.4881	78.3	32.4	0.07
$\mathscr{C}_{4,8}$	0.3834	0.4977	178.7	119.4	31.8
$\mathscr{C}_{5,10}$	0.3416	0.4994	267.4	208.1	112.2

The average number of pivots is for $n = 2\,048$.

criterion, the ensemble search results to be a fairly simple task: It is in fact sufficient to select a regular LDPC ensemble with sufficiently high variable and check node degrees to attain a near-optimal ML threshold (the results provided in Table 8.1 evidently confirm this claim). While this simple criterion results to be very effective from a threshold perspective, it hides a drawback in terms of decoding complexity. Regular ensembles characterized by large node degrees yield, under the efficient ML decoding algorithm described in Section 6.2, large pivot counts. The number of pivots is a major driver of the decoding complexity, since the (dense) system of equation solved in the final step of the structured Gaussian elimination method has as many unknowns as the number of pivots. Examples of the average numbers of pivots for some regular LDPC code ensembles are shown in Table 8.2. The results have been obtained under random pivoting (see Section 6.2.1). The (3, 6) regular ensemble attains by far the smallest average pivot counts.

A consequence of this result is that the choice of an LDPC code ensemble just based on the ML decoding threshold may yield codes with impractically high ML decoding complexity. Fortunately, a simple observation may be used to construct ensembles which attain simultaneously good ML decoding thresholds and manageable pivot counts. We start from the following intuition. For a given LDPC ensemble and for n large, when $\epsilon < \epsilon^{\star}_{\mathsf{BP}}$ we may expect to have (on average) a small number of pivots, since BP decoding would succeed with high probability. Assume now to operate over an erasure channel with $\epsilon^{\star}_{\mathsf{BP}} < \epsilon < \epsilon^{\star}_{\mathsf{ML}}$. BP decoding would often fail due to a number of erasures that exceeds $n\epsilon^{\star}_{\mathsf{BP}}$. Assume now that a genie reveals to the BP decoder the value of a few (randomly chosen) erased symbols, such that the BP decoder faces a new virtual erasure channel with a reduced erasure probability $\epsilon' < \epsilon^{\star}_{\mathsf{BP}}$. Thanks to the help of the genie, the BP decoder will succeed with high probability. When structured Gaussian elimination is applied, we may regard the pivots as symbols whose value will be revealed to us by a genie. The genie, represented here by the final solution of the pivot values, may fail in providing the solution. However, for the pure analysis of the pivot count, we shall consider the pivots as erased symbols whose value will be made available to the BP. It is obvious that, the closer the BP and the ML thresholds are, the fewer symbols have to be revealed by the genie when the channel erasure probability lies between the two thresholds. A first rough estimate of the expected number of pivots $\bar{\xi}$ required by a length-n code from a given ensemble, over an erasure channel with erasure probability $\epsilon^{\star}_{\mathsf{BP}} < \epsilon < \epsilon^{\star}_{\mathsf{ML}}$, can be found as

$$\bar{\xi} \approx n(\epsilon - \epsilon^{\star}_{\mathsf{BP}}). \tag{8.43}$$

Algorithm 3: LDPC ensemble design for ML decoding

1 Initialize
  a. The design rate R;
  b. The minimum and the maximum allowed variable node degrees;
  c. The minimum allowed critical fractional codeword weight;
  d. The maximum allowed gap ν between $\epsilon^{\star}_{\mathsf{ML}}$ and the Shannon limit $1 - \mathsf{R}$.

2 Choose an initial value for the maximum check node degree d_c.
3 For the current value of d_c, find a degree distribution pair λ, ρ with BP decoding threshold as close as possible to the Shannon limit, under the initialization constraints (a,b,c).
4 Compute the ML threshold associated to the degree distribution found at step 3.
5 If $(1 - \mathsf{R}) - \epsilon^{\star}_{\mathsf{ML}} \leq \nu$ exit. Otherwise, increment d_c by 1 and go to step 3.

We sketch Algorithm 3 as an example of LDPC code ensemble design which takes into account both the ML threshold and the BP threshold as a proxy to control the decoding complexity under structured Gaussian elimination.

To address the error floor performance, Algorithm 3 may be modified to take into account constraints on the degree distribution stemming from requirements on the ensemble critical fractional codeword weight. Once the ensemble has been optimized, the finite length code construction continues as usual, that is, by applying girth optimization techniques to the design of the parity-check matrix.

Example 8.3 Algorithm 3 has been used to design a (2048, 1024) LDPC code ensemble. The obtained degree distribution pair is

$$\begin{aligned} \lambda(x) &= 0.111111x + 0.430013x^2 + 0.279514x^{13} \\ &\quad + 0.120768x^{52} + 0.0585938x^{53} \\ \rho(x) &= x^8. \end{aligned} \tag{8.44}$$

This ensemble is characterized by $\epsilon^{\star}_{\mathsf{BP}} = 0.4792$ and $\epsilon^{\star}_{\mathsf{ML}} = 0.4988$. It was obtained by requiring a strictly positive critical fractional codeword weight, without imposing any specific positive lower bound on it. The critical fractional codeword weight is $6.6 \cdot 10^{-4}$. The performance of a code from the ensemble is compared with the one of a (3,6) regular LDPC code under BP and ML decoding in Figure 8.5. The parity-check matrices of both codes have been designed by means of the PEG algorithm. At an erasure probability of 0.48, ML decoding of the irregular code required on average $\bar{\xi} \approx 62.5$ pivots, while for the (3,6) code the average number of pivots have been estimated (via a Monte Carlo simulation) to be $\bar{\xi} \approx 78.3$.

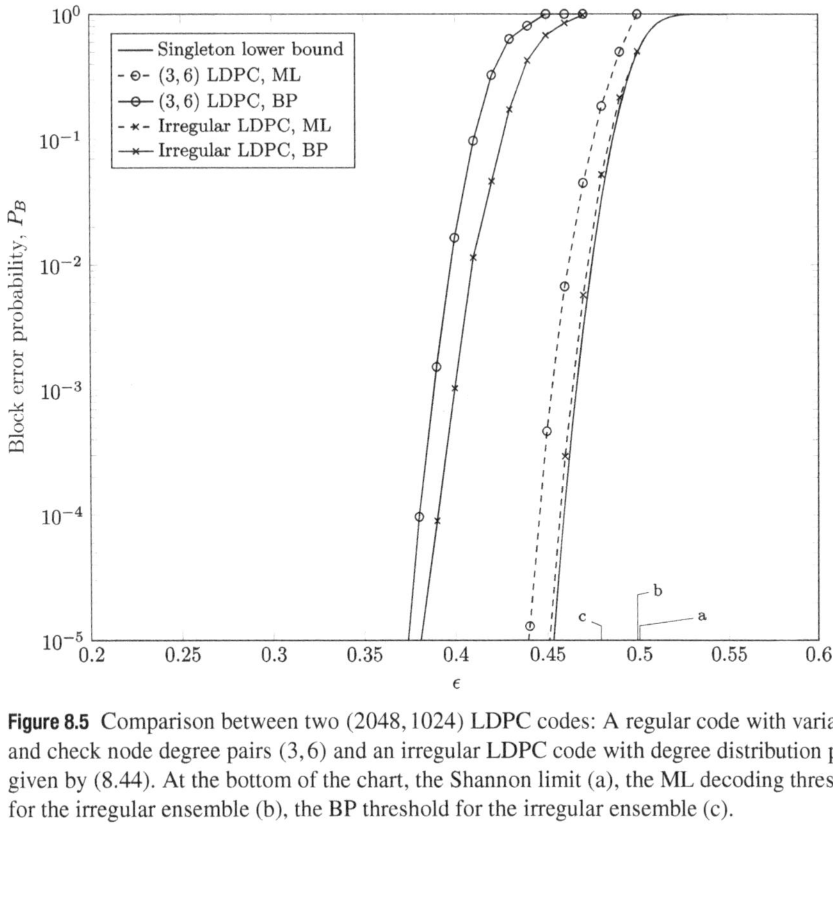

Figure 8.5 Comparison between two (2048, 1024) LDPC codes: A regular code with variable and check node degree pairs (3,6) and an irregular LDPC code with degree distribution pairs given by (8.44). At the bottom of the chart, the Shannon limit (a), the ML decoding threshold for the irregular ensemble (b), the BP threshold for the irregular ensemble (c).

8.4 Code Design for Channels with Memory

We introduce next code design strategies for erasure channels with memory, namely, for the SBUEC, the BLEC, and the GEC.

8.4.1 Single Burst Erasure Channels

Consider next the transmission with a given (n,k) LDPC code over an acyclic SBUEC. A possible way to optimize the code MTBL follows from the greedy approach presented in Section 7.6.1 with the subtlety of testing a given symbol permutation with an ML decoder. However, as soon as ML decoding is employed, it is possible to show that an optimum MTBL can be attained by means of a constructive approach under relatively mild conditions on the code structure. To show this, we focus on the simple case of an (n,k) binary linear code with $n-k$ that divides n. The code parity-check matrix can be partitioned as

$$\boldsymbol{H} = [\boldsymbol{H}_1 \quad \boldsymbol{H}_2 \quad \cdots \quad \boldsymbol{H}_N] \tag{8.45}$$

with $\boldsymbol{H}_\ell$, $\ell = 1,\ldots,N$ being $(n-k)\times(n-k)$ matrices. We will provide next a sequence of $N-1$ alternative parity-check matrices $\boldsymbol{H}^{(1)},\boldsymbol{H}^{(2)},\ldots,\boldsymbol{H}^{(N-1)}$ for the code, which can be used to form an $n \times n$ parity-check matrix with redundant rows (see Section 2.3) whose structure gives an immediate proof of the optimality in terms of MTBL.

A first requirement to achieve the result is to have the code parity-check matrix in the form (8.45), where each submatrix $\boldsymbol{H}_\ell$ is nonsingular. To meet the requirement, a permutation of the symbols in the codeword may be required, yielding a first contribution to the final interleaver design. While the requirement is quite easily matched by codes drawn from the random parity-check code ensemble, it is not obvious to state that a code drawn from a given LDPC code ensemble satisfies the wished property with high probability. However, as we will see, specific LDPC code constructions may facilitate the task of obtaining a parity-check matrix in the required form. The construction of the ℓth alternative matrix $\boldsymbol{H}^{(\ell)}$ proceeds as follows. Let us denote by $\boldsymbol{H}_\ell^{-1}$ the inverse of the ℓth submatrix in (8.45). An intermediate alternative parity-check matrix is obtained as

$$\boldsymbol{H}' = \boldsymbol{H}_\ell^{-1}\boldsymbol{H} \tag{8.46}$$

$$= \begin{bmatrix}\boldsymbol{H}'_1 & \boldsymbol{H}'_2 & \cdots & \boldsymbol{H}'_{\ell-1} & \boldsymbol{I} & \boldsymbol{H}'_{\ell+1} & \cdots & \boldsymbol{H}'_N\end{bmatrix}. \tag{8.47}$$

An additional requirement now comes into play: The submatrix $\boldsymbol{H}'_{\ell+1}$ must be transformable into an upper triangular form by row operations, with row additions allowed from top to bottom only. To achieve the result, permutations of the columns of $\boldsymbol{H}'_{\ell+1}$ are allowed (it easy to see that if $\boldsymbol{H}'_{\ell+1}$ is full rank then a permutation exists that allows the transformation), giving rise to a second contribution to the final symbol interleaver. Denoting by $\boldsymbol{U}^{(\ell)}$ the upper triangular form of $\boldsymbol{H}'_{\ell+1}$, the requirement implies that the linear transformation applied to (the suitably permuted version of) $\boldsymbol{H}'_{\ell+1}$ is defined by a lower-triangular matrix $\boldsymbol{L}^{(\ell)}$, that is,

$$\boldsymbol{U}^{(\ell)} = \boldsymbol{L}^{(\ell)}\boldsymbol{H}'_{\ell+1}. \tag{8.48}$$

When the transformation is applied to the intermediate matrix $\boldsymbol{H}'$, one obtains

$$\boldsymbol{H}^{(\ell)} = \boldsymbol{L}^{\ell}\boldsymbol{H}' \tag{8.49}$$

$$= \begin{bmatrix}\boldsymbol{H}_1^{(\ell)} & \boldsymbol{H}_2^{(\ell)} & \cdots & \boldsymbol{H}_{\ell-1}^{(\ell)} & \boldsymbol{L}^{(\ell)} & \boldsymbol{U}^{(\ell)} & \cdots & \boldsymbol{H}_N^{(\ell)}\end{bmatrix}. \tag{8.50}$$

The process is iterated for $\ell = 1,\ldots,N-1$. The final parity-check matrix (with redundant rows) is obtained by stacking the alternative parity-check matrices as

$$\boldsymbol{H}_{\text{BCF}} = \begin{bmatrix}\boldsymbol{H}^{(1)} \\ \boldsymbol{H}^{(2)} \\ \vdots \\ \boldsymbol{H}^{(N-1)}\end{bmatrix} \tag{8.51}$$

$$= \begin{bmatrix}\boldsymbol{L}^{(1)} & \boldsymbol{U}^{(1)} & \boldsymbol{H}_3^{(1)} & \cdots & \boldsymbol{H}_{N-1}^{(1)} & \boldsymbol{H}_N^{(1)} \\ \boldsymbol{H}_1^{(2)} & \boldsymbol{L}^{(2)} & \boldsymbol{U}^{(2)} & \cdots & \boldsymbol{H}_{N-1}^{(2)} & \boldsymbol{H}_N^{(2)} \\ \vdots & \vdots & \vdots & \ddots & \vdots & \vdots \\ \boldsymbol{H}_1^{(N-1)} & \boldsymbol{H}_2^{(N-1)} & \boldsymbol{H}_3^{(N-1)} & \cdots & \boldsymbol{L}^{(N-1)} & \boldsymbol{U}^{(N-1)}\end{bmatrix}. \tag{8.52}$$

The matrix $\boldsymbol{H}_{\mathrm{BCF}}$ is referred to as burst correction form parity-check matrix. By direct inspection of (8.52), it is possible to verify that any single (acyclic) erasure burst consisting of $n-k$ or less consecutive erasures can be recovered by exploiting the burst correction form parity-check matrix. BP decoding over the Tanner graph associated with (8.52) is sufficient to resolve the erasure burst. In fact, by denoting with i the initial position of an erasure burst of length L, it is sufficient to extract from $\boldsymbol{H}_{\mathrm{BCF}}$ the submatrix consisting of the L consecutive rows starting with index i to obtain a system of equations that can be solved by back substitution, as soon as L does not exceed $n-k$.

Example 8.4 Consider the $(8,4)$ binary linear block code with parity-check matrix given by

$$\boldsymbol{H} = \left[\begin{array}{cccc|cccc} 1 & 0 & 1 & 0 & 1 & 0 & 0 & 0 \\ 1 & 1 & 0 & 1 & 0 & 1 & 0 & 0 \\ 1 & 1 & 1 & 0 & 0 & 0 & 1 & 0 \\ 1 & 0 & 1 & 1 & 0 & 0 & 0 & 1 \end{array}\right]. \tag{8.53}$$

Observe that the left and the right submatrices of $\boldsymbol{H}$ are both full-rank. We proceed with the method illustrated in Section 8.4.1 to identify the permutation of the code coordinates that yields the maximum MTBL. By multiplying (8.53) by the inverse of the left submatrix, we obtain the form of (8.47) as

$$\boldsymbol{H}' = \left[\begin{array}{cccc|cccc} 1 & 0 & 0 & 0 & 0 & 1 & 1 & 1 \\ 0 & 1 & 0 & 0 & 1 & 0 & 1 & 0 \\ 0 & 0 & 1 & 0 & 1 & 1 & 1 & 1 \\ 0 & 0 & 0 & 1 & 1 & 0 & 0 & 1 \end{array}\right]. \tag{8.54}$$

The right submatrix of $\boldsymbol{H}'$ should be now put in upper triangular form by row operations, with row additions allowed from top to bottom only. It is easy to observe that it is not possible to perform this transformation without permuting some columns of the submatrix. A suitable permutation that yields a valid solution is obtained by swapping the fifth and the sixth columns of $\boldsymbol{H}'$. By introducing a slight abuse of notation, we keep referring to the permuted version of $\boldsymbol{H}'$ as $\boldsymbol{H}'$, which now is

$$\boldsymbol{H}' = \left[\begin{array}{cccc|cccc} 1 & 0 & 0 & 0 & 1 & 0 & 1 & 1 \\ 0 & 1 & 0 & 0 & 0 & 1 & 1 & 0 \\ 0 & 0 & 1 & 0 & 1 & 1 & 1 & 1 \\ 0 & 0 & 0 & 1 & 0 & 1 & 0 & 1 \end{array}\right]. \tag{8.55}$$

The (single) alternative parity-check matrix is hence given by

$$\boldsymbol{H}^{(1)} = \boldsymbol{L}^{(1)}\boldsymbol{H}' \tag{8.56}$$

with

$$\boldsymbol{L}^{(1)} = \left[\begin{array}{cccc} 1 & 0 & 0 & 0 \\ 0 & 1 & 0 & 0 \\ 1 & 1 & 1 & 0 \\ 1 & 0 & 1 & 1 \end{array}\right], \tag{8.57}$$

which gives

$$
\boldsymbol{H}^{(1)} = \left[\begin{array}{cccc|cccc} 1 & 0 & 0 & 0 & 1 & 0 & 1 & 0 \\ 0 & 1 & 0 & 0 & 0 & 1 & 1 & 1 \\ 1 & 1 & 1 & 0 & 0 & 0 & 1 & 0 \\ 1 & 0 & 1 & 1 & 0 & 0 & 0 & 1 \end{array}\right]. \tag{8.58}
$$

Since in this example $N = 2$, (8.58) turns to be the burst correction form of the parity-check matrix.

We emphasize that the method can be applied in two equivalent ways. One may store the burst correction form of the code parity-check matrix (after applying the necessary permutation of the code coordinates) and use it directly to obtain the system of equations in triangular form for any erasure burst of $n - k$ or less erasures. Alternatively, the method can be used to derive the optimal interleaving rule and then apply the general (efficient) ML decoding methods discussed in Section 6.2 over the suitably permuted LDPC code parity-check matrix (being ML decoding optimum, the guaranteed MTBL is achieved also in this case).

Example 8.5 Following Example 8.4, we can observe that the largest possible MTBL is achieved by permuting the fifth and the sixth columns of $\boldsymbol{H}$, that is, by applying the permutation described by the matrix

$$
\boldsymbol{\Pi} = \begin{bmatrix} 1 & 0 & 0 & 0 & 0 & 0 & 0 & 0 \\ 0 & 1 & 0 & 0 & 0 & 0 & 0 & 0 \\ 0 & 0 & 1 & 0 & 0 & 0 & 0 & 0 \\ 0 & 0 & 0 & 1 & 0 & 0 & 0 & 0 \\ 0 & 0 & 0 & 0 & 0 & 1 & 0 & 0 \\ 0 & 0 & 0 & 0 & 1 & 0 & 0 & 0 \\ 0 & 0 & 0 & 0 & 0 & 0 & 1 & 0 \\ 0 & 0 & 0 & 0 & 0 & 0 & 0 & 1 \end{bmatrix} \tag{8.59}
$$

to obtain the parity-check matrix of an equivalent code in the form $\boldsymbol{H}\boldsymbol{\Pi}$.

As we will see next, while this approach gives the best possible permutation for a SBUEC, we must avoid claiming any optimality for more sophisticated correlated erasure channels.

8.4.2 Block and Gilbert Erasure Channels

When transmitting over a correlated erasure channel, one may be tempted of using the MTBL as a proxy to estimate the resilience of the code with respect to bursty erasures. While the approach is reasonable for highly correlated channels, the blind use of an MTBL-driven design can yield fairly suboptimum results in the most general case, as anticipated in Section 7.6. This is especially true under the MTBL maximization strategy outlined in Section 8.4.1. In fact, the greedy optimization algorithm discussed in Section 7.6.3 tends to randomize column permutations, yielding an inherent robustness with respect to general bursty erasure channel models. On the contrary, the technique described in Section 8.4.1 may leverage on structured parity-check matrices to achieve the optimum MTBL, limiting the amount of permutations. This may come at the expense of a reduced robustness to multiple erasure bursts. To illustrate this, we show next the example of an LDPC code (designed to achieve the maximum possible MTBL), when used to transmit over various BLECs and GECs. For BLECs, the exact code block error probability can be easily obtained when L is moderate-small (e.g., $L \leq 20$). To show the result, we first provide some relevant definitions.

Definition 8.2 (Block cover) Consider an (n, k) linear block code where each codeword is partitioned into L sub-blocks of equal length. A block cover is a set of subblocks that contains the entire support of a codeword.

Definition 8.3 (Block cover enumerator) The block cover enumerator of a code is the number B_w of cardinality-w block covers.

Example 8.6 Consider a (6, 3) binary linear block code with generator matrix

$$\boldsymbol{G} = \begin{bmatrix} 1 & 1 & 0 & 1 & 0 & 0 \\ 1 & 0 & 1 & 0 & 1 & 0 \\ 1 & 1 & 1 & 0 & 0 & 1 \end{bmatrix}. \tag{8.60}$$

The set of codewords is given by the rows of the following array

$$\begin{array}{cc|cc|cc} 0 & 0 & 0 & 0 & 0 & 0 \\ 1 & 1 & 0 & 1 & 0 & 0 \\ 1 & 0 & 1 & 0 & 1 & 0 \\ 1 & 1 & 1 & 0 & 0 & 1 \\ 0 & 1 & 1 & 1 & 1 & 0 \\ 0 & 0 & 1 & 1 & 0 & 1 \\ 0 & 1 & 0 & 0 & 1 & 1 \\ 1 & 0 & 0 & 1 & 1 & 1 \end{array}$$

where the codewords have been partitioned into $L = 3$ blocks of two bits each. We denote by $\mathcal{B} = \{b_1, b_2, b_3\}$ the set of blocks. There are three block covers with cardinality 2 that we denote as $\mathcal{B}_{12} = \{b_1, b_2\}$, $\mathcal{B}_{13} = \{b_1, b_3\}$, and $\mathcal{B}_{23} = \{b_2, b_3\}$. In fact, $\mathcal{B}_{12}$ contains the entire support of the second codeword from the top, $\mathcal{B}_{13}$ contains the entire support of the second-to-last codeword, and $\mathcal{B}_{23}$ contains the entire support of the sixth codeword from the top. Additionally, there is one block cover with cardinality 3, which coincides with $\mathcal{B}$. Thus, $B_2 = 3$ and $B_3 = 1$.

Based on Definitions 8.2 and 8.3, we can express the block error probability over an L-blocks BLEC with erasure probability ϵ as

$$P_B = \sum_{w=1}^{L} \binom{L}{w} \epsilon^w (1 - \epsilon)^{L-w} B_w. \tag{8.61}$$

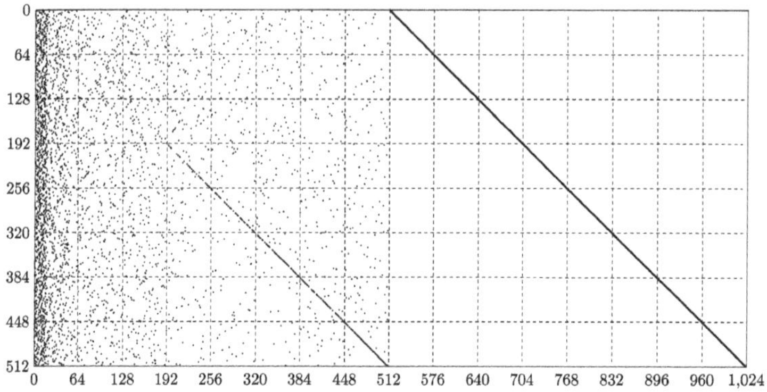

Figure 8.6 Parity-check matrix of the IRA code from the Example 8.7, already permuted according to the interleaving rule provided by the burst correcting form parity-check matrix construction.

The derivation of the block cover enumerator can be obtained by brute-force search when L is moderate or small. For example, one may enumerate all the 2^L possible realizations of an L-blocks BLEC. By performing ML decoding for each realization, it is then possible to count how many realizations with exactly w erased blocks yield a decoding failure.

Example 8.7 Consider now the example of an $(1024, 512)$ binary IRA code. IRA codes are natural candidates for the method discussed in Section 8.4.1 since the rightmost $(n-k) \times (n-k)$ submatrix of the parity-check matrix is nonsingular by construction. In our example, to enable the achievement of the optimum MTBL we must guarantee that the leftmost $(n-k) \times k$ submatrix of the parity-check matrix has rank k (observe that, for the given code parameters, $k = n - k$). This requirement can be fulfilled only if the variable nodes associated to the k rightmost columns do not all possess an even degree. As soon as this is ensured by the choice of an appropriate degree distribution, the construction of the right part of the parity-check matrix can be performed by trial and error, till the requirement on the rank is satisfied. As an example, the process has been applied to an IRA code with edge-oriented degree distribution pairs given by

$$\begin{aligned} \psi(x) &= 0.277902x^2 + 0.037109x^6 + 0.247768x^7 + 0.065290x^{17} \\ &\quad + 0.100725x^{18} + 0.271205x^{53} \\ \phi(x) &= 0.015067x^5 + 0.964844x^6 + 0.020089x^7 . \end{aligned} \tag{8.62}$$

The ensemble possesses a BP decoding threshold $\epsilon^\star_{\mathsf{BP}} = 0.4954$ and a ML decoding threshold $\epsilon^\star_{\mathsf{ML}} = 0.4986$. The code parity-check matrix is depicted in Figure 8.6. Figures 8.7 and 8.8 report the performance of the code on a BLEC with $L = 4$ and $L = 8$ for the code after permuting the columns to achieve the maximum possible MTBL (i.e., achieving $L_{\max} = 512$) and after applying a random columns

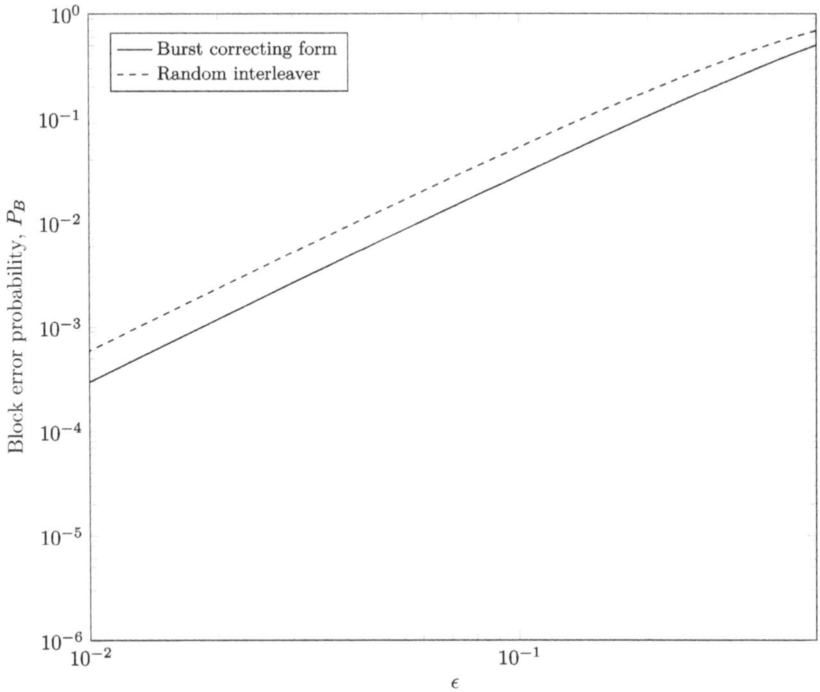

Figure 8.7 Block error probability for the IRA code from Example 8.7 over the BLEC with $L = 4$. The result is provided for the matrix permuted according to the interleaving rule provided by the burst correcting form parity-check matrix construction, as well as for random interleaving.

permutation (in this case a slightly suboptimum MTBL is achieved, with $L_{\text{max}} = 503$). For the case of $L = 4$, the block cover enumerator for the case of interleaving with maximum MTBL is given by

$$B_2 = 3 \qquad B_3 = 4 \qquad B_4 = 1,$$

while for the specific random interleaver generated in this example,

$$B_2 = 6 \qquad B_3 = 4 \qquad B_4 = 1.$$

From inspection of Figure 8.7, one can observe how the interleaving strategies provide a similar performance, and the two curves follow the same slope as the erasure probability grows (this is due to the fact both interleavers yield, in our example, a minimum block cover cardinality equal to 2). For the case of $L = 8$, the (truncated) block cover enumerator for the case of interleaving with maximum MTBL is given by

$$B_2 = 1 \qquad B_3 = 45 \qquad B_4 = 453 \qquad B_5 = 2151 \qquad \ldots,$$

while for the specific random interleaver generated in this example,

$$B_7 = 1 \qquad B_8 = 11958 \qquad B_9 = 11440 \qquad B_{10} = 8008 \qquad \ldots.$$

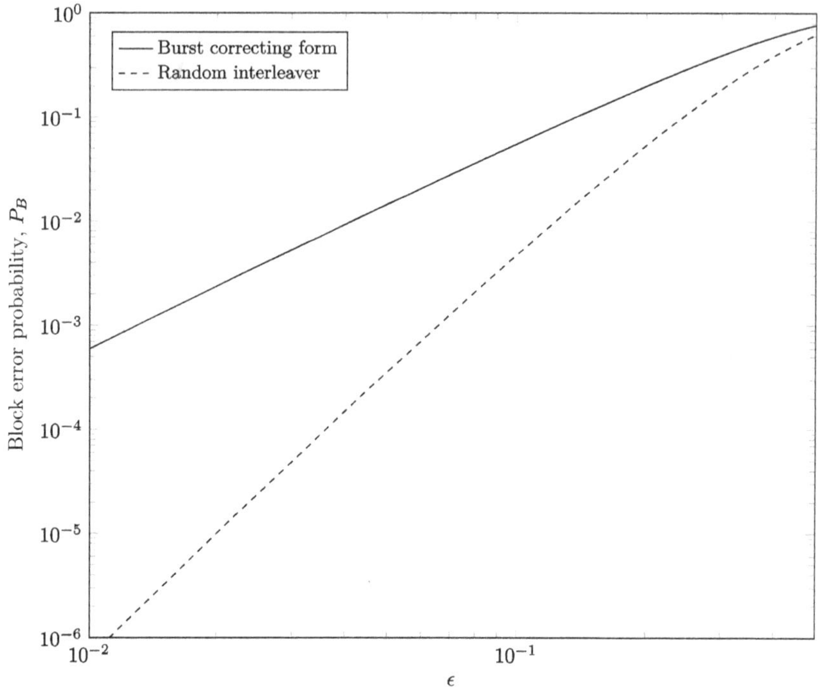

Figure 8.8 Block error probability for the IRA code from Example 8.7 over the BLEC with $L = 8$. The result is provided for the matrix permuted according to the interleaving rule provided by the burst correcting form parity-check matrix construction, as well as for random interleaving.

It is immediately visible that the MTBL maximizing interleaving strategy yields, in this case, a particularly small minimum block cover cardinality (equal to 2), whereas for the case of random interleaving the minimum block cover cardinality is fairly large (it equals 7). The gap shows with full evidence in Figure 8.8: The MTBL maximizing interleaving strategy performs remarkably worse than the random interleaving strategy. For further evidence, the performance of the various interleaving strategies is provided on two GECs, one characterized by a rather long average bad state duration $\delta = 100$ (Figure 8.9) and one characterized by a lower correlation, with $\delta = 20$ (Figure 8.10). As for the BLEC case, the MTBL maximizing interleaving strategy guarantees a relatively robust performance only in presence of strong channel correlation.

Example 8.7 provides an insightful case study and allows to claim that maximizing a code's MTBL is not sufficient for cope with more sophisticated types of bursty erasure channels. In fact, while the behavior observed in Example 8.7 may be considered a special case that stems from the specific structure of IRA codes (and to the specific random interleaver that was generated as competitor), it is still enough to warn engineers from using the MTBL as design criterion over bursty erasure channels. While

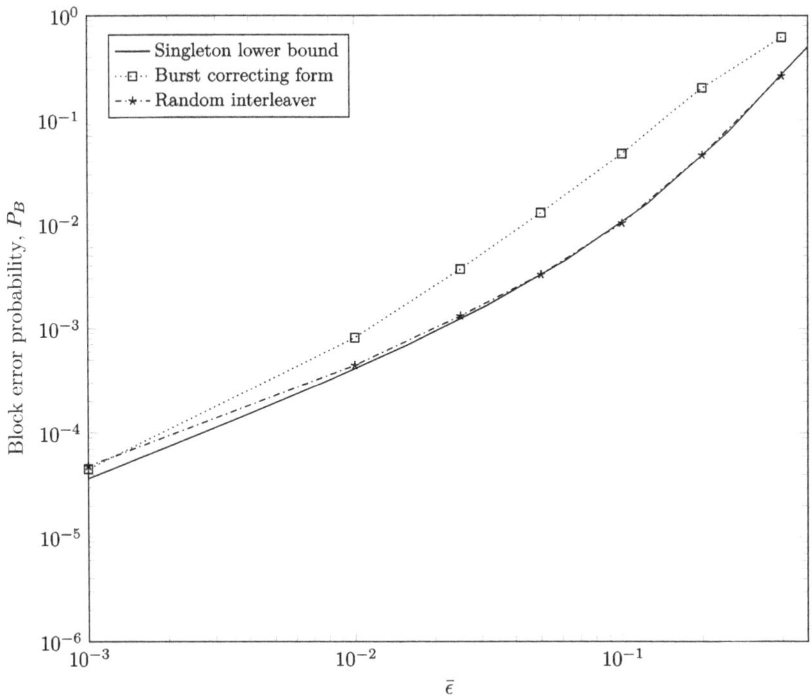

Figure 8.9 Block error probability for the IRA code from the Example 8.7 over the GEC with $\delta = 100$. The result is provided for the matrix permuted according to the interleaving rule provided by the burst correcting form parity-check matrix construction, as well as for random interleaving.

the approach may succeed for highly correlated erasure channel (as for the BLEC with $L = 4$), it may fail on erasure channels featuring faster variations (as for the BLEC with $L = 8$).

8.5 Historical Notes

The Area Theorem was originally introduced in [1]. The proof provided in Section 8.1 follows the derivation proposed in [2] (see also [3, Chapter 3]). The application of the Area Theorem to a (7,4) Hamming codes of Example 8.2 was illustrated in [1], where the derivation of a code EXIT function is provided in terms of split information functions. An alternative method for the derivation of the EXIT function of a binary linear block code was proposed in [4]. The approach of [4] relies on the trellis description of the code and on a Markov chain analysis of the ML decoder metrics. The adoption of the Area Theorem to upper bound the ML decoding threshold of LDPC code ensembles was proposed in [5], where the tightness of the threshold upper bound is thoroughly discussed. The use of the design rate in (8.27) is addressed in [5,6]. The approach of [5] was adapted in [7] to analyze the MAP threshold of joint detection-and-decoding schemes over intersymbol interference channels

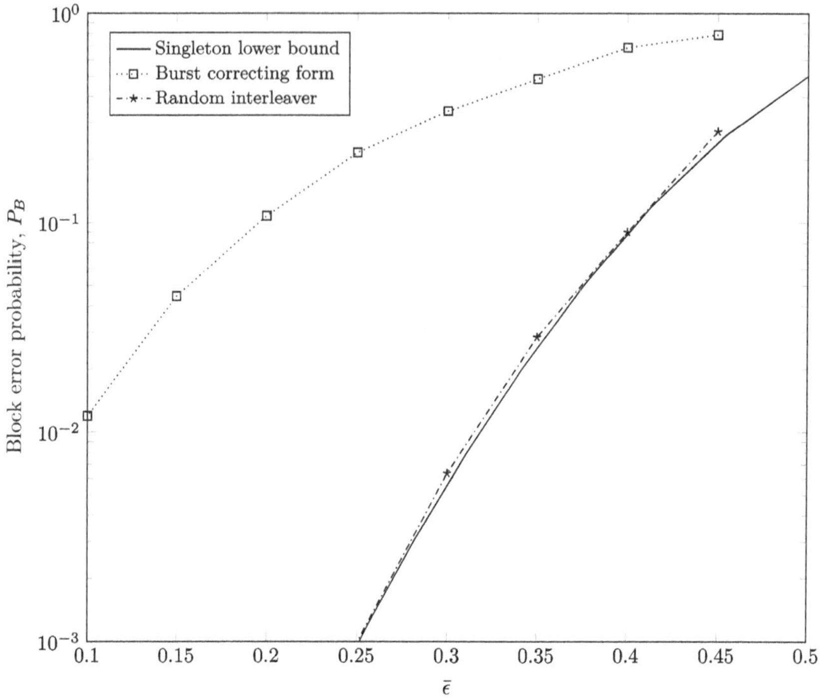

Figure 8.10 Block error probability for the IRA code from the Example 8.7 over the GEC with $\delta = 20$. The result is provided for the matrix permuted according to the interleaving rule provided by the burst correcting form parity-check matrix construction, as well as for random interleaving.

with erasure noise. The validity of the Area Theorem was extended beyond the BEC case by generalizing the notion of EXIT functions as the derivative of the normalized conditional entropy $\mathsf{H}(\boldsymbol{X}|\boldsymbol{Y})/n$, taken over a parameter that defines the channel noise level [8]. When the underlying model involved transmission over an AWGN channel, the derivative of the normalized conditional entropy with respect to the channel signal-to-noise ratio is proportional to the minimum mean-square error (MSE) [9]. This result leads immediately to a definition of the Area Theorem for MSE transfer charts [10]. The impact the Area Theorem had on the development of modern coding theory cannot be overstated. Among its most remarkable applications, it plays a central role in proving the capacity achieving properties of Reed–Muller codes over the BEC [11]. A detailed historical account on the development of the Area Theorem is provided in [12, Appendix 3.C].

The particular type of union bound used in Theorem 8.2 was derived in [13] for binary LDPC codes used to transmit over the BEC. The extensions to nonbinary code ensembles provided in Theorems 8.3 and 8.4 were studied in [14]. An alternative argument that yields a similar conclusion on the expected number of pivots given by (8.43) is based on the analysis of the guessing decoder (also known as Maxwell decoder) illustrated in Section 6.2.2 [5]. In [5], the guessing decoder is analyzed in the asymptotic setting for specific LDPC code ensembles. Suppose that a code out of the (3,6) regular LDPC ensemble is used to transmit on a BEC with $\epsilon = \epsilon^{\star}_{\mathsf{ML}}$. In [5] it was shown that as n tends to infinity, the average number of guesses $\bar{\xi}(n)$, normalized to n, converges to the area between the EXIT function and the BP EXIT function in Figure 8.2a. Again, a BP decoding threshold close to the ML one helps in keeping the average number of guesses (and hence the average number of pivots) small.

The construction illustrated in Section 8.4.1 for maximizing the MTBL follows the approach proposed in [15], and it is applicable to any linear block code for which the parity-check matrix can be placed in the form (8.45), where each submatrix $\boldsymbol{H}_\ell$ is nonsingular. Note that construction can be generalized to the case where $n - k$ does not divide n (the derivation is somehow more tedious, and we refer the reader to [15] for the details).

The behavior of the average ML block error probability of LDPC code ensembles as a function of the blocklength was analyzed in [16]. In particular, it was shown that the error probability can decrease at most polynomially in n. Based on the observation that a polynomial decay of the error probability is often dominated by a (polynomially small) fraction of bad codes in the ensemble, an exponential decay of the error probability can be achieved for some ensembles by expurgation. Lower bounds on the error exponents of expurgated binary LDPC code ensembles over the BEC were introduced in [17]. The result was generalized to nonbinary code ensembles (not necessarily LDPC) in [18].

8.6 Exercises

Exercise 8.1 Show that the area under the ML EXIT function of a $(n,1)$ repetition code is $1/n$.

Exercise 8.2 Show that the area under the ML EXIT function of a $(n,n-1)$ SPC code is $(n-1)/n$.

Exercise 8.3 Complete the derivation of the ML EXIT function for the (4,2) code of Example 8.1.

Exercise 8.4 Derive the EXIT function of the dual code of the (4,2) code of Example 8.1. By means of the duality property of EXIT functions (see Section 7.2), use the derived EXIT function to obtain the EXIT function of the (4,2) code of Example 8.1.

Exercise 8.5 Derive the EXIT function for the Hamming code of Example 8.2. *Hint*: Observe that both the Hamming code and its dual code are cyclic codes, thus the problem of deriving the EXIT function can be reduced to the problem of deriving the term associated to single coordinate. Moreover, recall that given a set of ℓ events $S_1, S_2, \ldots, S_\ell$, the general form of the inclusion–exclusion principle states

$$P\left(\bigcup_{i=1^\ell} S_i\right) = \sum_{k=1}^{n} (-1)^{k-1} \sum_{\mathcal{I}_k} P(S_{\mathcal{I}_k}), \tag{8.63}$$

where the inner sum is over all possible ordered subsets $\mathcal{I}_k$ of $\{1,2,\ldots,\ell\}$ with cardinality k, and

$$S_{\mathcal{I}_k} = \bigcap_{i \in \mathcal{I}_k} S_i. \tag{8.64}$$

Exercise 8.6 Using the results of the previous exercises, prove that an LDPC ensemble $\mathscr{C}^q_{\lambda,\rho}$ cannot attain a BP threshold, over a QEC, larger than $1 - R$.

Exercise 8.7 Consider a linear block code $\mathcal{C}$ over GF(q) and denote by $\boldsymbol{x}$ a nonzero codeword. Show that $|\mathcal{S}_{\boldsymbol{x}}| \geq q - 1$.

Exercise 8.8 Prove Theorem 8.3. *Hint*: Given a random code $\mathcal{C}$, write

$$P_B(\mathcal{C}) = \sum_{e=n-k+1}^{n} \binom{n}{e} \epsilon^e (1-\epsilon)^{n-e} + \sum_{e=1}^{n-k} \binom{n}{e} \epsilon^e (1-\epsilon)^{n-e} \times P\left(\bigcup_{\boldsymbol{x} \in \mathcal{C} \setminus \boldsymbol{x}_1} \{\text{supp}(\boldsymbol{x}) \subseteq \mathscr{I}_E\} \middle| |\mathscr{I}_E| = e \right) \tag{8.65}$$

and observe that

$$\mathsf{E}[P_B(\mathcal{C})] = \sum_{e=n-k+1}^{n} \binom{n}{e} \epsilon^e (1-\epsilon)^{n-e} + \sum_{e=1}^{n-k} \binom{n}{e} \epsilon^e (1-\epsilon)^{n-e} \times \mathsf{E}\left[P\left(\bigcup_{\boldsymbol{x} \in \mathcal{C} \setminus \boldsymbol{x}_1} \{\text{supp}(\boldsymbol{x}) \subseteq \mathscr{I}_E\} \middle| |\mathscr{I}_E| = e \right) \right]. \tag{8.66}$$

The proof continues by analyzing the inner expectation following the approach of (8.38).

Exercise 8.9 Prove Theorem 8.4. *Hint*: Given a random code $\mathcal{C}$, observe that the first sum in (8.35) can be replaced by the Singleton bound for a parity-check ensemble leading to

$$P_B(\mathcal{C}) = \sum_{e=m+1}^{n} \binom{n}{e} \epsilon^e (1-\epsilon)^{n-e} + \sum_{e=1}^{m} \binom{n}{e} \epsilon^e (1-\epsilon)^{n-e} \times P\left(\bigcup_{\boldsymbol{x} \in \mathcal{C} \setminus \boldsymbol{x}_1} \{\text{supp}(\boldsymbol{x}) \subseteq \mathscr{I}_E\} \middle| |\mathscr{I}_E| = e \right). \tag{8.67}$$

The proof continues as in Exercise 8.8.

Exercise 8.10 Derive the block cover enumerator B_w for an $(8,4)$ extended Hamming code for $L = 4$.

References

[1] A. Ashikhmin, G. Kramer, and S. ten Brink, "Extrinsic information transfer functions: Model and erasure channel properties," *IEEE Trans. Inf. Theory*, vol. 50, no. 11, pp. 2657–2673, Nov. 2004.

[2] C. Measson, A. Montanari, T. Richardson, and R. Urbanke, "Life above threshold: From list decoding to area theorem and MSE," in *Proc. IEEE Inf. Theory Workshop*, San Antonio, TX, USA, Oct. 2004.

[3] T. Richardson and R. Urbanke, *Modern coding theory*. New York, NY, USA: Cambridge University Press, 2008.

[4] M. Lentmaier, M. B. S. Tavares, and G. P. Fettweis, "Exact erasure channel density evolution for protograph-based generalized LDPC codes," in *Proc. IEEE Int. Symp. Inf. Theory*, Seoul, South Korea, June/July 2009, pp. 566–570.

[5] C. Measson, A. Montanari, and R. Urbanke, "Maxwell construction: The hidden bridge between iterative and maximum a posteriori decoding," *IEEE Trans. Inf. Theory*, vol. 54, no. 12, pp. 5277–5307, Dec. 2008.

[6] C. Measson, A. Montanari, and R. Urbanke, "Asymptotic rate versus design rate," in *Proc. IEEE Int. Symp. Inf. Theory*, Nice, France, June 2007.

[7] C. Wang and H. D. Pfister, "Upper bounds on the MAP threshold of iterative decoding systems with erasure noise," in *Proc. IEEE Int. Symp. Turbo Codes and Rel. Topics*, Lausanne, Switzerland, Sep. 2008, pp. 7–12.

[8] C. Measson, A. Montanari, T. J. Richardson, and R. Urbanke, "The generalized area theorem and some of its consequences," *IEEE Trans. Inf. Theory*, vol. 55, no. 11, pp. 4793–4821, Nov. 2009.

[9] D. Guo, S. Shamai, and S. Verdu, "Mutual information and minimum mean-square error in Gaussian channels," *IEEE Trans. Inf. Theory*, vol. 51, no. 4, pp. 1261–1282, Apr. 2005.

[10] K. Bhattad and K. R. Narayanan, "An MSE-based transfer chart for analyzing iterative decoding schemes using a Gaussian approximation," *IEEE Trans. Inf. Theory*, vol. 53, no. 1, pp. 22–38, Jan. 2007.

[11] S. Kudekar, S. Kumar, M. Mondelli, H. D. Pfister, E. Şaşoğlu, and R. L. Urbanke, "Reed–Muller codes achieve capacity on erasure channels," *IEEE Trans. Inf. Theory*, vol. 63, no. 7, pp. 4298–4316, July 2017.

[12] C. Measson, "Conservation laws for coding," PhD dissertation, École Polytechnique Fédérale de Lausanne, Lausanne, Switzerland, 2006.

[13] C. Di, D. Proietti, I. Telatar, T. Richardson, and R. Urbanke, "Finite-length analysis of low-density parity-check codes on the binary erasure channel," *IEEE Trans. Inf. Theory*, vol. 48, no. 6, pp. 1570–1579, June 2002.

[14] G. Liva, E. Paolini, and M. Chiani, "Bounds on the error probability of block codes over the q-ary erasure channel," *IEEE Trans. Commun.*, vol. 61, no. 6, pp. 2156–2165, June 2013.

[15] M. Fossorier, "Universal burst error correction," in *Proc. IEEE Int. Symp. Inf. Theory*, 2006, pp. 1969–1973.

[16] G. Miller and D. Burshtein, "Bounds on the maximum-likelihood decoding error probability of low-density parity-check codes," *IEEE Trans. Inf. Theory*, vol. 47, no. 11, pp. 2696–2710, Nov. 2001.

[17] D. Burshtein and G. Miller, "Asymptotic enumeration methods for analyzing LDPC codes," *IEEE Trans. Inf. Theory*, vol. 50, no. 6, pp. 1115–1131, June 2004.

[18] E. Paolini and G. Liva, "A lower bound on the error exponent of linear block codes over the erasure channel," in *Proc. IEEE Int. Symp. Inf. Theory*, Paris, France, July 2019.

9 Low-Density Parity-Check Codes: Generalizations

This chapter presents code ensembles that represent special instances, generalization, or modifications of LDPC code ensembles introduced in Chapter 4. The chapter starts with spatially coupled LDPC (SC-LDPC) codes, introduced within a protograph-based framework, then addresses generalized LDPC codes, where some of the CNs impose multiple linear constraints, and finally describes low-density generator matrix (LDGM) codes. Erasure decoding algorithms are described for all code classes.

9.1 Spatially Coupled Low-Density Parity-Check Codes

The LDPC codes addressed in Chapter 4 are linear block codes drawn from ensembles of codes characterized by common properties. Codes in the ensemble may, for example, share the same degree distribution pair (in case of unstructured LDPC code ensembles) or the same base matrix (in case of protograph LDPC code ensembles). By connecting the Tanner graphs associated with block LDPC code ensembles, larger Tanner graphs are obtained describing new codes. If performed according to certain rules, the procedure through which the Tanner graphs of block LDPC code ensembles are connected with each other is known as *spatial coupling*, and it leads to "convolutional" LDPC codes with remarkable properties under BP decoding, in terms of erasure correction performance and error floor. Such codes are usually referred to as spatially coupled low-density parity-check (SC-LDPC) codes. These codes are hereafter introduced starting from protograph LDPC codes. For the sake of clarity, the expression "block LDPC code" will be used to refer to ordinary LDPC codes discussed in Chapter 4. Binary SC-LDPC codes will be considered, the extension to nonbinary ones being a minor modification.

9.1.1 Protograph-Based Spatially Coupled LDPC Code Ensembles

Consider a block LDPC code ensemble based on a protograph with n_v VNs and n_c CNs, respectively. The number n_v of VNs in the protograph is given by $n_v = n_u + n_p$, where n_u and n_p are the numbers of transmitted and punctured protograph VNs, respectively. The n_v VNs are indexed by the elements of $\mathcal{J} = \{1,2,\ldots,n_v\}$ and the n_c CNs by the elements of $\mathcal{I} = \{1,2,\ldots,n_c\}$. Let $\boldsymbol{B} = \{B_{ij}\}$ be the $n_c \times n_v$ ensemble base matrix, where the nonnegative integer B_{ij} represents the number of edges connecting the VN of index $j \in \mathcal{J}$ to the CN of index $i \in \mathcal{I}$.

Take L identical copies of the protograph, where the positive integer L is referred to as the *chain length*, and index the L copies with $t \in \{1,2,\ldots,L\}$. For example, $L = 8$ copies of a protograph with one CN of degree 6 and two VNs, each of degree 3 (base matrix $\boldsymbol{B} = \begin{bmatrix} 3 & 3 \end{bmatrix}$), are illustrated in Figure 9.1a. After copying the protograph L times, we have L unconnected protographs, each of which defines the same block protograph LDPC code ensemble. The n_v VNs in the tth protograph are denoted by $\mathsf{v}_1(t), \mathsf{v}_2(t), \ldots, \mathsf{v}_{n_v}(t)$ and are referred to as protograph VNs at "spatial position" t; the corresponding n_c CNs are denoted by $\mathsf{c}_1(t), \mathsf{c}_2(t), \ldots, \mathsf{c}_{n_c}(t)$ and are referred to as protograph CNs at "spatial position" t.

Next define a *syndrome former memory*, that is, take a positive integer w fulfilling

$$w \le \left\lceil \frac{L(n_v - n_c)}{n_c} \right\rceil - 1 . \tag{9.1}$$

Moreover, take $w + 1$ *component base matrices*, that is, $w + 1$ matrices $\boldsymbol{B}_0, \boldsymbol{B}_1, \ldots, \boldsymbol{B}_w$ of size $n_c \times n_v$ such that: (i) for all $h \in \{0,1,\ldots,w\}$, the elements of $\boldsymbol{B}_h$ are nonnegative integers and (ii) the sum of the $w + 1$ matrices fulfills

$$\sum_{h=0}^{w} \boldsymbol{B}_h = \boldsymbol{B}, \tag{9.2}$$

where $\boldsymbol{B}$ is the base matrix of the protograph defining the original block protograph LDPC code ensemble.

The syndrome former memory w and the $w + 1$ component base matrices $\boldsymbol{B}_0$, $\boldsymbol{B}_1, \ldots, \boldsymbol{B}_w$ define an *edge spreading rule*, that is, a rule according to which the L unconnected protographs are connected with each other to obtain a spatially coupled protograph. More specifically, the n_c CNs of the original protograph are copied another w times, for a total of additional $w\, n_c$ protograph CNs. The new n_c CNs at spatial position $t \in \{L+1, L+2, \ldots, L+w\}$ are again denoted by $\mathsf{c}_1(t), \mathsf{c}_2(t), \ldots, \mathsf{c}_{n_c}(t)$. Next, for all $t \in \{1,2,\ldots,L\}$, all edges in the tth unconnected protograph are detached from the corresponding CNs. Finally, for all $t \in \{1,2,\ldots,L\}$, the edges from VNs $\mathsf{v}_1(t), \mathsf{v}_2(t), \ldots, \mathsf{v}_{n_v}(t)$ are connected to CNs at spatial positions $t, t+1, \ldots, t+w$ according to the following rule: The component base matrix $\boldsymbol{B}_h$ defines the connectivity between the n_v VNs at spatial position t and the n_c CNs at spatial position $t + h$.

After application of the edge spreading rule, we obtain a *spatially coupled protograph* whose base matrix $\boldsymbol{B}_{\mathsf{sc}}$ has size $(L + w)n_c \times Ln_v$ and structure

$$\boldsymbol{B}_{\mathsf{sc}} = \begin{bmatrix} \boldsymbol{B}_0 & & & \\ \boldsymbol{B}_1 & \boldsymbol{B}_0 & & \\ \vdots & \boldsymbol{B}_1 & \ddots & \\ \boldsymbol{B}_w & \vdots & \ddots & \boldsymbol{B}_0 \\ & \boldsymbol{B}_w & & \boldsymbol{B}_1 \\ & & \ddots & \vdots \\ & & & \boldsymbol{B}_w \end{bmatrix}. \tag{9.3}$$

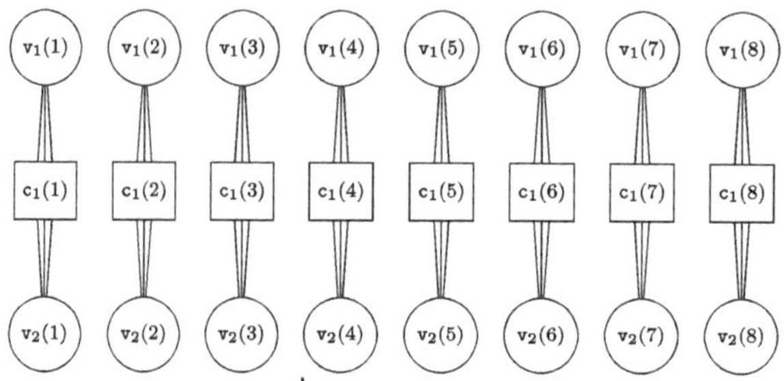

(a) A protograph with $n_v = 2$, $n_c = 1$, and base matrix $\boldsymbol{B} = \begin{bmatrix} 3 & 3 \end{bmatrix}$ is copied $L = 8$ times.

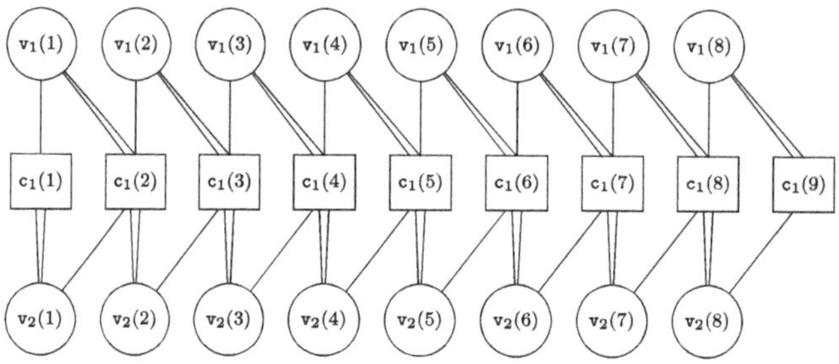

(b) For a syndrome former memory $w = 1$, the $L = 8$ unconnected protographs are coupled using the edge spreading rule defined by $\boldsymbol{B}_0 = \begin{bmatrix} 1 & 2 \end{bmatrix}$, $\boldsymbol{B}_1 = \begin{bmatrix} 2 & 1 \end{bmatrix}$.

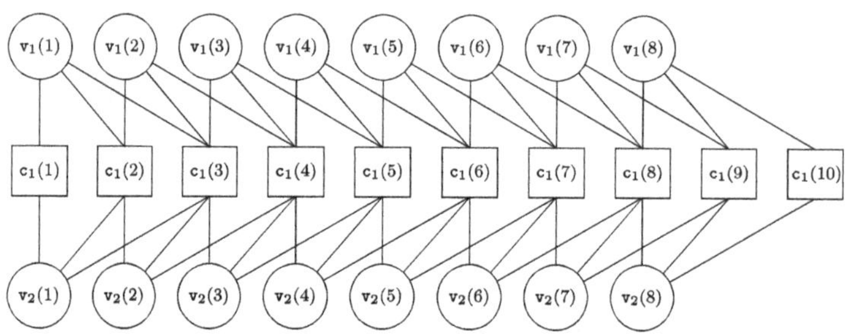

(c) For a syndrome former memory $w = 2$, the $L = 8$ unconnected protographs are coupled using the edge spreading rule defined by $\boldsymbol{B}_0 = \boldsymbol{B}_1 = \boldsymbol{B}_1 = \begin{bmatrix} 1 & 1 \end{bmatrix}$.

Figure 9.1 Two different edge spreading rules yielding two spatially coupled protographs from the same chain of unconnected protographs.

All nodes in the spatially coupled protograph retain the same degree they had before application of the edge spreading rule, with the exception of the protograph CNs in the two *terminations*. Each of these CNs is in fact characterized by a reduced degree compared with the degree it had in the unconnected chain. The newly introduced protograph CNs at the end of the chain are also characterized by reduced degrees. Shortly, the positive effect of reduced CN degrees at the two ends of the spatially coupled protograph will be illustrated.

The protograph with base matrix $\boldsymbol{B}_{\mathrm{sc}}$ given by (9.3) defines a protograph block code ensemble as described in Section 4.2.2. Hereafter this ensemble will be referred to as *protograph-based SC-LDPC block code ensemble*, and each code belonging to it as a protograph-based SC-LDPC block code. This ensemble is completely specified by the chain length L, the syndrome former memory w, and the w component base matrices $\boldsymbol{B}_0, \boldsymbol{B}_1, \ldots, \boldsymbol{B}_w$. It is constructed through the copy-and-permute procedure described in Section 4.2.2. We recall that this procedure consists of: copying the spatially coupled protograph Q times; taking all possible E_{p}-tuples of $Q \times Q$ permutation matrices (where E_{p} is the number of edges in the spatially coupled protograph) not leading to multiple edges between the same VN–CN pair; for each choice of the E_{p} permutation matrices, permuting the Q edges of the same type (i.e., the Q edges being copies of the same edge in the spatially coupled protograph) according to the corresponding permutation matrix. The probability distribution of a random code in the ensemble is the one induced by a random choice of the E_{p} permutation matrices, according to a uniform probability measure.

Recalling (4.32) the protograph SC-LDPC block code ensemble design rate is

$$\begin{aligned} \mathrm{R} &= \frac{L\,n_v - (L+w)n_c}{L\,n_u} \\ &= \frac{n_v - n_c}{n_u} - \frac{w\,n_c}{L\,n_u}, \end{aligned} \tag{9.4}$$

where the first term in the right-hand side of (9.4) is the design rate of the original block LDPC code ensemble, while the second term represents a rate loss due to the introduction of additional $w\,n_c$ protograph CNs in the coupled chain.

Example 9.1 Let $n_v = n_u = 2$, $n_p = 0$, $n_c = 1$, and $\boldsymbol{B} = \begin{bmatrix} 3 & 3 \end{bmatrix}$. Moreover, let $L = 8$. The protograph at spatial position $t \in \{1,2,\ldots,8\}$ is characterized by two VNs of degree 3, denoted by $\mathsf{v}_1(t)$ and $\mathsf{v}_2(t)$ and one CN of degree 6, denoted by $\mathsf{c}_1(t)$ (Figure 9.1a). Consider at first the edge spreading rule defined by $w = 1$, $\boldsymbol{B}_0 = \begin{bmatrix} 1 & 2 \end{bmatrix}$, and $\boldsymbol{B}_1 = \begin{bmatrix} 2 & 1 \end{bmatrix}$. With reference to Figure 9.1b, $w = 1$ implies that one new spatial position of CNs (represented by a unique protograph CN, as $n_c = 1$) is added to the chain. That $\boldsymbol{B}_0 = \begin{bmatrix} 1 & 2 \end{bmatrix}$ implies that, for all $t \in \{1,2,\ldots,8\}$, VN $\mathsf{v}_1(t)$ is connected to CN $\mathsf{c}_1(t)$ by one edge and VN $\mathsf{v}_1(t)$ is connected to the same CN by two parallel edges. Finally, that $\boldsymbol{B}_1 = \begin{bmatrix} 2 & 1 \end{bmatrix}$ implies that, for all $t \in \{1,2,\ldots,8\}$, VN $\mathsf{v}_1(t)$ is connected to CN $\mathsf{c}_1(t+1)$ by two parallel edges and VN $\mathsf{v}_1(t)$ is connected to the same CN by one edge. The resulting spatially coupled protograph has base matrix

$$\boldsymbol{B}_{\mathrm{sc}} = \begin{bmatrix} 1 & 2 & & & & & & & & \\ 2 & 1 & 1 & 2 & & & & & & \\ & & 2 & 1 & 1 & 2 & & & & \\ & & & & & & \ddots & & & \\ & & & & & & & 2 & 1 & 1 & 2 \\ & & & & & & & & & 2 & 1 \end{bmatrix} \tag{9.5}$$

of size 9×16. From (9.4) with $L = 8$, $w = 1$, $n_v = n_u = 2$, and $n_c = 1$, the design rate of the corresponding SC-LDPC block code ensemble is $\mathrm{R} = \frac{1}{2} - \frac{1}{16} = \frac{7}{16}$. Next, for the same initial chain of unconnected protographs depicted in Figure 9.1a, consider a second edge spreading rule defined by $w = 2$, $\boldsymbol{B}_0 = \begin{bmatrix} 1 & 1 \end{bmatrix}$, $\boldsymbol{B}_1 = \begin{bmatrix} 1 & 1 \end{bmatrix}$, and $\boldsymbol{B}_2 = \begin{bmatrix} 1 & 1 \end{bmatrix}$. In this case, $w = 2$ spatial positions of protograph CNs are added to the chain, each one represented again by one protograph CN only since $n_c = 1$. These two new CNs are denoted by $\mathsf{c}_1(8)$ and $\mathsf{c}_1(9)$ in Figure 9.1c. Since the three component base matrices are all equal to $[1, 1]$, for all $t \in \{1, 2, \ldots, 8\}$ each of the two VNs at spatial position t is connected to CNs $\mathsf{c}_1(t)$, $\mathsf{c}_1(t+1)$, and $\mathsf{c}_1(t+2)$ by one edge. Since in this case we have $L = 8$, $w = 2$, $n_v = n_u = 2$, and $n_c = 1$, the SC-LDPC block code ensemble design rate is $\mathrm{R} = \frac{1}{2} - \frac{1}{8} = \frac{3}{8}$.

At the end of the copy-and-permute procedure, the bipartite graph is characterized by $QL\,n_v$ VNs, of which $QL\,n_u$ are transmitted and $QL\,n_p$ are punctured, and $Q(L+w)n_c$ CNs. The parity-check matrix of any code in the ensemble defined by the base matrix $\boldsymbol{B}_{\mathsf{SC}}$ has the structure

$$\boldsymbol{H} = \begin{bmatrix} \boldsymbol{H}_0(1) & & & \\ \boldsymbol{H}_1(1) & \boldsymbol{H}_0(2) & & \\ \vdots & \boldsymbol{H}_1(2) & \ddots & \\ \boldsymbol{H}_w(1) & \vdots & \ddots & \boldsymbol{H}_0(L) \\ & \boldsymbol{H}_w(2) & & \boldsymbol{H}_1(L) \\ & & \ddots & \vdots \\ & & & \boldsymbol{H}_w(L) \end{bmatrix} \tag{9.6}$$

in which we have explicitly remarked, through the spatial position index in brackets, that the same $n_c \times n_v$ component matrix $\boldsymbol{B}_h$ may be lifted to different $Qn_c \times Qn_v$ matrices at different spatial positions of the chain.

Remark 9.1. Spatial coupling of the L unconnected protographs, through application of the edge spreading rule, may be interpreted as introduction of *memory* in the encoding process. Thinking of the spatial position index t as time and recalling that each unconnected protograph represents an LDPC code ensemble of length Qn_u, before coupling we have transmission of one codeword of length Qn_u bits at time, where each codeword is individually generated by encoding k bits, where $k \geq Q(n_v - n_c)$. By contrast, due to the structure of the parity-check matrix in (9.6), the Qn_u bits at time instant t (equivalently, at spatial position t) may be involved in parity-check equations at time $t, t+1, \ldots, t+w$. The quantity $\nu = Q(w+1)n_v$, representing the maximum number of symbols involved in each parity-check equation, is referred to as the *constraint length*.

Remark 9.2. As explained earlier, for a finite chain length L the edge spreading rule yields a block code of length $QL\,n_u$ symbols. An infinite chain length may also be considered, in which case the spatial position index t may grow unbounded. The (semi-infinite) spatially coupled protograph obtained after application of the edge spreading rule is called a *convolutional protograph* and the corresponding ensemble is called a *time-variant protograph-based SC-LDPC convolutional code ensemble*. An SC-LDPC block code ensemble may therefore be interpreted as a terminated SC-LDPC convolutional code ensemble; terminating a convolutional protograph consists of removing from it all VNs at any spatial position $t > L - 1$, all CNs that are connected solely to them, and all edges connecting such VNs to such CNs.

9.1.2 Belief Propagation Decoding and Threshold Saturation

Decoding of a protograph-based SC-LDPC block code over an erasure channel may be performed applying the BP decoder described in Section 6.1.2 over the whole bipartite graph. In each BP decoding iteration, all $Q(L + w)n_c$ CNs and then all $QL\,n_v$ VNs process messages in parallel.

In the limit as the copy factor Q tends to infinity, a BP decoding threshold may be computed by applying protograph density evolution analysis, presented in Section 7.3.2, to the spatially coupled protograph. Hereafter the BP decoding threshold will be denoted by $\epsilon^\star(L)$ to emphasize that, given an initial protograph with base matrix $\boldsymbol{B}$ and an edge spreading rule, it is a function of the chain length L. Threshold values $\epsilon^\star(L)$, rounded to the fourth decimal place, are displayed in Table 9.1 for an initial base matrix $\boldsymbol{B} = \begin{bmatrix} 3 & 3 \end{bmatrix}$ and the two edge spreading rules depicted in Figure 9.1b and c. Looking at the values in the table, we observe the same behavior in both cases: As the chain length L increases, the threshold initially decreases and then settles to a value that is the same for both edge spreading rules. Moreover, this value is larger than the BP decoding threshold of the protograph LDPC block code ensemble before spatial coupling, 0.4881 versus 0.4294. The threshold decrease as L increases can be attributed to an increase of the ensemble design rate, as per (9.4). As L gets large, the rate loss vanishes, the design rate tends to that of the original block code ensemble, and the threshold settles to a specific value. The fact that this value is reached earlier by the ensemble characterized by the edge spreading rule depicted in Figure 9.1b is related to its smaller syndrome former memory ($w = 1$ versus $w = 2$), corresponding to a smaller rate loss for the same value of L.

Table 9.1 BP decoding thresholds of protograph-based SC-LDPC block code ensembles. Uncoupled base matrix $\boldsymbol{B} = [3 \quad 3]$. $w = 1$: edge spreading rule $\boldsymbol{B}_0 = [1 \quad 2]$, $\boldsymbol{B}_1 = [2 \quad 1]$. $w = 2$: edge spreading rule $\boldsymbol{B}_0 = \boldsymbol{B}_1 = \boldsymbol{B}_2 = [1 \quad 1]$

L	$\epsilon^\star(L)$	
	$w = 1$	$w = 2$
8	0.4903	0.5223
16	0.4881	0.4892
32	0.4881	0.4882
48	0.4881	0.4881
64	0.4881	0.4881

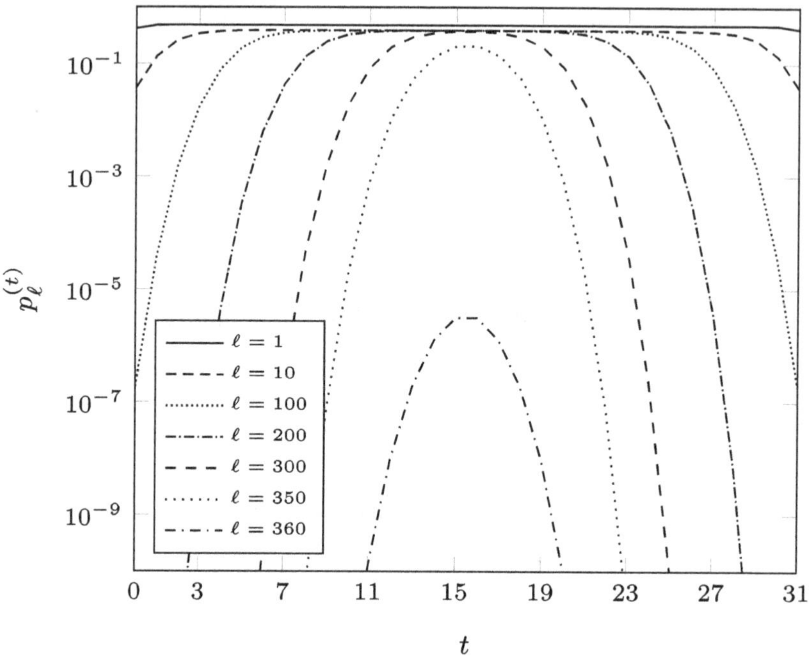

Figure 9.2 Profile of the a posteriori erasure probability for the VNs of the spatially coupled defined by $w = 2$ and $\boldsymbol{B}_0 = \boldsymbol{B}_1 = \boldsymbol{B}_2 = [1 \quad 1]$. Each curve corresponds to a specific number of density evolution iterations.

That the BP decoding threshold of the SC-LDPC protograph ensemble surpasses that of the original uncoupled ensemble finds its explanation in the beneficial effect of the chain *terminations*, as discussed hereafter. This effect is illustrated in Figure 9.2, for the ensemble defined by $L = 32$, $\boldsymbol{B} = \begin{bmatrix} 3 & 3 \end{bmatrix}$, $w = 2$, and $\boldsymbol{B}_0 = \boldsymbol{B}_1 = \boldsymbol{B}_2 = \begin{bmatrix} 1 & 1 \end{bmatrix}$. The figure shows the profile of the a posteriori erasure probability $p_\ell^{(t)}$ for the protograph VNs $\mathsf{v}_1(t)$, $t \in \{1,2,\ldots,L\}$, for different values of the number ℓ of density evolution iterations and for a channel erasure probability $\epsilon = 0.48$.[1] Note that this value of ϵ cannot be achieved by any of the original uncoupled ensembles in Figure 9.1a, whose BP decoding threshold, as above recalled, is equal to 0.4294. As the figure illustrates, for a given number ℓ of decoding iterations, the a posteriori erasure probability $p_\ell^{(t)}$ decreases when we move from the center of the chain ($t = 16, 17$) toward the two endpoints ($t = 1, 32$). Moreover, as the number of density evolution iterations increases, $p_\ell^{(t)}$ for VNs close to the endpoints converges to zero first, favoring convergence to zero of $p_\ell^{(t)}$ for nearby VNs. This "wave," originated at the two endpoints of the chain, travels through the chain and eventually reaches its center, fostering convergence to zero of the a posteriori erasure probability of all VNs in the spatially coupled protograph. A key question at this point is: What triggers the convergence process at the endpoints? The answer to this question is found in

[1] Due to symmetry in the spatially coupled protograph, for all $t \in \{1, \ldots, L\}$ the a posteriori erasure probability of $\mathsf{v}_2(t)$ coincides with that of $\mathsf{v}_1(t)$.

the CNs at spatial positions $t \in \{1,\ldots,w\}$ and $t \in \{L+1,\ldots,L+w\}$, that is, at the *terminations* of the chain. With reference to the presented example, CNs at spatial positions $t = 1$ and $t = 34$ are characterized by degree 2, CNs at spatial positions $t = 2$ and $t = 33$ by degree 4, and all of the other CNs by degree 6. The CNs with a lower degree feature a better erasure correction capability than the other CNs, which allows convergence to zero of the a posteriori erasure probability of VNs at the endpoints of the chain even for ϵ values larger that the BP threshold of the underlying block code ensemble.

A fundamental observation can be made by comparing the numerical values in Table 9.1 with those in Table 8.1. As the chain length L grows, the value of the BP decoding threshold of both SC-LDPC block ensembles becomes numerically indistinguishable, at least up to the fourth decimal place, from the *ML decoding threshold* of the original, uncoupled LDPC block code ensemble with $\mathsf{d}_v = 3$ and $\mathsf{d}_c = 6$. In other words, as L gets large the BP decoding threshold of the spatially coupled ensembles settles to a value that is not only larger than the BP decoding threshold of the underlying block ensemble but that is numerically very close to its ML decoding threshold. This celebrated phenomenon is usually referred to as *threshold saturation* and has been observed for protograph SC-LDPC code ensembles constructed from both regular and irregular block LDPC protograph ensembles, with different edge spreading rules. Attention should be paid to the fact that, in order to observe a coincidence (up to some decimal place) of the BP decoding threshold of the coupled ensemble with the ML decoding threshold of the uncoupled one, a sufficiently large syndrome former memory should be used.

Example 9.2 Let the initial uncoupled protograph be defined by the base matrix $\boldsymbol{B} = \begin{bmatrix} 4 & 4 \end{bmatrix}$. If we perform spatial coupling using the edge spreading rule defined by $w = 3$ and $\boldsymbol{B}_0 = \boldsymbol{B}_1 = \boldsymbol{B}_2 = \boldsymbol{B}_3 = \begin{bmatrix} 1 & 1 \end{bmatrix}$, and we round thresholds to the fourth decimal place, we observe that the threshold $\epsilon^*(L)$ settles to the value = 0.4977 as L increases. To the extent of this rounding, this value coincides with the ML decoding threshold of the uncoupled protograph ensemble. The threshold settles to the same value using the edge spreading rule defined by $w = 2$ and $\boldsymbol{B}_0 = \begin{bmatrix} 2 & 1 \end{bmatrix}$, $\boldsymbol{B}_1 = \begin{bmatrix} 1 & 2 \end{bmatrix}$, and $\boldsymbol{B}_2 = \begin{bmatrix} 1 & 1 \end{bmatrix}$. However, if we reduce the syndrome former memory to $w = 1$ and we take $\boldsymbol{B}_0 = \begin{bmatrix} 1 & 3 \end{bmatrix}$ and $\boldsymbol{B}_0 = \begin{bmatrix} 3 & 1 \end{bmatrix}$, the threshold $\epsilon^*(L)$ settles to the value 0.4962. Threshold saturation is therefore not observed in this case.

Remark 9.3. The threshold saturation phenomenon has been proved rigorously, although for a spatially coupled ensemble featuring less structure than the one based on protographs. The ensemble is still characterized by a chain of coupled block LDPC code ensembles but with a higher level of randomization in the connections between the VNs and CNs. Assume that there are T_v VNs at every spatial position $t \in \{1,\ldots,L\}$ and $T_c = T_v \mathsf{d}_v / \mathsf{d}_c$ CNs at every spatial position $t \in \{1,\ldots,L+w\}$. The ensemble is defined as follows. For every VN at spatial position t, a spatial position in the set $s \in \{t, \ldots, t+w-1\}$ is first picked uniformly at random; a CN at spatial position s is then picked uniformly at random, out of T_c CNs; finally, the VN is connected to the CN. For this "unstructured" SC-LDPC code ensemble, we have

$$\lim_{L\to\infty} \lim_{w\to\infty} \epsilon^{\star}_{\mathsf{BP}}(L, w, \mathsf{d}_v, \mathsf{d}_c) = \epsilon^{\star}_{\mathsf{ML}}(\mathsf{d}_v, \mathsf{d}_c), \tag{9.7}$$

where $\epsilon^{\star}_{\mathsf{BP}}(L, w, \mathsf{d}_v, \mathsf{d}_c)$ is the iterative BP decoding threshold of the coupled chain and $\epsilon^{\star}_{\mathsf{ML}}(\mathsf{d}_v, \mathsf{d}_c)$ is the ML decoding threshold of an unstructured regular block LDPC code ensemble $\mathscr{C}_{\mathsf{d}_v, \mathsf{d}_c}$. Note that, since the ML decoding threshold of an ensemble $\mathscr{C}_{(1-\mathrm{R})\mathsf{d}_c, \mathsf{d}_c}$ approaches capacity as d_c grows (see examples in Table 8.1), so does the BP decoding threshold $\epsilon^{\star}_{\mathsf{BP}}(L, w, \mathsf{d}_v, \mathsf{d}_c)$ (under the two limits in (9.7)).

9.1.3 Windowed Decoding

Windowed decoding of a protograph-based SC-LDPC block code consists of performing BP erasure decoding over a portion of the spatially coupled bipartite graph with the aim of recovering a subset of the symbols called the target symbols. When decoding over the current window position terminates, the window slides and BP decoding is performed again on the next window position.

More in detail, let the integer W be the window size, fulfilling $w+1 \leq W \leq w+L$. In the beginning, the window comprises all CNs at spatial position $t \in \{1, 2, \ldots, W\}$ and all VNs at spatial position $t \in \{1, 2, \ldots, \min\{W, L\}\}$, for a total of WQn_c CNs and $\min\{W, L\}Qn_v$ VNs. In this first window position, the target symbols are represented by the Qn_v VNs at spatial position $t = 1$. Iterative BP decoding is performed over the window, with the aim of recovering the target symbols. Edges connecting VNs in the window to CNs not belonging to it are not used in the message passing procedure.

When BP decoding performed on the first window configuration terminates, the window is slid by one spatial position along the spatially coupled chain and BP decoding is performed within the new window. Target symbols are now represented by the Qn_v VNs at spatial position $t = 1$. This process is iterated until the target VNs coincide with the Qn_v VNs at spatial position $t = L$. For any $t \in \{1, 2, \ldots, L\}$, when the target symbols are the ones at spatial position t, the window comprises all CNs whose spatial position belongs to $\{t, t+1, \ldots, \min\{t+W-1, L+w\}\}$ and all VNs whose spatial position belongs to $\{t, t+1, \ldots, \min\{t+W-1, L\}\}$.

Remark 9.4. The upper limit to the decoding window size is imposed by the total number of CN spatial positions in the spatially coupled protograph, equal to $L+w$. The lower limit is instead imposed by the need to include in the window all parity-check equations in which a target symbol may be involved for a code drawn from the ensemble. This number is equal to $Q(w+1)n_c$, from which the constraint $W \geq w+1$ follows.

Example 9.3 Figure 9.3 exemplifies windowed decoding, with window size $W = 4$, for any SC-LDPC code from the ensemble defined by the spatially coupled

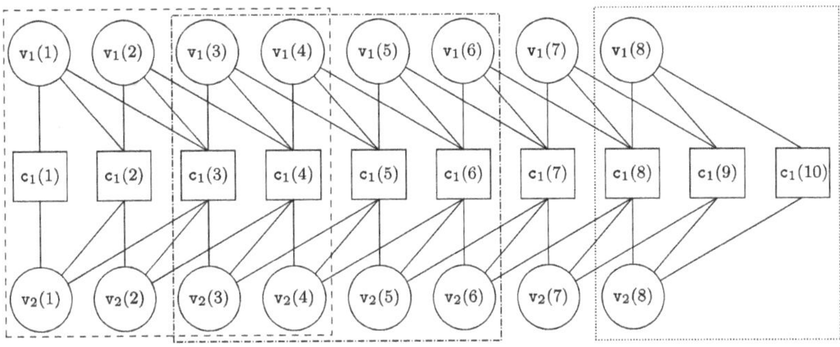

Figure 9.3 Windowed decoding for protograph-based SC-LDPC codes. Window size $W = 4$. Window positions starting at spatial positions $t = 1$ (dashed, initial position), $t = 3$ (dash-dotted), and $t = 8$ (dotted, final position).

protograph in Figure 9.1c. The ensemble is defined by $n_v = n_u = 2$, $n_c = 1$, $L = 8$, $w = 2$, $\boldsymbol{B}_0 = \boldsymbol{B}_1 = \boldsymbol{B}_2 = \begin{bmatrix} 1 & 1 \end{bmatrix}$. The first window position, represented by the dashed box, comprises the $4Q$ CNs being copies of $\mathsf{c}_1(t)$ for all $t \in \{1,2,3,4\}$. Moreover, since $\min\{W,L\} = W = 4$ it comprises the $8Q$ VNs being copies of $\mathsf{v}_1(t)$ and $\mathsf{v}_2(t)$ for all $t \in \{1,2,3,4\}$. Target symbols in the first window configuration are represented by the $2Q$ VNs being copies of $\mathsf{v}_1(1)$ and $\mathsf{v}_2(1)$. Iterative BP decoding is performed by exploiting all CNs and VNs within the window, with the goal of recovering the $2Q$ target symbols. Note that any edge being the copy of a protograph edge connecting a VN in the window (e.g., $\mathsf{v}_1(3)$) to a CN not in the window (e.g., $\mathsf{c}_1(5)$) is not used for BP decoding. There are $6Q$ such edges in the first window position. Equivalently, messages incoming toward the involved VNs along these edges are permanently set to E, that is, to "erasure."

Figure 9.3 also illustrates an intermediate window position comprising all VNs and CNs at spatial position $t \in \{3,4,5,6\}$ (dash-dotted box) and the final window position (dotted box). In this latter position, the maximum VN spatial position in the window is $\min\{L + W - 1, L\} = L = 8$. Similarly, the maximum CN spatial position in the window is $\min\{L + W - 1, L + w\} = L + w = 10$.

Let the starting spatial position of the decoding window be any $t \in \{1,2,\ldots,L-1\}$. When BP decoding over this window terminates, the messages outgoing from each target VN at the last iteration remain unchanged when the window is slid to any subsequent window position and BP decoding is performed on it. For example, looking at Figure 9.3, in every iteration of BP decoding over the window starting at spatial position 3 (dash-dotted box), any message outgoing from a VN at spatial position 1 or 2, toward a CN at spatial position 3 or 4, constantly keeps the last value it assumed when the VN was a target one.

Let again the starting spatial position of the decoding window be any $t \in \{1,2,\ldots,L-1\}$ and assume to perform BP decoding over this window. When decoding terminates, consider the last values of the messages outgoing from the VNs that belong to the window but whose spatial position is larger than t. For example, with reference again to the decoding window starting at spatial position 3 in Figure 9.3, consider the last values of the messages outgoing from any VN being a copy of $\mathsf{v}_1(t)$ or $\mathsf{v}_2(t)$, for $t \in \{4,5,6\}$. When the window is slid, a first strategy consists of using the values of these messages as the initial values for BP decoding in the next window position. We will refer to this approach as "strategy 1." Another strategy, hereafter referred to as "strategy 2," consists of reinitializing all of these messages based on the channel output for the corresponding VN. Over a channel introducing erasures, strategy 1 is always more efficient than strategy 2 and cannot perform worse than it. Nevertheless, strategy 2 may be useful on error channels, where a BP decoder failure may yield message deterioration phenomena.

9.1.3.1 Windowed Decoding Threshold

An asymptotic decoding threshold may be defined for protograph-based SC-LDPC codes under windowed decoding. The threshold is defined as the supremum of the set of all BEC erasure probabilities ϵ such that ϵ is *achievable* in all window positions. A BEC erasure probability ϵ is said to be achievable in window position starting at spatial position $t \in \{1,2,\ldots,L\}$ when protograph EXIT analysis, performed on the portion of the spatially coupled protograph that falls into the window, returns an a posteriori MI $\mathsf{I}_{\mathsf{APP}}$ fulfilling $\mathsf{I}_{\mathsf{APP}} > 1-\delta$ for every target protograph VN $\mathsf{v}_1(t)$, $\mathsf{v}_2(t)$, $\ldots$, $\mathsf{v}_{n_v}(t)$. Here, $0 < \delta < 1$ is a small positive number whose impact on the windowed decoding threshold will be illustrated through numerical examples.

Hereafter, the windowed decoding threshold will be denoted by $\epsilon^{\star}_{\mathsf{WD}}(\delta, W)$. As explicitly indicated, for a given spatially coupled protograph the threshold depends on both the quantity δ and the window size W.

The achievability of a specific value ϵ of the BEC erasure probability may be tested as follows. Consider the spatially coupled protograph and, for each edge of it, initialize to the value $1-\epsilon$ the message from the corresponding VN to the corresponding CN. Place the decoding window in its initial position, starting at spatial position $t = 1$, and perform protograph EXIT analysis within the window, iterating until a fixed point is reached for all messages. When EXIT analysis terminates, calculate the a posteriori MI $\mathsf{I}_{\mathsf{APP}}$ for each of the n_v target protograph VNs at spatial position 1. If any of the so computed n_v a posteriori MI values is less than $1-\delta$, declare the channel erasure probability ϵ as not achievable and terminate. Otherwise, slide the window by one position, letting it start at spatial position $t = 2$. If strategy 1 is used for decoding, perform protograph EXIT analysis on the new window position without reinitializing the variable-to-check node messages. Else, if strategy 2 is employed, reinitialize to the value $1-\epsilon$ the variable-to-check node message along every protograph edge in the new window position and perform protograph EXIT analysis. Again, compute the a posteriori MI $\mathsf{I}_{\mathsf{APP}}$ for each of the n_v target protograph VNs at spatial position 1, declaring non-achievability and terminating if any of them is less than $1-\delta$ and moving to the next window position otherwise. Iterate the described procedure, until ϵ is declared to be not achievable in some window position, or until all window positions have been successfully tested, in which case ϵ is declared as achievable.

Example 9.4 Consider the spatially coupled protograph depicted in Figure 9.1b and let the BEC erasure probability be $\epsilon = 0.430664$. Moreover, let $W = 4$ and $\delta = 10^{-6}$. Using strategy 1 we find that ϵ is declared to be not achievable in the window position starting at spatial position $t = 2$. More in detail, protograph EXIT analysis reaches a fixed point on the first window position, the one starting at spatial position $t = 1$, after 143 iterations; the a posteriori MI values for $\mathsf{v}_1(1)$ and $\mathsf{v}_2(1)$ are $1 - 9.68493 \times 10^{-7}$ and $1 - 5.03686 \times 10^{-7}$, respectively. In the window position starting at spatial position $t = 2$, a fixed point is reached after 136 iterations; the a posteriori MI values for $\mathsf{v}_1(2)$ and $\mathsf{v}_2(2)$ are $1 - 1.07056 \times 10^{-6}$ and $1 - 5.11670 \times 10^{-7}$, respectively. The same outcome, with the same final a posteriori MI values in both window positions, in obtained using strategy 2, with the only difference that 144 EXIT analysis iterations are performed in the window position starting at $t = 2$ before reaching the fixed point.

Concerning the dependence of the threshold on the quantity δ, a monotonicity property follows immediately from the definition of achievability of an erasure probability ϵ. In particular, for any δ' and δ'' such that $0 < \delta' < \delta'' < 1$, and for any $w + 1 \leq W \leq w + L$, we have

$$\epsilon^{\star}_{\mathsf{WD}}(\delta', W) \leq \epsilon^{\star}_{\mathsf{WD}}(\delta'', W)\,. \tag{9.8}$$

A monotonicity property also holds as regards dependence of the threshold on the window size W. In particular, for any W such that $w + 1 \leq W < w + L - 1$, and for any $0 < \delta < 1$, we have

$$\epsilon^{\star}_{\mathsf{WD}}(\delta, W) \leq \epsilon^{\star}_{\mathsf{WD}}(\delta, W + 1)\,. \tag{9.9}$$

The proof of this latter monotonicity property is proposed as an exercise at the end of the chapter. Note that (9.8) and (9.9) hold under both strategies 1 and 2.

Table 9.2 reports some windowed decoding threshold values $\epsilon^{\star}_{\mathsf{WD}}(\delta, W)$, rounded to the fifth decimal place, for a protograph-based SC-LDPC block code ensemble characterized by a chain length $L = 64$ and by the edge spreading rule depicted in Figure 9.1b. The ensemble design rate is $\mathsf{R} = \frac{1}{2} - \frac{1}{128}$. Each threshold value in the table corresponds to a specific choice of the window size $W \in \{4,5,6,7,8,16\}$ and of

Table 9.2 Windowed decoding thresholds for the protograph-based SC-LDPC block code ensemble characterized by $L = 64$, $w = 1$, $\boldsymbol{B}_0 = [1 \quad 2]$, and $\boldsymbol{B}_1 = [2 \quad 1]$

W	$\delta = 10^{-6}$	$\delta = 10^{-12}$	$\delta = 10^{-18}$
4	0.43018	0.33950	0.25690
5	0.46739	0.43702	0.39346
6	0.47271	0.46449	0.45542
7	0.47379	0.46782	0.46376
8	0.48794	0.48738	0.48641
16	0.48813	0.48813	0.48813

Table 9.3 Windowed decoding thresholds for the protograph-based SC-LDPC block code ensemble characterized by $L = 64$, $w = 2$, $\boldsymbol{B}_0 = \boldsymbol{B}_1 = \boldsymbol{B}_2 = [1 \quad 1]$

W	$\delta = 10^{-6}$	$\delta = 10^{-12}$	$\delta = 10^{-18}$
4	0.21321	0.07790	0.02877
5	0.31352	0.17292	0.09593
6	0.38795	0.27208	0.19002
7	0.43610	0.35565	0.28615
8	0.46320	0.41495	0.36534
16	0.48812	0.48804	0.48791

Table 9.4 Windowed decoding thresholds for protograph-based SC-LDPC block code ensembles characterized by $w = 1$, $\boldsymbol{B}_0 = [1 \quad 2]$, and $\boldsymbol{B}_1 = [2 \quad 1]$

L	5	6	7	16	32	64
$\epsilon^{\star}_{\mathrm{WD}}$	0.43024	0.43019	0.43018	0.43018	0.43018	0.43018

the parameter $\delta \in \{10^{-6}, 10^{-12}, 10^{-18}\}$. During the threshold search, achievability of each value of ϵ under test was analyzed adopting strategy 2, with a maximum allowed number of protograph EXIT iterations equal to 10^6; in all cases, a fixed point was reached before this maximum number. Achievement of a fixed point was declared in case none of the variable-to-check node messages had changed by more than 10^{-22} from one iteration to the next one. The numerical analysis confirms threshold monotonicity with respect to both δ and W. In particular, we can see how, for given δ, the threshold improves monotonically as the window size gets larger; for a window size sufficiently larger than the syndrome former memory w, the windowed decoding threshold exhibits a very small loss with respect to the BP decoding threshold. The analysis also reveals a pronounced sensitivity of the windowed decoding threshold to the parameter δ when for relatively small window size (e.g., $W = 4$ in the table).

The same remarks can be made as regards Table 9.3, reporting windowed decoding threshold values for the ensemble characterized by the same chain length ($L = 64$) and by the edge spreading rule depicted in Figure 9.1c. The ensemble design rate is in this case $\mathrm{R} = \frac{1}{2} - \frac{1}{64}$. In this case, we can observe some sensitivity of the threshold to the parameter δ even for $W = 16$, due to the larger syndrome former memory $w = 2$. A comparison of Tables 9.2 and 9.3 reveals how, for the same initial ensemble base matrix $\boldsymbol{B}$ and the same chain length L, the edge spreading rule may have a heavy impact on the windowed decoding threshold in case of relatively small window sizes.

For given edge spreading rule and window size, the windowed decoding threshold exhibits a very mild dependence on the chain length L. For example, Table 9.4 displays threshold values for ensembles characterized by the same edge spreading rule, the one illustrated in Figure 9.1b ($w = 1$) and by different chain lengths. In all cases, the parameter δ is set to 10^{-6} and the window size is $W = 4$. In this specific example, the threshold takes the same value (rounded to the fifth decimal place) for all chain lengths $L \geq 7$, while for $L = 5$ and $L = 6$, it only exhibits minor variations.

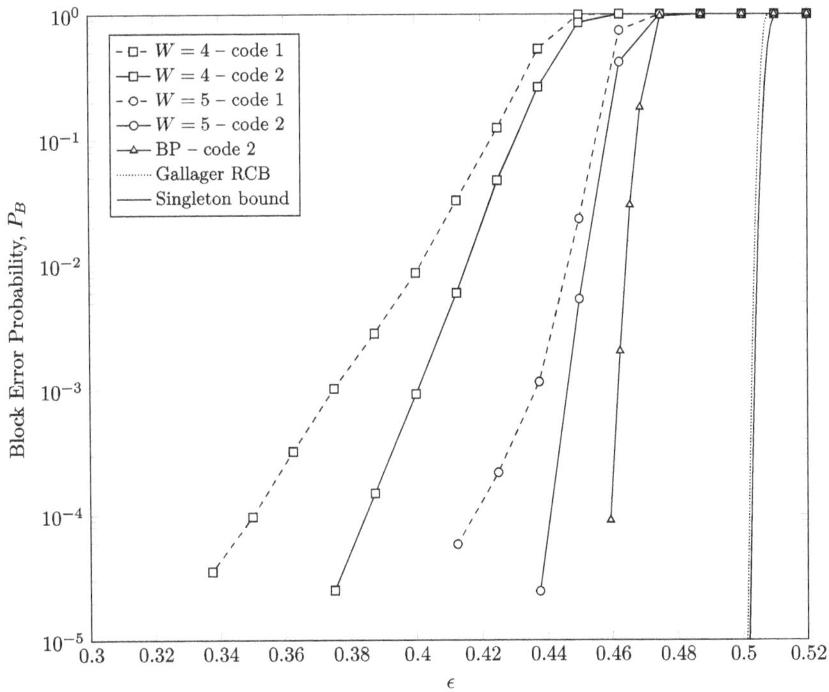

Figure 9.4 Block error rate versus the BEC erasure probability for two SC-LDPC block codes with $n = 131072$ and $R = 63/128$. The Gallager random coding bound and the Singleton bound are shown as benchmarks.

9.1.4 Finite-Length Performance

This subsection discusses the performance of some protograph-based SC-LDPC block codes in the finite-length regime. Figure 9.4 shows the block error rate of two such codes, both characterized by chain length $L = 64$, syndrome former memory $w = 1$, and edge spreading rule $\boldsymbol{B}_0 = \begin{bmatrix} 1 & 2 \end{bmatrix}$, $\boldsymbol{B}_1 = \begin{bmatrix} 2 & 1 \end{bmatrix}$. The two codes have been constructed by lifting the same spatially coupled protograph (the one depicted in Figure 9.1b with $L = 64$) using the same copy factor $Q = 1024$ but following two different strategies. To obtain the parity-check matrix $\boldsymbol{H}$ for the first code, hereafter referred to as "code 1," each element of $\boldsymbol{B}_{\text{SC}}$ equal to 1 was replaced with a $Q \times Q$ circulant permutation matrix drawn randomly, and each element of $\boldsymbol{B}_{\text{SC}}$ equal to 2 with the sum of two randomly drawn (but different) $Q \times Q$ circulant permutation matrices. Each null entry of $\boldsymbol{B}_{\text{SC}}$ was replaced by a $Q \times Q$ null matrix. Also the second code, referred to as "code 2" has been constructed by lifting $\boldsymbol{B}_{\text{SC}}$ with circulant permutation matrices. However, in this latter case the expansion was aimed at increasing the girth and at letting the shorter cycles be connected to the rest of the graph in a judicious way.

Both codes 1 and 2 are characterized by blocklength $n = 131\,072$, dimension $k = 645\,12$, and code rate $R = 63/128 \approx 0.492$. They were simulated over a BEC, both under BP decoding and windowed decoding with window size $W = 4$ and $W = 5$.

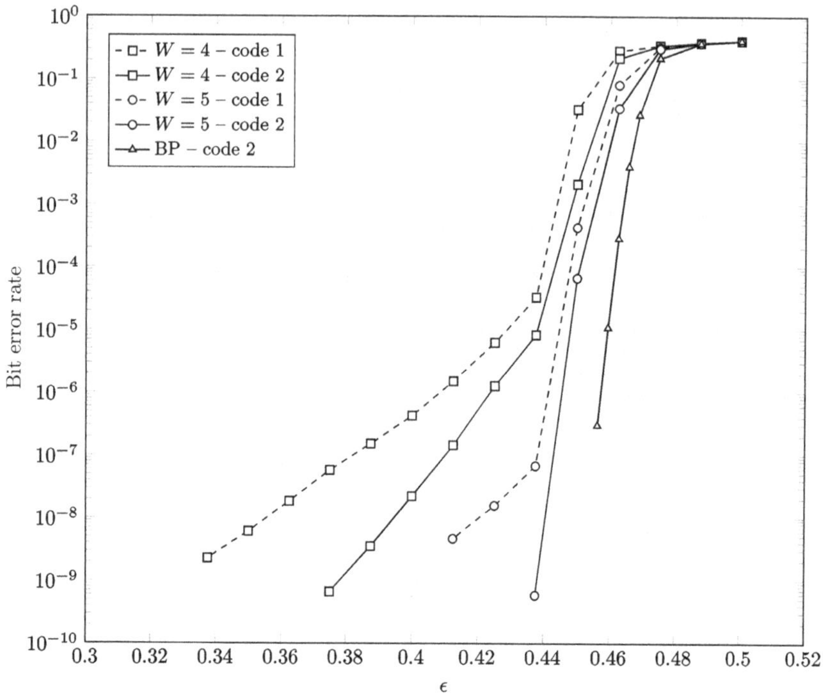

Figure 9.5 Bit error rate versus the BEC erasure probability for two SC-LDPC block codes with $n = 131072$ and $R = 63/128$.

As it is possible to observe from Figure 9.4, a judicious code design allows achieving a remarkably better performance, especially at low block error rate values where code 1 exhibits a marked slope change. We can also observe how the finite-length block error rate curves are in agreement with asymptotic threshold analysis. The windowed decoding with window size $W = 5$ performs quite close to the BP decoder. The corresponding bit error rate curves are shown in Figure 9.5.

9.2 Generalized Low-Density Parity-Check Codes

As illustrated in Chapters 7 and 8, LDPC codes can achieve excellent erasure correction performance under both BP iterative decoding and (with some additional complexity) ML decoding. As a matter of fact, a finite-length LDPC code, belonging to an ensemble whose iterative decoding threshold $\epsilon^{\star}_{\mathsf{BP}}$ is close to the erasure channel capacity, besides exhibiting an excellent waterfall performance may be characterized by a high error floor. This observation has motivated the investigation of more complex coding schemes, such as generalized LDPC (GLDPC) codes.

In GLDPC codes, generic linear block codes are used as CNs in replacement of (or in addition to) SPC codes. Linear block codes used as CNs are called *component codes* of the GLDPC code. In this book, we will address GLDPC codes constructed over the Galois field GF(2), although the results can be easily extended to GLDPC codes over GF(q) with $q > 2$.

9.2.1 Tanner Graph and Parity-Check Matrix

As discussed in Chapter 4, an LDPC code characterized by a sparse parity-check matrix $\boldsymbol{H}$ with m rows and n columns may be graphically represented by means of its Tanner graph, comprising n VNs and m CNs. In the Tanner graph, the VNs have a one-to-one correspondence with the encoded bits of the codeword, and the CNs represent the m linear equations, each involving a certain number of codeword bits.

A degree-d CN of an LDPC code constructed over GF(2) can be interpreted as a binary length-d SPC code, that is, as a $(d, d-1)$ binary linear block code. Analogously, a degree-d VN can be interpreted as a length-d binary repetition code, that is, as a $(d, 1)$ binary linear block code. In GLDPC codes we let some of (possibly, all of) the CNs be (d, h) linear block codes of dimension $h < d - 1$ and we refer to them as generalized CNs. A (d, h) generalized CN is still connected to d VNs, but it now corresponds to $d - h > 1$ independent parity-check equations. The number of CNs in the Tanner graph of a GLDPC code will be denoted by m_c, while we will still denote by m the number of rows in the parity-check matrix $\boldsymbol{H}$. As opposed to the LDPC code case, in GLDPC codes the two quantities are not equal to each other as we now have $m > m_c$. A GLDPC code represented by some Tanner graph is said to be variable-regular if all of the VNs in the Tanner graph have the same degree and variable-irregular otherwise. It is said to be check-regular if all of the CNs in the Tanner graph have the same degree and check-irregular otherwise. Finally, it is said to be check-homogeneous when all CNs are associated with the same component code and check-hybrid otherwise. A check-homogeneous GLDPC code is necessarily check-regular.

Similarly to the LDPC code case, also in the Tanner graph of a GLDPC code each CN should be thought as equipped with sockets to which edges are connected. In this respect, Definition 5.1 (CN socket assignment), Definition 5.2 (VN assignment), and Definition 5.3 (active edges) remain valid; moreover, there still exists a bijective mapping between the nonzero codewords of a GLDPC code and the VN assignments in its Tanner graph, whose induced CN socket assignment is valid.

Example 9.5 Figure 9.6 illustrates the Tanner graph of a GLDPC code with $n = 10$ VNs and $m_c = 5$ CNs. The four left-most CNs in the Tanner graph (the ones indexed from 1 to 4) are SPC codes, while the rightmost CN, c_5, is a (7,4) Hamming code. Let the parity-check matrix of the Hamming code associated with CN c_5 be the one specified in (2.64). Consider the VN assignment $\{\mathsf{v}_1, \mathsf{v}_2, \mathsf{v}_3, \mathsf{v}_4, \mathsf{v}_9\}$ and assign value $1 \in \mathrm{GF}(2)$ to each such VN. Moreover, assign value $0 \in \mathrm{GF}(2)$ to each VN not in the assignment. It is easy to verify that the induced CN socket assignment is valid. This is because each SPC CN has an even number of $1 \in \mathrm{GF}(2)$ values associated with its sockets, while the 7 sockets of the Hamming CN are associated with the 7-tuple $\begin{bmatrix} 1 & 1 & 0 & 0 & 0 & 1 & 0 \end{bmatrix}$, which is a valid local Hamming codeword.
Therefore

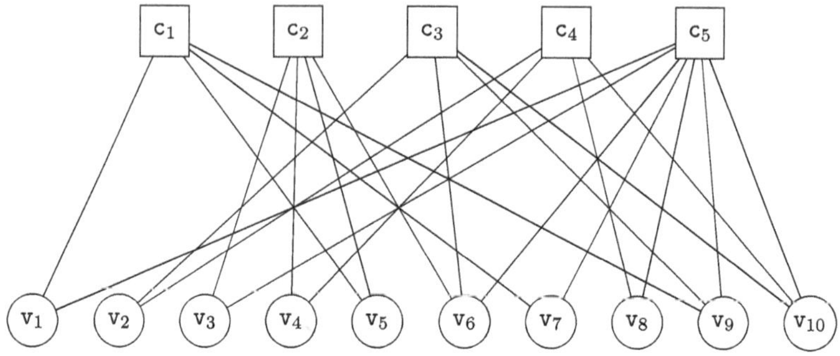

Figure 9.6 Tanner graph of a GLDPC code with blocklength $n = 10$ bits and 5 CNs. The CNs c_1, c_2, c_3, and c_4 are SPC codes. The CN c_5 is a (7,4) binary Hamming code.

$$\boldsymbol{c} = \begin{bmatrix} 1 & 1 & 1 & 1 & 0 & 0 & 0 & 0 & 1 & 0 \end{bmatrix} \tag{9.10}$$

is a codeword for the overall GLDPC code.

We denote by N_c the number of different component codes and by $\mathscr{C}^{(\mathrm{c})} = \{\mathcal{C}_1, \mathcal{C}_2, \ldots, \mathcal{C}_{N_c}\}$ the set of the component codes. The length of component code $\mathcal{C}_i$, $i \in \{1,2,\ldots,N_c\}$, is denoted by d_i, its dimension by h_i, and its code rate by $R_i = h_i/d_i$. Furthermore, we will indicate by $\boldsymbol{H}_i$ the $r_i \times d_i$ parity-check matrix used to represent code $\mathcal{C}_i$ (we have $r_i = d_i - h_i$ if $\boldsymbol{H}_i$ is full rank). A CN whose associated component code is $\mathcal{C}_i$ is said to be a type-i CN. An edge-oriented VN degree distribution $\lambda(x)$ is defined in the same way as for LDPC codes and is still given by (4.3), where d_v is the maximum degree of VNs. A CN degree distribution $\rho(x)$ may also be defined but in a slightly different way. In particular, we have

$$\rho(x) = \sum_{i=1}^{N_c} \rho_i x^{d_i - 1}, \tag{9.11}$$

where ρ_i is the fraction of edges connected to CNs of type i. The number m_c of CNs may be expressed as

$$m_c = n \frac{\int \rho}{\int \lambda}. \tag{9.12}$$

The fraction of VNs of degree d is $\Lambda_d = \lambda_d/d/\int\lambda$, while the fraction of CNs of type-i is $P_i = \rho_i/d_i/\int\rho$. The total number m of linear constraints imposed by the m_c CNs is given by

$$\begin{aligned} m &= \sum_{i=1}^{N_c} m_c\, P_i (1 - R_i) d_i \\ &= m_c \frac{\sum_{i=1}^{N_c} \rho_i (1 - R_i)}{\int \rho}, \end{aligned} \tag{9.13}$$

and therefore, recalling (9.12), the code rate of the GLDPC code fulfills

$$R \geq 1 - \frac{m}{n} = 1 - \frac{\sum_{i=1}^{N_c} \rho_i (1 - R_i)}{\int \lambda}, \tag{9.14}$$

where equality holds if and only if the m linear constraints are independent. Importantly, from (9.14) we see that the effect of using component codes of low rate is an overall decrease of the GLDPC code rate.

Example 9.6 For the GLDPC code whose Tanner graph is depicted in Figure 9.6 we have $N_c = 2$ and $\mathscr{C}^{(c)} = \{\mathcal{C}_1, \mathcal{C}_2\}$, where $\mathcal{C}_1$ is a (4,3) SPC code ($d_1 = 4$, $h_1 = 3$, $R_1 = 3/4$) and $\mathcal{C}_2$ is a (7,4) Hamming code ($d_2 = 7$, $h_2 = 4$, and $R_2 = 4/7$) with $\boldsymbol{H}_2$ expressed by (2.64). The edge-oriented degree distribution pair is

$$\lambda(x) = \frac{14}{23}x + \frac{9}{23}x^2 \quad \text{and} \quad \rho(x) = \frac{16}{23}x^3 + \frac{7}{23}x^6, \tag{9.15}$$

and the overall code rate fulfills $R \geq 3/10$.

The $m_c \times n$ binary matrix whose element (i, j) is equal to 1 if the ith check node is connected to the jth variable node, and is equal to 0 otherwise, is called *biadjacency matrix* of the Tanner graph, denoted by $\boldsymbol{A}$. The biadjacency matrix has a one-to-one correspondence with the Tanner graph. A parity-check matrix $\boldsymbol{H}$ for the GLDPC code may be obtained from the biadjacency matrix, using a *row expansion* rule. To describe it we let the CNs, and therefore the rows of $\boldsymbol{A}$, be indexed by $l \in \{1, 2, ..., m_c\}$; moreover, we let $\mathcal{C}_{i(l)}$ be the component code associated with the lth CN (represented by an $r_{i(l)} \times d_{i(l)}$ parity-check matrix $\boldsymbol{H}_{i(l)}$) and the $d_{i(l)}$ nonzero entries in the lth row of $\boldsymbol{A}$ be indexed (from left to right) by $t \in \{1, 2, \ldots, d_{i(l)}\}$.

The row expansion rule is the following. For $l = 1, 2, \ldots, m_c$, in row l of $\mathbf{A}$

- Substitute each 0 entry with an all-zero column vector of length $r_{i(l)}$;
- For $t = 1, 2, \ldots, d_{i(l)}$, substitute the tth nonzero entry with column t of $\boldsymbol{H}_{i(l)}$.

The above procedure yields an $m \times n$ matrix, $m = \sum_{l=1}^{m_c} r_{i(l)}$, which is a parity-check matrix $\boldsymbol{H}$ for the overall GLDPC code. The GLDPC code dimension fulfills $k \geq n - m$ with equality if and only if the m parity-check equations are linearly independent.

Example 9.7 The biadjacency matrix of the Tanner graph depicted in Figure 9.6 is

$$\boldsymbol{A} = \begin{bmatrix} 1 & 0 & 0 & 0 & 1 & 0 & 1 & 0 & 1 & 0 \\ 0 & 0 & 1 & 1 & 1 & 1 & 0 & 0 & 0 & 0 \\ 0 & 1 & 0 & 0 & 0 & 1 & 0 & 0 & 1 & 1 \\ 0 & 1 & 0 & 1 & 0 & 0 & 0 & 1 & 0 & 1 \\ 1 & 0 & 1 & 0 & 0 & 1 & 1 & 1 & 1 & 1 \end{bmatrix}. \tag{9.16}$$

Since the parity-check matrix of component code $\mathcal{C}_1$ (a binary SPC code of length $d_1 = 4$ and dimension $h_1 = 3$) is $\boldsymbol{H}_1 = [\ 1 \quad 1 \quad 1 \quad 1\ ]$, the row expansion procedure has no effect on the rows of $\boldsymbol{A}$ with indexes $l \in \{1,2,3,4\}$. The bottom row of $\boldsymbol{A}$, the one with index $l = 4$, is associated with component code $\mathcal{C}_2$ (Hamming code) with parity-check matrix given by (2.64). Applying the row expansion rule we obtain the 7×10 parity-check matrix

$$\boldsymbol{H} = \begin{bmatrix} 1 & 0 & 0 & 0 & 1 & 0 & 1 & 0 & 1 & 0 \\ 0 & 0 & 1 & 1 & 1 & 1 & 0 & 0 & 0 & 0 \\ 0 & 1 & 0 & 0 & 0 & 1 & 0 & 0 & 1 & 1 \\ 0 & 1 & 0 & 1 & 0 & 0 & 0 & 1 & 0 & 1 \\ 1 & 0 & 1 & 0 & 0 & 1 & 0 & 1 & 0 & 0 \\ 0 & 0 & 1 & 0 & 0 & 1 & 1 & 0 & 1 & 0 \\ 1 & 0 & 1 & 0 & 0 & 0 & 1 & 0 & 0 & 1 \end{bmatrix}. \tag{9.17}$$

It is easy to verify that $\boldsymbol{H}$ is full rank ($\text{rank}(\boldsymbol{H}) = 7$). Therefore the dimension of the GLDPC code is $k = 3$ and its code rate is $R = 3/10$ (while from (9.14) the value $3/10$ was a lower bound on the actual rate).

9.2.2 Decoding of GLDPC Codes on Erasure Channels

Iterative BP decoding of LDPC codes has been described in Section 6.1.2. Restricting the analysis to binary codes, it has been discussed how BP decoding on binary erasure channels may formulated in a very simple way compared to BP decoding for general error channels. In particular it has been discussed how, over a BEC, a simple equivalent formulation of BP decoding is possible in which any message m_i, $m_{i\to j}$, or $m_{i\leftarrow j}$ takes values in $\{0, 1, \mathsf{E}\}$. This is licit because over the BEC any such message is a two-dimensional p.m.f. (whose first and second elements correspond to $0 \in \text{GF}(2)$ and $1 \in \text{GF}(2)$, respectively) taking only the three values $\begin{bmatrix}1 & 0\end{bmatrix}$, $\begin{bmatrix}0 & 1\end{bmatrix}$, and $\begin{bmatrix}\frac{1}{2} & \frac{1}{2}\end{bmatrix}$. In this paragraph we extend BP decoding to binary GLDPC codes transmitted over the BEC.

The BP erasure decoder for GLDPC codes remains the same as the one formally described in Algorithm 1, except for rows from 8 to 13, describing the calculation of messages $m_{i\leftarrow j}$ from the generic CN c_j to its generic neighboring VN v_i. According to rows 8–13, if any of the messages incoming to the CN *except* the one from v_i is equal to E, then $m_{i\leftarrow j} = \mathsf{E}$; else $m_{i\leftarrow j}$ equals the additive inverse of the sum of all incoming messages *except* the one from v_i. To understand how rows 8–13 of Algorithm 1 should be generalized in the GLDPC code setting, we start by reformulating the corresponding processing.

Let the CN sockets be indexed by $t \in \{1, \ldots, d\}$ and let $\boldsymbol{l} = \begin{bmatrix} l_1 & l_2 & \ldots & l_d \end{bmatrix} \in \{0, 1, \mathsf{E}\}^d$ be the vector whose element l_t is the message received by the CN along its tth socket. Let $\boldsymbol{l}_{\sim i} \in \{0, 1, \mathsf{E}\}^d$ be a vector whose elements are all equal to the corresponding ones in $\boldsymbol{l}$, except (possibly) the element representing the message incoming from VN v_i. The value E is assigned to this latter element, regardless of its original value in $\boldsymbol{l}$. Then, the processing described in rows 8–13 of Algorithm 1 may be recast in compact form as

$$m_{i\leftarrow j} = \arg\max_{b\in\{0,1\}} P(b \mid \boldsymbol{l}_{\sim i}) \tag{9.18}$$

if $P(0 \mid \boldsymbol{l}_{\sim i}) \neq P(1 \mid \boldsymbol{l}_{\sim i})$ and

$$m_{i\leftarrow j} = \mathsf{E} \tag{9.19}$$

otherwise.

By recalling that a degree-d CN in the Tanner graph of an LDPC code is associated with a $(d, d-1)$ SPC component code, we can give the following interpretation to (9.18). A codeword of the SPC code is transmitted over an erasure channel whose output is the vector $\boldsymbol{l}_{\sim i}$ and the message $m_{i\leftarrow j}$ is calculated by applying incomplete symbol-wise MAP decoding to this channel observation. From a practical point of view, ML decoding is performed, locally at the CN, on $\boldsymbol{l}_{\sim i}$. If ML decoding is capable to uniquely determine a nonerasure value of the encoded bit corresponding to the socket connected to v_i then the obtained value is assigned to the message $m_{i\leftarrow j}$; otherwise an erasure (E) value is assigned to the message. Note that the channel observation $\boldsymbol{l}_{\sim i}$ carries no a priori information about the symbol $m_{i\leftarrow j}$ which therefore, after application of (9.18), represents the extrinsic information outgoing from the CN socket connected to v_i.

The described symbol-wise MAP calculation of the extrinsic messages outgoing from a CN does not rely on the component code being a SPC code and can therefore be extended to any (d, h) linear block component code: Once vector $\boldsymbol{l}$ has been received by the CN, vector $\boldsymbol{l}_{\sim i}$ is constructed for each neighboring VN v_i and ML decoding is applied to compute the extrinsic message $m_{i\leftarrow j}$. This is formalized in Algorithm 4.

Example 9.8 Consider the GLDPC code whose Tanner graph is depicted in Figure 9.6. Let the vector $\boldsymbol{l}$ of messages received by CN c_4, whose component code is a (7,4) Hamming code with $d_{\min} = 3$ and parity-check matrix specified in (2.64), be $\boldsymbol{l} = \begin{bmatrix} 1 & 0 & 0 & 0 & \mathsf{E} & \mathsf{E} & \mathsf{E} \end{bmatrix}$. Then, the vector of outgoing extrinsic messages after CN processing is $\begin{bmatrix} m_{1\leftarrow 5} & m_{3\leftarrow 5} & m_{6\leftarrow 5} & m_{7\leftarrow 5} & m_{8\leftarrow 5} & m_{9\leftarrow 5} & m_{10\leftarrow 5} \end{bmatrix} = \begin{bmatrix} \mathsf{E} & \mathsf{E} & \mathsf{E} & \mathsf{E} & 1 & 0 & 1 \end{bmatrix}$. In fact, to calculate $m_{8\leftarrow 5}$, $m_{9\leftarrow 5}$, and $m_{10\leftarrow 5}$ ML decoding is performed on $\boldsymbol{l}_{\sim 8} = \boldsymbol{l}_{\sim 9} = \boldsymbol{l}_{\sim 10} = \begin{bmatrix} 1 & 0 & 0 & 0 & \mathsf{E} & \mathsf{E} & \mathsf{E} \end{bmatrix}$; since the set of indexes $\{1,2,3,4\}$ is an information set (see Definition 2.23) for the Hamming code with generator matrix (2.52), in each of the three cases the value of the local encoded bit can be uniquely determined. Moreover, to calculate $m_{1\leftarrow 5}$ ML decoding is performed on $\boldsymbol{l}_{\sim 1} = \begin{bmatrix} \mathsf{E} & 0 & 0 & 0 & \mathsf{E} & \mathsf{E} & \mathsf{E} \end{bmatrix}$; since the value of the local encoded bit corresponding to the CN socket of index 0 cannot be recovered a message $m_{1\leftarrow 5} = \mathsf{E}$ is generated. The same applies to $m_{3\leftarrow 5}$, $m_{6\leftarrow 5}$, and $m_{7\leftarrow 5}$.

In Chapter 6 a stopping set of an LDPC code was defined as a subset of its VNs, such that if a CN is connected to $\mathcal{S}$ it is connected to it at least twice (Definition 6.2).

Algorithm 4: BP decoding of GLDPC codes on erasure channels

---

while *maximum number of iterations not reached* **do**
  compute outgoing messages for each VN v_i to each neighboring CN c_j as
  if *all incoming messages* $m_{i \leftarrow j'}$*, except the message from* c_j *have value* E **then**
    $m_{i \rightarrow j} = \mathsf{E}$
  else
    $m_{i \rightarrow j}$ is chosen to be any $m_{i \leftarrow j'}$ or m_i different from E
  end
  compute outgoing messages for each CN c_j to each neighboring VN v_i as
  if *ML decoding applied to* $\boldsymbol{l}_{\sim i}$ *uniquely determines a nonerasure value for the local encoded bit corresponding to* v_i **then**
    $m_{i \leftarrow j}$ is chosen to be the uniquely determined value
  else
    $m_{i \leftarrow j} = \mathsf{E}$
  end
end
compute values of code symbols $\hat{c}_i$ as
if *all incoming messages* $m_{i \leftarrow j}$*, have value* E **then**
  c_i cannot be determined and signal decoding failure
else
  c_i is any $m_{i \leftarrow j}$ or m_i different from E
end

It was also remarked that, if BP erasure decoding applied to an LDPC code stops prematurely, then the residual set of erasures forms a stopping set. In other words, BP erasure decoding of LDPC codes does not succeed whenever the set of VNs erased by the channel contains a stopping set.

Stopping sets may be generalized to GLDPC codes decoded via Algorithm 4. Given a local nonzero codeword of the component code associated with a CN, we say that a subset of the d CN sockets *covers* the codeword if every nonzero codeword bit corresponds to a socket in the subset. This leads us to the following definition.

Definition 9.1 (Stopping set of a GLDPC code) A stopping set of a GLDPC code is any subset of its VNs such that, if a CN is connected to it, the subset of the CN sockets connected to it covers at least one nonzero local component codeword.

When a GLDPC code transmitted over an erasure channel is decoded via Algorithm 4, decoding stops prematurely whenever the set of VNs erased by the channel contains a stopping set. Importantly, while for LDPC codes whether or not a subset of the VNs forms a stopping set depends only on the connectivity between the VNs and the CNs, in the GLDPC code case it also depends on the inner structure of each CN through its component code book. It is easy to see that, if all component codes

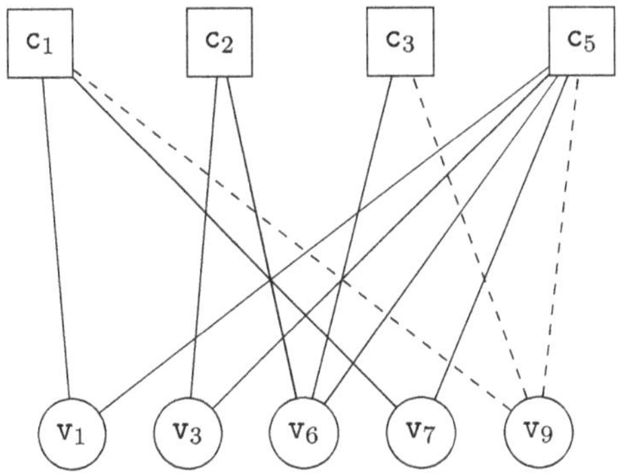

Figure 9.7 Subgraph induced by the stopping set $\{v_1, v_3, v_6, v_7, v_9\}$ of the GLDPC code whose Tanner graph is depicted in Figure 9.6.

of a GLDPC code are SPC codes, then Definition 9.1 becomes equivalent to Definition 6.2. In this case, in fact, for any pair of sockets of a CN there exists a local component codeword of Hamming weight 2 which is covered by the two sockets.

Example 9.9 The subset $\{v_1, v_3, v_6, v_7, v_9\}$ of the VNs of the GLDPC code whose Tanner graph is depicted in Figure 9.6 is a stopping set. The subgraph induced by this stopping set is depicted in Figure 9.7. In fact: the three CN sockets of c_1 connected to the stopping set cover the three local component codewords $\begin{bmatrix}1 & 0 & 1 & 0\end{bmatrix}$, $\begin{bmatrix}1 & 0 & 0 & 1\end{bmatrix}$, and $\begin{bmatrix}0 & 0 & 1 & 1\end{bmatrix}$; the two CN sockets of c_2 cover the local component codeword $\begin{bmatrix}1 & 0 & 0 & 1\end{bmatrix}$; the two CN sockets of c_3 cover the local component codeword (0 1 1 1); the five CN sockets of c_5 cover the two local component codewords $\begin{bmatrix}1 & 1 & 0 & 0 & 0 & 1 & 0\end{bmatrix}$ and $\begin{bmatrix}1 & 0 & 1 & 1 & 0 & 0 & 0\end{bmatrix}$. Removing v_9 from the stopping set yields the subset of VNs $\{v_1, v_3, v_6, v_7\}$, whose induced subgraph is obtained from the one in Figure 9.7 by canceling v_9 along with its edges (depicted as dashed lines). The new subset of VNs is not a stopping set anymore because no nonzero c_3 component codeword is now covered by the unique socket of c_3 connected to the subset. It is also easy to show that the subset $\{v_1, v_3, v_6, v_7\}$ does not contain any subset of smaller size.

The BP decoder described in Algorithm 4 admits a practical implementation which may be thought as the generalization to the GLDPC code framework of the peeling decoder (Section 6.1.3). Accordingly, at each decoding iteration the CNs are processed serially, from c_1 to c_{m_c}. For each CN c_j, the d_j VNs connected to c_j are considered and a local observable $\boldsymbol{l} \in \{0, 1, \mathsf{E}\}^{d_j}$ for the component code is formed with the values of these VNs. Local ML decoding is applied to the observation $\boldsymbol{l}$; the value (0, 1, or E) of each of the d_t component codeword bits after ML decoding is assigned to the corresponding VN. Decoding terminates successfully when the value of each VN erased by the channel has been recovered; it terminates with failure when a maximum number of iterations has been reached and there still are erased VNs.[2]

9.2.3 GLDPC Code Ensembles

Ensembles of GLDPC codes are defined in a similar way as LDPC code ones. This paragraph addresses unstructured and protograph GLDPC code ensembles.

9.2.3.1 Unstructured Ensembles

To construct an unstructured GLDPC code ensemble, we start from a set $\mathscr{C}^{(c)} = \{\mathcal{C}_1, \mathcal{C}_2, \ldots, \mathcal{C}_{N_c}\}$ of component codes and an from edge-oriented degree distribution pair (λ, ρ), where $\rho(x)$ is defined according to (9.11). Next we consider n VNs and m_c CNs, where m_c is given by (9.12). Each VN has a number of sockets so that, overall, the VN degree distribution $\lambda(x)$ is obtained. Each CN is associated with one of the N_c component codes in $\mathscr{C}^{(c)}$, and therefore its number of sockets equals the component code length, so that the overall CN distribution $\rho(x)$ is obtained.

Similar to LDPC codes, let E be the number of VN and CN sockets and $\boldsymbol{\Pi}$ be an $E \times E$ permutation matrix defining the connections between the E VN sockets and the E CN sockets. The ensemble construction consists of generating all possible $E!$ permutation matrices $\boldsymbol{\Pi}$ and of connecting the VN sockets to the CN sockets according to each such matrix; for each generated matrix $\boldsymbol{\Pi}$, a biadjacency matrix $\boldsymbol{A}$ is obtained which corresponds to some parity-check matrix $\boldsymbol{H}$ representing a GLDPC code. This unstructured GLDPC code ensemble will be denoted by $\mathscr{C}^n_{\lambda,\rho,\mathscr{C}^c}$. The probability distribution of a random code $\mathcal{C} \in \mathscr{C}^n_{\lambda,\rho,\mathscr{C}^c}$ is the one induced by a random choice of the permutation matrix $\boldsymbol{\Pi}$ according to a uniform probability measure. The design rate R of the ensemble $\mathscr{C}^n_{\lambda,\rho,\mathscr{C}^c}$ is defined as

$$\mathrm{R} = 1 - \frac{\sum_{i=1}^{N_c} \rho_i (1 - R_i)}{\int \lambda}, \tag{9.20}$$

which reduces to $\mathrm{R} = 1 - \mathsf{d}_v(1 - R_1)$ for ensembles of variable-regular and check-homogeneous GLDPC codes, where R_1 is the code rate of the unique component code. Note that every generalized CN corresponds to multiple parity-check equations, and therefore, it introduces a rate loss that is more pronounced for low-rate component codes.

Remark 9.5. Some $\boldsymbol{\Pi}$ matrices yield multiple edges between a VN and a CN. When this is the case, the corresponding element of the biadjacency matrix $\boldsymbol{A}$ is assigned the value 1, regardless of the number of multiple connections. Let the CN be of type $i \in \{1, 2, \ldots, N_c\}$ with parity-check matrix $\boldsymbol{H}_i$, and let u be the number of edges connecting the VN to the CN. Then, during the row expansion procedure which allows generating the GLDPC code parity-check matrix $\boldsymbol{H}$, the nonzero entry of $\boldsymbol{A}$ is replaced

[2] The following early termination strategy may also be applied: stop decoding and declare a decoding failure when no new VN has been recovered at the end of some iteration.

by the element-wise sum in GF(2) of the u columns of $\boldsymbol{H}_i$ corresponding to the u sockets of the CN connected to the VN.

In Section 5.4, the average weight distribution of unstructured LDPC code ensembles has been developed exploiting Lemmas 5.1 and 5.3. The former is a general result about generating functions; the latter, which provides an expression for the weight distribution of replicated codes, is valid for any base code of the replicated code. Therefore, the proof technique developed in Section 5.4 may be applied without modifications to develop the average weight distribution of an unstructured GLDPC code ensemble $\mathscr{C}^n_{\lambda,\rho,\mathscr{C}^c}$. Let $\bar{A}_w$ be the expected number of codewords of Hamming weight w of a random code $\mathcal{C} \in \mathscr{C}^n_{\lambda,\rho,\mathscr{C}^c}$. Let $A_i(W)$ be the weight enumerating function of component code $\mathcal{C}_i \in \mathscr{C}^{(c)}$, $\boldsymbol{w} = \begin{bmatrix} w_2 & w_3 & \ldots & w_{\mathsf{d}_v} \end{bmatrix}$ be a vector weight whose jth element, w_j, represents the portion of the total Hamming weight w apportioned to VNs of degree j, and $\mathcal{W}_w$ be the set of all vector weights $\boldsymbol{w}$ such that $0 \le w_j \le \Lambda_j n$ for all $j \in \{2,3,\ldots,\mathsf{d}_v\}$ and such that $\sum_{j=2}^{\mathsf{d}_v} w_j = w$. Then we have

$$\bar{A}_w = \sum_{\boldsymbol{w} \in \mathcal{W}_w} \frac{\operatorname{coeff}\left(\prod_{i=1}^{N_c} (A_i(W))^{P_i m_c}, W^{\sum_{j=2}^{\mathsf{d}_v} w_j j}\right)}{\binom{\sum_{j=2}^{\mathsf{d}_v} \Lambda_j j n}{\sum_{j=2}^{\mathsf{d}_v} w_j j}} \prod_{j=2}^{\mathsf{d}_v} \binom{\Lambda_j n}{w_j}, \tag{9.21}$$

which simplifies to

$$\bar{A}_w = \binom{n}{w} \frac{\operatorname{coeff}\left((A(W))^{m_c}, W^{w\mathsf{d}_v}\right)}{\binom{n\mathsf{d}_v}{w\mathsf{d}_v}} \tag{9.22}$$

for a random code whose realizations belong to a variable-regular and check-homogeneous GLDPC code ensemble, d_v and $A(W)$ being the constant VN degree and the weight distribution of the unique component code, respectively.

Since $\bar{A}_w$ in (9.21) and $\bar{A}_w$ in (5.30) (irregular LDPC code ensembles) have the same mathematical structure, the same approach can be used to find the weight spectral shape function. For a variable-irregular and check-hybrid GLDPC code ensemble sequence $\{\mathscr{C}^n_{\lambda,\rho,\mathscr{C}^c}\}$, we obtain

$$\begin{aligned} G(\omega) = \frac{1}{\ln 2} \Bigg[& \frac{\int \rho}{\int \lambda} \sum_{i=1}^{N_c} P_i \ln A_i(z_0) - \omega \ln x_0 \\ & + \sum_{j=2}^{\mathsf{d}_v} \Lambda_j \ln B_j(x_0, y_0) + \frac{1}{\int \lambda} \ln(1 - \beta \textstyle\int \lambda) \Bigg], \end{aligned} \tag{9.23}$$

where $B_j(X,Y) = 1 + XY^j$ for all $j \in \{2,3,\ldots,\mathsf{d}_v\}$ and where x_0, y_0, z_0, and β are the unique positive real roots of the four simultaneous equations given by

$$\frac{\int \rho}{\int \lambda} z_0 \sum_{i=1}^{N_c} P_i \left. \frac{\mathrm{d} \ln A_i(Z)}{\mathrm{d}Z} \right|_{z_0} = \beta \tag{9.24}$$

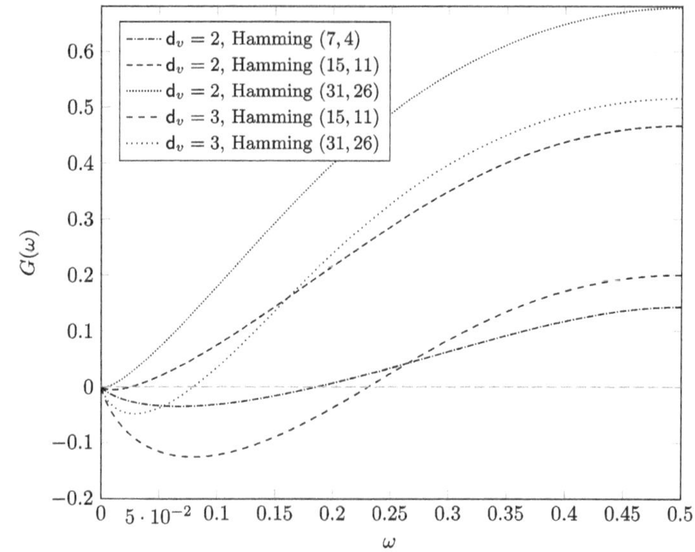

Figure 9.8 Weight spectral shape functions for unstructured variable-regular and check-homogeneous GLDPC code ensemble sequences with Hamming component codes.

and by (5.77), (5.81), and (5.82). The weight spectral shape function (9.23) can be simplified for sequences of variable-regular and check-hybrid GLDPC code ensembles. In particular, we have

$$G(\omega) = \frac{1}{\ln 2}\left[\mathsf{d}_v \left(\textstyle\int \rho\right) \sum_{i=1}^{N_c} P_i \ln A_i(z_0) - \omega\, \mathsf{d}_v \ln z_0 - (\mathsf{d}_v - 1)\mathsf{H}_e(\omega)\right], \tag{9.25}$$

where z_0 is the unique positive root to the equation

$$\left(\textstyle\int \rho\right) z_0 \sum_{i=1}^{N_c} P_i \left.\frac{\mathrm{d} \ln A_i(Z)}{\mathrm{d}Z}\right|_{z_0} = \omega\,. \tag{9.26}$$

Example 9.10 Let us consider variable-regular and check-homogeneous unstructured GLDPC ensemble sequences in which the unique component code is a binary Hamming code. A plot of the weight spectral shape curves is illustrated in Figure 9.8 for all admissible combinations of $\mathsf{d}_v \in \{2, 3\}$ and Hamming component codes of lengths 7, 15, and 31. Each weight spectral shape curve, obtained from (9.25) and (9.26), is plotted in the range $[0, \frac{1}{2}]$ of ω. Letting d and h be the length and the dimension of the Hamming component, respectively, the design rate of each ensemble is $\mathsf{R} = 1 - \frac{(d-h)\mathsf{d}_v}{d}$. At $\omega = \frac{1}{2}$, the value of each weight spectral shape function equals the corresponding design rate.

In Chapter 7, the performance of an LDPC code ensemble under BP decoding has been characterized, in the limit where the blocklength tends to infinity, in terms of an asymptotic decoding threshold $\epsilon^\star$. Under the Tanner graph local tree condition (see also Theorem 7.1), an asymptotic decoding threshold may be defined also for GLDPC code ensembles. Hereafter, we address the calculation of $\epsilon^\star$ for unstructured $\mathscr{C}^n_{\lambda,\rho,\mathscr{C}^c}$ ensembles under Algorithm 4. We first let $n \to \infty$ and then we track the evolution of the average (over the edges of a Tanner graph randomly drawn from the ensemble, upon a random choice of $\mathbf{\Pi}$) extrinsic erasure probability at the output of a VN. We can equivalently track the average MI between an extrinsic message and the associated codeword bit.

Let us start by considering again density evolution for unstructured LDPC code ensembles (Section 7.1.2). Recall that $p_\ell = \epsilon\lambda(q_\ell)$ expresses the average erasure probability p_ℓ of an extrinsic message outgoing from a VN at iteration ℓ, as a function of the channel erasure probability ϵ and of the average erasure probability q_ℓ of an extrinsic message outgoing from a CN at iteration ℓ. Since the processing performed by VNs in BP erasure decoding of LDPC codes (Algorithm 1) and BP erasure decoding of GLDPC codes (Algorithm 4) is the same, this relationship remains valid also in the unstructured GLDPC ensemble framework. Next, let us recall the equation $q_\ell = 1 - \rho(1 - p_{\ell-1})$, expressing q_ℓ as a function of $p_{\ell-1}$. To understand how this equation should be updated when SPC check component codes are replaced with generic linear block codes, we can observe that it may be recast as

$$\begin{aligned} q_\ell &= 1 - \sum_{d=2}^{\mathsf{d}_c} \rho_d (1 - p_{\ell-1})^{d-1} \\ &= \sum_{d=2}^{\mathsf{d}_c} \rho_d \, h_d(p_{\ell-1}) \end{aligned} \tag{9.27}$$

in which, with reference to (7.15), we have denoted by $h_d(\epsilon) = 1 - (1 - \epsilon)^{d-1}$ the ML EXIT function of a $(d, d-1)$ binary SPC code on a BEC with erasure probability ϵ. Recalling the definition and meaning of the ML EXIT function of a linear block code leads us to conclude that the validity of (9.27) is not confined to SPC CNs. For an unstructured GLDPC code ensemble, we may then use the pair of equations

$$p_\ell = \epsilon\, \lambda(q_\ell) \tag{9.28}$$

$$q_\ell = \sum_{i=1}^{N_c} \rho_i \, h_i(p_{\ell-1}) \tag{9.29}$$

to track the evolution of p_ℓ (or q_ℓ) over BP decoding iterations, where $h_i(\epsilon)$ represents the ML EXIT function of component code $C_i \in \mathscr{C}^{(c)}$. Equations (9.28) and (9.29) may be used to calculate the asymptotic decoding threshold of the ensemble in two different (but equivalent) ways.

The first approach consists of constructing a density evolution recursion. Accordingly, we can incorporate the right-hand side of (9.29) into (9.28) obtaining

$$p_\ell = \epsilon\, \lambda\!\left(\sum_{i=1}^{N_c} \rho_i \, h_i(p_{\ell-1}) \right) \tag{9.30}$$

with starting point $p_0 = \epsilon$. Density evolution recursion (9.30) is the generalization of (7.3) to unstructured GLDPC code ensembles. The asymptotic BP decoding threshold of the ensemble is defined as $\epsilon^\star = \sup\{\epsilon > 0 : p_\ell \to 0 \text{ as } \ell \to \infty\}$.

The second approach consists of constructing an EXIT chart. Owing to $\mathsf{I}_{\mathrm{AV}} = \mathsf{I}_{\mathrm{EC}}$ and $\mathsf{I}_{\mathrm{AC}} = \mathsf{I}_{\mathrm{EV}}$, for fixed ϵ we plot on the same chart the functions $\mathsf{I}_{\mathrm{EV}} = \mathsf{T}_{\mathrm{V}}(\mathsf{I}_{\mathrm{AV}}, \epsilon)$ and $\mathsf{I}_{\mathrm{AC}} = \mathsf{T}_{\mathrm{C}}^{-1}(\mathsf{I}_{\mathrm{EC}})$, where $\mathsf{T}_{\mathrm{V}}(\mathsf{I}_{\mathrm{AV}}, \epsilon) = 1 - \epsilon\lambda(1 - \mathsf{I}_{\mathrm{AV}})$ and $\mathsf{T}_{\mathrm{C}}(\mathsf{I}_{\mathrm{AC}}) = \sum_{i=1}^{N_c} \rho_i \, h_i(1 - \mathsf{I}_{\mathrm{AC}})$. The asymptotic BP decoding threshold is the supremum of the set of all $\epsilon > 0$ such that $\mathsf{T}_{\mathrm{V}}(x, \epsilon) > \mathsf{T}_{\mathrm{C}}^{-1}(x)$ for all $x \in [0, 1)$. As for LDPC codes, the larger the area gap $\int_0^1 T_{\mathrm{V}}(x, \epsilon^\star)\mathrm{d}x - \int_0^1 \mathsf{T}_{\mathrm{C}}^{-1}(x)\mathrm{d}x$, the larger the gap $(1 - \mathrm{R}) - \epsilon^\star$.

Example 9.11 Consider a variable-regular and check-homogeneous ($N_c = 1$) unstructured GLDPC code ensemble in which all VNs have degree 2 and the component code of every CNs is the (7,4) binary Hamming codes. From $\lambda(x) = x$ and Example 8.2, the density evolution recursion (9.30) for this ensemble is

$$p_\ell = \epsilon \, p_{\ell-1}^2 \left(3 + 4p_{\ell-1} - 15p_{\ell-1}^2 + 12p_{\ell-1}^3 - 3p_{\ell-1}^4\right) \tag{9.31}$$

with asymptotic BP decoding threshold $\epsilon^\star = 0.75645$.

9.2.3.2 Protograph Ensembles

Similar to the protograph LDPC code ensembles introduced in Section 4.2.3, protograph GLDPC code ensembles are defined by a small bipartite graph called the protograph. The numbers of CNs and VNs in the protograph are still denoted by n_c and $n_v = n_u + n_p$, respectively, where n_u and n_p are the numbers of protograph VNs that are transmitted and punctured, respectively. As opposed to the LDPC case, however, some protograph CNs represent linear block codes that are not SPC codes and that, therefore, impose multiple constraints on the codeword symbols. The sets of indexes of the n_c CNs and n_v VNs in the protograph are still denoted by $\mathcal{I}$ and $\mathcal{J} = \mathcal{J}_u \cup \mathcal{J}_p$, respectively, where $\mathcal{J}_u$ and $\mathcal{J}_p$ are the sets of indexes of the transmitted and punctured VNs, respectively. The number of edges in the protograph is E_{p} and $\Phi = \{1, 2, \ldots, E_{\mathrm{p}}\}$ is the set of indexes of the protograph edges. The subset of Φ including the edge indexes connected to c_i and v_j are Γ_i and Ξ_j, respectively. Considering the protograph CN c_i, the corresponding linear block component code is denoted by $\mathcal{C}_i$, its length by d_i, its dimension by h_i, and its $(d_i - h_i) \times h_i$ parity-check matrix by $\boldsymbol{H}_i$. The sockets of c_i are assumed to be indexed from 1 to d_i.

Remark 9.6. As opposed to protograph LDPC code ensembles, for which a base matrix was sufficient to define the protograph connectivity, a finer description level is necessary for protograph GLDPC ones. In this latter case, in fact, besides specifying to which protograph CN a protograph edge is connected, it is also necessary to specify to which CN *socket* the edge is tied. To allow this higher degree of connectivity description, and to facilitate density evolution analysis presented in the following, we define the four functions

$$\begin{aligned}
&\beta : g \in \Phi \mapsto \beta(g) \in \mathcal{J} \\
&\gamma : g \in \Phi \mapsto \gamma(g) \in \mathcal{I} \\
&\sigma : g \in \Phi \mapsto \sigma(g) \in \{1, \ldots, d_i\} \\
&\eta : (i, s) \in \mathcal{I} \times \{1, \ldots, d_i\} \mapsto \eta(i, s) \in \Phi,
\end{aligned} \tag{9.32}$$

where $\beta(g)$ and $\gamma(g)$ are the indexes of the protograph VN and CN to which the edge of index $g \in \Phi$ is connected, $\sigma(g)$ is the index of the socket of $\mathsf{c}_{\gamma(g)}$ to which the edge is tied, and $\eta(i,s)$ is the index of the edge connected to socket s of c_i.

A GLDPC code is drawn from the ensemble by performing a copy-and-permute operation analogous to the one that was described in Section 4.2.3. It consists of first copying Q times each VN v_j, $j \in \mathcal{J}$, each CN c_i, $i \in \mathcal{I}$, and each edge in the protograph. The Q copies of v_j and c_i are called VNs of type j and CNs of type i, respectively. Note that all Q type-i CNs are associated with the same linear block component code $\mathcal{C}_i$.[3] Consider any edge in the protograph and assume it connects v_j to c_i, for some $j \in \mathcal{J}$ and $i \in \mathcal{I}$. The edge is copied Q times, yielding Q edges each connecting a specific copy of v_j to a specific copy of c_i. These Q edges are then permuted, where the permutation is picked uniformly at random, to obtain a connected Tanner graph with Qn_v VNs and Qn_c CNs. Edge permutation is performed independently for each protograph edge.

Remark 9.7. Although the described procedure may appear identical to the one described in Section 4.2.3 for ordinary protograph binary LDPC codes, as previously mentioned in the case of protograph GLDPC codes a particular attention should be paid to node sockets when performing edge permutation. Let the protograph edge of index $g \in \Phi$ be connected to the socket of index $s \in \{1,\dots,d_i\}$ of c_i (i.e., $\eta(i,s) = g$). Then, after edge permutation, each of the Q copies of the considered edge must be connected to socket s of the corresponding type-i CN.

Remark 9.8. Protograph-based spatially coupled GLDPC code ensembles can be constructed using an approach similar to the one described in Section 9.1.1 for spatially coupled LDPC ones. The approach consists again of starting from unconnected copies of a block GLDPC protograph, of connecting them via an appropriate edge spreading rule, and of lifting the obtained spatially coupled GLDPC protograph. Most of the properties pointed out for protograph-based spatially coupled LDPC codes can be extended to GLDPC ones.

The biadjacency matrix $\boldsymbol{A}$ of the obtained Tanner graph is characterized by $m_c = Qn_c$ rows and $n = Qn_v$ columns. The parity-check matrix $\boldsymbol{H}$ of the drawn binary GLDPC code can be obtained by applying the row expansion rule described at the end of Section 9.2.1; the $\boldsymbol{H}$ matrix has dimensions $m \times n$, where $m = Q\sum_{i\in\mathcal{I}}(d_i - h_i)$. The design rate of the protograph GLDPC code ensemble is therefore

$$\mathsf{R} = \frac{n_v - \sum_{i\in\mathcal{I}}(d_i - h_i)}{n_u}. \tag{9.33}$$

Example 9.12 A GLDPC protograph is depicted in Figure 9.9. It is composed of the two CNs c_1 and c_2, hence $\mathcal{I} = \{1,2\}$, and of the seven transmitted VNs v_1 to v_7, hence $\mathcal{J} = \mathcal{J}_u = \{1,2,\dots,7\}$ and $\mathcal{J}_p = \{\}$. Each VN in the protograph has degree 2

[3] Two different protograph CNs may be associated with the same component code, that is, there may exist $i_1 \in \mathcal{I}$ and $i_2 \in \mathcal{I}$ such that $\mathcal{C}_{i_1} = \mathcal{C}_{i_2}$.

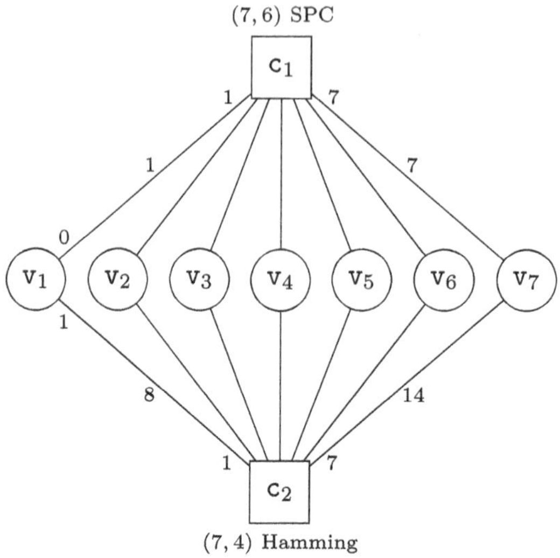

Figure 9.9 Example of GLDPC protograph. The check component code of c_1 is an ordinary SPC code of length 7, while the check component code of c_2 is a (7,4) Hamming code.

and is connected to both CNs. The two CNs have both degree 7, but they are associated with different component codes. Specifically, $\mathcal{C}_1$ is a (7,6) SPC code while $\mathcal{C}_2$ is a (7,4) Hamming code. The sockets of each VN are indexed as 1 and 2, according to the following rule: The socket index coincides with the index of the CN the corresponding edge is connected to. The sockets of each CN are indexed from 1 to 7, according to the same convention. The protograph edges are indexed from 1 to 14 (where only the indexes 1, 7, 8, and 14 are shown in the figure). Using (9.33), the protograph ensemble design rate is R = 3/7.

In the remainder of this section, we discuss how to extend density evolution of protograph LDPC ensembles, described in Section 7.3.2, to the GLDPC framework. We use the symbols $p_\ell^{(g)}$ and $q_\ell^{(g)}$ to denote the extrinsic erasure probability along the protograph edge of index $g \in \Phi$ from $\mathsf{v}_{\beta(g)}$ to $\mathsf{c}_{\gamma(g)}$ and from $\mathsf{c}_{\gamma(g)}$ to $\mathsf{v}_{\gamma(g)}$, respectively, at density evolution iteration ℓ. We adopt the compact notation

$$\begin{aligned} \boldsymbol{p}_\ell &= \left[p_\ell^{(1)} \quad p_\ell^{(2)} \quad \dots \quad p_\ell^{(E_\mathrm{p})}\right] \\ \boldsymbol{q}_\ell &= \left[q_\ell^{(1)} \quad q_\ell^{(2)} \quad \dots \quad q_\ell^{(E_\mathrm{p})}\right] \end{aligned} \tag{9.34}$$

and illustrate a method to express $\boldsymbol{p}_{\ell+1}$ as a function of $\boldsymbol{q}_\ell$ and $\boldsymbol{q}_\ell$ as a function of $\boldsymbol{p}_\ell$, thus allowing tracking of the extrinsic erasure probabilities along each edge of the protograph. For the sake of notational compactness, for any two real vectors $\boldsymbol{a}$ and $\boldsymbol{b}$ of equal length d we let $\boldsymbol{a}^{\boldsymbol{b}} = a_1^{b_1} a_2^{b_2} \cdots a_d^{b_d}$.

Concerning the problem of expressing $\boldsymbol{p}_{\ell+1}$ as a function of $\boldsymbol{q}_\ell$, (7.61) still holds but needs to be recast to the new ensemble description. To this aim we can simply write

$$p_{\ell+1}^{(g)} = \alpha(g) \prod_{e \in \Xi_{\beta(g)} \setminus \{g\}} q_{\ell}^{(e)}, \tag{9.35}$$

where $\alpha(g) = \epsilon$ for all $g \in \Phi$ such that $\beta(g) \in \mathcal{J}_u$ (the bits corresponding to the VN are transmitted) and $\alpha = 1$ for all $g \in \Phi$ such that $\beta(g) \in \mathcal{J}_p$ (the bits corresponding to the VN are punctured).

Regarding the problem of expressing $\boldsymbol{q}_\ell$ as a function of $\boldsymbol{p}_\ell$, we need to develop an expression for the extrinsic erasure probability outgoing from any socket of a CN whose component code is a generic linear block code. The expression we provide is valid under the hypothesis that ML decoding is performed locally at the CN. The proof of the claimed expression is proposed as an exercise at the end of the chapter.

Let $\boldsymbol{G}_i$ be an $h_i \times d_i$ generator matrix of $\mathcal{C}_i$. Let

$$\boldsymbol{t}_{i,\sim s} = \begin{bmatrix} t_{i,\sim s,1} & \dots & t_{i,\sim s,s-1} & t_{i,\sim s,s+1} & \dots & t_{i,\sim s,d_i} \end{bmatrix} \tag{9.36}$$

be a length-$(d_i - 1)$ binary vector, each element of which is associated with a socket of c_i, excluding the one of index s, and define $\boldsymbol{G}_i(\boldsymbol{t}_{i,\sim s})$ as the $h_i \times (\sum_l t_{i,\sim s,l})$ submatrix of $\boldsymbol{G}_i$ composed of the columns of $\boldsymbol{G}_i$ that correspond to nonzero elements of $\boldsymbol{t}_{i,\sim s}$. Let $\boldsymbol{G}_i(s)$ be the sth column of $\boldsymbol{G}_i$ and $\boldsymbol{G}_i(s)|\boldsymbol{G}_i(\boldsymbol{t}_{i,\sim s})$ be the $h_i \times (1 + \sum_l t_{i,\sim s,l})$ matrix obtained by prepending the column $\boldsymbol{G}_i(s)$ to $\boldsymbol{G}_i(\boldsymbol{t}_{i,\sim s})$. Moreover, let

$$\boldsymbol{p}_{i,\sim s,\ell} = \begin{bmatrix} p_\ell^{(\eta(i,1))} & \dots & p_\ell^{(\eta(i,s-1))} & p_\ell^{(\eta(i,s+1))} & p_\ell^{(\eta(i,d_i))} \end{bmatrix} \tag{9.37}$$

be the vector of erasure probabilities $p_\ell^{(g)}$ such that $\gamma(g) = i$ and $\sigma(g) \neq s$. Then, for all $g \in \Phi$ we have

$$q_\ell^{(g)} = \sum_{\boldsymbol{t}_{i,\sim s}} \boldsymbol{p}_{i,\sim s,\ell}^{\boldsymbol{1}-\boldsymbol{t}_{i,\sim s}} (\boldsymbol{1} - \boldsymbol{p}_{i,\sim s,\ell})^{\boldsymbol{t}_{i,\sim s}} \left[\operatorname{rank}(\boldsymbol{G}_i(s)|\boldsymbol{G}_i(\boldsymbol{t}_{i,\sim s})) - \operatorname{rank}(\boldsymbol{G}_i(\boldsymbol{t}_{i,\sim s}))\right], \tag{9.38}$$

where $i = \gamma(g)$, $s = \sigma(g)$, the sum is over all possible $\boldsymbol{t}_{i,\sim s} \in \{0,1\}^{d_i - 1}$ and $\boldsymbol{1}$ is an all-1 vector of length $d_i - 1$.

In summary, density evolution for protograph GLDPC code ensembles consists of initializing the gth element of $\boldsymbol{p}_1$, $g \in \{1, \dots, E_{\mathrm{p}}\}$, to $\alpha(g)$ and, for each $\ell \geq 1$, of computing the elements of $\boldsymbol{q}_\ell$ through (9.38) at first and then the elements of $\boldsymbol{p}_{\ell+1}$ through (9.35). The channel erasure probability value ϵ is said to be achievable when the a posteriori erasure probability of each protograph VN tends to zero as ℓ tends to infinity. The BP decoding threshold of the protograph GLDPC code ensemble (with ML decoding performed locally at the CNs), again denoted by $\epsilon^\star$, is defined as the supremum of the set of all achievable channel erasure probabilities.

Remark 9.9. Equation (9.38) is independent of the specific generator matrix $\boldsymbol{G}_i$ considered for the check component code $\mathcal{C}_i$. In other words, it depends solely on the set of codewords of $\mathcal{C}_i$, that is, on its code book.

Example 9.13 Consider again the ensemble defined by the GLDPC protograph depicted in Figure 9.9, with design rate 3/7. Applying the above-described density evolution analysis we obtain a threshold (rounded to the fifth decimal place) equal to $\epsilon^\star = 0.49602$. Note that, for any ϵ and at any density evolution iteration ℓ, we have $p_\ell^{(1)} = \cdots = p_\ell^{(7)}, q_\ell^{(1)} = \cdots = q_\ell^{(7)}, p_\ell^{(8)} = \cdots = p_\ell^{(14)}, q_\ell^{(8)} = \cdots = q_\ell^{(14)}$, so that we can only track, for example, the extrinsic erasure probabilities along the edges 0 and 8, thus reducing complexity considerably.

Example 9.14 The ensemble analyzed in Example 9.13 is characterized by a regular Tanner graph, since all VNs have degree $\mathsf{d}_v = 2$ and all CNs have degree $\mathsf{d}_c = 7$. Although its iterative decoding threshold is remarkably better than that of a regular LDPC code ensemble with the same code rate ($\epsilon^\star = 0.43659$ for an $\mathsf{R} = 3/7$ LDPC ensemble with CN degree $\mathsf{d}_c = 7$ and VN degree $\mathsf{d}_v = 4$), it is considerably far from the Shannon limit $\epsilon^\star_{\mathsf{Sh}} = 4/7 \approx 0.57143$. A simple technique to improve the ensemble threshold substantially for the same code rate consists of: (i) performing a standard extension of the Hamming component code, turning it into an (8,4) component code with minimum distance $d_{\min} = 4$, and therefore yielding a CN of degree 8; (ii) introducing a punctured degree-1 VN in the protograph, connected to the new socket of the generalized CN. The iterative decoding threshold of the new ensemble, again with ML decoding performed locally at the generalized CNs and rounded to the fifth decimal place, is $\epsilon^\star = 0.55314$, remarkably closer to capacity.

9.3 Low-Density Generator Matrix Codes

This section introduces LDGM codes, a class of linear block codes that may be seen as the dual of LDPC codes and that, similarly to LDPC codes, admit a graphical representation through a sparse graph. The difference between the Tanner graph of an LDPC code introduced in Section 4.1 and the sparse graph of an low-density generator matrix (LDGM) code is that the former is a representation of the code parity-check matrix, while the latter is a representation of the code *generator matrix*. Hence, while LDPC codes admit a sparse parity-check matrix representation and are decoded exploiting it, LDGM codes can be encoded through a sparse generator matrix that can also be employed for iterative decoding. Due to this property, LDGM codes are characterized by low-complexity encoding and decoding. Their distance properties, however, are generally poor. As for SC-LDPC codes and GLDPC codes, we again address the binary case, the extension to the nonbinary one being straightforward.

9.3.1 Code Structure

The generator matrix of a linear block code has been defined in Section 2.3. Letting $\mathcal{C}(n,k)$ be a binary linear block code with length n and dimension k, a generator matrix for $\mathcal{C}$ is any full-rank $k \times n$ matrix $\boldsymbol{G}$ whose k rows are codewords of $\mathcal{C}$. Through $\boldsymbol{G}$ we can map bijectively any vector $\boldsymbol{u} = \begin{bmatrix} u_1 & u_2 & \dots & u_k \end{bmatrix}$ of message bits to a codeword $\boldsymbol{c} = \begin{bmatrix} c_1 & c_2 & \dots & c_n \end{bmatrix}$ through the linear transformation $\boldsymbol{c} = \boldsymbol{u}\boldsymbol{G}$.

The binary linear block code $\mathcal{C}(n,k)$ is said to be an LDGM code when it admits at least one generator matrix $\boldsymbol{G}$ in *sparse* form, that is, with relatively few nonzero entries. The Tanner graph of an LDGM code[4] with sparse generator matrix $\boldsymbol{G}$ is a graph that consists of a set $\{\mathsf{v}_1^{\mathsf{u}}, \mathsf{v}_2^{\mathsf{u}}, \dots, \mathsf{v}_k^{\mathsf{u}}\}$ of k message VNs, a set $\{\mathsf{v}_1^{\mathsf{c}}, \mathsf{v}_2^{\mathsf{c}}, \dots, \mathsf{v}_n^{\mathsf{c}}\}$ of n coded VNs, and a set $\{\mathsf{c}_1, \mathsf{c}_2, \dots, \mathsf{c}_n\}$ of n CNs. For all $i \in \{0,1,\dots,k\}$ the message VN $\mathsf{v}_i^{\mathsf{u}}$ is bijectively associated with bit u_i of $\boldsymbol{u}$; moreover, for all $j \in \{1,2,\dots,n\}$ the coded VN $\mathsf{v}_j^{\mathsf{c}}$ is bijectively associated with the encoded bit c_j of $\boldsymbol{c}$. Every coded VN is connected to one and only one CN; moreover, no two CNs are connected to the same coded VN. The ith message VN $\mathsf{v}_i^{\mathsf{u}}$, $i \in \{1,2,\dots,k\}$, and the jth CN c_j, $j \in \{1,2,\dots,n\}$, are connected by an edge if and only if the entry of $\boldsymbol{G}$ in position (i,j) is nonzero, that is, if u_i participates in the generation of c_j. Note that the Tanner graph of an LDGM code defines the encoding function, that is, the mapping between messages and codewords.

Consider an LDGM code $\mathcal{C}(n,k)$ defined by a sparse matrix $\boldsymbol{G}$ in nonsystematic form. An edge-oriented degree distribution pair may be introduced for the corresponding Tanner graph to describe the degree profile of message VNs and CNs. The degree of a CN is here defined as the number of edges connecting the CN to the message VNs (i.e., the edge connecting the CN to its coded VN is not counted in the definition of CN degree). The fraction of edges connected to message VNs of degree d is denoted by λ_d; moreover, the fraction of edges connected to CNs of degree d is denoted by ρ_d. The polynomials $\lambda(x)$ and $\rho(x)$ are defined as in (4.3) and (4.4), respectively. The number of edges connecting message VNs and CNs is $k/\int\lambda = n/\int\rho$, form which the rate of $\mathcal{C}$ is

$$R = \frac{\int\lambda}{\int\rho} \tag{9.39}$$

with equality as $\boldsymbol{G}$ is full rank by definition.

Example 9.15 Consider a binary linear block code $\mathcal{C}$ with blocklength $n = 10$, dimension $k = 4$, and let

$$\boldsymbol{G} = \begin{bmatrix} 1 & 0 & 0 & 0 & 1 & 0 & 1 & 0 & 1 & 0 \\ 0 & 0 & 1 & 1 & 1 & 1 & 0 & 0 & 0 & 0 \\ 0 & 1 & 0 & 0 & 0 & 1 & 0 & 0 & 1 & 1 \\ 0 & 1 & 0 & 1 & 0 & 0 & 0 & 1 & 0 & 1 \end{bmatrix} \tag{9.40}$$

be a nonsystematic generator matrix for this code. The Tanner graph corresponding to $\boldsymbol{G}$ is depicted in Figure 9.10. It is characterized by 4 message VNs $\{\mathsf{v}_1^{\mathsf{u}}, \dots, \mathsf{v}_4^{\mathsf{u}}\}$, 10 coded VNs $\{\mathsf{v}_1^{\mathsf{c}}, \dots, \mathsf{v}_{10}^{\mathsf{c}}\}$, and 10 CNs $\{\mathsf{c}_1, \dots, \mathsf{c}_{10}\}$, each connected to one coded VN.

[4] For the sake of adopting a uniform notation, we use the name Tanner graph also for the sparse graph representing the generator matrix of an LDGM code.

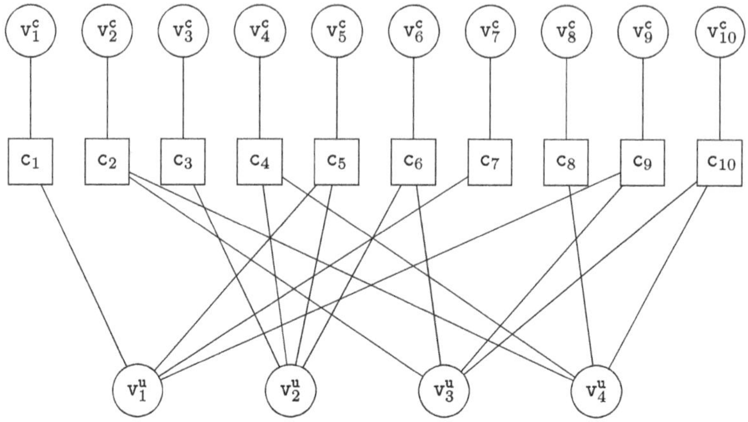

Figure 9.10 Tanner graph corresponding to the generator matrix (9.40) of a binary linear block code with $n = 10$ and $k = 4$.

The edge-oriented degree distribution pair is $\lambda(x) = x^3$ and $\rho(x) = \frac{4}{16} + \frac{12}{16}x$. The rate of $\mathcal{C}$ is $R = \int\lambda / \int\rho = \frac{2}{5}$.

Next, consider an LDGM code $\mathcal{C}(n, k)$ defined by a generator matrix $\boldsymbol{G} = [\boldsymbol{I}_k | \boldsymbol{P}]$ in systematic form, $\boldsymbol{P}$ being a sparse $k \times (n - k)$ matrix. The first k CNs have degree 1 and are connected to the k message VNs in a one-to-one fashion; the last $n - k$ CNs, hereafter called parity CNs, are associated with the $n - k$ parity bits. The degree of a CN is still defined as the number of edges connecting the CN to the message VNs. For a parity CN, this equals the Hamming weight of the corresponding column of $\boldsymbol{P}$. For systematic $\boldsymbol{G}$, we define the parity-check degree of a message VN as the number of edges connecting the VN to parity CNs. This equals the Hamming weight of the corresponding row of $\boldsymbol{P}$. An edge-oriented degree distribution pair $(\lambda_\mathsf{p}, \rho_\mathsf{p})$, where the subscript p reminds us "parity-check," may be used to describe the degree profile of the $n - k$ parity CNs and the parity-check degree profile of message VNs. The corresponding polynomials, defined as in (4.3) and (4.4), are denoted by $\lambda_\mathsf{p}(x)$ and $\rho_\mathsf{p}(x)$. The number of edges connecting message VNs and parity-check CNs is $k / \int\lambda_\mathsf{p} = n / (\int\rho_\mathsf{p} + \int\lambda_\mathsf{p})$, hence the rate of $\mathcal{C}$ is

$$R = \frac{1}{1 + \beta}, \tag{9.41}$$

where, by definition,

$$\beta = \frac{\int\rho_\mathsf{p}}{\int\lambda_\mathsf{p}}. \tag{9.42}$$

We have $R = 1/2$ when $\beta = 1$, $R > 1/2$ when $\beta < 1$, and $R < 1/2$ when $\beta > 1$.

Ensembles of LDGM codes may be defined with an approach similar to the one followed for LDPC (or GLDPC) codes. For example, an unstructured ensemble of LDGM codes in nonsystematic form, of length n and degree distribution pair (λ, ρ), is constructed as follows. There are n CNs, each connected to its own coded VN, and

$(\int\lambda/\int\rho)n$ message VNs. Degrees are assigned to message VNs and to CNs (e.g., with degrees increasing from left to right) to match the degree distribution pair; message VN sockets and CN sockets are indexed from 1 to $n/\int\rho$ (e.g., with indexes increasing from left to right). The ensemble is constructed by generating all $(n/\int\rho)!$ permutation matrices $\mathbf{\Pi}$ of dimension $n/\int\rho$ and of connecting the message VN sockets to the CN sockets according to each such matrix, to obtain a sparse generator matrix representing a binary linear block code. The probability distribution of a random code $\mathcal{C}$ in the ensemble is the one induced by a random choice of $\mathbf{\Pi}$, according to a uniform probability measure. The design rate of the unstructured ensemble is

$$\mathsf{R} = \frac{\int\lambda}{\int\rho}. \tag{9.43}$$

Letting $\mathcal{C}$ be any code in the ensemble, the rate of $\mathcal{C}$ fulfills $R(\mathcal{C}) \leq \mathsf{R}$ with equality if and only if the dimension of $\mathcal{C}$ equals $(\int\lambda/\int\rho)n$, that is, if and only if the $(\int\lambda/\int\rho)n\times n$ binary matrix associated with the Tanner graph is full rank.

More structured ensembles may also be constructed. Partially structured ensembles include, among others, ensembles of LDGM codes in systematic form. An ensemble of LDGM codes in systematic form, of length n and parity-check degree distribution pair $(\lambda_{\mathsf{p}}, \rho_{\mathsf{p}})$, is constructed from n CNs and $n/(1+\beta)$ message VNs, where β is defined in (9.42). There are $n/(1+\beta)$ permanent edges connecting the messages VNs to the first $n/(1+\beta)$ CNs in a one-to-one fashion, while the edges connecting the message VNs to the $n\beta/(1+\beta)$ parity CNs are permuted according to all possible permutation matrices of dimension $n/(\int\rho_{\mathsf{p}} + \int\lambda_{\mathsf{p}})$. The ensemble design rate is

$$\mathsf{R} = \frac{1}{1+\beta}, \tag{9.44}$$

which coincides with the rate $R(\mathcal{C})$ of any code $\mathcal{C}$ in the ensemble.

9.3.2 LDGM Decoding

9.3.2.1 Iterative Decoding

As mentioned earlier, LDGM codes may be decoded iteratively exploiting their sparse generator matrix representation, that is, working over their Tanner graph. A peeling decoder for LDGM codes is described in Algorithm 5. Recall that the channel observation corresponding to coded bit c_i is $y_i = c_i$ if the bit was not erased by the channel and $y_i = \mathsf{E}$ otherwise.

Step 2 in the algorithm is equivalent to removing from the generator matrix all columns that correspond to coded bits erased by the channel; we will refer to it as *column removal* step. Any of these columns is useless for iterative decoding, since its associated CN is not in a position to recover any of its neighboring information VNs.

Stopping sets can be defined for LDGM codes as they were defined for LDPC codes in Section 6.1.4. A stopping set of an LDGM code is any subset of the information VNs such that all neighboring CNs are connected to it at least twice. Iterative decoding of an LDGM code fails whenever, after the column removal step, the residual graph contains at least one stopping set.

Algorithm 5: Peeling decoder for LDGM codes

```
all information VNs v_i^u are marked as erased; all coded VNs v_j^c for which y_i = E
  are marked as erased, all others take the value as suggested by y_i
remove from the graph any coded VN v_j^c marked as erased, along with the CN
  c_j connected to it and all edges incident on c_j
while there exist CNs c_j with one neighboring erased VN v_i^u do
    recover value of v_i^u as the bit-wise sum of v_j^c and of all other VNs
    v_l^u ∈ N(c_j)\{v_j^c, v_i^u}
end
if no erased information VNs left then
    declare decoding to be successful
else
    decoding failed
end
```

Remark 9.10. From the definition of stopping set it follows that, in order for the iterative decoder to have a chance to succeed, the Tanner graph must include CNs with only one edge toward the information VNs (e.g., c_1, c_3, c_7, or c_8 in Figure 9.10). If the Tanner graph does not possess this property, iterative decoding cannot be triggered even over an ideal channel. An important consequence of this fact is that the concept of iterative decoding threshold, introduced for LDPC code ensembles and extended to SC-LDPC and GLDPC codes in this chapter, falls apart in the LDGM case since the density evolution recursion has always a fixed point not corresponding to successful decoding.

9.3.2.2 Maximum Likelihood Decoding

The Tanner graph of an LDGM code defines a linear system $\boldsymbol{G}^{\top}\boldsymbol{u}^{\top} = \boldsymbol{c}^{\top}$ of n equations in the $k < n$ unknowns $u_1, u_2, \ldots, u_k$. Let e be the number of coded bits erased by the channel, $\boldsymbol{c}_{\mathrm{K}}$ be the vector of $n - e$ nonerased coded bits, and $\boldsymbol{G}_{\mathrm{K}}$ be the corresponding $k \times (n - e)$ submatrix of $\boldsymbol{G}$. The peeling decoder addressed above attempts to solve the linear system

$$\boldsymbol{G}_{\mathrm{K}}^{\top}\boldsymbol{u}^{\top} = \boldsymbol{c}_{\mathrm{K}}^{\top} \tag{9.45}$$

iteratively, searching for equations involving one and only one unknown, solving for the unknown, and substituting the obtained value in the other equations.

An optimum ML decoder may also be adopted. With reference to Section 2.2, an ML decoder for LDGM codes makes the decision $\hat{\boldsymbol{u}}$ on $\boldsymbol{u}$ according to

$$\hat{\boldsymbol{u}} = \operatorname*{arg\,max}_{\boldsymbol{u} \in \mathrm{GF}(2)^k} P(\boldsymbol{y}|\boldsymbol{u}), \tag{9.46}$$

if there exists a unique message $\boldsymbol{u}$ maximizing $P(\boldsymbol{y}|\boldsymbol{u})$. If multiple such messages exist, the decoder may output a randomly chosen one or may flag the error (incomplete decoder). The ML decoder may be practically implemented by solving the linear system (9.45) via Gaussian elimination (GE). Note that the same column removal step, performed before iterative decoding, is performed before GE. This is because any column of $\boldsymbol{G}$ corresponding to a bit erased by the channel is associated with a linear equation, involving a subset of the information bits, whose constant term is unknown; as it can be easily proved (via the Rouché–Capelli theorem), such an equation turns out to be useless even under GE. A unique solution $\hat{\boldsymbol{u}}$ is found, and therefore decoding is successful, when $\text{rank}(\boldsymbol{G}_{\mathsf{K}}) = k$. This holds if and only if the erasure pattern does not include the support of any nonzero codeword. By contrast, when $\text{rank}(\boldsymbol{G}_{\mathsf{K}}) < k$ (which holds if and only if the support of any nonzero codeword is contained in the erasure pattern), multiple solutions to the system (9.45) are found and decoding fails.

Remark 9.11. Since an LDGM code possesses codewords with low Hamming weight by construction, performance in the error floor region is generally poor even under ML decoding.

Remark 9.12. When ML erasure decoding is adopted, the constraint on the LDGM code Tanner graph, expressed by Remark 9.10, is not required. In other words, ML erasure decoding does require CNs with only one edge toward the information VNs.

Owing to the sparsity of $\boldsymbol{G}$, an efficient ML decoder may be used that implements the structured GE described in Section 6.2. In practice, structured GE is applied to $\boldsymbol{G}_{\mathsf{K}}^{\mathsf{T}}$ with the goal to resolve a small fraction of unknowns by standard GE, all of the other unknowns being resolved by iterative decoding.

9.3.3 Tornado Codes

A Tornado encoder is obtained by cascading a number of LDGM encoders, each defined by a sparse generator matrix in systematic form. More specifically, the message $\boldsymbol{u} = \begin{bmatrix} c_1 & c_2 & \dots & c_n \end{bmatrix}$ is first encoded by multiplication with a $k \times n_1$ sparse generator matrix $\boldsymbol{G}_1$ in systematic form, yielding a codeword $\boldsymbol{u}\boldsymbol{G}_1 = \begin{bmatrix} \boldsymbol{u} & \boldsymbol{p}_1 \end{bmatrix}$, where $\boldsymbol{p}_1$ is the vector of $n_1 - k$ parity bits. The vector $\boldsymbol{p}_1$ is further encoded by a second LDGM encoder defined by a sparse generator matrix $\boldsymbol{G}_2$, again in systematic form, of dimension $(n_1 - k) \times n_2$, $n_2 > n_1 - k$. Encoding of $\boldsymbol{p}_1$ yields a codeword $\boldsymbol{p}_1\boldsymbol{G}_2 = \begin{bmatrix} \boldsymbol{p}_1 & \boldsymbol{p}_2 \end{bmatrix}$, where $\boldsymbol{p}_2$ is the vector of $n_2 - (n_1 - k)$ parity bits. This process, consisting of starting from $\boldsymbol{p}_0 = \boldsymbol{u}$ and performing, for all $l \in \{1,2,\dots,L-1\}$, systematic LDGM encoding of the parity bits $\boldsymbol{p}_{l-1}$, is repeated $L-1$ times. Vector $\boldsymbol{p}_{L-1}$ is possibly encoded using a classical (e.g., algebraic) systematic linear block encoder, generating $\begin{bmatrix} \boldsymbol{p}_{L-1} & \boldsymbol{p}_L \end{bmatrix}$. The final Tornado codeword, transmitted on the communication channel, is $\boldsymbol{c} = \begin{bmatrix} \boldsymbol{u} & \boldsymbol{p}_1 & \dots & \boldsymbol{p}_{L-1} & \boldsymbol{p}_L \end{bmatrix}$ if the last (classical) encoder is used and $\boldsymbol{c} = \begin{bmatrix} \boldsymbol{u} & \boldsymbol{p}_1 & \dots & \boldsymbol{p}_{L-1} \end{bmatrix}$ otherwise. A Tornado code may be constructed employing systematic LDGM encoders that are characterized by the same degree distribution pair $(\lambda_{\mathsf{p}}, \rho_{\mathsf{p}})$ and the same code rate, or by different ones.

Assuming no final classical encoder is employed, the Tornado code ensemble is defined by the message length k and by the $L-1$ degree distribution pairs $(\lambda_{\mathsf{p}}^{(l)}, \rho_{\mathsf{p}}^{(l)})$, $l \in \{1,2,\dots,L-1\}$. Recalling (9.42), the ensemble design rate is

$$R = (1 + \beta_1 + \beta_1\beta_2 + \cdots + \beta_1\beta_2 \cdots \beta_{L-1})^{-1}, \tag{9.47}$$

which reduces to

$$R = (1 + \beta + \cdots + \beta^{L-1})^{-1} = \frac{1-\beta}{1-\beta^L} \tag{9.48}$$

if all systematic LDGM encoders employ the same degree distribution pair $(\lambda_{\mathrm{p}}, \rho_{\mathrm{p}})$. The second equality in (9.48) holds if and only if $\beta \neq 1$.

Again in the case of no classical final encoder, the graph representing a Tornado code is a cascade of $L-1$ systematic LDGM Tanner graphs. Conventionally, in each such Tanner graph, only the VNs corresponding to parity bits are drawn on the coded VN side. The VNs associated with message bits $\boldsymbol{u}$ form the VN layer 0, the ones associated with bits $\boldsymbol{p}_1$ form the VN layer 1, and so on up to the VN layer $L-1$, for a total of L layers. An efficient way to decode Tornado codes consists of starting from the LDGM Tanner graph involving VNs at layer $L-2$ and $L-1$ and of performing BP erasure decoding on this Tanner graph. Once BP decoding terminates, the next LDGM Tanner graph, involving VN layers $L-3$ and $L-2$, is considered and BP decoding is performed on this graph. The process is repeated until BP decoding is performed on the first graph, the one involving the k VNs associated with the message bits.

9.4 Historical Notes

Early works on convolutional LDPC codes date back to Jiménez Felström and Zigangirov [1] (1999) and to Tanner *et al.* [2] (2004). Time-varying convolutional LDPC codes were developed in [1], along with efficient encoding and decoding algorithms; quasi-cyclic algebraic convolutional LDPC codes were investigated in [2] and compared with block counterparts. More recently, threshold analysis of terminated LDPC convolutional codes was carried out in 2010 by Lentmaier *et al.* in [3] and a solid theoretical justification of the threshold saturation phenomenon was provided in 2011 by Kudekar *et al.* [4]. The ensemble considered in [4] is the randomized ensemble that is mentioned in Remark 9.3 in this chapter. Threshold saturation was further investigated by Andriyanova and Graell i Amat in [5] for nonbinary SC-LDPC code ensembles.

Spatially coupled LDPC codes constructed from protographs were thoroughly investigated by Mitchell *et al.* in the 2015 work [6], which addressed several aspects of them, including constructions, edge spreading rules, minimum distance and threshold analysis, and also irregular constructions. The treatment provided in this chapter draws heavily from [6]. Windowed decoding of spatially coupled LDPC codes was analyzed in a number of works, among which we number Iyengar *et al.* [7], Iyengar *et al.* [8], and Wei *et al.* [9]. The monotonicity properties (9.8) and (9.9), mentioned in this chapter, are established in [7]. Several windowed decoding threshold properties and bounds for the randomized spatially coupled ensemble are available in [8], while windowed decoding of nonbinary spatially coupled LDPC code ensembles is investigated in [9].

The idea of generalized LDPC coding, that is, of error-correcting codes built from short codes and bipartite graphs, appears in Tanner [10] and later in Sipser and Spielman [11]. This idea was further developed in 1999 by Lentmaier and Zigangirov [12], who investigated GLDPC codes based on Hamming component codes, and by Boutros *et al.* [13], who developed several properties of regular GLDPC codes, including average weight distribution and minimum distance bounds. Remarkable GLDPC code constructions, mainly addressing error channels, were also proposed by Xu *et al.* [14], Liva *et al.* [15], Miladinovic and Fossorier [16]. Works of generalized LDPC codes and related constructions over the erasure channel include [17–25].

The first LDGM code construction was proposed in 2003 by Garcia-Frias and Zhong [26], where the performance of systematic LDGM code constructions was assessed over error channels. From a code-theoretical viewpoint, LDGM codes can be seen as the dual of LDPC codes (see also Forney's duality framework for codes on graphs [27]). Tornado codes, composed of a cascade of several bipartite graphs, appear in the 2001 work by Luby *et al.* [28], along with iterative decoding over the erasure channel and threshold analysis.

Finally, we point out that the result, mentioned in this chapter, about the ML ensemble threshold approaching capacity as the CN degree increases for the same design rate, founds its justification in [29].

9.5 Exercises

Exercise 9.1 Consider a protograph-based SC-LDPC block code ensemble with base matrix $\boldsymbol{B} = \begin{bmatrix} 4 & 4 \end{bmatrix}$. Consider the two values of the syndrome former memory $w = 1$ and $w = 2$. For each of them, compute the design rate of the corresponding ensemble for $L = 8, 16, 32, 48, 64$, evaluating the rate loss with respect to that of the original block LDPC ensemble, and enumerate all possible edge spreading rules.

Exercise 9.2 For two edge spreading rules identified in Exercise 9.1, one with $w = 3$ and one with $w = 4$, numerically compute the asymptotic decoding thresholds under BP decoding for $L = 8, 16, 32, 48, 64$ (see Table 9.1), checking the threshold saturation phenomenon.

Exercise 9.3 Section 5.6.4 describes a technique to calculate the weight spectral shape and the critical fractional codeword weight ω^* of a protograph LDPC code ensemble. Applying that approach, compute numerically the value of ω^* for the protograph-based SC-LDPC block code ensemble whose spatially coupled protograph is depicted in Figure 9.1c. Observe the behavior of the parameter $\omega^* L/(w+1)$ as L increases, showing numerically how it converges to a constant value. Provide an interpretation to the result. *Hint*: To compute numerically ω^* for some L, exploit symmetry in the spatially coupled protograph to reduce the size of the system.

Exercise 9.4 Prove inequality (9.9), establishing monotonicity of the windowed decoding threshold with the window size W. *Hint*: Augment the window of size W to one of size $W + 1$ by adding one extra level on the right, setting to constant 0, for every extra edge, the MI between any message over it and the corresponding codeword symbol. Compare protograph EXIT analysis for the window of size $W+1$ with protograph EXIT analysis for the augmented window. Use monotonicity of the expression in (7.61) with respect to any argument I_{AV} and of the expression in (7.62) with respect to any argument I_{AC}.

Exercise 9.5 Consider the density evolution equations (9.28) and (9.29) for an unstructured GLDPC code ensemble. Prove that (9.29) may be recast in the equivalent form

$$q_\ell = \sum_{i=1}^{N_c} \rho_i \frac{1}{d_i} \sum_{t=0}^{d_i-1} p_\ell^t (1-p_\ell)^{d_i-t-1} \left[(d_i - t)\tilde{e}^{(i)}_{d_i-t} - (t+1)\tilde{e}^{(i)}_{d_i-t-1} \right],$$

where $\tilde{e}^{(i)}_g$ is the gth unnormalized information function of the ith component code. This parameter equals the sum of the ranks of all matrices obtained selecting g columns in $\boldsymbol{G}_i$, where $\boldsymbol{G}_i$ is any generator matrix of the component code.

Exercise 9.6 Consider an unstructured GLDPC code ensemble constructed on GF(q). The ensemble definition is the same as the one given in Section 4.2.1 for LDPC codes, with the difference represented by the presence of generalized CNs constructed on GF(q). Extend to this framework the expression (9.21) of the expected number of codewords of weight w.

Exercise 9.7 Prove equation (9.38), which establishes an expression for the erasure probability $q_\ell^{(g)}$ along the gth protograph edge in the framework of density evolution for protograph-based GLDPC code ensembles.

Exercise 9.8 Example 9.14 presents a technique to improve the threshold of a protograph GLDPC code ensemble by extension and puncturing of a generalized CN. Apply the same technique to a protograph similar to the one depicted in Figure 9.9 in which, however, both CNs are generalized Hamming ones. Numerically evaluate the threshold of the initial ensemble, of the ensemble obtained by applying extension and puncturing to one of the CN, and of the ensemble obtained applying extension and puncturing to both CNs.

Exercise 9.9 Section 5.5 in Chapter 5 addressed an expression of the average weight distribution for a protograph LDPC code ensemble. Elaborating on it, Section 5.6.4 provided a technique to efficiently evaluate the ensemble weight spectral shape. Extend the analysis in Chapter 5 to protograph GLDPC code ensembles. Then, numerically evaluate the critical fractional codeword weight ω^* of the ensembles mentioned in Exercise 9.8.

Exercise 9.10 Write density evolution equations for nonsystematic LDGM code ensembles. Show that the asymptotic BP decoding threshold is zero, thus justifying Remark 9.10.

References

[1] A. Jiménez Felström and K. S. Zigangirov, "Time-varying periodic convolutional codes with low-density parity-check matrix," *IEEE Trans. Inf. Theory*, vol. 45, no. 6, pp. 2181–2191, Sep. 1999.

[2] R. M. Tanner, D. Sridhara, A. Sridharan, T. E. Fuja, and D. J. Costello, "LDPC block and convolutional codes based on circulant matrices," *IEEE Trans. Inf. Theory*, vol. 50, no. 12, pp. 2966–2984, Dec. 2004.

[3] M. Lentmaier, A. Sridharan, D. J. Costello, and K. S. Zigangirov, "Iterative decoding threshold analysis for LDPC convolutional codes," *IEEE Trans. Inf. Theory*, vol. 56, no. 10, pp. 5274–5289, Oct. 2010.

[4] S. Kudekar, T. J. Richardson, and R. L. Urbanke, "Threshold saturation via spatial coupling: Why convolutional LDPC ensembles perform so well over the BEC," *IEEE Trans. Inf. Theory*, vol. 57, no. 2, pp. 803–834, Feb. 2011.

[5] I. Andriyanova and A. Graell i Amat, "Threshold saturation for nonbinary SC-LDPC codes on the binary erasure channel," *IEEE Trans. Inf. Theory*, vol. 62, no. 5, pp. 2622–2638, May 2016.

[6] D. G. M. Mitchell, M. Lentmaier, and D. J. Costello, "Spatially coupled LDPC codes constructed from protographs," *IEEE Trans. Inf. Theory*, vol. 61, no. 9, pp. 4866–4889, Sep. 2015.

[7] A. R. Iyengar, M. Papaleo, P. H. Siegel, J. K. Wolf, A. Vanelli-Coralli, and G. E. Corazza, "Windowed decoding of protograph-based LDPC convolutional codes over erasure channels," *IEEE Trans. Inf. Theory*, vol. 58, no. 4, pp. 2303–2320, Apr. 2012.

[8] A. R. Iyengar, P. H. Siegel, R. L. Urbanke, and J. K. Wolf, "Windowed decoding of spatially coupled codes," *IEEE Trans. Inf. Theory*, vol. 59, no. 4, pp. 2277–2292, Apr. 2013.

[9] L. Wei, D. G. M. Mitchell, T. E. Fuja, and D. J. Costello, "Design of spatially coupled LDPC codes over GF(q) for windowed decoding," *IEEE Trans. Inf. Theory*, vol. 62, no. 9, pp. 4781–4800, Sep. 2016.

[10] M. Tanner, "A recursive approach to low complexity codes," *IEEE Trans. Inf. Theory*, vol. 27, no. 5, pp. 533–547, Sep. 1981.

[11] M. Sipser and D. Spielman, "Expander codes," *IEEE Trans. Inf. Theory*, vol. 42, no. 6, pp. 1710–1722, Nov. 1996.

[12] M. Lentmaier and K. S. Zigangirov, "On generalized low-density parity-check codes based on Hamming component codes," *IEEE Commun. Lett.*, vol. 3, no. 8, pp. 248–250, 1999.

[13] J. Boutros, O. Pothier, and G. Zemor, "Generalized low density (Tanner) codes," in *Proc. IEEE Int. Conf. Commun.*, vol. 1, Vancouver, Canada, June 1999, pp. 441–445.

[14] J. Xu, L. Chen, L. Zeng, L. Lan, and S. Lin, "Construction of low-density parity-check codes by superposition," *IEEE Trans. Commun.*, vol. 53, no. 2, pp. 243–251, Feb. 2005.

[15] G. Liva, W. E. Ryan, and M. Chiani, "Quasi-cyclic generalized LDPC codes with low error floors," *IEEE Trans. Commun.*, vol. 56, no. 1, pp. 49–57, Jan. 2008.

[16] N. Miladinovic and M. P. C. Fossorier, "Generalized LDPC codes and generalized stopping sets," *IEEE Trans. Commun.*, vol. 56, no. 2, pp. 201–212, Feb. 2008.

[17] A. Ashikhmin, G. Kramer, and S. ten Brink, "Extrinsic information transfer functions: Model and erasure channel properties," *IEEE Trans. Inf. Theory*, vol. 50, no. 11, pp. 2657–2673, Nov. 2004.

[18] M. Lentmaier, M. B. S. Tavares, and G. P. Fettweis, "Exact erasure channel density evolution for protograph-based generalized LDPC codes," in *Proc. IEEE Int. Symp. Inf. Theory*, Seoul, South Korea, June/July 2009, pp. 566–570.

[19] E. Paolini, M. P. C. Fossorier, and M. Chiani, "Doubly-generalized LDPC codes: Stability bound over the BEC," *IEEE Trans. Inf. Theory*, vol. 55, no. 3, pp. 1027–1046, Mar. 2009.

[20] E. Paolini, M. P. C. Fossorier, and M. Chiani, "Generalized and doubly generalized LDPC codes with random component codes for the binary erasure channel," *IEEE Trans. Inf. Theory*, vol. 56, no. 4, pp. 1651–1672, Apr. 2010.

[21] M. F. Flanagan, E. Paolini, M. Chiani, and M. P. C. Fossorier, "On the growth rate of the weight distribution of irregular doubly generalized LDPC codes," *IEEE Trans. Inf. Theory*, vol. 57, no. 6, pp. 3721–3737, June 2011.

[22] M. F. Flanagan, E. Paolini, M. Chiani, and M. P. C. Fossorier, "Spectral shape of doubly-generalized LDPC codes: Efficient and exact evaluation," *IEEE Trans. Inf. Theory*, vol. 59, no. 11, pp. 7212–7228, Nov. 2013.

[23] P. M. Olmos, D. G. M. Mitchell, and D. J. Costello, "Analyzing the finite-length performance of generalized LDPC codes," in *Proc. 2015 IEEE Int. Symp. Inf. Theory*, Hong Kong, China, June 2015, pp. 2683–2687.

[24] C. Häger, H. D. Pfister, A. Graell i Amat, and F. Brännström, "Density evolution for deterministic generalized product codes on the binary erasure channel at high rates," *IEEE Trans. Inf. Theory*, vol. 63, no. 7, pp. 4357–4378, July 2017.

[25] Y. Liu, P. M. Olmos, and T. Koch, "A probabilistic peeling decoder to efficiently analyze generalized LDPC codes over the BEC," *IEEE Trans. Inf. Theory*, vol. 65, no. 8, pp. 4831–4853, Aug. 2019.

[26] J. Garcia-Frias and Wei Zhong, "Approaching Shannon performance by iterative decoding of linear codes with low-density generator matrix," *IEEE Commun. Lett.*, vol. 7, no. 6, pp. 266–268, June 2003.

[27] G. D. Forney, "Codes on graphs: Duality and MacWilliams identities," *IEEE Trans. Inf. Theory*, vol. 57, no. 3, pp. 1382–1397, Mar. 2011.

[28] M. Luby, M. Mitzenmacher, A. Shokrollahi, and D. Spielman, "Efficient erasure correcting codes," *IEEE Trans. Inf. Theory*, vol. 47, no. 2, pp. 569–584, Feb. 2001.

[29] C. Hsu and A. Anastasopoulos, "Capacity-achieving codes with bounded graphical complexity and maximum likelihood decoding," *IEEE Trans. Inf. Theory*, vol. 56, no. 3, pp. 992–1006, Mar. 2010.

10 Polar Codes

This chapter describes polar codes, which are block codes designed for simplifying the implementation of the encoder and the decoder. They are closely related to Reed–Muller (RM) codes, an important family of codes introduced in the '60s. Polar codes have the property that their code rate can be adjusted from 0 to 1 without changing the encoder and decoder, and have a recursive structure, which makes them highly suitable for implementation in hardware. Specifically, polar codes are designed assuming a successive cancellation (SC) decoder, and could achieve capacity over symmetric binary DMC with encoding and decoding methods of complexity $\mathcal{O}(n \log n)$. Polar codes are the first provably capacity-achieving class of codes with polynomial-time algorithms for encoding and decoding and with an explicit code construction.

10.1 Reed–Muller Codes

For $n = 2^m$ define the $n \times n$ matrix $\boldsymbol{G}_{\mathrm{RM}}(n,n)$ as the m-fold Kronecker product

$$\boldsymbol{G}_{\mathrm{RM}}(n,n) = \boldsymbol{F}_2 \otimes \boldsymbol{F}_2 \otimes \cdots \otimes \boldsymbol{F}_2 = \boldsymbol{F}_2^{\otimes n}, \tag{10.1}$$

where $\boldsymbol{F}_2$ is the 2×2 binary Hadamard matrix

$$\boldsymbol{F}_2 = \begin{bmatrix} 1 & 0 \\ 1 & 1 \end{bmatrix} \tag{10.2}$$

and all operations are in GF(2). For example, with $n = 8$ we have

$$\boldsymbol{G}_{\mathrm{RM}}(8,8) = \boldsymbol{F}_2 \otimes \boldsymbol{F}_2 \otimes \boldsymbol{F}_2 = \begin{bmatrix} 1 & 0 & 0 & 0 & 0 & 0 & 0 & 0 \\ 1 & 1 & 0 & 0 & 0 & 0 & 0 & 0 \\ 1 & 0 & 1 & 0 & 0 & 0 & 0 & 0 \\ 1 & 1 & 1 & 1 & 0 & 0 & 0 & 0 \\ 1 & 0 & 0 & 0 & 1 & 0 & 0 & 0 \\ 1 & 1 & 0 & 0 & 1 & 1 & 0 & 0 \\ 1 & 0 & 1 & 0 & 1 & 0 & 1 & 0 \\ 1 & 1 & 1 & 1 & 1 & 1 & 1 & 1 \end{bmatrix}. \tag{10.3}$$

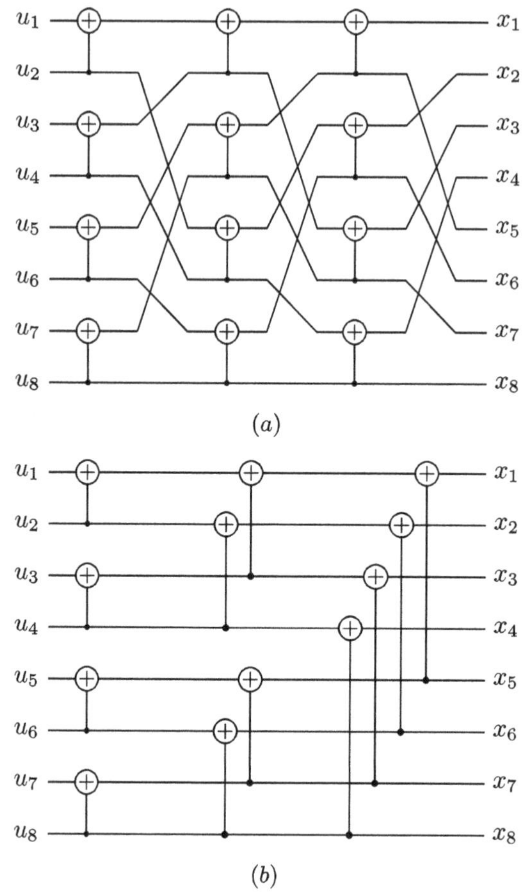

Figure 10.1 Two equivalent graphical representations of the automorphism defined by $\boldsymbol{G}_{\text{RM}}(8,8) = \boldsymbol{F}_2 \otimes \boldsymbol{F}_2 \otimes \boldsymbol{F}_2$. (a) Three identical stages, composed of 2×2 transforms followed by permutation. (b) FFT-like structure. © AEIT. Reproduced with permission.

The weights of all rows of $\boldsymbol{G}_{\text{RM}}(n,n)$ are powers of two, and in particular there are $\binom{m}{t}$ rows of weight 2^t, $t \in \{0,1,\ldots,m\}$. The matrix is full rank and self-inverse, and defines the automorphism $\boldsymbol{x} = \boldsymbol{u}\,\boldsymbol{G}_{\text{RM}}(n,n)$. A schematic diagram illustrating such a linear transformation for the $n = 8$ case is depicted in Figure 10.1. An (n,k) linear block code can be built by choosing any k rows of $\boldsymbol{G}_{\text{RM}}(n,n)$ as generator matrix.

The RM code of length $n = 2^m$ and order $0 \leq r \leq m$ is the code generated by the $k = \sum_{t=0}^{r} \binom{m}{t}$ rows of $\boldsymbol{G}_{\text{RM}}(n,n)$ whose Hamming weight is not less than 2^{m-r}. The corresponding generator matrix is hereafter denoted by $\boldsymbol{G}_{\text{RM}}(n,k)$. It can be shown that the minimum distance $d_{\min}$ of the resulting code is 2^{m-r}, that is, it coincides with the minimum Hamming weight of the selected rows.

For example, with $n = 8$ and $r = 1$ we get $d_{\min} = 4, k = 4$ with generator matrix

$$\boldsymbol{G}_{\mathrm{RM}}(8,4) = \begin{bmatrix} 1 & 1 & 1 & 1 & 0 & 0 & 0 & 0 \\ 1 & 1 & 0 & 0 & 1 & 1 & 0 & 0 \\ 1 & 0 & 1 & 0 & 1 & 0 & 1 & 0 \\ 1 & 1 & 1 & 1 & 1 & 1 & 1 & 1 \end{bmatrix}, \tag{10.4}$$

while for $d_{\min} = 2$ we have $k = 7$ and generator matrix

$$\boldsymbol{G}_{\mathrm{RM}}(8,7) = \begin{bmatrix} 1 & 1 & 0 & 0 & 0 & 0 & 0 & 0 \\ 1 & 0 & 1 & 0 & 0 & 0 & 0 & 0 \\ 1 & 1 & 1 & 1 & 0 & 0 & 0 & 0 \\ 1 & 0 & 0 & 0 & 1 & 0 & 0 & 0 \\ 1 & 1 & 0 & 0 & 1 & 1 & 0 & 0 \\ 1 & 0 & 1 & 0 & 1 & 0 & 1 & 0 \\ 1 & 1 & 1 & 1 & 1 & 1 & 1 & 1 \end{bmatrix}. \tag{10.5}$$

The encoding rule of Reed–Muller (RM) codes may simply be defined through the generator matrix obtained with the selected rows. An alternative way to define the encoding rule is to use the full matrix $\boldsymbol{G}_{\mathrm{RM}}(n,n)$, forcing to zero all elements of the input vector $\boldsymbol{u}$ of length n corresponding to the discarded rows, and leaving free all the other elements of $\boldsymbol{u}$. The former elements of $\boldsymbol{u}$, corresponding to the rows of $\boldsymbol{G}_{\mathrm{RM}}(n,n)$ of weight less than 2^{m-r}, are referred to as *frozen bits*, while the latter ones represent the information bits.

For example, for the (8, 4) RM code above, we have

$$\begin{bmatrix} x_1 & x_2 & \dots & x_8 \end{bmatrix} = \begin{bmatrix} 0 & 0 & 0 & u_4 & 0 & u_6 & u_7 & u_8 \end{bmatrix} \boldsymbol{G}_{\mathrm{RM}}(8,8). \tag{10.6}$$

RM codes are easy to encode and admit several decoding strategies, including SC decoding, ML, and BP algorithms. In selecting the rows of the full $\boldsymbol{G}_{\mathrm{RM}}(n,n)$, the criterion adopted to generate RM codes is that of the largest weights, yielding the largest possible minimum distance. In contrast, polar codes are obtained by selecting the rows of $\boldsymbol{G}_{\mathrm{RM}}(n,n)$ in order to achieve the best performance under SC decoding on a given channel.

10.2 Decomposing Multi-input Multi-output Channels for Successive Cancellation Decoding

Consider the multi-input multi-output (MIMO) DMC depicted in Figure 10.2, where $\boldsymbol{u} = \begin{bmatrix} u_1 & \dots & u_n \end{bmatrix}$ and $\boldsymbol{y} = \begin{bmatrix} y_1 & \dots & y_n \end{bmatrix}$ are the input and the output vectors, and where $\boldsymbol{U} = \begin{bmatrix} U_1 & \dots & U_n \end{bmatrix}$ and $\boldsymbol{Y} = \begin{bmatrix} Y_1 & \dots & Y_n \end{bmatrix}$ are the corresponding random vectors. From the chain rule for entropy, $\mathsf{H}(X_1,\dots,X_n) = \sum_{i=1}^{n} \mathsf{H}(X_i|X_{i-1},X_{i-2},\dots,X_1)$, the mutual information between the vectors $\boldsymbol{U}$ and $\boldsymbol{Y}$ can be written as

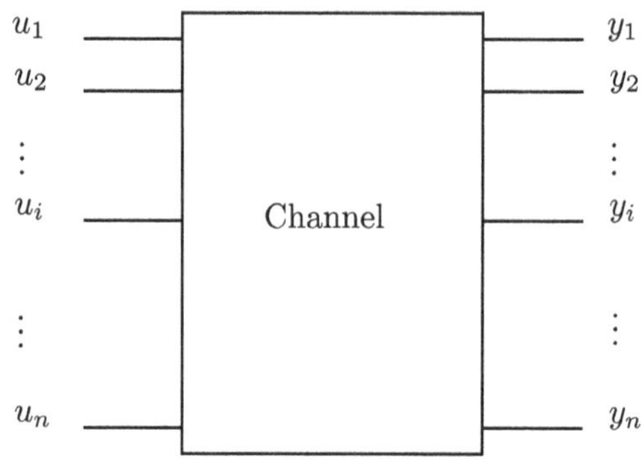

Figure 10.2 A MIMO DMC channel. © AEIT. Reproduced with permission.

$$\begin{aligned} \mathsf{I}(\boldsymbol{U};\boldsymbol{Y}) &= \mathsf{H}(\boldsymbol{U}) - \mathsf{H}(\boldsymbol{U}|\boldsymbol{Y}) \\ &= \sum_{i=1}^{n} \mathsf{H}(U_i|U_{i-1},\ldots,U_2,U_1) - \mathsf{H}(U_i|\boldsymbol{Y},U_{i-1},\ldots,U_2,U_1). \end{aligned} \tag{10.7}$$

If we further assume that the symbols $U_1, U_2, \ldots$ are independent, then

$$\mathsf{H}(U_i|U_{i-1},\ldots,U_2,U_1) = \mathsf{H}(U_i), \tag{10.8}$$

and therefore

$$\begin{aligned} \mathsf{I}(\boldsymbol{U};\boldsymbol{Y}) &= \sum_{i=1}^{n} \mathsf{H}(U_i) - \mathsf{H}(U_i|\boldsymbol{Y},U_{i-1},\ldots,U_2,U_1) \\ &= \sum_{i=1}^{n} \mathsf{I}(U_i;\boldsymbol{Y},U_{i-1},\ldots,U_2,U_1) = \sum_{i=1}^{n} \mathsf{I}_i, \end{aligned} \tag{10.9}$$

where

$$\mathsf{I}_i = \mathsf{I}(U_i;\boldsymbol{Y},U_{i-1},\ldots,U_2,U_1) \tag{10.10}$$

can be interpreted as the mutual information of a "virtual channel" Q_i with input U_i and, as output, $\boldsymbol{Y}$ together with all previous input symbols $U_{i-1},\ldots,U_2,U_1$. This virtual channel is depicted in Figure 10.3. This simple analysis reveals that a MIMO channel with independent input symbols can be decomposed into n channels $Q_1,\ldots,Q_n$, where Q_i has input u_i and, as observables, the vector $\boldsymbol{y}$ and all input symbols $u_{i-1},\ldots,u_1$. We may think of Q_i as the channel we deal with when decoding u_i in the framework of an SC decoder (also known as a multistage decoder), assuming all previous symbols have been correctly recovered (genie-aided SC decoder).

In the particular case in which each y_i depends solely on the corresponding u_i (i.e., if the MIMO channel is composed of n parallel channels, as exemplified in Figure 10.4(a) for $n = 3$), then we trivially have $\mathsf{I}_i = \mathsf{I}(U_i;Y_i)$ and (10.9) does not provide any insight. However, if $\boldsymbol{u}$ is first mapped onto $\boldsymbol{x}$ by an automorphism defined by a full-rank matrix $\boldsymbol{G}$, and then the elements of $\boldsymbol{x}$ are transmitted over the n parallel

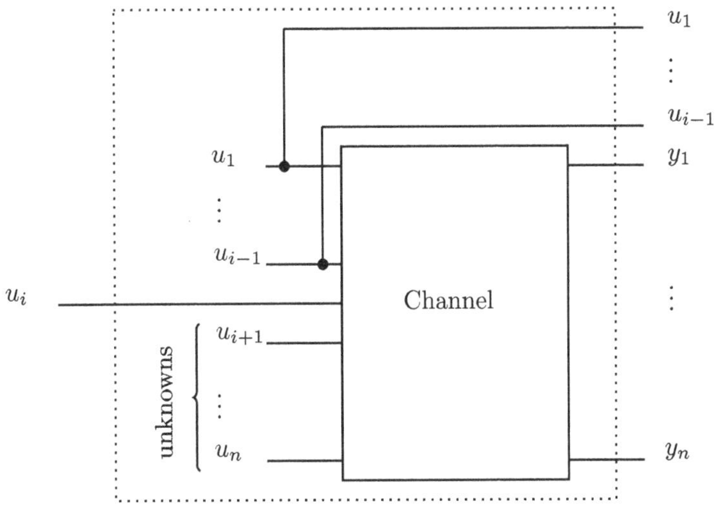

Figure 10.3 Representation of virtual channel Q_i in the decomposition (10.9). © AEIT. Reproduced with permission.

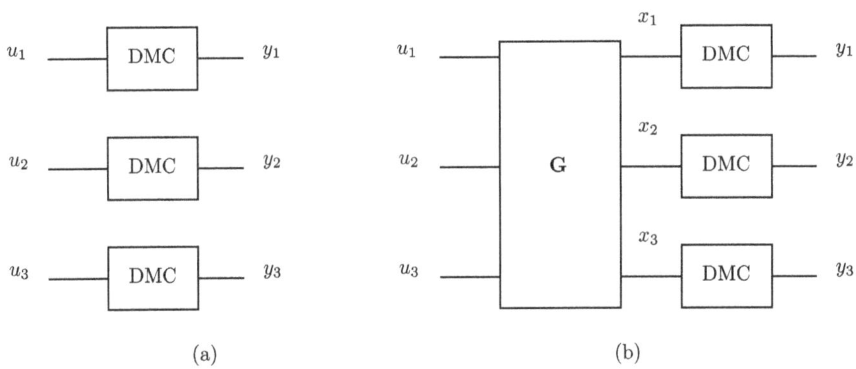

Figure 10.4 Parallel DMC: (a) use in parallel and (b) use preceded by a transformation. © AEIT. Reproduced with permission.

DMCs (Figure 10.4(b), again for $n = 3$), then each y_i will depend in general on several input symbols u_i. In this case, the virtual channels Q_i obtained by (10.9) will have different mutual information values.

For large n and binary symmetric DMCs, many automorphisms lead to the celebrated channel polarization phenomenon, that is, to virtual channels Q_i whose mutual information I_i approaches either 1 or 0, apart from a vanishing fraction of virtual channels. Since the total mutual information must be $n\,\mathsf{I}(U_i;Y_i)$, it turns out that under the polarization effect, there are approximately $n\,\mathsf{I}(U_i;Y_i)$ *noiseless* virtual channels, the remaining ones being *useless* (and therefore not used, enforcing frozen values $u_i = 0$). Thus, one may transmit $k \approx n\,\mathsf{I}(U_i;Y_i)$ data symbols over the noiseless virtual channels; at the decoder side, use SC to decode successively each data symbol from the channel outputs and previously decoded data symbols. The virtual channels require perfect knowledge of $u_1,\ldots,u_{i-1}$. Instead, in practice, the SC decoder only provides an estimate. For the analysis we resort to the genie-aided SC decoder, arguing that the block error probability is the same (see Exercise 10.7).

10.2.1 The Linear Transformation $F_2^{\otimes m}$ as a Way to Introduce Channel Polarization

As a particular case of Figure 10.4, assume we have n parallel BECs, all with the same erasure probability ϵ and mutual information $\mathsf{I}(\epsilon) = 1 - \epsilon$. Each such BEC is denoted by W. Note that $\mathsf{I}(\epsilon)$ coincides with the probability that the channel output is the same as the input.

10.2.1.1 Analysis of a 2 × 2 Channel Defined by F_2

Consider the elementary scheme in Figure 10.5, where $n = 2$ and where the invertible mapping $\boldsymbol{x} = \boldsymbol{u}\boldsymbol{F}_2$ is applied to two bits u_1 and u_2 to obtain the transformed bits $x_1 = u_1 \oplus u_2$ and $x_2 = u_2$ that are transmitted over two parallel W channels. The overall 2×2 MIMO channel is denoted by W_2. The (invertible) transformation can be written as

$$\boldsymbol{x} = \boldsymbol{u}\boldsymbol{F}_2, \tag{10.11}$$

where operations are in GF(2), and $\boldsymbol{F}_2$ is the 2×2 binary Hadamard matrix in (10.2). Each y_i is or exactly coincident with x_i (with probability $\mathsf{I}(\epsilon)$), or an erasure (with probability ϵ). Owing to decomposition (10.9), let us first consider decoding of symbol u_1, assuming only the observable $\boldsymbol{y} = \begin{bmatrix} y_1 & y_2 \end{bmatrix}$ is available. In this case, since $u_1 = x_1 \oplus x_2$, we have that u_1 can be recovered only if both x_1 and x_2 are known, that is, if and only if neither y_1 nor y_2 is erased. Therefore

$$\begin{aligned} P(u_1 \text{ correctly decoded } | y_1, y_2) &= P(x_1 \text{ and } x_2 \text{ correctly received } | y_1, y_2) \\ &= \mathsf{I}(\epsilon)\mathsf{I}(\epsilon) = \mathsf{I}^2(\epsilon)\,. \end{aligned} \tag{10.12}$$

The second virtual channel in (10.9) has input u_2 and observables (y_1, y_2, u_1). For this subchannel, assuming u_1 is known (e.g., by a previous decision, assuming no error propagation), it is possible to recover u_2 from either x_1 or x_2. Thus,

$$\begin{aligned} P(u_2 \text{ correctly decoded } | y_1, y_2, u_1) &= P(x_1 \text{ or } x_2 \text{ correctly received } | y_1, y_2) \\ &= \mathsf{I}(\epsilon) + \mathsf{I}(\epsilon) - \mathsf{I}^2(\epsilon) = 2\,\mathsf{I}(\epsilon) - \mathsf{I}^2(\epsilon)\,. \end{aligned} \tag{10.13}$$

Observing that over the BEC mutual information values coincide with probabilities of correct decoding, we may conclude that the mutual information is transformed, from

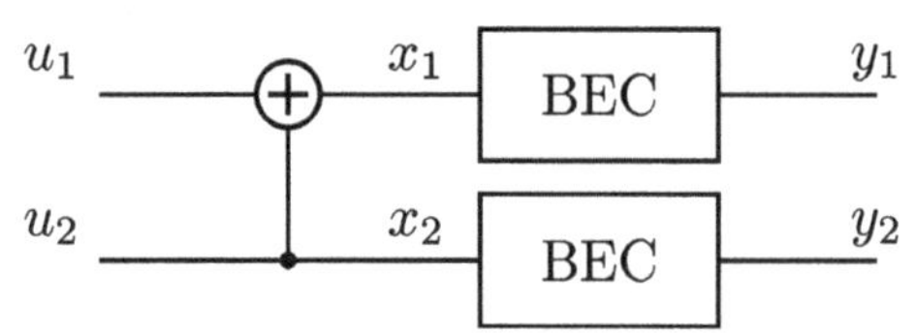

Figure 10.5 The channel W_2. © AEIT. Reproduced with permission.

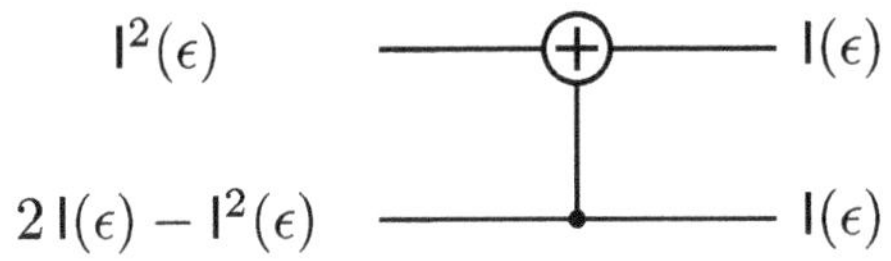

Figure 10.6 Mutual information transformation for the erasure channel and $\boldsymbol{F}_2$ automorphism. Mutual information values on the left-hand side are referred to as SC decoding, that is, they apply when u_1 is decoded first, and then u_2 is decoded assuming u_1 is known. © AEIT. Reproduced with permission.

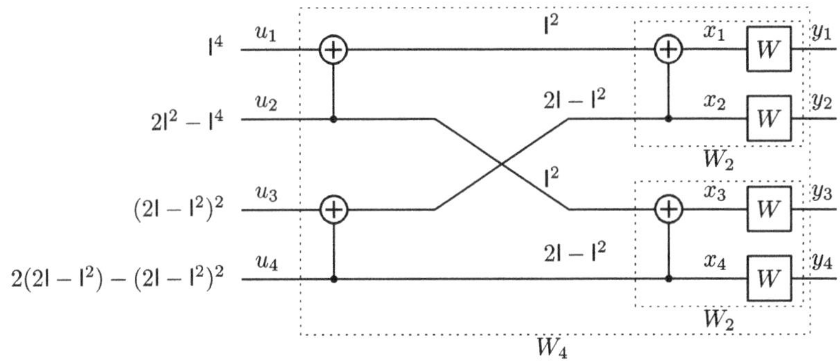

Figure 10.7 Transformation of the mutual information from the channels to the symbols. The BEC is indicated as W. © AEIT. Reproduced with permission.

right to left, as illustrated in Figure 10.6. Note that the mutual information, $2\mathsf{I}(\epsilon)$, is preserved. With $0 < \mathsf{I}(\epsilon) < 1$ we have $\mathsf{I}^2(\epsilon) < \mathsf{I}(\epsilon)$ and $2\mathsf{I}(\epsilon) - \mathsf{I}^2(\epsilon) > \mathsf{I}(\epsilon)$. Thus, under SC decoding, the transformation produces two virtual channels, the upper with a degraded and the lower with a better quality than that of the two independent W channels, both with mutual information $\mathsf{I}(\epsilon)$. The elementary structure in Figure 10.6 is hereafter referred to as "butterfly".

10.2.1.2 Extension to an $n \times n$ Channel Defined by $\boldsymbol{F}_2^{\otimes m}$

We can extend the previous 2×2 MIMO channel construction and analysis to the case of $n = 2^m$ parallel W channels, each characterized by mutual information $\mathsf{I}(\epsilon)$. Specifically, we can construct an $n \times n$ MIMO channel, denoted by W_n, by taking the input vector $\boldsymbol{u}$, mapping it onto $\boldsymbol{x} = \boldsymbol{u}\,\boldsymbol{F}_2^{\otimes m}$ using the automorphism $\boldsymbol{F}_2^{\otimes m} = \boldsymbol{G}_{\mathrm{RM}}(n,n)$, and letting each x_i be the input of a parallel BEC W with output y_i. (Note that this construction closely resembles Fast Fourier and Hadamard transforms.)

The propagation of the mutual information from right to left (from the channels to the data u_i) for the transformation $\boldsymbol{F}_2^{\otimes m}$ is hence the reiteration of that described for $\boldsymbol{F}_2$ in Figure 10.6: When moving from right to left, a summation node has the effect to square the mutual information from I to I^2, while a copying node transforms from I to $2\mathsf{I} - \mathsf{I}^2$ at the left of the node. An example for $n = 4$ is reported in Figure 10.7.

The following theorem provides an expression of the mutual information I_i of virtual channel Q_i in the decomposition (10.9) applied to W_n.

Theorem 10.1 *Consider the $n \times n$ MIMO channel W_n, $n = 2^m$, with input $\boldsymbol{u} = \begin{bmatrix} u_1 & u_2 & \dots & u_n \end{bmatrix}$ and output $\boldsymbol{y} = \begin{bmatrix} y_1 & y_2 & \dots & y_n \end{bmatrix}$, such that $\boldsymbol{x} = \boldsymbol{u}\,\boldsymbol{F}_2^{\otimes m}$ and each y_i is the output of a parallel BEC W, with input x_i and mutual information $\mathsf{I}(\epsilon) = 1 - \epsilon$. Let $r = (b(r,m-1) \cdots b(r,0))_2$ where $b(r,\ell) \in \{0,1\}$ is the ℓth digit in the binary representation of integer r. Then, the mutual information I_i of virtual channel Q_i, defined in (10.10), is*

$$\begin{aligned} \mathsf{I}_i &= P(u_i \text{ correctly decoded } | \boldsymbol{y}, u_{i-1}, \dots, u_1) \\ &= \left(f^{b(i-1,0)} \circ f^{b(i-1,1)} \circ \dots \circ f^{b(i-1,m-1)}\right)(\mathsf{I}(\epsilon)) \\ &= f^{b(i-1,0)}\left(f^{b(i-1,1)}\left(\cdots f^{b(i-1,m-1)}(\mathsf{I}(\epsilon))\right)\right), \end{aligned} \tag{10.14}$$

where

$$\begin{aligned} f^0(x) &= x^2 \\ f^1(x) &= 2x - x^2 \,. \end{aligned} \tag{10.15}$$

For example, the probability of correctly decoding u_3 in Figure 10.7 when using SC genie-aided decoding is obtained by observing that the binary representation of 2 is $(10)_2$, thus

$$\mathsf{I}_3 = f^0 \circ f^1(\mathsf{I}(\epsilon)) = (2\mathsf{I}(\epsilon) - \mathsf{I}^2(\epsilon))^2 \,. \tag{10.16}$$

Graphically, the evolution of the mutual information for the virtual channel Q_i in the MIMO channel W_n may be described through a *transformation chart*, that is, by a path within the region of the plane between the two curves $y = f^0(x) = x^2$ and $y = f^1(x) = 2x - x^2$, $0 \le x \le 1$. Starting from the point of coordinates $(\mathsf{I}(\epsilon), \mathsf{I}(\epsilon))$, a copying node determines a vertical movement to the $y = f^1(x)$ curve followed by a horizontal movement back to the curve $y = x$. Analogously, a summation node determines a vertical movement to the $y = f^0(x)$ curve followed by a horizontal movement back to the curve $y = x$.

Example 10.1 The network corresponding to channel W_8 is depicted in Figure 10.1. The path across the network from x_7 to u_7 touches two copying $\perp$ nodes and then a summation $\oplus$ node, as correctly predicted by the bits of the binary representation $6 = (110)_2$. The probability of correctly decoding u_7 under genie-aided SC decoding is

$$\begin{aligned} \mathsf{I}_7 &= f^0 \circ f^1 \circ f^1(\mathsf{I}(\epsilon)) \\ &= \left(2(2I(\epsilon) - 1) - (2I(\epsilon) - 1)^2\right)^2 . \end{aligned} \tag{10.17}$$

and the corresponding mutual information evolution is depicted in the transformation chart of Figure 10.8 for $\mathsf{I}(\epsilon) = 0.6$, where the two curves $y = f^0(x)$ and $y = f^1(x)$ are displayed in grey. As another example, in case of u_2, since $1 = (001)_2$, we obtain $\mathsf{I}_1 = 2I^4(\epsilon) - I^8(\epsilon)$ which also represents the probability of correct decoding over the BEC under SC decoding with no error propagation. The corresponding mutual information evolution is also represented in Figure 10.8, again for $\mathsf{I}(\epsilon) = 0.6$. Another example is depicted in Figure 10.9 for $n = 2^{10}$, $i = 710$, and $\mathsf{I}(\epsilon) = 0.5$.

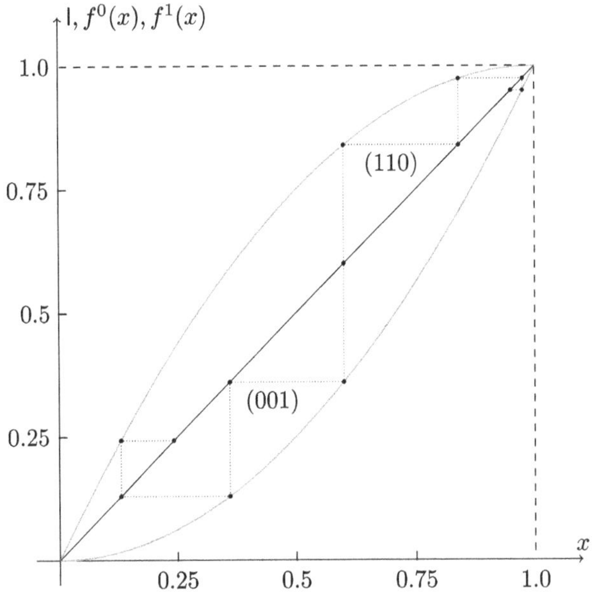

Figure 10.8 Mutual information transformation from the channel to the symbols $\boldsymbol{u} = \boldsymbol{x}\boldsymbol{G}_{\mathrm{RM}}(8,8)$. Evolution for u_7 ($6 = (110)_2$) and for u_2 ($1 = (001)_2$), starting from BEC with $\mathsf{I}(\epsilon) = 0.6$. © AEIT. Reproduced with permission.

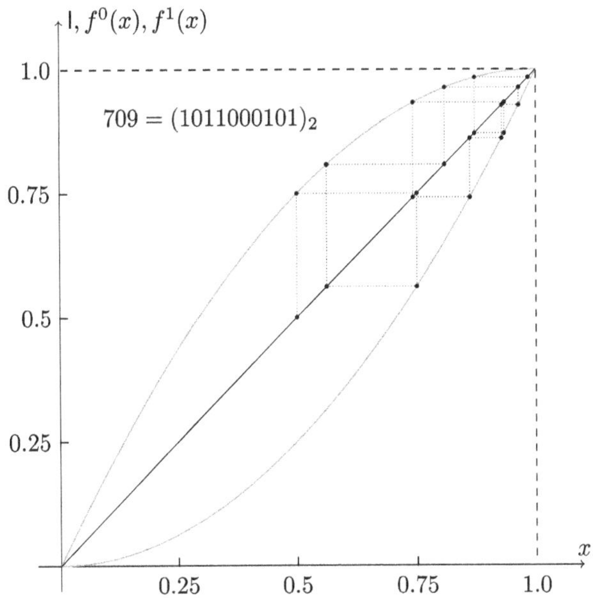

Figure 10.9 Example of mutual information evolution for $n = 2^{10}$, starting from BEC with $\mathsf{I}(\epsilon) = 0.5$, node u_{710}. The final mutual information of the node is $\mathsf{I}_{710} = 0.9834$. © AEIT. Reproduced with permission.

Remark 10.1. It is interesting to investigate if there are cycles in the transformation chart, starting from a given $0 < \mathsf{I} < 1$ and returning to the same mutual information I after a path in the map corresponding to a given binary sequence $\mathbf{b}$. We call *eigeninformation* the value $\mathsf{I}_\mathbf{b}$ for which this happens. For example, a cycle with the binary sequence $\mathbf{b} = (1,0)$ is obtained if $(2\mathsf{I} - \mathsf{I}^2)^2 = \mathsf{I}$, that solved gives $\mathsf{I}_\mathbf{b} = (3 - \sqrt{5})/2 \approx 0.382$. It is interesting to note that $(3 - \sqrt{5})/2 = 1/\varphi^2$, where φ is the Golden Ratio. Therefore, if the channel mutual information is $1/\varphi^2$ and the number of stages m is even, the input indexed by $(1010\ldots10)_2$ has the same mutual information as its corresponding output. Similarly, a cycle with the binary sequence $\mathbf{b} = (0,1)$ is obtained if $2\mathsf{I}^2 - \mathsf{I}^4 = \mathsf{I}$, that solved gives $\mathsf{I}_\mathbf{b} = (\sqrt{5} - 1)/2 = 1/\varphi \approx 0.618$. Therefore, if the channel mutual information is $1/\varphi$ and the number of stages m is even, the input indexed by $(0101\ldots01)_2$ has again the same mutual information as its corresponding output. These two cycles are the same in the transformation chart, alternating between $1/\varphi$ and $1/\varphi^2$, as shown in Figure 10.10 (a). It can be checked that for any input u_i, $i \in \{2,3,\ldots,n-1\}$, there is a value of the BEC mutual information $\mathsf{I}(\epsilon)$ such that the mutual information of the ith virtual channel is $\mathsf{I}_i = \mathsf{I}(\epsilon)$; this value is found by solving

$$f^{b(i-1,0)} \circ f^{b(i-1,1)} \circ \cdots \circ f^{b(i-1,m-1)}(\mathsf{I}(\epsilon)) = \mathsf{I}(\epsilon) \tag{10.18}$$

in the interval $(0,1)$. An example is reported in Figure 10.10(b) for the input u_{420} of a polar code with $n = 2^{10}$. For a given $n = 2^m$, there are several distinct eigeninformation values (at most $n-2$). If the BEC mutual information equals any of these, there is one input for which the transformed mutual information remains $I(\epsilon)$.

10.2.1.3 Channel Polarization and Subchannel Ranking

Consider the W_2 channel in Figure 10.6 with the output probability of correct decoding $\mathsf{I} = \mathsf{I}(\epsilon)$. The two values of the correct decoding probability at the channel input are I^2 and $\mathsf{I}(2-\mathsf{I})$, so that the variation of mutual information is $\Delta\mathsf{I} = \pm(\mathsf{I} - \mathsf{I}^2)$. The function $|\Delta\mathsf{I}| = \mathsf{I}(1-\mathsf{I})$ is reported in Figure 10.11. As previously remarked, $\Delta\mathsf{I}$ is always negative for the upper node and positive for the lower node. The value of $|\Delta\mathsf{I}|$ decreases whenever I moves toward 0 or toward 1. Hence, these two points play the role attractors for the paths in the transformation chart. For any BEC mutual information $\mathsf{I}(\epsilon)$, for large n almost all possible paths in the map converge to one of the two extreme points $\mathsf{I}_i = 0$ or $\mathsf{I}_i = 1$.[1] As previously pointed out, since the sum of all I_i must equal $n\,\mathsf{I}(\epsilon)$ it follows that the fraction of subchannels in (10.9) with $\mathsf{I}_i = 1$ is $\mathsf{I}(\epsilon)$. In other words, for large n, there are approximately $n\,\mathsf{I}(\epsilon)$ inputs u_i with $\mathsf{I}_i \approx 1$, and the remaining have $\mathsf{I}_i \approx 0$. This is the so-called channel polarization phenomenon, which has been proved to apply for $n \to \infty$ not just for BEC, but for any kind of symmetric binary-input DMCs, when the precoding matrix $\boldsymbol{G}_{\mathrm{RM}}(n,n)$ is used in Figure 10.4(b).

[1] If I is one of the eigeninformation values, there is one transformed channel with $\mathsf{I}_i = \mathsf{I}(\epsilon)$.

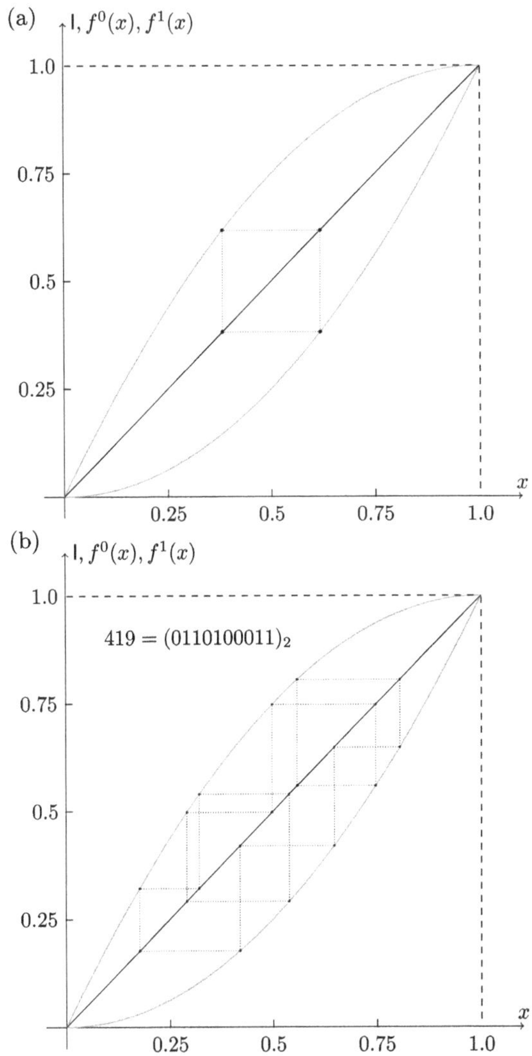

Figure 10.10 Eigeninformation and cycle sequences. (a) The shortest cycle (length 2), alternating between mutual information values $1/\varphi^2 \approx 0.382$ and $1/\varphi \approx 0.618$. (b) Sequence with eigeninformation $\mathsf{I} \approx 0.539366$, cycle of length 10. The cycle gives the eigeninformation values for all sequences obtained by cyclic shift and bit reversal of the indicated eigensequence. © AEIT. Reproduced with permission.

By applying (10.14) we can find the correct decoding probability, under genie-aided SC decoding, for any u_i. This allows ranking virtual channels Q_i to retain only the best ones. For a given code dimension k, encoding of an (n,k) polar code is performed as $\boldsymbol{x} = \boldsymbol{u}\, \boldsymbol{G}_{\mathrm{RM}}(n,n)$, by associating the k information bits to the elements of $\boldsymbol{u}$ corresponding to the best virtual channels and by setting to a frozen (0) value the remaining elements of $\boldsymbol{u}$. Generally, the rows of $\boldsymbol{G}_{\mathrm{RM}}(n,n)$ to retain are chosen based on the channel parameter $\mathsf{I}(\epsilon)$, so the code is optimized for a given channel quality.

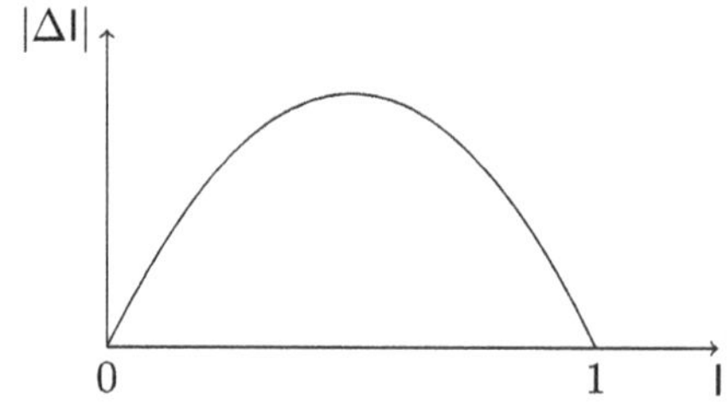

Figure 10.11 Variation of the mutual information per stage, $|\Delta \mathsf{I}| = \mathsf{I}(1 - \mathsf{I})$.

Example 10.2 For $n = 8$ and $\mathsf{I}(\epsilon) = 0.5$, by (10.14) we obtain the decoding success probabilities

$$\begin{array}{llll} \mathsf{I}_1 = 0.004 & \mathsf{I}_2 = 0.121 & \mathsf{I}_3 = 0.191 & \mathsf{I}_4 = 0.684 \\ \mathsf{I}_5 = 0.316 & \mathsf{I}_6 = 0.809 & \mathsf{I}_7 = 0.879 & \mathsf{I}_8 = 0.996. \end{array} \tag{10.19}$$

Thus, for a code with $k = 4$ we would freeze to zero u_1, u_2, u_3, u_5, and use u_4, u_6, u_7, u_8 as information bits. In this case, the resulting $(8,4)$ block code coincides with the $(8,4)$ RM code in (10.6), which is designed to maximize the code minimum distance. For larger n, polar codes and RM codes are generally different.

10.3 Decoding of Polar Codes

Due to the structure of the transformation $\boldsymbol{F}_2^{\otimes m}$, polar codes have $\mathcal{O}(n \log n)$ encoding and SC decoding complexity, and a recursive structure that makes them fit for hardware implementations.

In SC decoding, each symbol u_i in the decomposition represented in Figure 10.3 is decoded sequentially, by observing the channel output $\begin{bmatrix} y_1 & y_2 & \dots & y_n \end{bmatrix}$ assuming all previous symbols $\begin{bmatrix} u_1 & u_2 & \dots & u_{i-1} \end{bmatrix}$ have been decoded successfully. Thus, indicating with the hat the decoded symbols, the SC decoder decides the next symbol as

$$\hat{u}_i = \arg\max_u P(U_i = u | y_1 \dots y_n, u_1, u_2, \dots, u_{i-1}), \tag{10.20}$$

where for the previous symbols we use the decisions, that is, (10.20) is calculated with $u_1 = \hat{u}_1, u_2 = \hat{u}_2, \dots, u_{i-1} = \hat{u}_{i-1}$. Equivalently, the decision is taken based on the LLR

$$\ln \frac{P(U_i = 1 | y_1 \dots y_n, u_1, u_2, \dots, u_{i-1})}{P(U_i = 0 | y_1 \dots y_n, u_1, u_2, \dots, u_{i-1})}. \tag{10.21}$$

These LLRs can be calculated iteratively by exploiting the structure of the code, leading to a new decision $\hat{u}_i$, which is used in subsequent decisions. The basic idea is reported in Figure 10.12 (see also Exercises 10.4, 10.5, and 10.6 at the end of the chapter). Starting from the channel outputs $\begin{bmatrix} y_1 & y_2 & \dots y_8 \end{bmatrix}$ the four upper symbols are

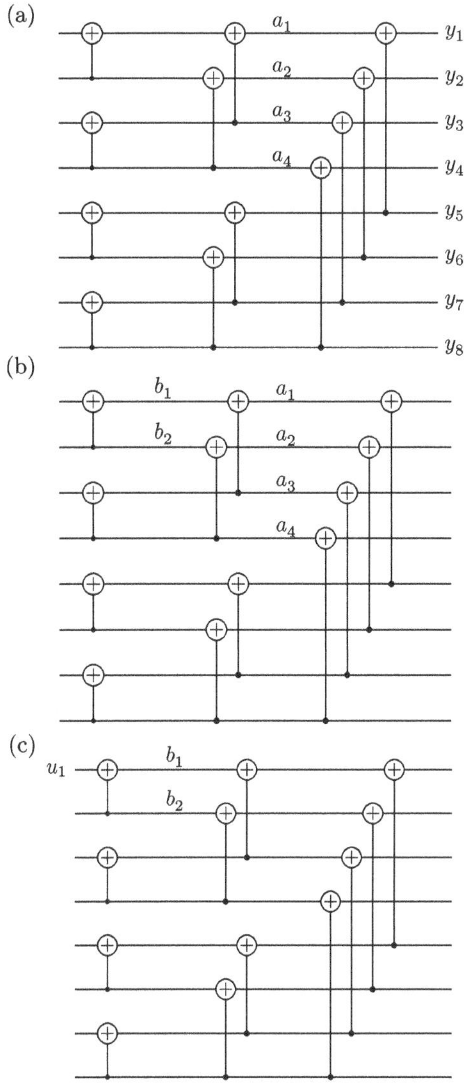

Figure 10.12 Example for $n = 8$, decoding the bit u_1.

first decoded by calculating the likelihoods for the symbols $\begin{bmatrix} a_1 & a_2 & a_3 & a_4 \end{bmatrix}$ (with just the $\begin{bmatrix} y_1 & \dots & y_8 \end{bmatrix}$ known, Figure 10.12(a)). In turn, the likelihoods of $\begin{bmatrix} b_1 & b_2 \end{bmatrix}$ are calculated from those of $\begin{bmatrix} a_1 & a_2 & a_3 & a_4 \end{bmatrix}$ (see Figure 10.12(b)). Finally, the bit u_1 is estimated based on $\begin{bmatrix} b_1 & b_2 \end{bmatrix}$ or set to zero if it is a frozen bit (see Figure 10.12(c)). Then, u_2 is estimated based on u_1, b_1, b_2, and the values of (b_1, b_2) are updated. The procedure is then propagated to the other nodes until the symbols $\begin{bmatrix} a_1 & a_2 & a_3 & a_4 \end{bmatrix}$ are estimated. These estimates are then used to decode the lower four symbols.

For the performance over DMC with mutual information I, it is possible to prove that, for every $R < \mathsf{I}$ there exists for sufficiently large n a rate R polar code for which the block error rate under an SC decoder is $\mathcal{O}\left(n^{-1/(2+\delta)}\right)$, where $\delta > 0$ is arbitrarily small.

Despite this asymptotic performance, studies indicate that for short and moderate blocklengths, SC decoding of polar codes does not perform as well as LDPC codes. To improve the performance, different decoders were suggested, for example, belief propagation decoding, improved SC decoding, and successive cancellation list (SCL) decoding. In particular, with SCL, the input bits are successively decoded one-by-one, but keeping track of several decoding results instead of just one. In order to keep the best paths at each stage, a suitable pruning criterion must be adopted. The performance of SCL decoding is very close to that of a full maximum-likelihood decoding even for moderate values of the list size. Unfortunately, even with SCL decoding, the performances of polar codes are in general worse than turbo and LDPC codes of similar length. However, it has been observed that the performance could be improved by using an additional cyclic redundancy check (CRC) code. Combined with CRC precoding, the SCL decoder provides the best performance for polar codes to date. Over the BEC, efficient implementations are provided by SC inactivation decoding where, whenever an information bit decodes as erased, it is replaced by a dummy variable, that is, it is inactivated, and the decision on it is postponed to the end of decoding process.

10.4 SC Performance Bounds

Over the BEC with erasure probability ϵ, the quantity $1 - \mathsf{I}_k = 1 - \mathsf{I}_k(\epsilon)$ gives the probability of erasure for virtual channel Q_k, assuming genie-aided SC decoding. Let us order virtual channels as $Q_{o(1)}, Q_{o(2)}, \ldots, Q_{o(n)}$, where $o(1)$, $o(2)$, ... are the indexes of the best subchannel, the second-best subchannel, and so on. If we use the first $i \leq n$ virtual channels, by union bound the probability that the genie-aided SC decoder produces a wrong codeword that is upper bounded by

$$\begin{aligned} P_e &\leq \min\left\{1, \sum_{k=1}^{i} P(\text{erasure at } k\text{th best subchannel})\right\} \\ &= \min\left\{1, \sum_{k=1}^{i} (1 - \mathsf{I}_{o(k)})\right\} \end{aligned} \tag{10.22}$$

and the rate of the code is $R = i/n$. For the same code rate, the error probability may also be lower bounded as

$$P_e \geq \max_{k \in \{1,\ldots,i\}} (1 - \mathsf{I}_{o(k)}) \,. \tag{10.23}$$

In Figures 10.13–10.15, we report the above upper and lower bounds versus the coding rate, for length $n = 2^{10}, 2^{15}$, and 2^{20} and several values of the channel erasure probability ϵ. As expected, for increasing n the curves are more steep.

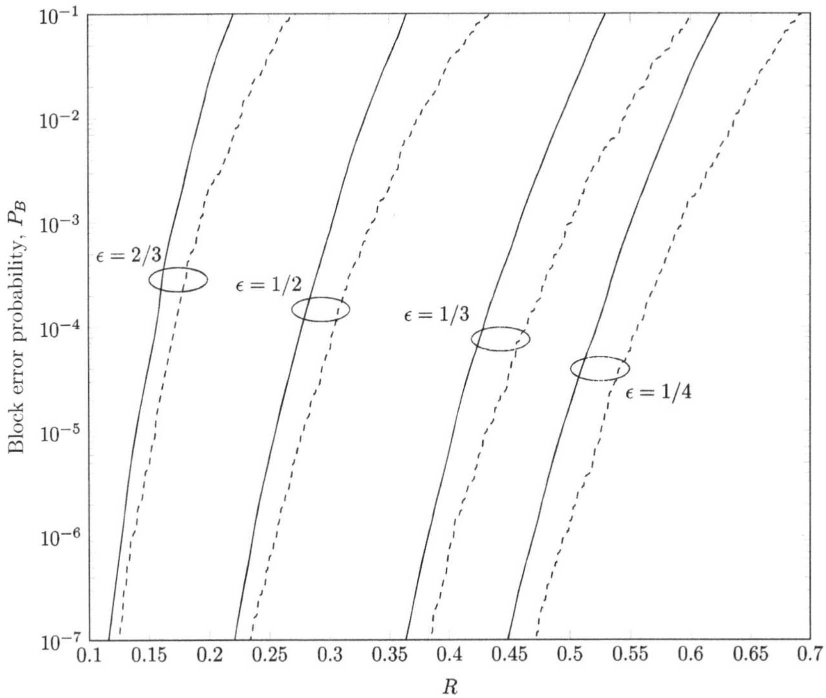

Figure 10.13 Bounds on the block error probability versus the code rate R for polar codes of length $n = 2^{10}$ under SC decoding. BEC with erasure probability $\epsilon = 2/3$, $1/2$, $1/3$, and $1/4$. Solid: Upper bound (10.22). Dashed: Lower bound (10.23). © AEIT. Reproduced with permission.

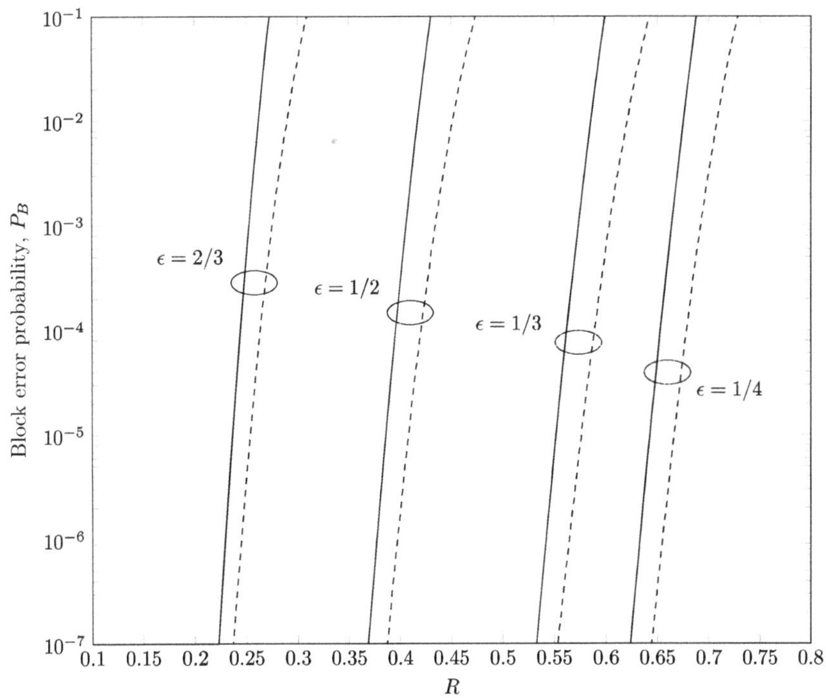

Figure 10.14 Bounds on the block error probability versus the code rate R for polar codes of length $n = 2^{15}$ under SC decoding. BEC with erasure probability $\epsilon = 2/3$, $1/2$, $1/3$, and $1/4$. Solid: Upper bound (10.22). Dashed: Lower bound (10.23).

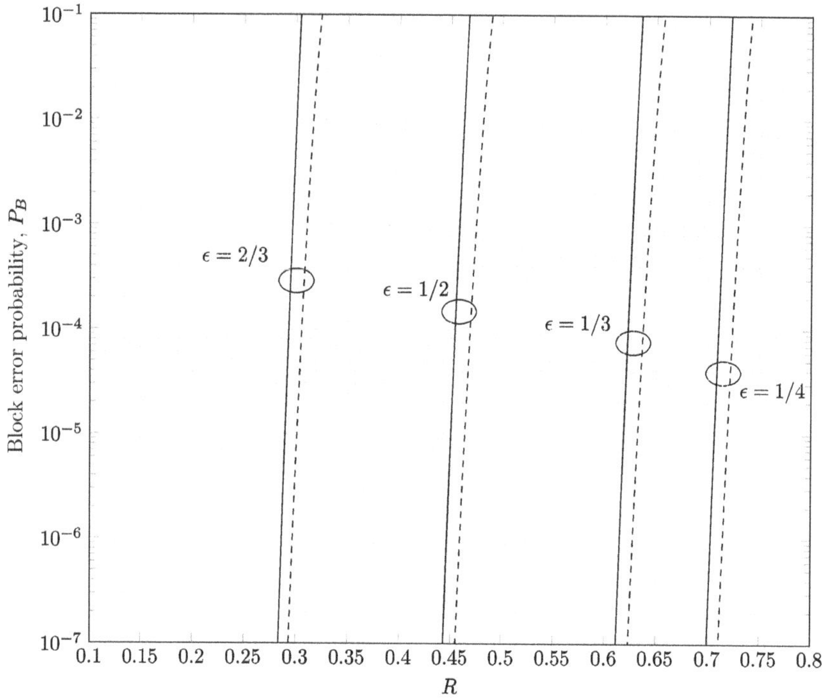

Figure 10.15 Bounds on the block error probability versus the code rate R for polar codes of length $n = 2^{20}$ under SC decoding. BEC with erasure probability $\epsilon = 2/3$, $1/2$, $1/3$, and $1/4$. Solid: Upper bound (10.22). Dashed: Lower bound (10.23).

10.5 Historical Notes

The proposal of polar codes and the proof that they achieve the symmetric capacity of a binary-input discrete memoryless channel can be found in the seminal paper by Arıkan [1]. The idea that Reed–Muller codes can be suitably modified for multistage decoding and for list decoding can also be found in the PhD thesis of N. Stolte [2]. In [3], it has been proved that Reed–Muller codes achieve capacity under MAP decoding over the BEC. In [4], it was shown that the class of BEC capacity-achieving codes includes certain sequences of cyclic codes. Extension to other channels and a recent survey on RM codes and their capacity-achieving properties can be found in [5,6].

Implementation issues about SC and other decoders can be found in [7–11]. In particular, list decoding was proposed in [11], the performance of polar codes in terms of scaling exponent and the analysis of the error floor regimes can be found in [12], and polarization is analyzed in [1, 13–15]. Notable results on channel ranking may be found, for instance, in [16, 17] and references therein. Polar subcodes, which allows some of the frozen symbols, called dynamic frozen symbols, to be data-dependent, have been introduced in [18]. Inactivation decoding for efficient implementation of SCL decoders for polar codes over the BEC has been proposed in [19].

10.6 Exercises

Exercise 10.1 (Reed–Muller codes) Verify for the RM $m = 4$ that the order $r = 0$ RM code is a repetition code, that the order 1 RM code is the dual of an extended Hamming code (also called biorthogonal code), and that the order $m - 2$ RM code is an extended Hamming code.

Exercise 10.2 Prove Theorem 10.1.

Hint: the network transforming $\boldsymbol{u}$ into $\boldsymbol{x}$ according to $\boldsymbol{x} = \boldsymbol{F}_2^{\otimes m}\boldsymbol{u}$ may be described as a cascade of $m = \log_2 n$ identical stages (see Figure 10.1(a)). The jth stage, $j \in \{0, 1, \ldots, m-1\}$, has input $\boldsymbol{x}_j$ and output $\boldsymbol{x}_{j+1}$; moreover, $\boldsymbol{x}_0 = \boldsymbol{u}$ and $\boldsymbol{x}_m = \boldsymbol{x}$. The jth stage transforms $\boldsymbol{x}_j$ into $\boldsymbol{x}_{j+1}$ by a bank of $n/2$ parallel butterflies, followed by a permutation.

Exercise 10.3 Prove that the eigeninformation values, that is, the values $x \in (0,1)$ giving $f^{b(i,0)}\left(f^{b(i,1)}\left(\cdots f^{b(i,m-1)}(x)\right)\right) = x$, are irrational.

Hint: use the rational root theorem of algebra for a monic polynomial $P(x) = c_n x^n + \cdots + c_1 x + c_0$ with integer coefficients.

Exercise 10.4 To better understand how SC works, let us consider the transformation in Figure 10.1. Starting from the received vector, the SC decoder can be build iteratively, by first decoding the four upper symbols, and then the four lower inputs assuming the decisions on the first four, as represented in Figure 10.12. The upper four can in turn be decoded by first decoding the two upper and then the two lowest. Therefore, the basic block to build an SC decoder for polar codes is the 2×2 butterfly reported in the figure, where a_1, a_2, b_1, and b_2 represent bits.

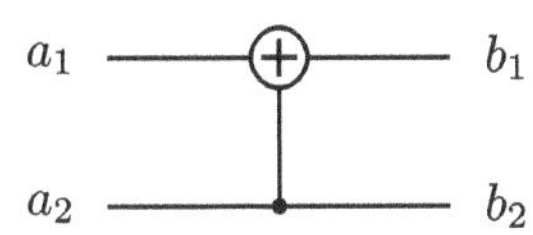

For the BEC the possible values of $P(b_i = 0)$ are $\{0, 1/2, 1\}$ corresponding to $b_i = \{1, \mathsf{E}, 0\}$. Find $P(a_1 = 0)$, $P(a_2 = 0|a_1 = 0)$ and $P(a_2 = 0|a_1 = 1)$ as a function of $r_1 = P(b_1 = 0)$ and $r_2 = P(b_2 = 0)$.

Exercise 10.5 More in general, given the output likelihood ratio (LR) $r_1 = P(b_1 = 0)/P(b_1 = 1)$ and $r_2 = P(b_2 = 0)/P(b_2 = 1)$, prove that the input ratios are

$$\frac{P(a_1 = 0)}{P(a_1 = 1)} = \frac{r_1 r_2 + 1}{r_1 + r_2}$$

and, conditional to the symbol in the upper arm,

$$\frac{P(a_2 = 0|a_1 = 0)}{P(a_2 = 1|a_1 = 0)} = r_1 r_2 \qquad \frac{P(a_2 = 0|a_1 = 1)}{P(a_2 = 1|a_1 = 1)} = \frac{r_2}{r_1}.$$

Clearly the expressions are similar for propagating the LR from left to right. These expressions can be used to propagate the LR in one direction or in the other.

Exercise 10.6 For the previous exercise, show that using the LLRs $R_i = \ln r_i$ we have

$$\ln \frac{P(a_1 = 0)}{P(a_1 = 1)} = \ln \frac{1 + e^{R_1+R_2}}{e^{R_1} + e^{R_2}} \approx \mathrm{sgn}(R_1)\mathrm{sgn}(R_2) \min\{|R_1|, |R_2|\}$$

and

$$\ln \frac{P(a_2 = 0|a_1 = 0)}{P(a_2 = 1|a_1 = 0)} = R_1 + R_2 \quad \ln \frac{P(a_2 = 0|a_1 = 1)}{P(a_2 = 1|a_1 = 1)} = R_2 - R_1.$$

Exercise 10.7 Prove equation (10.22).

Hint: first show that it holds for genie-aided decoding, then show that the decoder inputs $y_1, \ldots, y_n$ that make the genie-aided fail are the same that make the SC decoder fail, on the BEC.

Exercise 10.8 Compare SC versus ML erasure decoding for the (8,4) RM code.

Exercise 10.9 Draw the upper bound (10.22) for SC decoding of a (8,4) RM code, and compare it with the upper bound on ML decoding based on the weight enumeration function.

Exercise 10.10 Show that for a polar code $d_{\min} = \min w_H(g_i), i = 1, \ldots, k$.

References

[1] E. Arıkan, "Channel polarization: A method for constructing capacity-achieving codes for symmetric binary-input memoryless channels," *IEEE Trans. Inf. Theory*, vol. 55, no. 7, pp. 3051–3073, Jul. 2009.

[2] N. Stolte, "Rekursive codes mit der Plotkin-konstruktion und ihre decodierung (Recursive Codes with the Plotkin-Construction and Their Decoding)," Ph.D. dissertation, Technische Universität Darmstadt, 2002.

[3] S. Kudekar, S. Kumar, M. Mondelli, H. D. Pfister, E. Şaşoğlu, and R. L. Urbanke, "Reed–Muller codes achieve capacity on erasure channels," *IEEE Trans. Inf. Theory*, vol. 63, no. 7, p. 4298–4316, July 2017.

[4] S. Kumar, R. Calderbank, and H. D. Pfister, "Beyond double transitivity: Capacity-achieving cyclic codes on erasure channels," in *Proc. IEEE Inf. Theory Workshop*, Cambridge, UK, Sep. 2016, p. 241–245.

[5] G. Reeves and H. D. Pfister, "Reed–Muller codes on BMS channels achieve vanishing bit-error probability for all rates below capacity," *IEEE Trans. Inf. Theory*, vol. 70, no. 2, p. 920–949, Feb. 2024.

[6] E. Abbe, O. Sberlo, A. Shpilka, and M. Ye, "Reed-Muller codes," *Foundations and Trends® in Communications and Information Theory*, vol. 20, no. 1–2, p. 1–156, 2023.

[7] E. Arıkan, "A performance comparison of polar codes and Reed-Muller codes," *IEEE Commun. Lett.*, vol. 12, no. 6, 2008.

[8] C. Leroux, A. J. Raymond, G. Sarkis, I. Tal, A. Vardy, and W. J. Gross, "Hardware implementation of successive-cancellation decoders for polar codes," *J. Signal Processing Systems*, vol. 69, no. 3, pp. 305–315, 2012.

[9] P. Trifonov, "Efficient design and decoding of polar codes," *IEEE Trans. Commun.*, vol. 60, no. 11, pp. 3221–3227, 2012.

[10] K. Chen, K. Niu, and J. Lin, "Improved successive cancellation decoding of polar codes," *IEEE Trans. Commun.*, vol. 61, no. 8, pp. 3100–3107, 2013.

[11] I. Tal and A. Vardy, "List decoding of polar codes," *IEEE Trans. Inf. Theory*, vol. 61, no. 5, pp. 2213–2226, May 2015.

[12] M. Mondelli, S. H. Hassani, and R. L. Urbanke, "Unified scaling of polar codes: Error exponent, scaling exponent, moderate deviations, and error floors," *IEEE Trans. Inf. Theory*, vol. 62, no. 12, pp. 6698–6712, 2016.

[13] E. Arıkan and E. Telatar, "On the rate of channel polarization," in *Proc. IEEE Int. Symp. Inf. Theory*, Seoul, South Korea, 2009, pp. 1493–1495.

[14] M. Alsan and E. Telatar, "A simple proof of polarization and polarization for non-stationary memoryless channels," *IEEE Trans. Inf. Theory*, vol. 62, no. 9, pp. 4873–4878, 2016.

[15] J. Błasiok, V. Guruswami, P. Nakkiran, A. Rudra, and M. Sudan, "General strong polarization," in *Proc. ACM Symp. Theory of Comput.* Los Angeles, CA, USA, June 2018, pp. 485–492.

[16] M. Bardet, V. Dragoi, A. Otmani, and J. P. Tillich, "Algebraic properties of polar codes from a new polynomial formalism," in *Proc. IEEE Int. Symp. Inf. Theory*, Barcelona, Spain, July 2016, pp. 230–234.

[17] W. Wu, B. Fan, and P. H. Siegel, "Generalized partial orders for polar code bit-channels," in *Proc. Ann. Allerton Conf. on Commun., Control, and Comput.*, Monticello, IL, USA, Oct. 2017, pp. 541–548.

[18] P. Trifonov and V. Miloslavskaya, "Polar subcodes," *IEEE J. Sel. Areas Commun.*, vol. 34, no. 2, pp. 254–266, 2016.

[19] M. C. Coşkun, J. Neu, and H. D. Pfister, "Successive cancellation inactivation decoding for modified Reed-Muller and eBCH codes," in *Proc. IEEE Int. Symp. Inf. Theory*. Los Angeles, CA, USA, June 2020, pp. 437–442.

11 Fountain Codes

In this chapter, we address an alternative paradigm for erasure coding that exploits feedback from multiple receivers to the transmitter. The feedback is employed to adapt on the fly the coding rate to the channel conditions. This approach, which shares several similarities with the framework of rate compatible codes with hybrid ARQ, relies on so-called *fountain codes*. The chapter starts by introducing the main concepts of fountain coding. Important classes of fountain codes, namely LT and Raptor codes, are then discussed and analyzed.

11.1 Preliminary Definitions

Fountain codes provide a practical solution to the following problem: A sender wishes to deliver an information message composed of k symbols (symbols here can be intended as bits or packets, that is, vectors of bits) to multiple receivers. Each sender-receiver link is assumed to be an erasure channel that erases symbols with some probability. Links experience independent erasures, and the erasure probability may not be the same for all links. To this purpose, a fountain encoder can be used: The encoder continuously produces codeword symbols, which are broadcast to the receivers. Due to the erasure patterns affecting the links, each receiver gets a subset of the transmitted symbols. The receivers try to decode the message at each new symbol reception. The fountain encoder stops generating new codeword symbols once all receivers have been able to decode with success. Each receiver communicates the success to the sender via a feedback link. Naming this approach "fountain coding" comes in handy in recalling the underlying principle: Receivers are seen as empty buckets, which are used to collect water (the symbols) at a fountain (the sender); once all buckets are full, the fountain can be stopped. Fountain codes are often referred to as *rateless* codes to emphasize the capability of the encoder to produce a potentially infinite number of codeword symbols for a given information message.

Let us consider the transmission of a message $\boldsymbol{u}$ of k symbols in $\mathrm{GF}(q)$ over a QEC. At each channel use, the encoder outputs a coded symbol given by a linear combination in $\mathrm{GF}(q)$ of the message symbols. The encoder output at time i is hence in the form

$$x_i = \sum_{j=1}^{k} g_{ji} u_j \qquad i = 1, 2, \ldots. \tag{11.1}$$

Encoding can be compactly described by resorting to the semi-infinite code generator matrix with coefficients in GF(q)

$$G = \begin{bmatrix} g_{11} & g_{12} & \cdots & g_{1i} & \cdots \\ g_{21} & g_{22} & \cdots & g_{2i} & \cdots \\ \vdots & \vdots & & \vdots & \\ g_{k1} & g_{k2} & \cdots & g_{ki} & \cdots \end{bmatrix} \tag{11.2}$$

and by performing the product

$$\boldsymbol{x} = \boldsymbol{u}\boldsymbol{G} \tag{11.3}$$

with $\boldsymbol{x} = \begin{bmatrix} x_1 & x_2 & \ldots \end{bmatrix}$ being semi-infinite. The symbols in $\boldsymbol{u}$ are called the *input symbols*, whereas the symbols in $\boldsymbol{x}$ are referred to as *output symbols*. At the receiver side, due to the erasures introduced by the channel, at a given time, the decoder has access to only a subset of the so far transmitted codeword symbols. Let us denote by $\boldsymbol{x}'$ the vector containing the nonerased symbols and by $\boldsymbol{G}'$ the submatrix of $\boldsymbol{G}$ composed of the columns of $\boldsymbol{G}$ associated with the nonerased symbols. Denote by n the number of nonerased symbols, that is, the length of $\boldsymbol{x}'$. Decoding consists of solving a system of n linear equations in k unknowns

$$\boldsymbol{u}\boldsymbol{G}' = \boldsymbol{x}'. \tag{11.4}$$

Decoding may succeed only if $\text{rank}(\boldsymbol{G}') = k$. Note that this condition is also sufficient under ML decoding, whereas it is generally only a necessary condition under any other sub-optimum decoding algorithm. In the jargon of fountain coding, the number of equations in excess with respect to the number of unknowns is denoted by $\delta = n - k$, and it is referred to as (absolute) *receiver overhead*. Sometimes, the overhead is normalized with respect to k. In this case, it is denoted by $\epsilon_o = (n-k)/k$, and it is referred to as relative receiver overhead. The performance of a fountain code $\mathcal{C}$ under a given decoding algorithm $\mathcal{D}$ is characterized by its decoding failure probability $P_B(\mathcal{C})$ as a function of the overhead.

Definition 11.1 (Reliable decoding algorithm) A sequence of fountain codes $\{\mathcal{C}_k\}$ is said to possess a reliable decoding algorithm $\mathcal{D}$ if there is an overhead sequence $\delta(\mathcal{C}_k)$ such that the decoding failure probability of $\mathcal{D}$ can be upper bounded as

$$P_B(\mathcal{C}_k) \leq \frac{1}{k^c} \tag{11.5}$$

for some positive constant c.

Definition 11.2 (Encoding cost) The encoding cost of a fountain code is the expected number of operations required to compute a codeword symbol.

Definition 11.3 (Decoding cost) The decoding cost of a fountain code with a given (reliable) decoding algorithm $\mathcal{D}$ is the expected number of operations required to successfully decode a message, normalized by the message length k.

A reliable decoding algorithm yields a failure probability that tends to zero at least polynomially fast in k for a given overhead sequence $\delta(\mathcal{C}_k)$.

Example 11.1 (Linear random fountain codes) Consider the transmission of a message $\boldsymbol{u}$ composed of k symbols in GF(q) over a QEC. At each channel use, the encoder outputs a coded symbol given by a random linear combination of the message symbols. In particular, the coefficients g_{ji} in (11.1) are drawn independently and uniformly at random from GF(q). The obtained code is referred to as a linear random fountain (LRF) code.

Let us assume an optimum ML decoder, which tries to solve (11.4) by Gaussian elimination. The decoding failure probability is here determined by the probability that a uniformly distributed $k \times n$ random matrix $\boldsymbol{G}'$ with coefficients in GF(q) has rank lower than k. The failure probability is hence given by

$$P_B(\delta) = 1 - \prod_{i=1}^{k}\left(1 - \frac{q^{i-1}}{q^{k+\delta}}\right) \tag{11.6}$$

(see Appendix B). The decoding failure probability provided by (11.6) admits tight lower and upper bounds as

$$q^{-\delta-1} \leq P_B(\delta) < \frac{1}{q-1}q^{-\delta}. \tag{11.7}$$

The upper and lower bounds are compared with the exact expression in Figure 11.1. As k grows large, uniform LRF codes allow decoding with an arbitrarily small relative overhead ϵ_o. More specifically, it is easy to show that LRF codes possess a

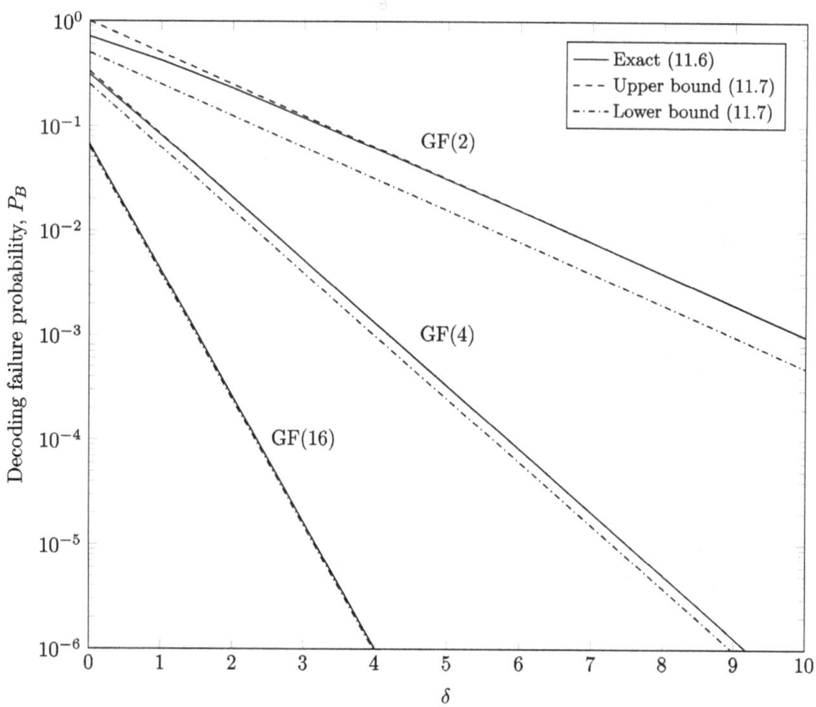

Figure 11.1 Decoding failure probability for LRF codes as a function of the overhead.

reliable decoding algorithm, that is, ML decoding, with relative overhead $\mathcal{O}(\log(k)/k)$. The result comes, anyhow, at the expense of an encoding cost that is $\mathcal{O}(k)$ and of a decoding cost that is $\mathcal{O}\left(k^2\right)$. LT codes (to be discussed next) attack the problem of designing fountain codes with reduced encoding and decoding complexity.

11.2 LT Codes

In this section, a basic description of binary LT codes is provided. We restrict to the binary case since LT codes are originally conceived to be decoded via a peeling decoding algorithm that bears no influence from the field order used in the code construction.[1] For a given source message length k, an LT code is fully specified by an *output symbols degree distribution* $\Omega = (\Omega_1, \Omega_2, \ldots, \Omega_k)$.

11.2.1 Encoding

Encoding proceeds as follows.

(i) For a generic output symbols x_i, an output symbol degree d is sampled according to the probability distribution Ω;

(ii) A subset $\mathcal{I}_i$ of d distinct indexes in $\{1, 2, \ldots, k\}$ is formed by selecting the indexes uniformly at random without replacement;

(iii) The output symbol x_i is generated as

$$x_i = \sum_{j \in \mathcal{I}_i} u_j. \tag{11.8}$$

The generation of the output symbols is performed independently, that is, for every output symbol, a new degree and a new subset $\mathcal{I}_i$ are sampled.

11.2.2 Peeling Decoding

Peeling decoding of LT codes follows closely the principles of peeling decoding applied to LDPC codes (see Section 6.1.3), and it can be described by introducing a bipartite graph representation of the relation between input symbols and nonerased output symbols. Denoting by n the number of nonerased output symbols, the graph consists of a set of k input symbol nodes, a set of n output symbol nodes, and a set of edges connecting input and output nodes. Specifically, for a generic nonerased output symbol x'_i, the corresponding output node is connected with the input nodes associated with the input symbols with indexes in $\mathcal{I}_i$, where $\mathcal{I}_i$ defines the input symbols

[1] We refer to Section 11.4 for the ML decoding analysis of LT codes, for which the decoding failure probability depends on the field order, contrary to what happens under peeling decoding.

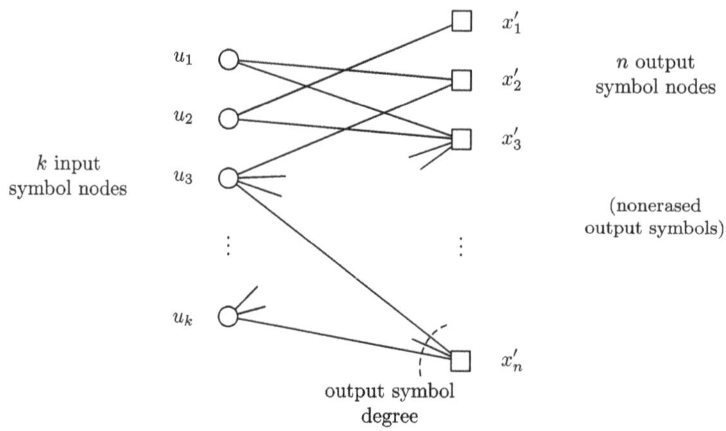

Figure 11.2 Example of bipartite graph for an LT code.

chosen to generate x'_i. An example is provided in Figure 11.2. We will refer to output (input) symbols and output (input) nodes without distinction in the following.[2]

Peeling decoding proceeds by recovering the value of an input symbol at each iteration. In particular, the decoder selects an output symbol at each iteration among the ones with degree 1. Denote by x'_i the selected output symbol, and denote by u_j its only neighbor. The input symbol u_j is set to x'_i, that is, its value is recovered. The output node is removed from the graph, together with the edge connecting u_j and x'_i. As the value of u is now available, the value of the residual output nodes connected to u_j is updated as

$$x' \leftarrow x' + u_j \tag{11.9}$$

for all $x' \in \mathcal{N}(u_j)$, being $\mathcal{N}(u_j)$ the set of residual neighbors of u_j. The edges connecting u_j to its neighbors are finally removed. Note that, as an outcome of the edge removal process, the number of output symbols with *residual* degree equal to one may increase. Decoding continues by resolving one input symbol at a time till all input symbols are resolved or no more degree-1 output symbol can be found. It follows that the decoding algorithm runs for k iterations in case of a decoding success and stops earlier in case of a decoding failure. Figure 11.3 shows an example of (successful) peeling decoding for an LT code with source message length $k = 5$, and with $n = 6$ nonerased output symbols.

11.2.3 Design of the Degree Distribution

It is immediately evident that the success of the peeling decoder is directly connected to the availability of degree-1 output symbols at each iteration. Due to the special role played by degree-1 output symbols in the decoding process, the notions of *input and output ripples* are introduced.

[2] It should be noted that LT codes are instances of rateless LDGM codes. It follows that their bipartite graph representation can be equivalently formulated as described in Section 9.3.

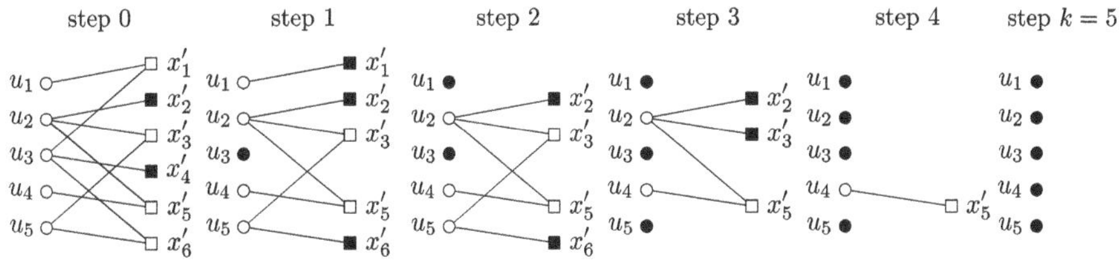

Figure 11.3 Example of peeling decoding for an LT code with $k = 5$ input symbols, at various iteration counts. Output (symbol) nodes with residual degree equal to 1 are depicted as dark squares, whereas dark circles denote the revealed input symbols.

Definition 11.4 (Input ripple) Given an LT code bipartite graph, the input ripple is the set of input nodes connected to output nodes with residual degree 1.

Definition 11.5 (Output ripple) Given an LT code bipartite graph, the output ripple is the set of output nodes with residual degree 1.

As an example, in Figure 11.3, the input ripple at the beginning of the decoding process is $\{u_2, u_3\}$, while the output ripple is $\{x'_2, x'_4\}$.

From a failure probability analysis viewpoint, we are interested in tracking the evolution of the cardinality of the ripple during the decoding process. In particular, we are interested in keeping the probability that the ripple gets empty small for any iteration index $\ell < k$ by selecting the degree distribution Ω in a judicious manner. A heuristic reasoning that provides a first design criterion for the degree distribution relies on the following observation: At a given iteration, the probability that an input symbol is connected to more than one output symbol in the output ripple should be kept small. In fact, if an input symbol is connected to several output symbols in the output ripple, the output ripple size would dangerously decrease when the input symbol is recovered. This implies that the size of the input ripple should be kept small at each decoding step, with ideally one input symbol entering the input ripple at every iteration. By doing so, the input ripple size would stay constant during the decoding process. Leaning on this reasoning, a first-degree distribution, referred to as the *ideal soliton distribution*, can be derived.

Example 11.2 (Ideal soliton distribution) The ideal soliton distribution is

$$\Omega_d = \begin{cases} \frac{1}{k} & \text{for } d = 1 \\ \frac{1}{d(d-1)} & \text{for } d = 2, 3, \ldots, k. \end{cases} \tag{11.10}$$

Some observations are in order. First, the average output degree is

$$\bar{d} = \frac{1}{k} + \sum_{d=2}^{k} \frac{d}{d(d-1)}. \tag{11.11}$$

Observing that

$$\log(k+1) < \sum_{d=1}^{k} \frac{1}{d} < \log k + 1, \tag{11.12}$$

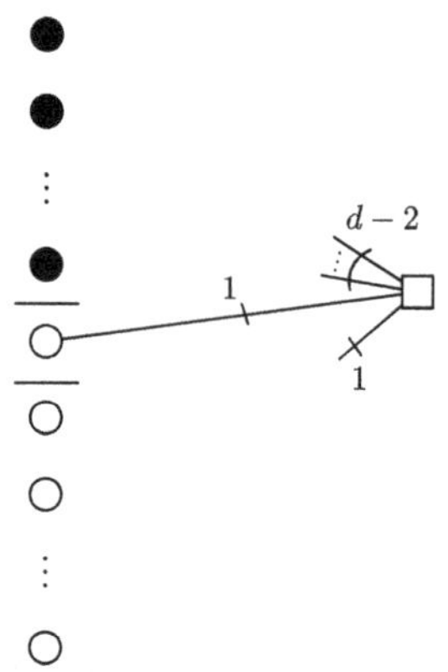

Figure 11.4 Condition for a degree-d output symbol to enter the ripple at the ℓth iteration.

we conclude that the average encoding cost is $\mathcal{O}(\log(k))$. Moreover, let us denote by q_ℓ the probability that a given output symbol enters the output ripple at iteration ℓ. We will show that the ideal soliton distribution yields $q_\ell = 1/k$ for $\ell = 1, \ldots, k-1$. To do so, we first express the probability that a given output symbol with degree d enters the output ripple at iteration ℓ. We denote this quantity as $q_{\ell,d}$. Figure 11.4 shows the condition that we are seeking: The output symbol, with degree d, must be connected with one edge to the input symbol that is revealed during the ℓth iteration. Moreover, the output symbol needs to have one and only one neighbor among the $k - \ell$ input symbols that are still unrevealed. We have

$$q_{\ell,d} = \frac{(k-\ell)\binom{\ell-1}{d-2}}{\binom{k}{d}}. \tag{11.13}$$

At the beginning of the decoding process, the output ripple is formed by output symbols with original degree 1. Hence, $q_0 = \Omega_1 = 1/k$. For $\ell = 1, \ldots, k-1$, by averaging (11.13) over Ω, we have

$$\begin{aligned} q_\ell &= \sum_{d=2}^{\ell+1} \frac{1}{d(d-1)} \frac{(k-\ell)\binom{\ell-1}{d-2}}{\binom{k}{d}} && (11.14)\\ &= \sum_{d=2}^{\ell+1} (k-\ell) \frac{(\ell-1)!\,(k-d)!}{k!\,(\ell-d+1)!} && (11.15)\\ &= \sum_{d=2}^{\ell+1} \binom{k-d}{k-\ell-1} \frac{(k-\ell)!\,(\ell-1)!}{k!} && (11.16)\\ &= \binom{k-1}{\ell-1}^{-1} \frac{1}{k} \sum_{d=2}^{\ell+1} \binom{k-d}{k-\ell-1} && (11.17)\\ &= \frac{1}{k}, && (11.18) \end{aligned}$$

where in (11.14) the lower limit in the sum is 2 since degree-1 output symbols are already in the ripple before the first iteration, (11.15) follows by simple algebraic operations, (11.16) is obtained by multiplying the numerator and the denominator in (11.15) by $(k - \ell - 1)!$, (11.17) follows by a trivial manipulation of the expression, and finally (11.18) is obtained through the identity

$$\sum_{i=a}^{b} \binom{i}{a} = \binom{b+1}{a+1}. \tag{11.19}$$

The result given in (11.18) implies that, at each iteration, on average n/k output symbols enter the ripple. If each realization of the decoding process would follow the average behavior, k output symbols would be sufficient to succeed (with an overhead of zero). In practice, statistical fluctuations around the average may happen, leading to a high decoding failure probability for the ideal soliton distribution: The expected ripple size is one, providing no margin with respect to possible deviations in the ripple size.

A variation of the ideal soliton distribution that solves this shortcoming is the *robust soliton distribution*, illustrated in the following example.

Example 11.3 (Robust soliton distribution) The robust soliton distribution is obtained by adding a bias to the ideal soliton distribution, which helps maintain the output ripple size larger than one throughout the decoding process. The bias term is

$$\beta_d = \begin{cases} \dfrac{\chi}{dk} & d = 1, \dots, \dfrac{k}{\chi} - 1 \\[2ex] \dfrac{\chi \log(\chi/p)}{k} & d = \dfrac{k}{\chi} \\[2ex] 0 & d = \dfrac{k}{\chi} + 1, \dots, k \end{cases}, \tag{11.20}$$

where

$$\chi = c \log \frac{k}{p} \sqrt{k}$$

for some positive constant c, and p is the target decoding failure probability. The robust soliton distribution is obtained by adding the bias β_d to the ideal soliton distribution Ω_d in (11.10). It can be shown that the robust soliton distribution achieves the target decoding failure probability with an absolute overhead that is $\mathcal{O}(\sqrt{k} \log^2(k/p))$. Since the average output symbol degree $\bar{d}$ is $\mathcal{O}(\log(k/\delta))$, the encoding and decoding costs are $\mathcal{O}(\log(k/\delta))$, too.

11.2.4 Limitations

An important question is whether degree distributions exist that allow reliable decoding of LT codes with a relative overhead close to zero, with linear encoding/decoding complexities, that is, with encoding and decoding costs independent of k. The answer is, unfortunately, negative. In particular, the following theorem provides a first hint to the fact that a necessary condition to achieve reliable decoding in the sense of Definition 11.1 is to have an encoding cost that is at least proportional to $\log k$.

Theorem 11.1 *Consider an LT code with output degree distribution Ω. The probability of decoding failure under ML decoding is lower bounded as*

$$P_B(\delta) \geq \left(1 - \frac{\bar{d}}{k}\right)^{k+\delta}, \tag{11.21}$$

where $\bar{d} = \sum_{d=1}^{k} d\,\Omega_d$ is the average output symbol degree and $k + \delta$ is the number of nonerased output symbols.

Proof The probability that an input symbol is not connected to any of the $k+\delta$ output symbols is

$$\left[\sum_{d=1}^{k} \Omega_d \left(1 - \frac{d}{k}\right)\right]^{k+\delta} = \left(1 - \frac{\bar{d}}{k}\right)^{k+\delta}. \tag{11.22}$$

If an input symbol is not connected to any output symbol, its value cannot be revealed, causing a decoding failure. □

A necessary condition for reliable decoding is to have

$$\left(1 - \frac{\bar{d}}{k}\right)^{n} \leq \frac{1}{k^c} \tag{11.23}$$

for some positive constant c. By observing that the left-hand side in (11.23) is lower bounded by

$$\exp\left(-\frac{\bar{d}n}{k - \bar{d}}\right),$$

we have that

$$\frac{\bar{d}n}{k - \bar{d}} \geq c \log k \tag{11.24}$$

implying that, for $n = k + \delta = k(1 + \epsilon_o)$

$$\begin{aligned}
\bar{d} &\geq \frac{1}{(1+\epsilon_o) + c/k \log k} c \log k \\
&\geq \frac{1}{(1+\epsilon_o) + c/3 \log 3} c \log k \\
&= \text{const.} \times \log k.
\end{aligned} \tag{11.25}$$

This result has, as a major consequence, the following fact: If we wish to achieve reliable decoding with a fixed relative overhead, the average output degree has to grow at least as $\log k$, yielding an encoding cost that is at least in the order of $\log k$.

11.3 Raptor Codes

Raptor codes allow to overcome the limitation of LT codes discussed in the previous section. The underlying principle is to adopt an LT code with constant encoding and decoding cost, characterized by a nonzero average symbol erasure probability at its peeling decoder output, and to concatenate it with a high-rate linear block code in charge of recovering the residual erasures. The LT code hence plays the role of the inner code in the concatenation, whereas the high-rate code is the outer code. A Raptor code is fully described by the triplet $(k, \mathcal{C}, \Omega)$, where k is the number of input symbols, $\mathcal{C}$ is the outer code and Ω the degree distribution used by the LT encoder. We may distinguish three classes of Raptor codes. Binary Raptor codes have both the outer code and the inner LT code defined over GF(2). Nonbinary Raptor codes are a trivial generalization of the previous class, with both outer code and inner LT code defined over GF(q) with $q = 2^m$ and $m > 1$. Finally, RaptorQ codes are based on a hybrid concatenation: The outer code is defined over GF(q) with $q = 2^m$ and $m > 1$, whereas the inner LT code is defined over the subfield GF(2).

11.3.1 Encoding

Encoding of a $(k, \mathcal{C}, \Omega)$ Raptor code works as follows. The input symbols $\boldsymbol{u} = \begin{bmatrix} u_1 & u_2 & \ldots & u_k \end{bmatrix}$ are first precoded via the outer code encoder, resulting in a vector of h intermediate symbols that we denote as $\boldsymbol{v} = \begin{bmatrix} v_1 & v_2 & \ldots & v_h \end{bmatrix}$. The intermediate symbols are then encoded via the inner LT encoder, as described in Section 11.2.1 (with the caveat that the LT encoder operates now with h symbols at its input), generating the output symbols $x_1, x_2, \ldots$.

11.3.2 Two-Stage Decoding

Raptor codes can be decoded by following the concatenation principle. In particular, upon observing a subset $\begin{bmatrix} x'_1 & x'_2 & \ldots & x'_n \end{bmatrix}$ of output symbols at the channel output, one may first apply peeling decoding to the inner LT code, enabling the recovery of (a subset of) the intermediate symbols. The outer decoder is then run to attempt the recovery of the residual missing intermediate symbols. For a given LT distribution, the complexity/performance trade-off is inherently linked to the specific choice of the outer code and its encoding/decoding algorithms. For example, the outer code may be chosen as a high-rate algebraic code with an efficient erasure decoding algorithm or a high-rate LDPC code. We refer to decoding algorithms that decode the inner and the outer code separately as *two-stage decoding* algorithms. The separation of the inner and the outer code decoders contrasts with the general approach presented in Section 11.4, which implies the joint decoding of the inner and outer code. Some examples of Raptor codes amenable to two-stage decoding are discussed in Section 11.3.3.

11.3.3 Performance Analysis

Under two-stage decoding, the complexity/performance trade-off for a Raptor code largely depends on the choice of the outer code. A sensible choice to keep the decoding complexity low is to adopt an LDPC code as outer code. This construction shows remarkable properties in the limit of large k. The construction relies on an inner (h, Ω) LT code with a modified Soliton distribution. The distribution is

$$\Omega_d = \begin{cases} \dfrac{\mu}{\mu+1} & d = 1 \\ \dfrac{1}{(\mu+1)(d-1)d} & d = 2, \ldots, D, \\ \dfrac{1}{(\mu+1)D} & d = D+1 \end{cases} \tag{11.26}$$

where, for a positive ε,

$$D = \left\lceil \frac{4(1+\varepsilon)}{\varepsilon} \right\rceil \tag{11.27}$$

and

$$\mu = \frac{\varepsilon}{2} + \left(\frac{\varepsilon}{2}\right)^2. \tag{11.28}$$

The distribution in (11.26) has the following property: As $h \to \infty$, the residual erasure probability of the input symbols under peeling decoding is upper bounded by $(\varepsilon/4)/(1+\varepsilon)$ for any relative overhead $\epsilon_o > \varepsilon/2$. An asymptocally good Raptor code can be constructed by using, as an outer code, an LDPC code with rate $R = (1+\varepsilon/2)/(1+\varepsilon)$. The code should enable successful decoding with high probability when the probability of erasure for the intermediate symbols is $(\varepsilon/4)/(1+\varepsilon)$, under peeling decoding, with a number of operations that is $\mathcal{O}(h)$. A properly designed IRA code can serve the purpose (by comparing the code rate and the required erasure correction capability, we observe that the outer code is not required to be capacity-achieving). This construction achieves reliable decoding with a relative overhead $\epsilon_o = \varepsilon$, at constant encoding and decoding costs (albeit dependent on ϵ_o).

Despite the remarkable asymptotic properties, Raptor code constructions of this kind tend to perform very poorly for moderate source message lengths (e.g., up to several thousand symbols). In order to construct Raptor codes that work well in practice, one has to relax the demand of a constant decoding cost. In particular, most of the Raptor codes used in practice are not designed to be decoded with the two-stage decoding approach outlined in this section but rather through an efficient ML decoding algorithm, which relies on the principles discussed in Section 6.2 for the case of LDPC codes.

11.4 Maximum-Likelihood Decoding of LT and Raptor Codes

Owing to their linearity, ML decoding of LT and Raptor codes can be attained by solving a system of linear equations on GF(q). For LT codes, the system to be solved is trivially

$$\boldsymbol{u}\boldsymbol{G}' = \boldsymbol{x}', \tag{11.29}$$

where $\boldsymbol{G}'$ is the matrix induced by the encoding rule of (11.8). In particular, $\boldsymbol{G}'$ is a $k \times n$ sparse matrix whose columns have Hamming weights distributed according to Ω. This observation enables the adoption of the efficient decoding based on structured Gaussian elimination, described in Section 6.2.1 in the context of ML decoding of LDPC codes.[3]

The case of Raptor codes requires more attention. Denote by $\boldsymbol{H}_O$ the $(h-k)\times h$ parity-check matrix of the outer code, and by $\boldsymbol{G}'$ the $h\times n$ matrix defining the encoding of the inner LT code. We have

$$\boldsymbol{v}\boldsymbol{H}_O^{\mathsf{T}} = \boldsymbol{0} \tag{11.30}$$

and

$$\boldsymbol{v}\boldsymbol{G}' = \boldsymbol{x}'. \tag{11.31}$$

It follows that the decoding of Raptor codes can be achieved by solving the vector of h intermediate symbols $\boldsymbol{v}$ through the linear system

$$\boldsymbol{v}\left[\boldsymbol{H}_O^{\mathsf{T}} \quad \boldsymbol{G}'\right] = \left[\boldsymbol{0} \quad \boldsymbol{x}'\right]. \tag{11.32}$$

After recovering $\boldsymbol{v}$, the k-symbols input vector $\boldsymbol{u}$ can be obtained. We may observe that the matrix $\left[\boldsymbol{H}_O^{\mathsf{T}} \quad \boldsymbol{G}'\right]$ contains an $h\times n$ sparse right submatrix $\boldsymbol{G}'$, whose columns have Hamming weights distributed according to Ω, and an $h\times(h-k)$ left submatrix $\boldsymbol{H}_O^{\mathsf{T}}$, which is in general nonsparse (it turns out to be sparse if the outer code is an LDPC code). Since the outer code rate is high, that is, since $(h-k)$ is typically much smaller than h and n the number of columns $\boldsymbol{G}'$ is predominant in the system matrix $\left[\boldsymbol{H}_O^{\mathsf{T}} \quad \boldsymbol{G}'\right]$ for any positive overhead, and the system can be regarded as sparse. As such, also (11.32) is well-suited for efficient decoding through the method illustrated in Section 6.2.1.

11.4.1 Performance Analysis

Since LT and Raptor codes can be decoded via the efficient ML decoding algorithm described in Section 6.2.1, we address next the analysis of their ML decoding failure probability. In particular, we provide tight upper and lower bounds on the ML decoding failure probability of (nonbinary) LT and Raptor codes as a function of the overhead δ.

The first result is a refinement of the lower bound on the decoding failure probability of an LT code provided by Theorem 11.1.

[3] In the fountain codes jargon, efficient ML decoding following the method of Section 6.2.1 is dubbed *inactivation decoding*.

Theorem 11.2 *Consider a* (k,Ω) *LT code over* $GF(q)$. *The decoding failure probability under ML decoding is lower bounded as*

$$P_B(\delta) \geq \sum_{i=1}^{k} (-1)^{i+1} \binom{k}{i} \left(\sum_{d=1}^{d_{max}} \Omega_d \frac{\binom{k-i}{d}}{\binom{k}{d}} \right)^{k+\delta}. \tag{11.33}$$

Proof We define two events V_i and W_j, where V_i is the event where a specific subset of i input nodes is not connected to any of the $k+\delta$ output nodes associated with unerased symbols, and W_j is the event where there are exactly j input nodes not connected to any of the $n = k + \delta$ output nodes associated with unerased symbols. We have

$$P_B(\delta) \geq P\left(\bigcup_{j=1}^{k} W_j\right) \tag{11.34}$$

$$= \sum_{j=1}^{k} P\left(W_j\right), \tag{11.35}$$

where (11.35) follows from the observation that the events $W_1, W_2, \ldots, W_k$ are mutually exclusive. We proceed by expressing $P(V_i)$ in terms of $P\left(W_j\right)$. In particular, for $j \geq i$, we have that

$$P(V_i) = \sum_{j=1}^{k} P\left(V_i | W_j\right) P\left(W_j\right) \tag{11.36}$$

$$= \sum_{j=1}^{k} \frac{\binom{k-i}{j-i}}{\binom{k}{j}} P\left(W_j\right) \tag{11.37}$$

$$= \frac{1}{\binom{k}{i}} \sum_{j=1}^{k} \binom{j}{i} P\left(W_j\right). \tag{11.38}$$

Owing to the identity

$$\sum_{i=1}^{k} (-1)^{i+1} \binom{k}{i} = 1, \tag{11.39}$$

we can write

$$\sum_{i=1}^{k} (-1)^{i+1} \binom{k}{i} P(V_i) = \sum_{i=1}^{k} (-1)^{i+1} \sum_{j=1}^{k} \binom{j}{i} P\left(W_j\right) \tag{11.40}$$

$$= \sum_{i=1}^{k} \sum_{j=1}^{k} \binom{j}{i} (-1)^{i+1} P\left(W_j\right) \tag{11.41}$$

$$= \sum_{j=1}^{k} \sum_{i=1}^{j} \binom{j}{i} (-1)^{i+1} P\left(W_j\right) \tag{11.42}$$

$$= \sum_{j=1}^{k} P\left(W_j\right). \tag{11.43}$$

We may observe that $P(V_i)$ can be expressed also as

$$P(V_i) = \left(\sum_{d=1}^{d_{\max}} \Omega_d \frac{\binom{k-i}{d}}{\binom{k}{d}} \right)^{k+\delta}. \tag{11.44}$$

By combining (11.43) with (11.34), (11.38), and (11.44), the statement follows. □

The lower bound provided in Theorem 11.2 is especially tight at low failure probabilities (i.e., for large overheads), where the failure probability is dominated by the probability of having input symbols that do not participate in the equation that generates any of the $k + \delta$ nonerased output symbols. Instead, the bound happens to be loose at low overheads, where the failure probability is dominated by the probability of a rank deficiency in the matrix $\boldsymbol{G}'$.

An upper bound on the decoding failure probability of an LT code under ML decoding is provided next.

Theorem 11.3 *Consider a* (k,Ω) *LT code over* $GF(q)$*. The probability of decoding failure under ML decoding is upper-bounded as*

$$P_B(\delta) \leq \sum_{i=1}^{k} (q-1)^{i-1} \binom{k}{i} \pi_i^{k+\delta}, \tag{11.45}$$

where π_i *is the probability that an output symbol is* 0 *given that* $\boldsymbol{u}$ *has Hamming weight* i*, and it is equal to*

$$\pi_i = \frac{1}{q} + \frac{q-1}{q} \sum_{j=1}^{d_{max}} \Omega_j \frac{\mathcal{K}_j(i;h,q)}{\mathcal{K}_j(0;h,q)} \tag{11.46}$$

where

$$\mathcal{K}_j(x;h,q) = \sum_{\ell=0}^{j} (-1)^{\ell} \binom{x}{\ell} \binom{h-x}{j-\ell} (q-1)^{j-\ell} \tag{11.47}$$

is the Krawtchouk polynomial of degree j *with parameters* h *and* q*.*

The proof of Theorem 11.3 follows as a special case of the proof of Theorem 11.4, provided next, which generalizes the upper bound to Raptor codes (see Exercise 11.7).

Theorem 11.4 *Consider a* $(k,\mathcal{C},\Omega)$ *Raptor code over* $GF(q)$*, where the* (h,k) *outer code* $\mathcal{C}$ *has weight enumerator* A_i*. The decoding failure probability under ML decoding is upper bounded as*

$$P_B(\delta) \leq \frac{1}{q-1} \sum_{i=1}^{h} A_i \pi_i^{k+\delta}, \tag{11.48}$$

where

$$\pi_i = \frac{1}{q} + \frac{q-1}{q} \sum_{j=1}^{d_{max}} \Omega_j \frac{\mathcal{K}_j(i;h,q)}{\mathcal{K}_j(0;h,q)} \tag{11.49}$$

is the probability that an output symbol is 0 *given that the intermediate vector* $\boldsymbol{v}$ *has Hamming weight* i.

Proof The input symbol vector $\boldsymbol{u}$ and the received output symbols vector $\boldsymbol{x}'$ are related as $\boldsymbol{x}' = \boldsymbol{u}\boldsymbol{G}_o\boldsymbol{G}'$. Here, $\boldsymbol{G}_o$ is the $k \times h$ generator matrix of the outer code, and $\boldsymbol{G}'$ is the matrix defining the relation between the intermediate symbol vector and the received output symbols vector as in (11.31). Let us define $\boldsymbol{G}_r = \boldsymbol{G}_o\boldsymbol{G}'$. Observe that decoding fails whenever the system does not admit a unique solution that is, if and only if $\text{rank}(\boldsymbol{G}_r) < k$. The rank deficiency implies the existence of a nonzero vector $\boldsymbol{u} \in \text{GF}(q)^k$ for which $\boldsymbol{u}\boldsymbol{G}_r = \boldsymbol{0}$. Following this observation, for any two vectors $\boldsymbol{u} \in \text{GF}(q)^k$ and $\boldsymbol{v} \in \text{GF}(q)^h$, we define $E_{\boldsymbol{u}}$ as the event $\boldsymbol{u}\boldsymbol{G}_r = \boldsymbol{0}$, and $F_{\boldsymbol{v}}$ as the event $\boldsymbol{v}\boldsymbol{G}' = \boldsymbol{0}$. Relying on the outer code linearity, observe that the all-zero intermediate word is only generated by the all-zero input vector. It follows that

$$P_B(\delta) = P\left(\bigcup_{\boldsymbol{u} \in \text{GF}(q)^k \setminus \{\boldsymbol{0}\}} E_{\boldsymbol{u}}\right) = P\left(\bigcup_{\boldsymbol{v} \in \mathcal{C} \setminus \{\boldsymbol{0}\}} F_{\boldsymbol{v}}\right). \tag{11.50}$$

Moreover, again due to the linearity of the outer code, if $\boldsymbol{v} \in \mathcal{C}$, then $\beta\boldsymbol{v} \in \mathcal{C}$ for any $\beta \in \text{GF}(q)\setminus\{0\}$. Furthermore, for any $\beta \in \text{GF}(q)\setminus\{0\}$, $\boldsymbol{v}\,\boldsymbol{G}' = \boldsymbol{0}$ if and only if $\beta\boldsymbol{v}\boldsymbol{G}' = \boldsymbol{0}$. Thus, for any two outer codewords $\boldsymbol{v}_1$ and $\boldsymbol{v}_2$ such that $\boldsymbol{v}_1 = \beta\boldsymbol{v}_2$ for some $\beta \in \text{GF}(q) \setminus \{0\}$, the events $F_{\boldsymbol{v}_1}, F_{\boldsymbol{v}_2}$ coincide. We now take a union bound on (11.50). Following the observation above, we can divide it by a factor $q-1$, yielding

$$P_B(\delta) \le \frac{1}{q-1} \sum_{\boldsymbol{v} \in \mathcal{C} \setminus \{\boldsymbol{0}\}} P(F_{\boldsymbol{v}}). \tag{11.51}$$

An alternative expression of the union bound can be attained by partitioning the code $\mathcal{C}$ in the subsets $\mathcal{C}_l = \{\boldsymbol{v} \in \mathcal{C} : w_{\text{H}}(\boldsymbol{v}) = l\}$, with $l = 1, \dots, h$, and re-writing (11.51) as

$$P_B(\delta) \le \frac{1}{q-1} \sum_{l=1}^{h} \sum_{\boldsymbol{v} \in \mathcal{C}_l} P(F_{\boldsymbol{v}}) = \frac{1}{q-1} \sum_{l=1}^{h} A_l P(F_{\boldsymbol{v}} | w_{\text{H}}(\boldsymbol{v}) = l). \tag{11.52}$$

Since the output symbols are drawn independently, we have

$$P(F_{\boldsymbol{v}} | w_{\text{H}}(\boldsymbol{v}) = l) = \pi_l^{k+\delta}, \tag{11.53}$$

where $\pi_l = P(x' = 0 | w_{\text{H}}(\boldsymbol{v}) = l)$. We introduce now the r.v.s J and I, where J denotes the number of intermediate symbols that are linearly combined to generate the generic output symbol x', and I is the number of nonzero intermediate symbols. Obviously, we have $I \le \min\{J, w_{\text{H}}(\boldsymbol{v})\}$. Observe also that

$$P(J = j | w_{\text{H}}(\boldsymbol{v}) = l) = P(J = j) = \Omega_j \tag{11.54}$$

that is, the output symbol degree does not depend on the weight of the intermediate vector and that

$$P(x' = 0|w_{\mathrm{H}}(\boldsymbol{v}) = l, J = j, I = i) = P(x' = 0|I = i). \tag{11.55}$$

We can then recast π_l as

$$\pi_l = \sum_{j=1}^{d_{\max}} P(x' = 0|w_{\mathrm{H}}(\boldsymbol{v}) = l, J = j) P(J = j|w_{\mathrm{H}}(\boldsymbol{v}) = l) \tag{11.56}$$

$$= \sum_{j=1}^{d_{\max}} \Omega_j P(x' = 0|w_{\mathrm{H}}(\boldsymbol{v}) = l, J = j) \tag{11.57}$$

$$= \sum_{j=1}^{d_{\max}} \Omega_j \sum_{i=0}^{\min\{j,l\}} P(x' = 0|I = i) P(I = i|w_{\mathrm{H}}(\boldsymbol{v}) = l, J = j). \tag{11.58}$$

Recalling that the j intermediate symbols are chosen uniformly at random, the right-most term in (11.58) takes the form

$$P(I = i|w_{\mathrm{H}}(\boldsymbol{v}) = l, J = j) = \frac{\binom{l}{i}\binom{h-l}{j-i}}{\binom{h}{j}}. \tag{11.59}$$

Moreover, we may observe that the nonzero elements of $\boldsymbol{G}'$ are i.i.d. and uniformly drawn in $\mathrm{GF}(q) \setminus \{0\}$. It follows that

$$P(x' = 0|I = i) = \frac{1}{q}\left(1 + \frac{(-1)^i}{(q-1)^{i-1}}\right). \tag{11.60}$$

The statement follows by substituting (11.59) and (11.60) in (11.58), and rewriting the resulting expression using Krawtchouk polynomials. □

Example 11.4 Consider a $(57, \mathcal{C}, \Omega)$ binary Raptor code where the LT code degree distribution is

$$\Omega_1 = 0.0098 \qquad \Omega_2 = 0.4590 \qquad \Omega_3 = 0.2110$$
$$\Omega_4 = 0.1134 \qquad \Omega_{10} = 0.1113 \qquad \Omega_{11} = 0.0799 \qquad \Omega_{40} = 0.0156$$

and the outer code is a (63, 57) Hamming code. Note that weight enumerator of a Hamming code of blocklength $h = 2^t - 1$ and dimension $k = h - t$ can be obtained recursively as $(i+1)\, A_{i+1} + A_i + (h - i + 1)\, A_{i-1} = \binom{h}{i}$ with $A_0 = 1$ and $A_1 = 0$. Using this results in Theorem 11.4, one can derive the corresponding upper bound on the ML decoding failure probability. The result, depicted together with simulations at various values of overhead, is provided in Figure 11.5.

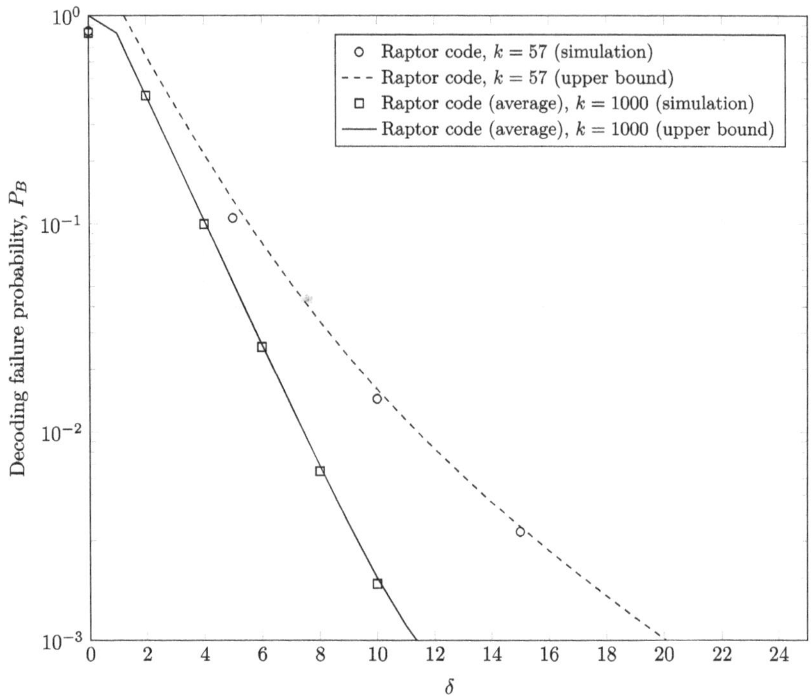

Figure 11.5 Decoding failure probability as a function of the overhead for the two Raptor codes of Examples 11.4 and 11.5.

Example 11.5 As a second example, we provide the performance in terms of ML decoding failure probability versus overhead for a Raptor code based on an LT code with the same distribution of the previous example and on an outer LDPC code. More specifically, we provide an expectation of the upper bound in Theorem 11.4 with respect to random LDPC code drawn from the (3, 15) regular LDPC ensemble with blocklength $h = 1250$ (and $k = 1000$). Due to linearity, the expectation of the upper bound in Theorem 11.4 yields

$$\mathsf{E}[P_B(\delta)] \leq \frac{1}{q-1} \sum_{i=1}^{h} \bar{A}_i \pi_i^{k+\delta}, \tag{11.61}$$

where $\bar{A}_i$ is the average weight enumerator. The bound, compared with simulation results, is depicted in Figure 11.5.

11.5 Historical Notes

A first, pioneering approach to improve the network throughput in broadcast scenarios relying on erasure correcting codes is due to J.J. Metzner [1]. The framework that lays the foundation of fountain codes appeared in [2,3]. LT codes were introduced as a first example of efficient, low-complexity fountain codes in [4], where their performance and complexity are analyzed for the first time. The analysis of LT codes has been subsequently refined, addressing their performance at finite source message length in [5] and [6]. Raptor codes [7] were invented by A. Shokrollahi in 2000, and since then, they have been included in a number of communication standards and recommendations (see Appendix A). A detailed survey on their design, including their most recent variants (RaptorQ codes) is contained in [8]. Nearly at the same time when Raptor codes were invented, a closely related class of rateless codes (dubbed *online codes*) was introduced by P. Maymounkov [9]. Inactivation decoding of Raptor codes was patented in [10]. Closely related decoding algorithms can be found in [11, Section 2.6], and more recently in [12–14]. Inactivation decoding for LT codes is studied in [15]. A first performance analysis of some classes on nonbinary fountain codes under ML decoding was provided in [16]. A thorough analysis of different classes of fountain codes under ML decoding, which includes the derivation of several upper and lower bounds on the decoding failure probability, is available in [17]. The derivation of the lower bound on the decoding failure probability given in Theorem 11.2 follows closely the one proposed in [17]. The upper bound on the decoding failure probability given in Theorem 11.3 was also originally derived in [17]. The extension of the bounds to the case of binary/nonbinary Raptor codes and RaptorQ codes was proposed in [18], where lower bounds on the error exponent of Raptor codes were derived. Raptor codes are used as fixed-rate codes (i.e., linear block erasure codes) in a few applications. The distance properties of fixed-rate Raptor codes have been analyzed in [19]. For a further, detailed account of the applications of Raptor codes, we point to Appendix A and to [8] [20].

11.6 Exercises

Exercise 11.1 A binary fountain code is used to encode a message of 5 bits. The vector containing the nonerased symbols at the channel output is

$$\boldsymbol{x}' = \begin{bmatrix} 1 & 0 & 0 & 1 & 0 & 1 \end{bmatrix}, \qquad (11.62)$$

whereas the matrix defining the relation between input symbols and the nonerased output symbols according to (11.4) is

$$\boldsymbol{G}' = \begin{bmatrix} 1 & 1 & 0 & 0 & 1 & 1 \\ 0 & 1 & 1 & 0 & 0 & 1 \\ 0 & 0 & 1 & 1 & 1 & 0 \\ 1 & 0 & 1 & 0 & 1 & 1 \\ 0 & 1 & 1 & 1 & 0 & 1 \end{bmatrix}. \qquad (11.63)$$

Perform ML decoding. Does decoding succeed?

Exercise 11.2 Consider using a binary fountain code to transmit over a QEC, where the channel input alphabet has cardinality $q = 2^m$ with $m > 1$. In particular, the binary fountain encoder generates blocks of m bits, which are mapped over the channel input symbol at every channel use. In this setting, we refer to the overhead δ in terms of channel uses, and we assume that the source message length k (in bits) divides m. Show that if the binary linear combinations are generated by drawing the coefficients g_{ji} of (11.1) independently and uniformly in GF(2), the decoding failure probability can be upper bounded by

$$P_B(\delta) < q^{-\delta}. \tag{11.64}$$

Exercise 11.3 Assume to know the decoding failure probability P_B of a fountain code as a function of the absolute overhead δ. What is the probability of failing at decoding after collecting $k + \delta - 1$ nonerased output symbols and succeeding after collecting an additional output symbol (i.e., with overhead δ)?

Exercise 11.4 A binary LT code is used to encode a message of 6 bits. The vector containing the nonerased symbols at the channel output is

$$\boldsymbol{x}' = \begin{bmatrix} 1 & 0 & 1 & 1 & 1 & 0 \end{bmatrix}, \tag{11.65}$$

whereas the matrix defining the relation between input symbols and the nonerased output symbols according to (11.4) is

$$\boldsymbol{G}' = \begin{bmatrix} 1 & 1 & 0 & 0 & 1 & 1 \\ 0 & 1 & 1 & 0 & 0 & 0 \\ 0 & 0 & 1 & 1 & 0 & 0 \\ 0 & 0 & 1 & 0 & 1 & 1 \\ 0 & 1 & 1 & 1 & 0 & 1 \\ 0 & 1 & 0 & 0 & 0 & 0 \end{bmatrix}. \tag{11.66}$$

Perform peeling decoding. Does decoding succeed?

Exercise 11.5 A binary LT code is used to encode a message of 6 bits. The vector containing the nonerased symbols at the channel output is

$$\boldsymbol{x}' = \begin{bmatrix} 1 & 0 & 1 & 1 & 1 & 0 \end{bmatrix}, \tag{11.67}$$

whereas the matrix defining the relation between input symbols and the nonerased output symbols according to (11.4) is

$$\boldsymbol{G}' = \begin{bmatrix} 1 & 1 & 0 & 0 & 1 & 1 \\ 0 & 1 & 1 & 0 & 0 & 0 \\ 0 & 0 & 1 & 1 & 1 & 0 \\ 0 & 0 & 1 & 0 & 1 & 1 \\ 0 & 1 & 1 & 1 & 0 & 1 \\ 0 & 1 & 0 & 0 & 0 & 0 \end{bmatrix}. \tag{11.68}$$

Note that the matrix $\boldsymbol{G}'$ differs from the matrix $\boldsymbol{G}'$ of Exercise 11.4 only by one element (third row, second to last column). Perform peeling decoding. Does decoding succeed?

Exercise 11.6 Perform ML decoding on the LT code of the previous example. Does decoding succeed?

Exercise 11.7 Using the result of Theorem 11.4, prove Theorem 11.3. *Hint*: What is the weight distribution of a rate-1 linear block code?

Exercise 11.8 Prove (11.60) for the case where $q = 2^m$. In particular, show that if X_1, X_2 ... X_l are discrete i.i.d. random variables uniformly distributed over $\mathrm{GF}(2^m)\backslash\{0\}$, then

$$P(X_1 + X_2 + \ldots + X_l = 0) = \frac{1}{q}\left(1 + \frac{(-1)^l}{(q-1)^{l-1}}\right), \qquad (11.69)$$

where $q = 2^m$. *Hint*: Observe that the additive group of $\mathrm{GF}(2^m)$ is isomorphic to the vector space $\mathbb{Z}_2^m$. Thus, one may let X_1, X_2 ... X_l be i.i.d random variables with uniform probability mass function over the vector space $\mathbb{Z}_2^m\backslash\{0\}$, and use the discrete Fourier transform to complete the proof.

Exercise 11.9 Extend the proof of the previous exercise to the case where $q = p^m$ with p being a prime.

Exercise 11.10 Reproduce the upper bound on the decoding failure probability under ML decoding for the Raptor code of Example 11.4.

References

[1] J. Metzner, "An improved broadcast retransmission protocol," *IEEE Trans. Commun.*, vol. 32, no. 6, pp. 679–683, June 1984.

[2] J. Byers, M. Luby, M. Mitzenmacher, and A. Rege, "A digital fountain approach to reliable distribution of bulk data," in *Proc. ACM SIGCOMM*, vol. 28, no. 4, Vancouver, Canada, Oct. 1998, pp. 56–67.

[3] J. Byers, M. Luby, and M. Mitzenmacher, "A digital fountain approach to asynchronous reliable multicast," *IEEE J. Sel. Areas Commun.*, vol. 20, no. 8, pp. 1528–1540, Oct. 2002.

[4] M. Luby, "LT codes," in *Proc. IEEE Symp. Found. Comp. Sci.*, Vancouver, Canada, Nov. 2002, pp. 271–282.

[5] R. Karp, M. Luby, and A. Shokrollahi, "Finite length analysis of LT codes," in *Proc. 2004 IEEE International Symp. on Inf. Theory*, Chicago, Illinois, US, June 2004.

[6] E. Maneva and A. Shokrollahi, "New model for rigorous analysis of LT-codes," in *Proc. 2006 IEEE International Symp. on Inf. Theory*. Seattle, Washington, US, 2006, pp. 2677–2679.

[7] A. Shokrollahi, "Raptor codes," *IEEE Trans. Inf. Theory*, vol. 52, no. 6, pp. 2551–2567, June 2006.

[8] A. Shokrollahi and M. Luby, "Raptor codes," *Foundations and Trends in Commun. and Inf. Theory*, vol. 6, no. 3-4, pp. 213–322, Jan. 2011.

[9] P. Maymounkov, "Online codes," Technical report, New York University, Technical Report, 2002.

[10] M. Shokrollahi, S. Lassen, and R. Karp, "Systems and processes for decoding chain reaction codes through inactivation," Feb. 2005, US Patent 6,856,263.

[11] E. Berlekamp, "The technology of error-correcting codes," *Proc. IEEE*, vol. 68, no. 5, pp. 564–593, May 1980.

[12] T. Richardson and R. Urbanke, "Efficient encoding of low-density parity-check codes," *IEEE Trans. Inf. Theory*, vol. 47, no. 2, pp. 638–656, Feb. 2001.

[13] H. Pishro-Nik and F. Fekri, "On decoding of low-density parity-check codes over the binary erasure channel," *IEEE Trans. Inf. Theory*, vol. 50, no. 3, pp. 439–454, Mar. 2004.

[14] D. Burshtein and G. Miller, "An efficient maximum likelihood decoding of LDPC codes over the binary erasure channel," *IEEE Trans. Inf. Theory*, vol. 50, no. 11, pp. 2837–2844, Nov. 2004.

[15] F. Lázaro, G. Liva, and G. Bauch, "Inactivation decoding of LT and Raptor codes: Analysis and code design," *IEEE Trans. Commun.*, vol. 65, no. 10, pp. 4114–4127, Oct. 2017.

[16] G. Liva, E. Paolini, and M. Chiani, "Performance versus overhead for fountain codes over $\mathbb{F}_q$," *IEEE Commun. Lett.*, vol. 14, no. 2, pp. 178–180, Feb. 2010.

[17] B. E. Schotsch, "Rateless coding in the finite length regime," Ph.D. dissertation, Inst. of Commun. Systems and Data Proc., RWTH Aachen, Aachen, Germany, July 2014.

[18] F. Lázaro, G. Liva, E. Paolini, and G. Bauch, "Bounds on the error probability of raptor codes under maximum likelihood decoding," *IEEE Trans. Inf. Theory*, vol. 67, no. 3, pp. 1537–1558, Mar. 2021.

[19] F. Lázaro, E. Paolini, G. Liva, and G. Bauch, "Distance spectrum of fixed-rate Raptor codes with linear random precoders," *IEEE J. Sel. Areas Commun.*, vol. 34, no. 2, pp. 422–436, Feb. 2016.

[20] M. Luby, M. Watson, T. Gasiba, T. Stockhammer, and W. Xu, "Raptor codes for reliable download delivery in wireless broadcast systems," in *Proc. IEEE Consumer Communications and Networking Conf.*, vol. 1, Las Vegas, NV, USA, Jan. 2006, pp. 192–197.

Appendix A Applications of Erasure Codes

In this appendix, a brief overview of erasure code applications in the context of data transmission, storage, computing, and privacy is presented. The list of use cases that is illustrated here is not to be considered as complete. Nevertheless, it covers some of the most popular and successful applications of erasure codes in current systems.

A.1 Reliable Data Transmission

In many communication systems, erasure codes are conceived as an additional, low-complexity protection mechanism that is applied on top of classical error correcting codes that protect, at the physical layer, individual packets from noise and errors introduced by the communication medium. In this context, erasure codes may serve as an alternative or a complement to ARQ protocols. In fact, while packet losses may be recovered by requiring the retransmission of the data units that have been erroneously decoded, such an approach may turn out to be unsuited in the presence of delay constraints or an unavailable feedback link between receiver and transmitter. Likewise, erasure codes are particularly appealing as a countermeasure to fading/outage events in wireless communication links since an erasure code encoding a sequence of packets can provide sufficient time diversity to cope with the impairments introduced by the fading channel.

From a classical coding perspective, a good erasure code aims at guaranteeing block error rates close to theoretical limits, such as the Singleton bound (see Chapter 3) at manageable encoding/decoding complexity. RS codes are a popular choice in many applications since they are MDS. Recall that an (n, k) MDS code can recover up to $n - k$ erasures, where $n - k$ corresponds to the redundancy introduced by the channel code. However, decoding complexity might become a bottleneck for increasing blocklengths. For instance, the complexity of the algorithms illustrated in Section 2.4 grows quadratically with the block length n (or the field order q). Longer codes can nevertheless be obtained by concatenation of shorter component codes.

As an example of application to wireless communication systems, RS codes have been standardized in Digital Video Broadcasting (DVB) [1] as network layer erasure recovery mechanism. In particular, they are used for protection of internet protocol (IP) datagrams, to increase reliability on mobile fading channels. To this end, so-called multi-protocol encapsulation (MPE)-FEC frames are generated, each frame

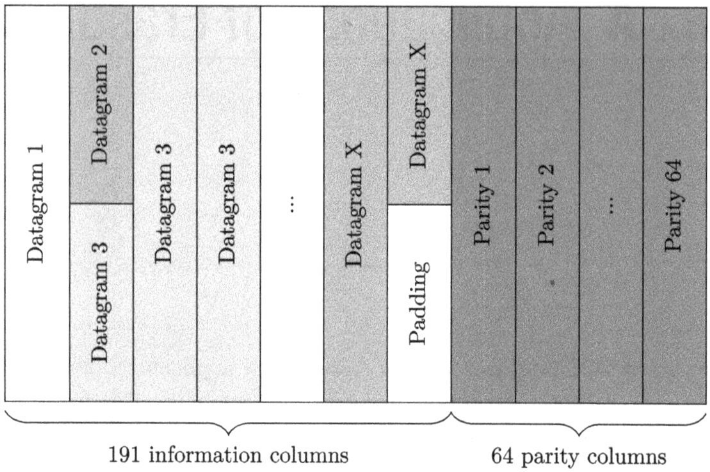

Figure A.1 Example of protection of IP datagrams using a (255, 191) RS code, according to the MPE-FEC construction.

being an array with 255 columns and a flexible number of rows (with a maximum of 1024), whose entries are from GF(256). Thereby, the first 191 leftmost columns contain the original IP datagrams (datagrams are placed in the array column-wise, with each IP datagram possibly spanning several columns). The 191 leftmost columns, hosting the data to be protected, are row-wise encoded by a (255, 191) RS code, generating 64 columns of redundancy symbols (see Figure A.1). The receivers can exploit the redundancy to recover symbol erasures in the RS codewords. An example of usage is the Digital Video Broadcasting – Handheld (DVB-H) standard, targeting video broadcasting to mobile handheld devices [2].

A family of LDPC erasure codes is part of an experimental Consultative Committee for Space Data Systems (CCSDS) recommendation for near-Earth and deep-space communications [3]. These codes are for use on upper layers (application, transport, network, or data link layer) of space communication systems as a complement to the physical layer FEC. Application scenarios are, for instance, optical low Earth orbit (LEO) satellite downlinks subject to long fading events. The LDPC codes of [3] are constructed as a concatenation of a binary (random) outer code with a binary inner IRA code. The latter one is obtained by a random permutation-based construction that allows to generate the code's parity-check matrix flexibly on the fly for a large number of code parameters [4]. Information lengths of 512, 2048, and 16384 and code rates of 2/3, 4/5, and 8/9 are covered in [3], but many other information lengths and code rates could be generated in a fast algorithmic manner. The code construction has been optimized for high-speed ML decoding through efficient structured GE (see Section 6.2).

Raptor codes are part of various standards for multimedia streaming and bulk data download services. As an example, consider the multimedia broadcast multicast services (MBMS) of the 3rd Generation Partnership Project (3GPP) long-term evolution (LTE) standard [5]. The data is first partitioned into k symbols of l bits each and encoded by a systematic (binary) Raptor encoder which produces additional $n - k$ redundancy (repair) symbols. Although the number of redundancy symbols can be chosen flexibly, transmissions of additional redundancy symbols upon decoding failure are not foreseen (fixed rate setting). Similarly, Raptor codes are also used for television services over IP networks in the framework of DVB [6]. They are also part of the DVB standard which defines the protection of IP datagrams [1]. Raptor codes on nonbinary fields, termed RaptorQ, are part of the North American Advanced Television Systems Committee (ATSC) standard [7] for streaming broadcast. They can be seen as an improvement of binary Raptor codes. By choosing the precode to be nonbinary over GF(256), but keeping the LT code to be binary, the erasure recovery properties are improved with respect to binary Raptor codes. The number of input symbols can range up to 56403, while for the binary Raptor codes in the aforementioned standards, it is at most 8192.

In the framework of the Internet Engineering Task Force (IETF), various solutions based on RS, LDPC, and Raptor codes are currently considered for reliable data delivery [8].

A.2 Distributed Storage

Low-cost, reliable storage of a tremendous amount of user data with instantaneous and ubiquitous availability is a key requirement of modern cloud-based storage systems. Distributed storage systems face this challenge by storing user data over multiple servers. These servers are subject to hardware failures which, without any remedy, would make user data temporarily (in the worst case, even permanently) unavailable. By adding redundancy to the stored data, erasure codes can be employed to reconstruct the content of unavailable servers from data stored on other servers. Already in the late '60s, the problem of reliable data storage was present. By observing that an expensive, reliable disk drive may be outperformed by an array of many low-cost (and less reliable) drives by adding redundancy, redundant array of independent disks (RAID) systems were introduced and quickly adopted by industry [9]. It was soon recognized that, although failures would rise in proportion to the number of drives in the array, the reliability of a redundant array can exceed that of any large single drive.

In the sequel, we will abstract a distributed system by a set of storage nodes (e.g., servers) in which the user's data is stored. In a simplified model, each storage node hosts one (or more) symbols of a codeword. Failures of nodes are modeled as code symbol erasures. This is illustrated in Figure A.2 for a network of three nodes. A $(3,2)$ single parity-check code is used to generate redundancy that is stored in the third node. Note that in the figure, the data on the nodes (e.g., servers) is further subdivided into 3 blocks to mimic partial node failures (e.g., hard disk failures). Whenever a node (partially) fails, the erased data can be restored from the remaining two nodes.

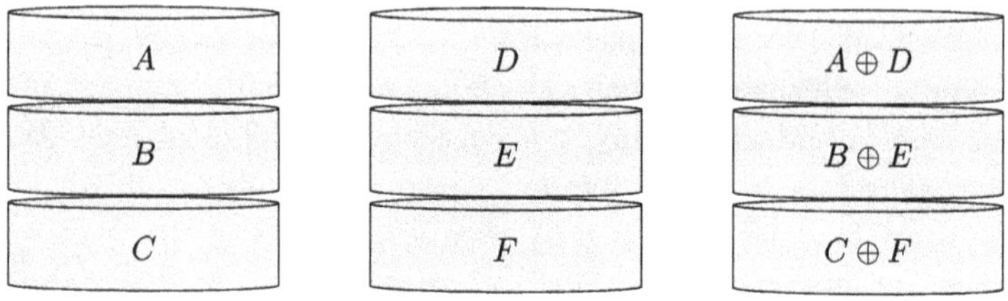

Figure A.2 Coding across nodes using a (3,2) single parity-check code. If any of the three nodes fail, the lost data can be recovered from the remaining two.

A.2.1 Availability and Storage Efficiency

Availability refers to the probability that a system is operational at a given time. High availability of a distributed storage system can be ensured by different forms of erasure coding where coding is performed across different nodes of the network. The simplest and thus the earliest coding strategy is repetition coding: Full copies of the data are stored on different nodes, and if any (but not all) of these nodes fail, the data is still accessible. Repetition codes have some appealing features, as will be made clear in the sequel, but also consume a large amount of storage space.

Example A.1 (Storage efficiency) Consider a $(3, 1)$ repetition code which is commonly used in distributed storage systems. It can recover from up to two erasures but requires to storing two redundancy symbols per data symbol. The resulting storage overhead, defined as $(n-k)/k$, is often given in percent. For the $(3, 1)$ repetition code, we have a storage overhead of 200%. As a comparison, a $(15, 11)$ Hamming code with minimum distance $d_{\min} = 3$ can also tolerate up to two erasures but requires only $\frac{4}{11} = 36\%$ of storage overhead. This reduction in overhead comes from the fact that several information symbols (or packets) are jointly processed to generate coded data. Increasing the information length may achieve lower overheads. However, there are practical constraints on the choice of the code parameters, e.g., number of disks, affordable decoding complexity, number of decoding operations, etc.

A.2.2 Decoding Complexity

RS codes are an appealing choice in storage systems: since RS codes are MDS, they ensure recovery of the data from any subset of k codeword symbols. Encoding and decoding operations for RS codes are performed on GF(q). Nowadays, many efficient implementations exist. However, for very high throughput applications, it might be desirable to resort to binary operations rather than GF(q) operations. Therefore, various families of MDS codes with this property have been developed in the literature. Examples are two-erasure correcting EVENODD codes [10, 11] or their evolution, Microsoft's three-erasure correcting STAR codes [12]. It is important to highlight that these codes are encoded and decoded as binary codes, but the aforementioned MDS property holds when considering their nonbinary representation, with code symbols which are assumed to be $(p-1)$-bit vectors. This is clarified in the subsequent example.

Example A.2 (EVENODD codes) EVENODD codes have parameters $(p+2,p)$, where p is a prime number and the parameters are expressed in symbols. Each nonbinary code symbol can be represented as a string of $p-1$ bits, and the length-$(p+2)$ codeword can be represented as a $(p-1)\times(p+2)$ binary array

$$A = \left[\begin{array}{cccc|c|c} a_{0,0} & a_{0,1} & \dots & a_{0,p-1} & a_{0,p} & a_{0,p+1} \\ a_{1,0} & a_{1,1} & \dots & a_{1,p-1} & a_{1,p} & a_{1,p+1} \\ \vdots & & & \vdots & \vdots & \vdots \\ a_{p-2,0} & a_{p-2,0} & \dots & a_{p-2,p-1} & a_{p-2,p} & a_{p-2,p+1} \end{array}\right]. \quad \text{(A.1)}$$

The leftmost p columns in A are initialized with the information bits. Encoding proceeds as follows. The first parity symbol, whose bitwise representation is $\begin{bmatrix} a_{0,p} & a_{1,p} & \dots & a_{p-2,p}\end{bmatrix}^{\mathsf{T}}$, is computed as bit-by-bit sum of the first p columns, i.e.,

$$a_{i,p} = a_{i,0} + a_{i,1} + \ldots + a_{i,p-1} \quad \text{(A.2)}$$

for $i = 0,1,\ldots,p-2$. The second parity symbol $\begin{bmatrix} a_{0,p+1} & a_{1,p+1} & \dots & a_{p-2,p+2}\end{bmatrix}^{\mathsf{T}}$ is obtained through the sum of (wrapped) diagonal elements of the p leftmost columns of A, and an additional term s (also referred to as adjuster), i.e.,

$$a_{i,p+1} = s + \sum_{\substack{j=0 \\ j\neq i+1}}^{p-1} a_{(i-j) \bmod p, j} \quad \text{(A.3)}$$

where

$$s = \sum_{j=1}^{p-1} a_{(p-j-1) \bmod p, j}. \quad \text{(A.4)}$$

It can be shown that this code can recover up to two symbols (i.e., column) erasures. By interpreting this code as a nonbinary code with symbols in $\{0,1\}^{p-1}$, this fact implies that it achieves the Singleton bound (see Theorem 2.4). Hence the code is MDS.

A.2.3 Node Repair Problem

Whenever a storage node in a distributed network fails, a new node (replacement node) tries to recover the lost information from working storage nodes in the network. The recovery of the data causes a considerable amount of disk access operations and network traffic. For an $(n,n-1)$ single parity-check code, we need to access data from $n-1$ nodes to recover a single erasure, while for an $(n,1)$ repetition code, it suffices to access the data on one storage node, and transfer it through the network to the replacement node. This illustrates that, apart from low storage overhead, also efficient repair of erased nodes is a key issue.

Example A.3 (Node repair problem) The Facebook warehouse cluster [13] stores hundreds of petabytes of data across a few thousand machines. Files are partitioned in blocks of size $l = 256$MB, and k blocks are encoded by an erasure code that generates $m = n - k$ additional redundancy blocks. Frequently asked data (also termed *hot* data) is encoded by a $(3, 1)$ repetition code, while data that is not been accessed for some time (*cold* data) is encoded by a $(14, 10)$ RS code. Code symbols (data blocks) of the same codeword are stored on different machines. Experiments show that a median of 180TB of data needs to be accessed and transferred through the network on a daily basis only to recover lost RS encoded data.

To mitigate the node repair problem, various types of erasure codes were proposed. Regenerating codes aim at minimizing the network traffic [14]. Usually, storage nodes send out linear combinations of their stored data following network coding concepts. Locally repairable codes, by contrast, require that any erased storage node can be recovered from no more than r other storage nodes [15]. We speak of codes with locality r. Such codes minimize the disk access operations and also reduce the network traffic. Regenerating codes with locality combines both aforementioned concepts.

Example A.4 (Code with locality) As an enhancement of the $(14, 10)$ RS code with systematic encoding, Facebook adopted a $(16, 10)$ locally repairable code in their Hadoop file system (HDFS), called Hadoop file system (HDFS)-Xorbas [13]. Consider a $(14, 10)$ shortened RS code over GF(16) with a systematic encoder. Let the codeword be $\boldsymbol{x} = \begin{bmatrix} x_1 & x_2 & \dots & x_{14} \end{bmatrix}$. Each codeword symbol of this code can be reconstructed from any subset of 10 other symbols, that is, $r = 10$. Enhancements can be found in [16]. We may modify the code as follows. We construct three additional parity-check symbols x_{15}, x_{16}, x_{17} as linear combinations

$$x_{15} = \sum_{i=1}^{5} a_i x_i \tag{A.5}$$

$$x_{16} = \sum_{i=6}^{10} a_i x_i \tag{A.6}$$

$$x_{17} = \sum_{i=11}^{14} a_i x_i \,. \tag{A.7}$$

We select the coefficients $a_i \in \mathrm{GF}(16) \setminus \{0\}$, such that $x_{15} + x_{16} = x_{17}$. Then, x_{17} does not need to be stored since it can be obtained from x_{15} and x_{16} if required. We obtain a $(16, 10)$ code with locality 5, i.e., each symbol can be recovered from 5 other symbols using (A.5), (A.6), (A.7). The network traffic and disk access are reduced by a factor of 2 compared to the $(14, 10)$ RS code. This comes at the price of increasing the storage overhead from 40% to 60% and shows an example of how to trade space efficiency for disk access/network load.

A.2.4 Access Latency

Beyond the node repair problem and storage efficiency, one may have further requirements for a distributed storage system. Among those is access latency. Here we may distinguish between decoding latency (associated with the time required to reconstruct data blocks upon need) and communication latency of the network if the requested data is stored in different geographical locations. Regarding decoding latency, systematic channel encoders might be appealing since the user data is part of the codeword and can be directly accessed without decoding. However, repair operations might be more complex than for nonsystematic encoders. The reason is that nonsystematic encoders allow functional repair, i.e., the content of a replacement node might be different from the failed node. This is in contrast to exact repair, i.e., the content of an erased node has to be exactly restored at the replacement node, which is a requirement for the systematic code symbols. A second aspect is latency due to geographically spread systems, i.e., user data stored on nodes in physically different locations. Access delay can be reduced by storing copies of frequently accessed data (hot data) in nodes close to the user (which can be accessed with low latency). This is referred to as caching and comes at the price of additional cache memory (decrease in storage efficiency) [17].

A.3 Further Applications

The next examples demonstrate additional applications of erasure codes.

A.3.1 Distributed Computing

Distributed computing systems can be understood as networks of computers that (in a coordinated manner) aim to achieve a common goal. Parallelization of certain computations (namely vector-matrix multiplications) can tremendously reduce execution time but also requires that the different computers deliver their results timely. A delay of a single computer in the computation phase (so-called *straggler*) will slow down the delivery of the final result. Similarly, delays during data transfer in a network (communication bottlenecks) will cause unwanted latency [18]. Here erasure codes can be used to improve the performance of distributed computing systems (see, e.g., [19–22]). In particular, redundancy is added to the computation so that the final output can be reconstructed even if some servers do not deliver the results in time. The output of such servers is simply treated as an erasure. Decoding complexity constraints are a key issue for such systems and need to be taken into account in the choice of the erasure code.

More in detail, assume that we are interested in computing vectors $\boldsymbol{u}$ as a vector-matrix multiplication

$$\boldsymbol{u} = \boldsymbol{d}\boldsymbol{A}. \tag{A.8}$$

Let the $r \times c$ matrix $\boldsymbol{A} \in \mathbb{F}^{r\times c}$, $\boldsymbol{d} \in \mathbb{F}^{1\times r}$, and $\boldsymbol{u} \in \mathbb{F}^{1\times c}$, where $\mathbb{F}$ is some field. Assume that we have k distributed nodes (workers) to solve the problem in (A.8). We may subdivide the $r \times c$ matrix $\boldsymbol{A}$ into k submatrices $\boldsymbol{A}_i$ of dimension $r \times (c/k)$.[1] A generic node i now performs the multiplication

$$\boldsymbol{u}_i = \boldsymbol{d}\boldsymbol{A}_i \tag{A.9}$$

and we obtain

$$\begin{bmatrix}\boldsymbol{u}_1 & \boldsymbol{u}_2 & \dots & \boldsymbol{u}_k\end{bmatrix} = \boldsymbol{d}\begin{bmatrix}\boldsymbol{A}_1 & \boldsymbol{A}_2 & \dots & \boldsymbol{A}_k\end{bmatrix}. \tag{A.10}$$

A central node assembles the vector $\boldsymbol{u} = \begin{bmatrix}\boldsymbol{u}_1 & \boldsymbol{u}_2 & \dots & \boldsymbol{u}_k\end{bmatrix}$ based on the inputs $\boldsymbol{u}_i$, $i \in \{1,2,\dots,k\}$. Then, $\boldsymbol{u}$ is available only when the last fragment $\boldsymbol{u}_i$ arrives at the central node. This can cause unwanted delays due to slow servers or bottlenecks in the network.

The problem above can be mitigated by erasure coding. Again we subdivide $\boldsymbol{A}$ into k submatrices $\boldsymbol{A}_i$ of dimension $r \times (c/k)$, yielding an information word $\begin{bmatrix}\boldsymbol{u}_1 & \boldsymbol{u}_2 & \dots & \boldsymbol{u}_k\end{bmatrix}$. Using a linear (n,k) channel code over $\mathbb{F}$ with generator matrix $\boldsymbol{G} = \left\{g_{i,j}\right\}_{i=1:k}^{j=1:n}$ and $n > k$, the information word is mapped to a length-n codeword $\boldsymbol{x} = \begin{bmatrix}\boldsymbol{x}_1 & \boldsymbol{x}_2 & \dots & \boldsymbol{x}_n\end{bmatrix}$ as

$$\boldsymbol{x} = \boldsymbol{u}\boldsymbol{G}. \tag{A.11}$$

For the code symbols, we obtain

$$\begin{aligned}\boldsymbol{x}_j &= \sum_{i=1}^{k} g_{i,j}\boldsymbol{u}_i && \text{(A.12)}\\ &= \sum_{i=1}^{k} g_{i,j}\boldsymbol{d}\boldsymbol{A}_i && \text{(A.13)}\\ &= \boldsymbol{d}\sum_{i=1}^{k} g_{i,j}\boldsymbol{A}_i && \text{(A.14)}\\ &= \boldsymbol{d}\boldsymbol{B}_j && \text{(A.15)}\end{aligned}$$

where (A.13) is due to the relation in (A.10). The terms $\boldsymbol{B}_j = \sum_{i=1}^{k} g_{i,j}\boldsymbol{A}_i$ can be precomputed in advance and shared with the distributed nodes.

Assume that a subset of code symbols $\boldsymbol{x}_j$ with $j \in \bar{\mathcal{E}}$, $\bar{\mathcal{E}}$ being the indexes of the not erased symbols, is available at the central node at a certain time t. The task is to recover $\boldsymbol{u}$ from the available symbols. In the case of MDS codes, we require $|\bar{\mathcal{E}}| \geq k$ for successful decoding. If at time t $|\bar{\mathcal{E}}| < k$, the central node may wait for more code

[1] We assume that k divides c.

symbols to arrive or declare decoding failure and not deliver any result (e.g., when t exceeds a maximum time).

Remark A.1. Observe that the number n of required nodes (workers) is greater than for the uncoded case since we have $n > k$. Also, the computational load at the central node is increased since it has to recover (decode) the original information from the received code symbols. Further, additional network traffic is generated.

Example A.5 (Distributed matrix multiplication) We are interested in computing

$$\boldsymbol{d}\boldsymbol{A} = \boldsymbol{u} \tag{A.16}$$

with $n = 3$ distributed nodes available and all operations are in GF(2). For coded distributed computing, we use a binary (3,2) single parity-check code with

$$\boldsymbol{G} = \begin{bmatrix} 1 & 0 & 1 \\ 0 & 1 & 1 \end{bmatrix}. \tag{A.17}$$

We have

$$\begin{bmatrix} \boldsymbol{x}_1 & \boldsymbol{x}_2 & \boldsymbol{x}_3 \end{bmatrix} = \boldsymbol{d}\begin{bmatrix} \boldsymbol{A}_1 & \boldsymbol{A}_2 & \boldsymbol{A}_1 \oplus \boldsymbol{A}_2 \end{bmatrix}. \tag{A.18}$$

Assume that till time t, we have only received $\boldsymbol{x}_1$ and $\boldsymbol{x}_3$. We can compute $\boldsymbol{x}_2$ as the sum of $\boldsymbol{x}_1$ and $\boldsymbol{x}_3$. Since the encoder is systematic, $\boldsymbol{u} = \begin{bmatrix} \boldsymbol{x}_1 & \boldsymbol{x}_2 \end{bmatrix}$.

A.3.2 Network Coding with Application to Coded Caching

The idea of network coding is to reduce network traffic by applying channel coding in the network nodes. One of the earliest works [23] investigated the communication of two ground terminals through a satellite and illustrated a saving of 50% of bandwidth in the downlink (i.e., in the transmission from the satellite to the users).

Example A.6 (Two-way relaying with network coding) Assume that user A wants to deliver to user B the data unit $\boldsymbol{u}_A$ and user B to user A the data unit $\boldsymbol{u}_B$. Communication takes place via a relay (e.g., a satellite). For simplicity, assume that the communication channel is error-free (which can be ensured by a proper physical layer code). Once the relay receives both data units, it may broadcast a linear combination $\boldsymbol{u}_A + \boldsymbol{u}_B$ in a single broadcast transmission, instead of using two separate transmissions (i.e., two separate time or frequency slots) to transmit $\boldsymbol{u}_A$ and $\boldsymbol{u}_B$, respectively. Then, user A may decode $\boldsymbol{u}_B$ by removing its message $\boldsymbol{u}_A$ from the broadcast message $\boldsymbol{u}_A + \boldsymbol{u}_B$. User B can proceed in an analogous way.

Network coding received considerable attention due to the seminal works in [24–26]. Network coding techniques help improve the throughput in a network but may also have other advantages, such as resilience to eavesdropping.

The ideas of network coding find a wide set of applications. One example is coded caching [27]. Here a central entity possesses a set of information units (e.g., files) $\mathcal{I} = \{\boldsymbol{u}_1, \boldsymbol{u}_2, \ldots, \boldsymbol{u}_k\}$ and serves d users which request part of the information (e.g., each user may request a certain file). Further, it is assumed that each user has some local cache, which can store a (small) subset of $\mathcal{I}$ of size M. In a so-called placement phase, the central entity can populate those caches with some data beforehand (e.g., in off-peak times, when the network traffic is low). Whenever the requested user information is not in the local cache, the central node transmits messages on a (for simplicity, error-free) broadcast link to all users in a delivery phase. Now erasure codes can be used for populating the users' caches and for the message broadcast by the central node with the hope of saving bandwidth during the delivery phase: A coded data unit sent by the central node can be useful for various users to extract (decode) the requested information with the help of their local cache content.

Remark A.2. Coded caching is a special instance of the index coding problem [28], which can be used to describe a broadcast problem with side information at the receivers.

A.3.3 Privacy

The problem of private information retrieval (PIR) deals with the download of a database entry (*message*) M from a database located on a remote server, where the database query must not reveal the identity of the message to the potentially mistrusted operator of the database [29]. When accessing only a single database server, it can be shown that the database operator has full uncertainty about the message identified if the user requests the complete database for download. By doing so, no information about the message is leaked. Obviously, this approach is impractical. Hence, a common assumption for PIR is that multiple copies of the same database are available on multiple servers, which are *noncolluding*. The goal is to find efficient retrieval schemes that maximize the ratio between the amount of desired information bits (the size of the message) to the total amount of downloaded bits while preserving privacy. Thereby, assuming that the number of servers is N, the user sends one query Q_i to each server $i \in \{1,2,\ldots N\}$. Queries specify some linear combination of the database items. The goal is to choose these linear combinations such that the requested message can be recovered. At the same time, the queries should not reveal any information about the message when accessing a given database. This requirement can be defined in information-theoretic terms by demanding that the mutual information between the requested message and each query is zero, i.e., $\mathsf{I}(Q_i, M) = 0$ for all $i \in \{1,2,\ldots N\}$. The definition of the queries can rely on well-known erasure code constructions.

Example A.7 (Two noncolluding servers) Consider a system of two noncolluding servers, each hosting an identical copy of a database. For simplicity, we assume that the database consists of L binary entries $u_1, u_2, \ldots, u_L$, and that a user wants to retrieve the message (i.e., the bit) with index i. The user sends to the first server a query Q_1 which specifies a set of indexes $\mathcal{S}_1$, where a generic index $j \in \{1,\ldots,L\}$ is included with the set with probability $1/2$ (random linear coding). Here we use a random linear coding approach: an index $j \in \{1,\ldots,L\}$ is selected with probability 0.5. The user sends to the second server a query Q_2 specifying a second set of indexes $\mathcal{S}_2$. Initially, $\mathcal{S}_2$ is set to be equal to $\mathcal{S}_1$. The set $\mathcal{S}_2$ is then modified as follows: if the index i of the message of interest is included in $\mathcal{S}_1$, it is removed from $\mathcal{S}_2$. Conversely, if $\mathcal{S}_1$ does not contain the index i, i is included in $\mathcal{S}_2$. Upon receiving the queries, the two servers generate a sum of the database items indexed by the respective queries. The first server generates the GF(2) linear combination

$$x_1 = \sum_{j \in \mathcal{S}_1} u_j \tag{A.19}$$

whereas the second server generates

$$x_2 = \sum_{j \in \mathcal{S}_2} u_j. \tag{A.20}$$

After downloading x_1 and x_2, the user can retrieve the required database entry via the GF(2) sum $u_i = x_1 + x_2$. Since the two queries are random (but not independent), each of the two servers cannot retrieve any information about the message, unless the servers *collude* to unmask the user request.

To reduce storage cost, coded storage of the data over several databases has been proposed [30–32]. The PIR capacity, defined as the maximum number of bits of the desired message per bit of total download, was analyzed in [33]. The case where the requirement of perfect privacy is relaxed was introduced and studied in [34].

References

[1] *Digital Video Broadcasting (DVB); Specification of Multi-Protocol Encapsulation – Inter-burst Forward Error Correction (MPE-iFEC)*, ETSI TS 102 772 V1.1.1, ETSI European Standard (Telecommunications series), Sep. 2009.

[2] *Digital Video Broadcasting (DVB); Transmission System for Handheld Terminals (DVB-H)*, ETSI EN 302 304 v1.1.1, ETSI European Standard (Telecommunications series), Nov. 2004.

[3] *Erasure correcting codes for use in near-Earth and deep-space communications*, Orange Book, Issue 1, Consultative Committee for Space Data Systems (CCSDS) Experimental Specification 131.5-O-1, Nov. 2014.

[4] G. Liva, P. Pulini, and M. Chiani, "On-line construction of irregular repeat accumulate codes for packet erasure channels," *IEEE Trans. Wireless Commun.*, vol. 12, no. 2, pp. 680–689, Feb. 2013.

[5] *Multimedia Broadcast/Multicast Service (MBMS); Protocols and codecs*, 3rd Generation Partnership Project (3GPP) Std. version 9.4.0, Mar. 2010.

[6] *Digital Video Broadcasting (DVB); Transport of MPEG-2 TS Based DVB Services over IP Based Networks*, ETSI TS 102 034 V1.4.1, ETSI European Standard (Telecommunications series), Aug. 2009.

[7] *ATSC Standard: Signaling, Delivery, Synchronization, and Error Protection*, Doc. A/331, ATSC Std., Jun. 2019.

[8] https://www.rfc-editor.org/standards.

[9] D. A. Patterson, G. Gibson, and R. H. Katz, "A case for redundant arrays of inexpensive disks (RAID)," in *Proc. ACM International Conference on Management of Data (SIGMOD)*, Chicago, IL, USA, Sep. 1988, pp. 109–116.

[10] M. Blaum, J. Brady, J. Bruck, and J. Menon, "EVENODD: An efficient scheme for tolerating double disk failures in RAID architectures," *IEEE Trans. Comput.*, vol. 44, no. 2, pp. 192–102, Feb. 1995.

[11] M. Blaum, J. Bruck, and A. Vardy, "MDS array codes with independent parity symbols," *IEEE Trans. Inf. Theory*, vol. 42, no. 2, pp. 529–542, Mar. 1996.

[12] C. Huang and L. Xu, "STAR: An efficient coding scheme for correcting triple storage node failures," *IEEE Trans. Comput.*, vol. 57, no. 7, pp. 889–901, May 2008.

[13] K. V. Rashmi, N. B. Shah, D. Gu, H. Kuang, D. Borthakur, and K. Ramchandran, "A solution to the network challenges of data recovery in erasure-coded distributed storage systems: A study on the Facebook warehouse cluster," in *Proc. USENIX Workshop on Hot Topics in Storage and File Systems*, San Jose, CA, USA, Jun. 2013, pp. 1–5.

[14] A. G. Dimakis, P. B. Godfrey, Y. Wu, M. J. Wainwright, and K. Ramchandran, "Network coding for distributed storage systems," *IEEE Trans. Inf. Theory*, vol. 56, no. 9, pp. 4539–4551, Sep. 2010.

[15] P. Gopalan, C. Huang, H. Simitci, and S. Yekhanin, "On the locality of codeword symbols," *IEEE Trans. Inf. Theory*, vol. 58, no. 11, pp. 6925–6934, Nov. 2012.

[16] V. Guruswami and M. Wootters, "Repairing Reed-Solomon codes," *IEEE Trans. Commun.*, vol. 63, no. 9, pp. 5684–5698, Sep. 2017.

[17] R. Halalai, P. Felber, A. Kermarrec, and F. Taïani, "Agar: A caching system for erasure-coded data," in *Proc. Int. Conf. Dist. Computing Sys. (ICDCS)*, Atlanta, GA, USA, Jun. 2017, pp. 23–33.

[18] S. Li, M. A. Maddah-Ali, Q. Yu, and A. S. Avestimehr, "A fundamental tradeoff between computation and communication in distributed computing," *IEEE Trans. Inf. Theory*, vol. 64, no. 1, p. 109–128, Jan. 2018.

[19] K. Lee, M. Lam, R. Pedarsani, D. Papailiopoulos, and K. Ramchandran, "Speeding up distributed machine learning using codes," *IEEE Trans. Inf. Theory*, vol. 64, no. 3, pp. 1514–1529, Mar. 2018.

[20] A. Severinson, A. Graell i Amat, and E. Rosnes, "Block-diagonal and LT codes for distributed computing with straggling servers," *IEEE Trans. Inf. Theory*, vol. 67, no. 3, p. 1739–1753, Mar. 2019.

[21] A. K. Pradhan, A. Heidarzadeh, and K. R. Narayanan, "Factored LT and factored Raptor codes for large-scale distributed matrix multiplication," *IEEE J. Sel. Areas Commun.*, vol. 2, no. 3, pp. 893–906, Sep. 2021.

[22] R. Schlegel, S. Kumar, E. Rosnes, and A. Graell i Amat, "CodedPaddedFL and CodedSecAgg: Straggler mitigation and secure aggregation in federated learning," *IEEE Trans. Commun.*, vol. 71, no. 4, pp. 2013–2027, Apr. 2023.

[23] M. Celebiler and G. Stette, "On increasing the down-link capacity of a regenerative satellite repeater in point-to-point communications," *Proc. IEEE*, vol. 66, no. 1, pp. 98–100, Jan. 1978.

[24] R. Ahlswede, N. Cai, S.-Y. R. Li, and R. W. Yeung, "Network information flow," *IEEE Trans. Inf. Theory*, vol. 46, no. 4, pp. 1204–1216, Jul. 2000.

[25] S.-Y. Li, R. Yeung, and N. Cai, "Linear network coding," *IEEE Trans. Commun.*, vol. 49, no. 2, pp. 371–381, Feb. 2003.

[26] R. Koetter and M. Medard, "An algebraic approach to network coding," *IEEE/ACM Trans. Netw.*, vol. 11, no. 5, pp. 782–795, 2003.

[27] M. A. Maddah-Ali and U. Niesen, "Fundamental limits of caching," *IEEE Trans. Inf. Theory*, vol. 60, no. 5, pp. 2856–2867, May 2014.

[28] Y. Birk and T. Kol, "Informed-source coding-on-demand (ISCOD) over broadcast channels," in *Proc. IEEE INFOCOM*, vol. 3, San Francisco, CA, USA, Aug. 1998, pp. 1257–1264 vol.3.

[29] B. Chor, O. Goldreich, E. Kushilevitz, and M. Sudan, "Private information retrieval," *J. of the ACM*, vol. 45, no. 6, pp. 965–982, Nov. 1998.

[30] N. B. Shah, K. V. Rashmi, and K. Ramchandran, "One extra bit of download ensures perfectly private information retrieval," in *Proc. IEEE Int. Symp. Inf. Theory*, Honolulu, HI, USA, Jun. 2014, pp. 856–860.

[31] A. Fazeli, A. Vardy, and E. Yaakobi, "Codes for distributed PIR with low storage overhead," in *Proc. IEEE Int. Symp. Inf. Theory*, Hong Kong, China, Jun. 2015, pp. 2852–2856.

[32] T. H. Chan, S.-W. Ho, and H. Yamamoto, "Private information retrieval for coded storage," in *Proc. IEEE Int. Symp. Inf. Theory*, Hong Kong, China, Jun. 2015, pp. 2842–2846.

[33] K. Banawan and S. Ulukus, "The capacity of private information retrieval from coded databases," *IEEE Trans. Inf. Theory*, vol. 64, no. 3, pp. 1945–1956, Jan. 2018.

[34] H.-Y. Lin, S. Kumar, E. Rosnes, A. Graell i Amat, and E. Yaakobi, "Multi-server weakly-private information retrieval," *IEEE Trans. Inf. Theory*, vol. 68, no. 2, pp. 1197–1219, Feb. 2022.

Appendix B Random Matrices over Finite Fields

Here we present some basic results for random matrices over finite fields.

Definition B.1 (Random Matrix) A random matrix $\boldsymbol{X}$ is a matrix whose entries are random variables. The statistical description of a random matrix requires therefore the joint distribution of all elements.

The types of random matrices of interest in information theory and physics include those with real or complex entries. For example, many problems are related to the distribution of the eigenvalues of Hermitian/symmetric matrices derived from Gaussian i.i.d. random variables, leading to the class of Gaussian Orthogonal Ensemble (GOE), Gaussian Unitary Ensemble (GUE), and Wishart matrices which describe the distribution of the sample covariance matrix in multivariate statistical analysis [3].

In coding theory, we are interested in random matrices with entries on finite fields.

Definition B.2 (Random Matrix over finite field $\mathrm{GF}(q)$) A random matrix over the finite field $\mathrm{GF}(q)$ is a random matrix $\boldsymbol{X} \in \mathrm{GF}(q)^{m \times n}$ whose elements are in $\mathrm{GF}(q)$.

One important question is what is the probability that a randomly generated matrix is of full rank. To find it, we will use the following theorem.

Theorem B.1 *The number of full rank matrices over* $\mathrm{GF}q$ *of dimension* $m \times n$*, with* $m \leq n$*, is*

$$(q^n - 1)(q^n - q)\left(q^n - q^2\right)\cdots\left(q^n - q^{m-1}\right). \tag{B.1}$$

Proof The matrix is of full rank if and only if the m rows are linearly independent. Therefore, we have to count all possible ways to arrange m linearly independent rows. The first row can be any n-tuple except that of all zeros; the second row can be any n-tuple except all q linear combinations of the first, and so on, until the mth row, which cannot be any linear combination of the previous $m - 1$. □

A particular case is constituted by full rank square matrices, which are therefore invertible. The set of full rank $n \times n$ matrices over $\mathrm{GF}(q)$, together with the operation of ordinary matrix multiplication, is a group, called the finite general linear group, $GL(n, q)$. In fact, the product of two invertible matrices is an invertible matrix, and the identity matrix acts as the identity element of the group. As already derived by Galois, this group has a number of elements given by (B.1) with $n = m$.

In this Appendix, we are interested in the probability of randomly picking a full-rank matrix. Since the total number of distinct random matrices is $q^{n\cdot m}$, we have the following result.

Theorem B.2 *Assume to generate a random matrix* $\boldsymbol{X} \in GF(q)^{m\times n}$ *with i.i.d. elements, each with uniform distribution over the field set* $GF(q)$*. Denoting by* $\delta = n - m \geq 0$ *the excess of columns, the probability that the matrix is of full rank is*

$$A(m,\delta) = P(rank(\boldsymbol{X}) = m) = \left(1 - \frac{1}{q^{m+\delta}}\right)\left(1 - \frac{1}{q^{m+\delta-1}}\right)\cdots\left(1 - \frac{1}{q^{\delta+1}}\right)$$
$$= \prod_{\ell=1}^{m}\left(1 - \frac{1}{q^{\delta+\ell}}\right). \tag{B.2}$$

Proof Just divide the number of full rank matrices (B.1) by the total number of matrices $q^{n\cdot m}$. □

From this theorem, we can derive simple bounds on the probability that a random matrix is not full rank, $1 - A(m,\delta)$, as follows [4]

$$\frac{1}{q^{\delta+1}} \leq 1 - A(m,\delta) < \frac{1}{q^{\delta}(q-1)} \tag{B.3}$$

where the equality holds only for $m = 1$. Note that both bounds do not depend on m. The lower bound derives immediately from (B.2), by keeping only the first factor. For the upper bound, observing that $A(m,\delta) < 1$ and the recursion

$$A(m+1,\delta) = A(m,\delta)\left(1 - \frac{1}{q^{\delta+1+m}}\right) = A(m,\delta) - \frac{A(m,\delta)}{q^{\delta+1+m}}, \tag{B.4}$$

we have

$$A(m+1,\delta) > A(m,\delta) - \frac{1}{q^{\delta+1+m}}$$
$$> \cdots > A(1,\delta) - \frac{1}{q^{\delta+1+1}} - \cdots - \frac{1}{q^{\delta+1+m}} =$$
$$= 1 - \frac{1}{q^{\delta+1}} - \frac{1}{q^{\delta+1}}\frac{1-q^{-m}}{q-1} = 1 - \frac{1}{q^{\delta}(q-1)} + \frac{1}{q^{\delta+1}}\frac{q^{-m}}{q-1}$$
$$> 1 - \frac{1}{q^{\delta}(q-1)}. \tag{B.5}$$

Also, an approximation can be obtained by observing that, since $1 - x < e^{-x}$, we can write

$$A(m,\delta) < \prod_{\ell=1}^{m} e^{-q^{-\delta-\ell}} = e^{-q^{-\delta}(1-q^{-m})/(q-1)} \approx e^{-q^{\delta}/(q-1)} \tag{B.6}$$

where we used the geometric sum for $\sum_\ell q^{-\ell}$. Actually, the bound and approximation in (B.6) are almost coincident, and very close to the exact value. Also, from (B.3), we see that the ratio of the upper to the lower bounds is $q/(q-1)$, which, for increasing q, tends to one. Thus we expect that for increasing q, the bounds in (B.3) become closer and closer. Indeed, the curves reported in Figure B.1 show that the probability that a

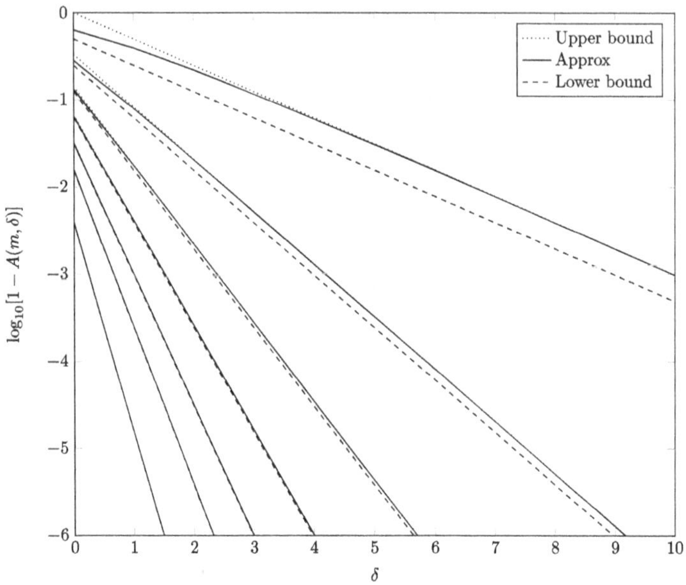

Figure B.1 Probability that a random matrix $X \in \mathrm{GF}(q)^{m \times n}$ is not full rank, as a function of the excess dimension $\delta = n - m$. The curves refer to (from upper to lower) $q = 2,4,8,32,256$. For each q, there are three curves: the upper and the lower bounds in (B.3), and the approximation (B.6). Note that the upper and lower bounds are very close, and are distinguishable only for small q.

random matrix is full rank depends essentially not on the matrix dimensions, but on the excess dimension δ and on the field order q.

References

[1] G. V. Balakin, "The distribution of the rank of random matrices over a finite field," *Theory Probab. Appl.*, vol. 13, no. 4, pp. 594–605, 1968.

[2] C. Cooper, "On the distribution of rank of a random matrix over a finite field," *Random Struct. Algor.*, vol. 17, no. 3-4, pp. 197–212, 2000.

[3] C. Tracy and H. Widom, "The distributions of random matrix theory and their applications," *New Trends Math. Phys.*, pp. 753–765, 2009.

[4] G. Liva, E. Paolini, and M. Chiani, "Performance versus overhead for fountain codes over $\mathbb{F}_q$," *IEEE Commun. Lett.*, vol. 14, no. 2, pp. 178–180, Feb. 2010.

Index